SPACECRAFT DYNAMICS

Thomas R. Kane
Stanford University

Peter W. Likins
Lehigh University

David A. Levinson
Lockheed Palo Alto Research Laboratory

McGraw-Hill Book Company

New York St. Louis San Francisco Auckland Bogotá Hamburg
London Madrid Mexico Montreal New Delhi
Panama Paris São Paulo Singapore Sydney Tokyo Toronto

This book was set in Times Roman by Andover Graphics.
The editors were Diane D. Heiberg and Madelaine Eichberg;
the production supervisor was Leroy A. Young.
New drawings were done by Danmark & Michaels, Inc.
The cover was designed by Mark Wieboldt.
Halliday Lithograph Corporation was printer and binder.

SPACECRAFT DYNAMICS

3 4 5 6 7 8 9 0 HDHD 8 9 8

ISBN 0-07-037843-6

Library of Congress Cataloging in Publication Data

Kane, Thomas R.
Spacecraft dynamics.

Includes index.
1. Space vehicles—Dynamics. I. Likins, Peter W.
II. Levinson, David A. III. Title.
TL1050.K36 629.47'01'53111 81-8431
ISBN 0-07-037843-6 AACR2

CONTENTS

PREFACE

This book is the outgrowth of courses taught at Stanford University and at the University of California, Los Angeles, and of the authors' professional activities in the field of spacecraft dynamics. It is intended both for use as a textbook in courses of instruction at the graduate level and as a reference work for engineers engaged in research, design, and development in this field.

The choice and arrangement of topics was dictated by the following considerations.

The process of solving a spacecraft dynamics problem generally necessitates the construction of a mathematical model, the use of principles of mechanics to formulate equations governing the quantities appearing in the mathematical model, and the extraction of useful information from the equations. Skill in constructing mathematical models of spacecraft is acquired best through experience and cannot be transmitted easily from one individual to another, particularly by means of the printed word. Hence, this subject is not treated formally in the book. However, through examples, the reader is brought into contact with a considerable number of mathematical models of spacecraft and, by working with the book, he can gain much experience of the kind required. By way of contrast, the formulation of equations of motion is a subject that *can* be presented formally, and it is essential that this topic be treated effectively, for there is no point in attempting to extract information from incorrect equations of motion. Now, every spacecraft dynamics analysis necessitates use of various kinematical relationships, some of which have played such a small role in the development of technology prior to the space age that they have been treated only cursorily, if at all, in the general mechanics literature. Accordingly, the book begins with what is meant to be a unified, modern treatment of the kinematical ideas that are most useful in dealing with spacecraft dynamics problems.

To place the topics to be treated in the book into perspective, we turn to the familiar relationship $F=ma$, here regarding it as a conceptual guideline rather than as the statement of a law of physics. Seen in this light, the a represents all kinematical quantities, the F all forces that come into play, the m all inertia properties, and the sign of equality the assertion that kinematical quantities, forces, and inertia properties are related to each other. It is then clear that one should deal with the topics of kinematics, forces, and inertia properties before taking up the study of a technique for formulating equations of motion.

The subject of inertia properties, that is, the finding of mass centers, moments and products of inertia, principal axes of inertia, and so on, is treated extensively in available textbooks and acquires no new facets in connection with spacecraft. Hence, we presume that the reader knows this material. Detailed information regarding forces that affect the behavior of spacecraft is not so readily accessible. Therefore, we address this topic in Chapter 2, confining attention to gravitational forces, which play a preeminent role in spacecraft dynamics. This brings us into position to attack specific problems in Chapters 3 and 4, these chapters differing from each other in one important respect: throughout Chapter 3, which deals with relatively simple spacecraft, we rely solely upon the angular momentum principle for the formulation of dynamical equations of motion, whereas in Chapter 4, where we are concerned with complex spacecraft, we first develop and then use a more powerful method for formulating equations of motion, one that is particularly well suited for problems involving multi-degrees-of-freedom spacecraft.

Most sections of the book are arranged as follows. They begin with the presentation of theoretical material in the form of categorical statements that either constitute definitions or that can be shown to be valid by means of deductive arguments. In the latter case, the arguments are presented under the heading "Derivations"; each section contains, in addition, an example in the application of the theory set forth in the section. In the classroom, an instructor can, therefore, focus attention wherever he pleases, that is, on the presentation of theory, on formal proofs, or on examples, leaving it to his students to devote time outside of the classroom to the matters not pursued in depth in class.

In addition to the examples appearing in the text, the book contains four sets of problems, each set associated intimately with one of the four chapters. To acquire mastery of the material in the book, a student should solve every one of these problems, for each covers material not covered by any other. Results are given wherever possible, so that one always can determine whether or not one has solved a problem correctly.

To cover all of the material in the book is rather difficult in the time generally available in a graduate curriculum. However, the material lends itself well to presentation in two courses dealing, respectively, with Sections

1.1–1.20, 2.1–2.8, 3.1–3.11, and Sections 1.21, 2.9–2.18, and 4.1–4.11. The first course, covering kinematical fundamentals, the vectorial treatment of gravitational forces and moments, and simple spacecraft, can prepare a student to work effectively on spacecraft dynamics problems and to pursue further studies in this field; successful completion of the second course, which concentrates upon new kinematical ideas, gravitational potential theory, a new approach to the formulation of equations of motion, and complex spacecraft problems, enables one to undertake work at the forefront of spacecraft dynamics.

The use of computers in the solution of spacecraft dynamics problems is so pervasive that familiarity of a worker in this field with programming concepts must be presumed. The writing and running of computer programs are meant to take place in connection with courses based on this book, but are not absolutely essential; that is, most problems are set up in such a way that considerable benefit can be obtained by studying them even without the aid of a computer.

Since it was our intention to produce a self-contained book, we have omitted all references to the extensive literature on spacecraft dynamics. However, we gratefully acknowledge our debt to the many authors whose work has influenced our thinking, and we wish to express our thanks to students who, with comments and suggestions, have aided us in writing this book, as well as to Stanford University and the University of California at Los Angeles, which supported this undertaking by generously making time and computational resources available to us.

Thomas R. Kane
Peter W. Likins
David A. Levinson

TO THE READER

Each of the four chapters of this book is divided into sections. A section is identified by two numbers separated by a decimal point, the first number referring to the chapter in which the section appears, and the second identifying the section within the chapter. Thus, the identifier 2.13 refers to the thirteenth section of the second chapter. A section identifier appears at the top of each page.

Equations are numbered serially within sections. For example, the equations in Secs. 2.12 and 2.13 are numbered (1)–(46) and (1)–(31), respectively. References to an equation may be made both within the section in which the equation appears and in other sections. In the first case, the equation number is cited as a single number, whereas, in the second case, the section number is included as part of a three-number designation. Thus, within Sec. 2.12, Eq. (1) of Sec. 2.12 is referred to as Eq. (1), whereas, in Sec. 2.13, the same equation is referred to as Eq. (2.12.1). To locate an equation cited in this manner, one may make use of the section identifiers appearing at the tops of the pages.

Figures appearing in the chapters are numbered so as to identify the sections in which the figures appear. For example, the two figures in Sec. 3.9 are designated Fig. 3.9.1 and Fig. 3.9.2. To avoid confusion of these figures with those in problem sets, the figure number is preceded by the letter P in the case of the latter; the double number that follows this letter refers to the problem statement in which the figure is introduced. For example, Fig. P3.8 is introduced in Prob. 3.8. Similarly, Table 1.15.2 is the designation for a table in Sec. 1.15, whereas Table P1.21 is associated with Prob. 1.21.

Thomas R. Kane
Peter W. Likins
David A. Levinson

CHAPTER

ONE

KINEMATICS

Every spacecraft dynamics problem involves considerations of kinematics. Indeed, the solution of most such problems begins with the formulation of kinematical relationships. Hence, this topic requires detailed attention.

The first nine sections of this chapter deal with changes in the orientation of a rigid body in a reference frame, without regard to the time that necessarily elapses when a real body experiences a change in orientation. Time enters the discussion in the remaining sections, in all of which the concept of angular velocity plays a central role. The concluding section contains the definitions of quantities employed to great advantage, in Chap. 4, in connection with the formulation of dynamical equations of motion of complex spacecraft. This section (and Probs. 1.27–1.31) may be omitted by readers concerned primarily with simple spacecraft, such as those treated in Chap. 3.

1.1 SIMPLE ROTATION

A motion of a rigid body or reference frame B relative to a rigid body or reference frame A is called a *simple rotation of B in A* if there exists a line L, called an *axis of rotation*, whose orientation relative to both A and B remains unaltered throughout the motion. This sort of motion is important because, as will be shown in Sec. 1.3, every change in the relative orientation of A and B can be produced by means of a simple rotation of B in A.

If $\mathbf{a}$ is any vector fixed in A (see Fig. 1.1.1), and $\mathbf{b}$ is a vector fixed in B and equal to $\mathbf{a}$ prior to the motion of B in A, then, when B has performed a simple rotation in A, $\mathbf{b}$ can be expressed in terms of the vector $\mathbf{a}$, a unit vector $\boldsymbol{\lambda}$ parallel to L, and the radian measure θ of the angle between two lines, L_A and L_B, which are fixed in A and B, respectively, are perpendicular to L, and are parallel to each other initially. Specifically, if θ is regarded as

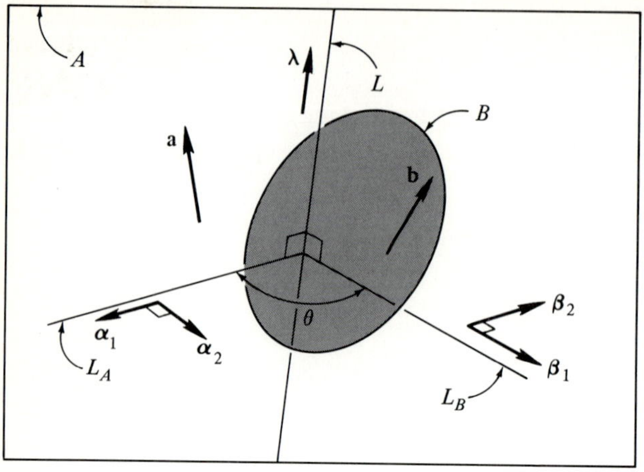

Figure 1.1.1

positive when the angle between L_A and L_B is generated by a $\boldsymbol{\lambda}$-rotation of L_B relative to L_A, that is, by a rotation during which a right-handed screw fixed in B with its axis parallel to $\boldsymbol{\lambda}$ advances in the direction of $\boldsymbol{\lambda}$ when B rotates relative to A, then

$$\mathbf{b} = \mathbf{a}\cos\theta - \mathbf{a} \times \boldsymbol{\lambda}\sin\theta + \mathbf{a} \cdot \boldsymbol{\lambda}\boldsymbol{\lambda}(1 - \cos\theta) \tag{1}$$

Equivalently, if a dyadic $\mathbf{C}$ is defined as

$$\mathbf{C} \triangleq \mathbf{U}\cos\theta - \mathbf{U} \times \boldsymbol{\lambda}\sin\theta + \boldsymbol{\lambda}\boldsymbol{\lambda}(1 - \cos\theta) \tag{2}$$

where $\mathbf{U}$ is the unit (or identity) dyadic, then

$$\mathbf{b} = \mathbf{a} \cdot \mathbf{C} \tag{3}$$

Derivations Let $\boldsymbol{\alpha}_1$ and $\boldsymbol{\alpha}_2$ be unit vectors fixed in A, with $\boldsymbol{\alpha}_1$ parallel to L_A and $\boldsymbol{\alpha}_2 = \boldsymbol{\lambda} \times \boldsymbol{\alpha}_1$; and let $\boldsymbol{\beta}_1$ and $\boldsymbol{\beta}_2$ be unit vectors fixed in B, with $\boldsymbol{\beta}_1$ parallel to L_B and $\boldsymbol{\beta}_2 = \boldsymbol{\lambda} \times \boldsymbol{\beta}_1$, as shown in Fig. 1.1.1. Then, if $\mathbf{a}$ and $\mathbf{b}$ are resolved into components parallel to $\boldsymbol{\alpha}_1, \boldsymbol{\alpha}_2, \boldsymbol{\lambda}$ and $\boldsymbol{\beta}_1, \boldsymbol{\beta}_2, \boldsymbol{\lambda}$, respectively, corresponding coefficients are equal to each other because $\boldsymbol{\alpha}_1 = \boldsymbol{\beta}_1$, $\boldsymbol{\alpha}_2 = \boldsymbol{\beta}_2$, and $\mathbf{a} = \mathbf{b}$ when $\theta = 0$. In other words, $\mathbf{a}$ and $\mathbf{b}$ can be expressed as

$$\mathbf{a} = p\boldsymbol{\alpha}_1 + q\boldsymbol{\alpha}_2 + r\boldsymbol{\lambda} \tag{4}$$

and

$$\mathbf{b} = p\boldsymbol{\beta}_1 + q\boldsymbol{\beta}_2 + r\boldsymbol{\lambda} \tag{5}$$

where p, q, and r are constants.

Expressed in terms of $\boldsymbol{\alpha}_1$ and $\boldsymbol{\alpha}_2$, the unit vectors $\boldsymbol{\beta}_1$ and $\boldsymbol{\beta}_2$ are given by

$$\boldsymbol{\beta}_1 = \cos\theta\, \boldsymbol{\alpha}_1 + \sin\theta\, \boldsymbol{\alpha}_2 \tag{6}$$

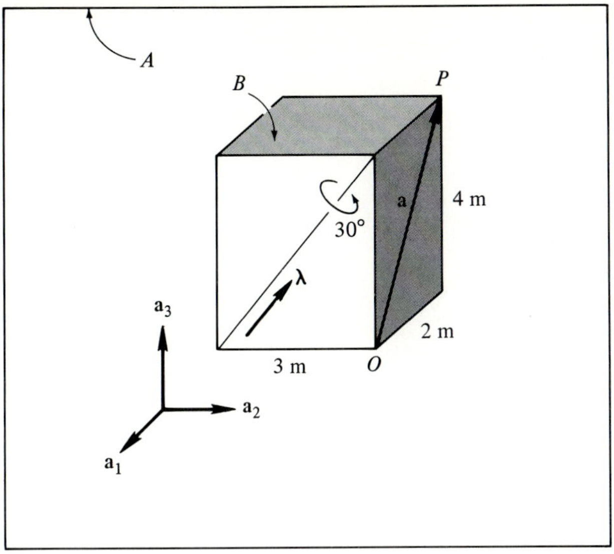

Figure 1.1.2

and
$$\boldsymbol{\beta}_2 = -\sin\theta\, \boldsymbol{\alpha}_1 + \cos\theta\, \boldsymbol{\alpha}_2 \tag{7}$$

so that, substituting into Eq. (5), one finds that

$$\mathbf{b} = (p\cos\theta - q\sin\theta)\boldsymbol{\alpha}_1 + (p\sin\theta + q\cos\theta)\boldsymbol{\alpha}_2 + r\boldsymbol{\lambda} \tag{8}$$

The right-hand member of Eq. (8) is precisely what one obtains by carrying out the operations indicated in the right-hand member of Eq. (1), using Eq. (4), and making use of the relationships $\boldsymbol{\lambda} \times \boldsymbol{\alpha}_1 = \boldsymbol{\alpha}_2$ and $\boldsymbol{\lambda} \times \boldsymbol{\alpha}_2 = -\boldsymbol{\alpha}_1$. Thus the validity of Eq. (1) is established, and Eq. (3) follows directly from Eqs. (1) and (2).

Example A rectangular block B having the dimensions shown in Fig. 1.1.2 forms a portion of an antenna structure mounted in a spacecraft A. This block is subjected to a simple rotation in A about a diagonal of one face of B, the sense and amount of the rotation being those indicated in the sketch. The angle ϕ between the line OP in its initial and final positions is to be determined.

If $\mathbf{a}_1$, $\mathbf{a}_2$, and $\mathbf{a}_3$ are unit vectors fixed in A and parallel to the edges of B prior to B's rotation, then a unit vector $\boldsymbol{\lambda}$ directed as shown in the sketch can be expressed as

$$\boldsymbol{\lambda} = \frac{3\mathbf{a}_2 + 4\mathbf{a}_3}{5} \tag{9}$$

And if $\mathbf{a}$ denotes the position vector of P relative to O prior to B's rotation, then

$$\mathbf{a} = -2\mathbf{a}_1 + 4\mathbf{a}_3 \tag{10}$$

$$\mathbf{a} \times \boldsymbol{\lambda} = \frac{-12\mathbf{a}_1 + 8\mathbf{a}_2 - 6\mathbf{a}_3}{5} \tag{11}$$

and

$$\mathbf{a} \cdot \boldsymbol{\lambda}\boldsymbol{\lambda} = \frac{48\mathbf{a}_2 + 64\mathbf{a}_3}{25} \tag{12}$$

Consequently, if $\mathbf{b}$ is the position vector of P relative to O subsequent to B's rotation,†

$$\mathbf{b} \underset{(1)}{=} (-2\mathbf{a}_1 + 4\mathbf{a}_3)\cos\frac{\pi}{6} + \frac{12\mathbf{a}_1 - 8\mathbf{a}_2 + 6\mathbf{a}_3}{5}\sin\frac{\pi}{6}$$

$$+ \frac{48\mathbf{a}_2 + 64\mathbf{a}_3}{25}\left(1 - \cos\frac{\pi}{6}\right)$$

$$= -0.532\mathbf{a}_1 - 0.543\mathbf{a}_2 + 4.407\mathbf{a}_3 \tag{13}$$

Since ϕ is the angle between $\mathbf{a}$ and $\mathbf{b}$,

$$\cos\phi = \frac{\mathbf{a} \cdot \mathbf{b}}{|\mathbf{a}|\,|\mathbf{b}|} \tag{14}$$

where $|\mathbf{a}|$ and $|\mathbf{b}|$ denote the (equal) magnitudes of $\mathbf{a}$ and $\mathbf{b}$. Hence

$$\cos\phi = \frac{(-2)(-0.532) + 4(4.407)}{(4 + 16)^{1/2}\,(4 + 16)^{1/2}} = 0.935 \tag{15}$$

and $\phi = 20.77°$.

1.2 DIRECTION COSINES

If $\mathbf{a}_1, \mathbf{a}_2, \mathbf{a}_3$ and $\mathbf{b}_1, \mathbf{b}_2, \mathbf{b}_3$ are two dextral sets of orthogonal unit vectors, and nine quantities C_{ij} $(i, j = 1, 2, 3)$, called *direction cosines*, are defined as

$$C_{ij} \triangleq \mathbf{a}_i \cdot \mathbf{b}_j \qquad (i, j = 1, 2, 3) \tag{1}$$

then the two row matrices $[\mathbf{a}_1 \quad \mathbf{a}_2 \quad \mathbf{a}_3]$ and $[\mathbf{b}_1 \quad \mathbf{b}_2 \quad \mathbf{b}_3]$ are related to each other as follows:

$$[\mathbf{b}_1 \quad \mathbf{b}_2 \quad \mathbf{b}_3] = [\mathbf{a}_1 \quad \mathbf{a}_2 \quad \mathbf{a}_3]\, C \tag{2}$$

† Numbers beneath signs of equality are intended to direct attention to equations numbered correspondingly.

where C is a square matrix defined as

$$C \triangleq \begin{bmatrix} C_{11} & C_{12} & C_{13} \\ C_{21} & C_{22} & C_{23} \\ C_{31} & C_{32} & C_{33} \end{bmatrix} \tag{3}$$

If a superscript T is used to denote transposition, that is, if C^T is defined as

$$C^T \triangleq \begin{bmatrix} C_{11} & C_{21} & C_{31} \\ C_{12} & C_{22} & C_{32} \\ C_{13} & C_{23} & C_{33} \end{bmatrix} \tag{4}$$

then Eq. (2) can be replaced with the equivalent relationship

$$[\mathbf{a}_1 \quad \mathbf{a}_2 \quad \mathbf{a}_3] = [\mathbf{b}_1 \quad \mathbf{b}_2 \quad \mathbf{b}_3]\, C^T \tag{5}$$

The matrix C, called a *direction cosine matrix*, can be employed to describe the relative orientation of two reference frames or rigid bodies A and B. In that context, it can be advantageous to replace the symbol C with the more elaborate symbol ${}^AC^B$. In view of Eqs. (2) and (5), one must then regard the interchanging of superscripts as signifying transposition; that is,

$${}^BC^A = ({}^AC^B)^T \tag{6}$$

The direction cosine matrix C plays a role in a number of useful relationships. For example, if $\mathbf{v}$ is any vector and ${}^Av_i$ and ${}^Bv_i$ $(i = 1, 2, 3)$ are defined as

$${}^Av_i \triangleq \mathbf{v} \cdot \mathbf{a}_i \qquad (i = 1, 2, 3) \tag{7}$$

and

$${}^Bv_i \triangleq \mathbf{v} \cdot \mathbf{b}_i \qquad (i = 1, 2, 3) \tag{8}$$

while ${}^Av$ and ${}^Bv$ denote the row matrices having the elements ${}^Av_1$, ${}^Av_2$, ${}^Av_3$ and ${}^Bv_1$, ${}^Bv_2$, ${}^Bv_3$, respectively, then

$${}^Bv = {}^AvC \tag{9}$$

Similarly, if $\mathbf{D}$ is any dyadic, and ${}^AD_{ij}$ and ${}^BD_{ij}$ $(i, j = 1, 2, 3)$ are defined as

$${}^AD_{ij} \triangleq \mathbf{a}_i \cdot \mathbf{D} \cdot \mathbf{a}_j \qquad (i, j = 1, 2, 3) \tag{10}$$

and

$${}^BD_{ij} \triangleq \mathbf{b}_i \cdot \mathbf{D} \cdot \mathbf{b}_j \qquad (i, j = 1, 2, 3) \tag{11}$$

while ${}^AD$ and ${}^BD$ denote square matrices having ${}^AD_{ij}$ and ${}^BD_{ij}$, respectively, as the elements in the ith rows and jth columns, then

$${}^BD = C^T\, {}^AD\, C \tag{12}$$

Use of the summation convention for repeated subscripts frequently makes it possible to formulate important relationships rather concisely. For example, if δ_{ij} is defined as

$$\delta_{ij} \triangleq 1 - \tfrac{1}{4}(i-j)^2[5-(i-j)^2] \qquad (i, j = 1, 2, 3) \tag{13}$$

so that δ_{ij} is equal to unity when the subscripts have the same value, and equal to zero when the subscripts have different values, then use of the summation convention permits one to express a set of six relationships governing direction cosines as

$$C_{ik}\, C_{jk} = \delta_{ij} \qquad (i, j = 1, 2, 3) \tag{14}$$

or an equivalent set as

$$C_{ki}\, C_{kj} = \delta_{ij} \qquad (i, j = 1, 2, 3) \tag{15}$$

Alternatively, these relationships can be stated in matrix form as

$$C\, C^T = U \tag{16}$$

and

$$C^T\, C = U \tag{17}$$

where U denotes the unit (or identity) matrix, defined as

$$U \triangleq \begin{bmatrix} 1 & 0 & 0 \\ 0 & 1 & 0 \\ 0 & 0 & 1 \end{bmatrix} \tag{18}$$

Each element of the matrix C is equal to its cofactor in the determinant of C; and, if $|C|$ denotes this determinant, then

$$|C| = 1 \tag{19}$$

Consequently, C is an orthogonal matrix, that is, a matrix whose inverse and whose transpose are equal to each other. Moreover,

$$|C - U| = 0 \tag{20}$$

Hence, unity is an eigenvalue of every direction cosine matrix. In other words, for every direction cosine matrix C there exist row matrices $[\kappa_1 \quad \kappa_2 \quad \kappa_3]$, called eigenvectors, which satisfy the equation

$$[\kappa_1 \quad \kappa_2 \quad \kappa_3]\, C = [\kappa_1 \quad \kappa_2 \quad \kappa_3] \tag{21}$$

Suppose now that $\mathbf{a}_i$ and $\mathbf{b}_i$ $(i = 1, 2, 3)$ are fixed in reference frames or rigid bodies A and B, respectively, and that B is subjected to a simple rotation in A (see Sec. 1.1); further, that $\mathbf{a}_i = \mathbf{b}_i$ $(i = 1, 2, 3)$ prior to the rotation, that $\boldsymbol{\lambda}$ and θ are defined as in Sec. 1.1, and that λ_i is defined as

$$\lambda_i \triangleq \boldsymbol{\lambda} \cdot \mathbf{a}_i = \boldsymbol{\lambda} \cdot \mathbf{b}_i \qquad (i = 1, 2, 3) \tag{22}$$

Then the elements of C are given by

$$C_{11} = \cos\theta + \lambda_1{}^2(1 - \cos\theta) \tag{23}$$

$$C_{12} = -\lambda_3 \sin\theta + \lambda_1\lambda_2(1 - \cos\theta) \tag{24}$$

$$C_{13} = \lambda_2 \sin\theta + \lambda_3\lambda_1(1 - \cos\theta) \tag{25}$$

$$C_{21} = \lambda_3 \sin\theta + \lambda_1\lambda_2(1 - \cos\theta) \tag{26}$$

$$C_{22} = \cos\theta + \lambda_2^2(1 - \cos\theta) \tag{27}$$

$$C_{23} = -\lambda_1 \sin\theta + \lambda_2\lambda_3(1 - \cos\theta) \tag{28}$$

$$C_{31} = -\lambda_2 \sin\theta + \lambda_3\lambda_1(1 - \cos\theta) \tag{29}$$

$$C_{32} = \lambda_1 \sin\theta + \lambda_2\lambda_3(1 - \cos\theta) \tag{30}$$

$$C_{33} = \cos\theta + \lambda_3^2(1 - \cos\theta) \tag{31}$$

Equations (23)–(31) can be expressed more concisely after defining ϵ_{ijk} as

$$\epsilon_{ijk} \triangleq \tfrac{1}{2}(i-j)(j-k)(k-i) \qquad (i, j, k = 1, 2, 3) \tag{32}$$

(The quantity ϵ_{ijk} vanishes when two or three subscripts have the same value; it is equal to unity when the subscripts appear in cyclic order, that is, in the order 1, 2, 3, the order 2, 3, 1, or the order 3, 1, 2; and it is equal to negative unity in all other cases.) Using the summation convention, one can then replace Eqs. (23)–(31) with

$$C_{ij} = \delta_{ij}\cos\theta - \epsilon_{ijk}\lambda_k \sin\theta + \lambda_i\lambda_j(1 - \cos\theta) \qquad (i, j = 1, 2, 3) \tag{33}$$

Alternatively, C_{ij} can be expressed in terms of the dyadic $\mathbf{C}$ defined in Eq. (1.1.2):

$$C_{ij} = \mathbf{a}_i \cdot (\mathbf{a}_j \cdot \mathbf{C}) \qquad (i, j = 1, 2, 3) \tag{34}$$

All of these results simplify substantially when $\boldsymbol{\lambda}$ is parallel to $\mathbf{a}_i$, and hence to $\mathbf{b}_i$ $(i = 1, 2, 3)$. If $C_i(\theta)$ denotes C for $\boldsymbol{\lambda} = \mathbf{a}_i = \mathbf{b}_i$, then

$$C_1(\theta) = \begin{bmatrix} 1 & 0 & 0 \\ 0 & \cos\theta & -\sin\theta \\ 0 & \sin\theta & \cos\theta \end{bmatrix} \tag{35}$$

$$C_2(\theta) = \begin{bmatrix} \cos\theta & 0 & \sin\theta \\ 0 & 1 & 0 \\ -\sin\theta & 0 & \cos\theta \end{bmatrix} \tag{36}$$

$$C_3(\theta) = \begin{bmatrix} \cos\theta & -\sin\theta & 0 \\ \sin\theta & \cos\theta & 0 \\ 0 & 0 & 1 \end{bmatrix} \tag{37}$$

It was mentioned previously that unity is an eigenvalue of every direction cosine matrix. If the elements of a direction cosine matrix C are given by Eqs. (23)–(31), then the row matrix $[\lambda_1 \quad \lambda_2 \quad \lambda_3]$ is one of the eigenvectors corresponding to the eigenvalue unity of C; that is,

$$[\lambda_1 \quad \lambda_2 \quad \lambda_3]\, C = [\lambda_1 \quad \lambda_2 \quad \lambda_3] \tag{38}$$

Equivalently, when **C** is the dyadic defined in Eq. (1.1.2), then

$$\boldsymbol{\lambda} \cdot \mathbf{C} = \boldsymbol{\lambda} \tag{39}$$

Derivations For any vector **v**, the following is an identity:

$$\mathbf{v} = (\mathbf{a}_1 \cdot \mathbf{v})\mathbf{a}_1 + (\mathbf{a}_2 \cdot \mathbf{v})\mathbf{a}_2 + (\mathbf{a}_3 \cdot \mathbf{v})\mathbf{a}_3 \tag{40}$$

Hence, letting $\mathbf{b}_1$ play the role of **v**, one can express $\mathbf{b}_1$ as

$$\mathbf{b}_1 = (\mathbf{a}_1 \cdot \mathbf{b}_1)\mathbf{a}_1 + (\mathbf{a}_2 \cdot \mathbf{b}_1)\mathbf{a}_2 + (\mathbf{a}_3 \cdot \mathbf{b}_1)\mathbf{a}_3 \tag{41}$$

or, by using Eq. (1), as

$$\mathbf{b}_1 = C_{11}\,\mathbf{a}_1 + C_{21}\,\mathbf{a}_2 + C_{31}\,\mathbf{a}_3 \tag{42}$$

Similarly,

$$\mathbf{b}_2 = C_{12}\,\mathbf{a}_1 + C_{22}\,\mathbf{a}_2 + C_{32}\,\mathbf{a}_3 \tag{43}$$

and

$$\mathbf{b}_3 = C_{13}\,\mathbf{a}_1 + C_{23}\,\mathbf{a}_2 + C_{33}\,\mathbf{a}_3 \tag{44}$$

These three equations are precisely what one obtains when forming expressions for $\mathbf{b}_1$, $\mathbf{b}_2$, $\mathbf{b}_3$ in accordance with Eq. (2) and with the rules for matrix multiplication; and a similar line of reasoning leads to Eq. (5).

To see that Eq. (9) is valid one needs only to observe that

$$\begin{aligned} {}^B v_i &\underset{(8,2)}{=} \mathbf{v} \cdot (\mathbf{a}_1 C_{1i} + \mathbf{a}_2 C_{2i} + \mathbf{a}_3 C_{3i}) \\ &= \mathbf{v} \cdot \mathbf{a}_1 C_{1i} + \mathbf{v} \cdot \mathbf{a}_2 C_{2i} + \mathbf{v} \cdot \mathbf{a}_3 C_{3i} \\ &\underset{(7)}{=} {}^A v_1 C_{1i} + {}^A v_2 C_{2i} + {}^A v_3 C_{3i} \end{aligned} \tag{45}$$

and to recall the definitions of ${}^A v$ and ${}^B v$. Similarly, Eq. (12) follows from

$$\begin{aligned} {}^B D_{ij} &\underset{(11,2)}{=} (\mathbf{a}_1 C_{1i} + \mathbf{a}_2 C_{2i} + \mathbf{a}_3 C_{3i}) \cdot \mathbf{D} \cdot (\mathbf{a}_1 C_{1j} + \mathbf{a}_2 C_{2j} + \mathbf{a}_3 C_{3j}) \\ &\underset{(10)}{=} C_{1i}\,({}^A D_{11} C_{1j} + {}^A D_{12} C_{2j} + {}^A D_{13} C_{3j}) \\ &\quad + C_{2i}\,({}^A D_{21} C_{1j} + {}^A D_{22} C_{2j} + {}^A D_{23} C_{3j}) \\ &\quad + C_{3i}\,({}^A D_{31} C_{1j} + {}^A D_{32} C_{2j} + {}^A D_{33} C_{3j}) \end{aligned} \tag{46}$$

and from the definitions of ${}^A D$ and ${}^B D$.

As for Eqs. (14) and (15), these are consequences of (using the summation convention)

$$\mathbf{a}_i \cdot \mathbf{a}_j \underset{(1,5)}{=} C_{ik}\,C_{jk} \qquad (i, j = 1, 2, 3) \tag{47}$$

and

$$\mathbf{b}_i \cdot \mathbf{b}_j = C_{ki}\,C_{kj} \qquad (i, j = 1, 2, 3) \tag{48}$$

respectively, because $\mathbf{a}_i \cdot \mathbf{a}_j$ is equal to unity when $i = j$, and equal to zero when $i \neq j$, and similarly for $\mathbf{b}_i \cdot \mathbf{b}_j$; and Eqs. (16) and (17) can be seen to be equivalent to Eqs. (14) and (15), respectively, by referring to Eqs. (3), (4), and (18) when carrying out the indicated multiplications.

To verify that each element of C is equal to its cofactor in the determinant of C, note that

$$\mathbf{b}_1 \underset{(2)}{=} C_{11}\,\mathbf{a}_1 + C_{21}\,\mathbf{a}_2 + C_{31}\,\mathbf{a}_3 \tag{49}$$

and

$$\mathbf{b}_2 \times \mathbf{b}_3 \underset{(2)}{=} (C_{22}C_{33} - C_{32}C_{23})\mathbf{a}_1 + (C_{32}C_{13} - C_{12}C_{33})\mathbf{a}_2 + (C_{12}C_{23} - C_{22}C_{13})\mathbf{a}_3 \tag{50}$$

so that, since $\mathbf{b}_1 = \mathbf{b}_2 \times \mathbf{b}_3$ because $\mathbf{b}_1, \mathbf{b}_2, \mathbf{b}_3$ form a dextral set of orthogonal unit vectors,

$$C_{11} = C_{22}C_{33} - C_{32}C_{23} \tag{51}$$

$$C_{21} = C_{32}C_{13} - C_{12}C_{33} \tag{52}$$

and

$$C_{31} = C_{12}C_{23} - C_{22}C_{13} \tag{53}$$

Thus each element in the first column of C [see Eq. (3)] is seen to be equal to its cofactor in C; and, using the relationships $\mathbf{b}_2 = \mathbf{b}_3 \times \mathbf{b}_1$ and $\mathbf{b}_3 = \mathbf{b}_1 \times \mathbf{b}_2$, one obtains corresponding results for the elements in the second and third columns of C. Furthermore, expanding C by cofactors of elements of the first row, and using Eq. (14) with $i = j = 1$, one arrives at Eq. (19).

The determinant of $C - U$ can be expressed as

$$|C - U| \underset{(3)}{=} |C| + C_{11} + C_{22} + C_{33} + C_{12}C_{21} + C_{23}C_{32} + C_{31}C_{13} - (C_{11}C_{22} + C_{22}C_{33} + C_{33}C_{11}) - 1 \tag{54}$$

Hence, replacing C_{11}, C_{22}, and C_{33} with their respective cofactors in $|C|$, one finds that

$$|C - U| = |C| - 1 \underset{(19)}{=} 0 \tag{55}$$

in agreement with Eq. (20); and the existence of row matrices $[\kappa_1 \quad \kappa_2 \quad \kappa_3]$ satisfying Eq. (21) is thus guaranteed.

The equality of $\boldsymbol{\lambda} \cdot \mathbf{a}_i$ and $\boldsymbol{\lambda} \cdot \mathbf{b}_i$ in Eq. (22) is a consequence of the fact that these two quantities are equal to each other prior to the rotation of B relative to A, that is, when $\mathbf{a}_i = \mathbf{b}_i$, and that neither $\boldsymbol{\lambda} \cdot \mathbf{a}_i$ nor $\boldsymbol{\lambda} \cdot \mathbf{b}_i$ changes during the rotation, since, by construction, $\boldsymbol{\lambda}$ is parallel to a line whose orientation in both A and B remains unaltered during the rotation.

With $\mathbf{a}$ and $\mathbf{b}$ replaced by $\mathbf{a}_j$ and $\mathbf{b}_j$, respectively, Eq. (1.1.3) becomes

$$\mathbf{b}_j = \mathbf{a}_j \cdot \mathbf{C} \tag{56}$$

Hence

$$C_{ij} \underset{(1)}{=} \mathbf{a}_i \cdot \mathbf{b}_j \underset{(56)}{=} \mathbf{a}_i \cdot (\mathbf{a}_j \cdot \mathbf{C}) \qquad (i, j = 1, 2, 3) \tag{57}$$

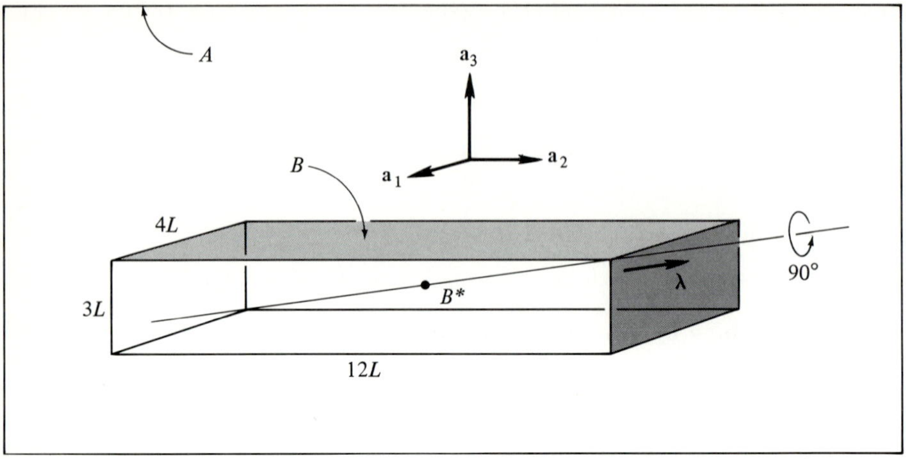

Figure 1.2.1

which is Eq. (34). Moreover, substituting for **C** the expression given in Eq. (1.1.2), one finds that

$$C_{ij} = \mathbf{a}_i \cdot \mathbf{a}_j \cos\theta - \mathbf{a}_i \cdot \mathbf{a}_j \times \boldsymbol{\lambda} \sin\theta + \mathbf{a}_i \cdot \boldsymbol{\lambda\lambda} \cdot \mathbf{a}_j(1 - \cos\theta) \qquad (i, j = 1, 2, 3) \quad (58)$$

and this, together with Eq. (22) leads directly to Eqs. (23)–(31), or, in view of Eq. (32), to Eq. (33).

Equation (35) is obtained by setting $\lambda_1 = 1$ and $\lambda_2 = \lambda_3 = 0$ in Eqs. (23)–(31) and then using Eq. (3). Similarly, $\lambda_1 = 0$, $\lambda_2 = 1$, and $\lambda_3 = 0$ lead to Eq. (36), and $\lambda_1 = \lambda_2 = 0$, $\lambda_3 = 1$ yield Eq. (37).

Finally, Eq. (39) is derived from the observation that†

$$\begin{aligned} \boldsymbol{\lambda} \cdot \mathbf{C} &\underset{(1.1.2)}{=} \boldsymbol{\lambda} \cdot \mathbf{U} \cos\theta - \boldsymbol{\lambda} \times \boldsymbol{\lambda} \sin\theta + \boldsymbol{\lambda}^2 \boldsymbol{\lambda}(1 - \cos\theta) \\ &= \boldsymbol{\lambda} \cos\theta + 0 + \boldsymbol{\lambda}(1 - \cos\theta) = \boldsymbol{\lambda} \end{aligned} \quad (59)$$

since $\boldsymbol{\lambda}$ is a unit vector, so that $\boldsymbol{\lambda}^2 \triangleq \boldsymbol{\lambda} \cdot \boldsymbol{\lambda} = 1$; and the equivalence of Eqs. (38) and (39) follows from Eqs. (22) and (34).

Example In Fig. 1.2.1, B designates a uniform rectangular block which is part of a scanning platform mounted in a spacecraft A. Initially, the edges of B are parallel to unit vectors $\mathbf{a}_1$, $\mathbf{a}_2$ and $\mathbf{a}_3$ which are fixed in A, and the platform is then subjected to a simple 90° rotation about a diagonal of B, as indicated in the sketch. If **I** is the inertia dyadic of B for the mass center B^* of B, and ${}^A I_{ij}$ is defined as

† When it is necessary to refer to an equation from an earlier section, the section number is cited together with the equation number. For example, (1.1.2) refers to Eq. (2) in Sec. 1.1.

$$ {}^{A}I_{ij} \triangleq \mathbf{a}_i \cdot \mathbf{I} \cdot \mathbf{a}_j \qquad (i, j = 1, 2, 3) \tag{60} $$

what is the value of ${}^{A}I_{ij}$ $(i, j = 1, 2, 3)$ subsequent to the rotation?

Let $\mathbf{b}_i$ $(i=1, 2, 3)$ be a unit vector fixed in B and equal to $\mathbf{a}_i$ $(i=1, 2, 3)$ prior to the rotation; and define ${}^{B}I_{ij}$ as

$$ {}^{B}I_{ij} \triangleq \mathbf{b}_i \cdot \mathbf{I} \cdot \mathbf{b}_j \qquad (i, j = 1, 2, 3) \tag{61} $$

Then $\mathbf{b}_1$, $\mathbf{b}_2$, and $\mathbf{b}_3$ are parallel to principal axes of inertia of B for B^*, so that

$$ {}^{B}I_{12} = {}^{B}I_{21} = {}^{B}I_{23} = {}^{B}I_{32} = {}^{B}I_{31} = {}^{B}I_{13} = 0 \tag{62} $$

and, if m is the mass of B,

$$ {}^{B}I_{11} = \frac{m}{12}(12^2 + 3^2)L^2 = \frac{153}{12} mL^2 \tag{63} $$

$$ {}^{B}I_{22} = \frac{m}{12}(3^2 + 4^2)L^2 = \frac{25}{12} mL^2 \tag{64} $$

and

$$ {}^{B}I_{33} = \frac{m}{12}(4^2 + 12^2)L^2 = \frac{160}{12} mL^2 \tag{65} $$

Hence, if ${}^{B}I$ denotes the square matrix having ${}^{B}I_{ij}$ as the element in the ith row and jth column, then

$$ {}^{B}I = \frac{mL^2}{12} \begin{bmatrix} 153 & 0 & 0 \\ 0 & 25 & 0 \\ 0 & 0 & 160 \end{bmatrix} \tag{66} $$

The unit vector $\boldsymbol{\lambda}$ shown in Fig. 1.2.1 can be expressed as

$$ \boldsymbol{\lambda} = \frac{4\mathbf{a}_1 + 12\mathbf{a}_2 + 3\mathbf{a}_3}{(4^2 + 12^2 + 3^2)^{1/2}} = \frac{4}{13}\mathbf{a}_1 + \frac{12}{13}\mathbf{a}_2 + \frac{3}{13}\mathbf{a}_3 \tag{67} $$

Consequently, λ_1, λ_2, and λ_3, if defined as in Eq. (22), are given by

$$ \lambda_1 = \tfrac{4}{13} \qquad \lambda_2 = \tfrac{12}{13} \qquad \lambda_3 = \tfrac{3}{13} \tag{68} $$

and, with $\theta = \pi/2$ rad, Eqs. (23)–(31) lead to the following expression for the direction cosine matrix C:

$$ C \underset{(3)}{=} \frac{1}{169} \begin{bmatrix} 16 & 9 & 168 \\ 87 & 144 & -16 \\ -144 & 88 & 9 \end{bmatrix} \tag{69} $$

If ${}^{A}I$ is now defined as the square matrix having ${}^{A}I_{ij}$ as the element in the ith row and jth column, then

$$ {}^{B}I \underset{(12)}{=} C^T \, {}^{A}I \, C \tag{70} $$

and simultaneous premultiplication with C and postmultiplication with C^T gives

$$C\,{}^BI\,C^T \underset{(70)}{=} C\,C^T\,{}^AI\,C\,C^T \underset{(16)}{=} U\,{}^AI\,U = {}^AI \tag{71}$$

Consequently,

$${}^AI \underset{(66,69,71)}{=} \frac{mL^2}{12 \times 169 \times 169} \begin{bmatrix} 16 & 9 & 168 \\ 87 & 144 & -16 \\ -144 & 88 & 9 \end{bmatrix} \begin{bmatrix} 153 & 0 & 0 \\ 0 & 25 & 0 \\ 0 & 0 & 160 \end{bmatrix} \begin{bmatrix} 16 & 87 & -144 \\ 9 & 144 & 88 \\ 168 & -16 & 9 \end{bmatrix}$$

$$= \frac{mL^2}{12 \times 169 \times 169} \begin{bmatrix} 4{,}557{,}033 & -184{,}704 & -90{,}792 \\ -184{,}704 & 1{,}717{,}417 & -1{,}623{,}024 \\ -90{,}792 & -1{,}623{,}034 & 3{,}379{,}168 \end{bmatrix} \tag{72}$$

and

$${}^AI_{11} = \frac{4{,}557{,}033\; mL^2}{12 \times 169 \times 169} \qquad {}^AI_{12} = -\frac{184{,}704\; mL^2}{12 \times 169 \times 169} \tag{73}$$

and so forth.

1.3 EULER PARAMETERS

The unit vector $\boldsymbol{\lambda}$ and the angle θ introduced in Sec. 1.1 can be used to associate a vector $\boldsymbol{\epsilon}$, called the *Euler vector*, and four scalar quantities, $\epsilon_1, \ldots, \epsilon_4$, called *Euler parameters*, with a simple rotation of a rigid body B in a reference frame A by letting

$$\boldsymbol{\epsilon} \overset{\triangle}{=} \boldsymbol{\lambda} \sin \frac{\theta}{2} \tag{1}$$

$$\epsilon_i \overset{\triangle}{=} \boldsymbol{\epsilon} \cdot \mathbf{a}_i = \boldsymbol{\epsilon} \cdot \mathbf{b}_i \qquad (i = 1, 2, 3) \tag{2}$$

and

$$\epsilon_4 \overset{\triangle}{=} \cos \frac{\theta}{2} \tag{3}$$

where $\mathbf{a}_1, \mathbf{a}_2, \mathbf{a}_3$ and $\mathbf{b}_1, \mathbf{b}_2, \mathbf{b}_3$ are dextral sets of orthogonal unit vectors fixed in A and B respectively, with $\mathbf{a}_i = \mathbf{b}_i$ $(i = 1, 2, 3)$ prior to the rotation. (Where a discussion involves more than two bodies or reference frames, notations such as ${}^A\boldsymbol{\epsilon}^B$ and ${}^A\epsilon_i^B$ will be used.)

The Euler parameters are not independent of each other, for the sum of their squares is necessarily equal to unity:

$$\epsilon_1^2 + \epsilon_2^2 + \epsilon_3^2 + \epsilon_4^2 = \boldsymbol{\epsilon}^2 + \epsilon_4^2 = 1 \tag{4}$$

An indication of the utility of the Euler parameters may be gleaned from the fact that the elements of the direction cosine matrix C introduced in Sec. 1.2 assume a particularly simple and orderly form when expressed in terms of $\epsilon_1, \ldots, \epsilon_4$: If C_{ij} is defined as

$$C_{ij} \triangleq \mathbf{a}_i \cdot \mathbf{b}_j \qquad (i, j = 1, 2, 3) \tag{5}$$

then

$$C_{11} = \epsilon_1^2 - \epsilon_2^2 - \epsilon_3^2 + \epsilon_4^2 = 1 - 2\epsilon_2^2 - 2\epsilon_3^2 \tag{6}$$

$$C_{12} = 2(\epsilon_1 \epsilon_2 - \epsilon_3 \epsilon_4) \tag{7}$$

$$C_{13} = 2(\epsilon_3 \epsilon_1 + \epsilon_2 \epsilon_4) \tag{8}$$

$$C_{21} = 2(\epsilon_1 \epsilon_2 + \epsilon_3 \epsilon_4) \tag{9}$$

$$C_{22} = \epsilon_2^2 - \epsilon_3^2 - \epsilon_1^2 + \epsilon_4^2 = 1 - 2\epsilon_3^2 - 2\epsilon_1^2 \tag{10}$$

$$C_{23} = 2(\epsilon_2 \epsilon_3 - \epsilon_1 \epsilon_4) \tag{11}$$

$$C_{31} = 2(\epsilon_3 \epsilon_1 - \epsilon_2 \epsilon_4) \tag{12}$$

$$C_{32} = 2(\epsilon_2 \epsilon_3 + \epsilon_1 \epsilon_4) \tag{13}$$

$$C_{33} = \epsilon_3^2 - \epsilon_1^2 - \epsilon_2^2 + \epsilon_4^2 = 1 - 2\epsilon_1^2 - 2\epsilon_2^2 \tag{14}$$

The Euler parameters can be expressed in terms of direction cosines in such a way that Eqs. (6)–(14) are satisfied identically. This is accomplished by taking

$$\epsilon_1 = \frac{C_{32} - C_{23}}{4\epsilon_4} \tag{15}$$

$$\epsilon_2 = \frac{C_{13} - C_{31}}{4\epsilon_4} \tag{16}$$

$$\epsilon_3 = \frac{C_{21} - C_{12}}{4\epsilon_4} \tag{17}$$

and

$$\epsilon_4 = \frac{1}{2}(1 + C_{11} + C_{22} + C_{33})^{1/2} \tag{18}$$

Since Eqs. (1) and (3) are satisfied if

$$\boldsymbol{\lambda} = \frac{\epsilon_1 \mathbf{a}_1 + \epsilon_2 \mathbf{a}_2 + \epsilon_3 \mathbf{a}_3}{(\epsilon_1^2 + \epsilon_2^2 + \epsilon_3^2)^{1/2}} \tag{19}$$

and

$$\theta = 2 \cos^{-1} \epsilon_4 \qquad 0 \le \theta \le \pi \tag{20}$$

one can thus find a simple rotation such that the direction cosines associated with this rotation as in Eqs. (1.2.23)–(1.2.31) are equal to corresponding elements of any direction cosine matrix C that satisfies Eq. (1.2.2). In other words, every change in the relative orientation of two rigid bodies or reference frames A and B can be produced by means of a simple rotation of B in A. This proposition is known as *Euler's theorem on rotation*.

As an alternative to Eqs. (1.1.1) and (1.1.3), the relationship between a vector **a** fixed in a reference frame A and a vector **b** fixed in a rigid body B and equal to **a** prior to a simple rotation of B in A can be expressed in terms of $\boldsymbol{\epsilon}$ and ϵ_4 as

$$\mathbf{b} = \mathbf{a} + 2[\epsilon_4\, \boldsymbol{\epsilon} \times \mathbf{a} + \boldsymbol{\epsilon} \times (\boldsymbol{\epsilon} \times \mathbf{a})] \tag{21}$$

Derivations The equality of $\boldsymbol{\epsilon} \cdot \mathbf{a}_i$ and $\boldsymbol{\epsilon} \cdot \mathbf{b}_i$ [see Eq. (2)] follows from Eqs. (1) and (1.2.22); Eqs. (4) are consequences of Eqs. (1)–(3) and of the fact that $\boldsymbol{\lambda}$ is a unit vector; and Eqs. (6)–(14) can be obtained from Eqs. (1.2.23)–(1.2.31) by replacing functions of θ with functions of $\theta/2$ and using Eq. (1.2.22) together with Eqs. (1)–(4). For example,

$$C_{11} \underset{(1.2.23)}{=} 2\cos^2\frac{\theta}{2} - 1 + 2\lambda_1^2 \sin^2\frac{\theta}{2} \tag{22}$$

and

$$\lambda_1 \sin\frac{\theta}{2} \underset{(1.2.22)}{=} \boldsymbol{\lambda} \cdot \mathbf{a}_1 \sin\frac{\theta}{2} \underset{(1)}{=} \boldsymbol{\epsilon} \cdot \mathbf{a}_1 \underset{(2)}{=} \epsilon_1 \tag{23}$$

while

$$\cos\frac{\theta}{2} \underset{(3)}{=} \epsilon_4 \tag{24}$$

Hence,

$$C_{11} = 2\epsilon_4^2 - 1 + 2\epsilon_1^2 \underset{(4)}{=} \epsilon_1^2 - \epsilon_2^2 - \epsilon_3^2 + \epsilon_4^2 \tag{25}$$

in agreement with Eq. (6).

The validity of Eqs. (15)–(18) can be established by showing that the left-hand members of Eqs. (6)–(14) may be obtained by substituting from Eqs. (15)-(18) into the right-hand members. For example,

$$\begin{aligned} 1 - 2\epsilon_2^2 - 2\epsilon_3^2 &\underset{(16\text{-}18)}{=} \frac{2(1 + C_{11} + C_{22} + C_{33}) - C_{13}^2 + 2C_{13}C_{31} - C_{31}^2 - C_{21}^2 + 2C_{12}C_{21} - C_{12}^2}{2(1 + C_{11} + C_{22} + C_{33})} \\ &\underset{(1.2.14)}{=} \frac{C_{11} + C_{22} + C_{33} + C_{13}C_{31} + C_{12}C_{21} + C_{11}^2}{1 + C_{11} + C_{22} + C_{33}} \end{aligned} \tag{26}$$

But, since each element of C is equal to its cofactor in $|C|$,

$$C_{13}C_{31} = C_{11}C_{33} - C_{22} \tag{27}$$

and

$$C_{12}C_{21} = C_{11}C_{22} - C_{33} \tag{28}$$

Consequently,

$$1 - 2\epsilon_2^2 - 2\epsilon_3^2 = \frac{C_{11} + C_{11}C_{33} + C_{11}C_{22} + C_{11}^2}{1 + C_{11} + C_{22} + C_{33}} = C_{11} \tag{29}$$

as required by Eq. (6).

To see that Eqs. (1) and (3) are satisfied if $\boldsymbol{\lambda}$ and θ are given by Eqs. (19) and (20), note that

$$\cos\frac{\theta}{2} \underset{(20)}{=} \epsilon_4 \tag{30}$$

which is Eq. (3), and that

$$\sin\frac{\theta}{2} \underset{(20)}{=} (1 - \epsilon_4^2)^{1/2} \underset{(4)}{=} (\epsilon_1^2 + \epsilon_2^2 + \epsilon_3^2)^{1/2} \tag{31}$$

so that

$$\boldsymbol{\lambda} \sin\frac{\theta}{2} \underset{(19)}{=} \epsilon_1 \mathbf{a}_1 + \epsilon_2 \mathbf{a}_2 + \epsilon_3 \mathbf{a}_3 \underset{(2)}{=} \boldsymbol{\epsilon} \tag{32}$$

as required by Eq. (1). Finally, Eq. (1.1.1) is equivalent to

$$\begin{aligned} \mathbf{b} &= \mathbf{a} + \boldsymbol{\lambda} \times \mathbf{a} \sin\theta + \boldsymbol{\lambda} \times (\boldsymbol{\lambda} \times \mathbf{a})(1 - \cos\theta) \\ &\underset{(1)}{=} \mathbf{a} + 2\boldsymbol{\epsilon} \times \mathbf{a} \cos\frac{\theta}{2} + 2\boldsymbol{\epsilon} \times (\boldsymbol{\epsilon} \times \mathbf{a}) \underset{(3)}{=} \mathbf{a} + 2\left[\epsilon_4 \boldsymbol{\epsilon} \times \mathbf{a} + \boldsymbol{\epsilon} \times (\boldsymbol{\epsilon} \times \mathbf{a})\right] \end{aligned} \tag{33}$$

in agreement with Eq. (21).

Example Triangle *ABC* in Fig. 1.3.1 can be brought into the position *A'B'C'* by moving point *A* to *A'*, without changing the orientation of the triangle, and then performing a simple rotation of the triangle while keeping *A* fixed at *A'*. To find $\boldsymbol{\lambda}$, a unit vector parallel to the axis of rotation, and to determine θ, the associated angle of rotation, let the unit vectors $\mathbf{a}_i$ and $\mathbf{b}_i$ $(i = 1, 2, 3)$ be directed as shown in Fig. 1.3.1, thus insuring that $\mathbf{a}_i = \mathbf{b}_i$ $(i = 1, 2, 3)$ prior to the rotation; determine C_{ij} by evaluating $\mathbf{a}_i \cdot \mathbf{b}_j$; and use Eqs. (15)–(18) to form ϵ_i $(i = 1, \ldots, 4)$:

$$\begin{aligned} \epsilon_4 &\underset{(18)}{=} \frac{1}{2}(1 + \mathbf{a}_1 \cdot \mathbf{b}_1 + \mathbf{a}_2 \cdot \mathbf{b}_2 + \mathbf{a}_3 \cdot \mathbf{b}_3)^{1/2} \\ &= \frac{1}{2}(1 + 0 + 0 + 0)^{1/2} = \frac{1}{2} \end{aligned} \tag{34}$$

$$\epsilon_1 \underset{(15)}{=} \frac{\mathbf{a}_3 \cdot \mathbf{b}_2 - \mathbf{a}_2 \cdot \mathbf{b}_3}{4(\frac{1}{2})} = \frac{1 - 0}{2} = \frac{1}{2} \tag{35}$$

$$\epsilon_2 \underset{(16)}{=} \frac{\mathbf{a}_1 \cdot \mathbf{b}_3 - \mathbf{a}_3 \cdot \mathbf{b}_1}{4(\frac{1}{2})} = \frac{-1 - 0}{2} = -\frac{1}{2} \tag{36}$$

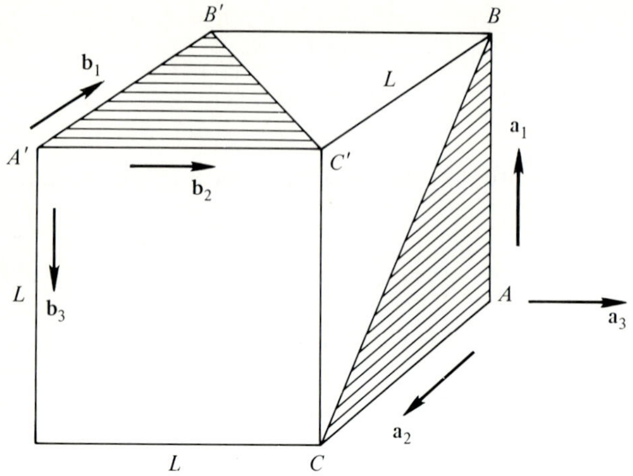

Figure 1.3.1

$$\epsilon_3 \underset{(17)}{=} \frac{\mathbf{a}_2 \cdot \mathbf{b}_1 - \mathbf{a}_1 \cdot \mathbf{b}_2}{4(\frac{1}{2})} = \frac{-1-0}{2} = -\frac{1}{2} \tag{37}$$

Then

$$\boldsymbol{\lambda} \underset{(19)}{=} \frac{\frac{1}{2}\mathbf{a}_1 - \frac{1}{2}\mathbf{a}_2 - \frac{1}{2}\mathbf{a}_3}{(\frac{3}{4})^{1/2}} = \frac{\mathbf{a}_1 - \mathbf{a}_2 - \mathbf{a}_3}{\sqrt{3}} \tag{38}$$

and

$$\theta \underset{(20)}{=} 2\cos^{-1}\frac{1}{2} = \frac{2\pi}{3} \text{ rad} \tag{39}$$

1.4 RODRIGUES PARAMETERS

A vector $\boldsymbol{\rho}$, called the *Rodrigues vector*, and three scalar quantities, ρ_1, ρ_2, ρ_3, called *Rodrigues parameters*, can be associated with a simple rotation of a rigid body B in a reference frame A (see Sec. 1.1) by letting

$$\boldsymbol{\rho} \triangleq \boldsymbol{\lambda} \tan\frac{\theta}{2} \tag{1}$$

and

$$\rho_i \triangleq \boldsymbol{\rho} \cdot \mathbf{a}_i = \boldsymbol{\rho} \cdot \mathbf{b}_i \qquad (i = 1, 2, 3) \tag{2}$$

where $\boldsymbol{\lambda}$ and θ have the same meaning as in Sec. 1.1, and $\mathbf{a}_1$, $\mathbf{a}_2$, $\mathbf{a}_3$ and $\mathbf{b}_1$, $\mathbf{b}_2$, $\mathbf{b}_3$ are dextral sets of orthogonal unit vectors fixed in A and B, respectively, with $\mathbf{a}_i = \mathbf{b}_i$ $(i = 1, 2, 3)$ prior to the rotation. (When a discussion involves more than two bodies or reference frames, notations such as ${}^A\boldsymbol{\rho}^B$ and ${}^A\rho_i^B$ will be used.)

The Rodrigues parameters are intimately related to the Euler parameters (see Sec. 1.3):

$$\rho_i = \frac{\epsilon_i}{\epsilon_4} \qquad (i = 1, 2, 3) \tag{3}$$

An advantage of the Rodrigues parameters over the Euler parameters is that they are fewer in number; but this advantage is at times offset by the fact that the Rodrigues parameters can become infinite, whereas the absolute value of any Euler parameter cannot exceed unity.

Expressed in terms of Rodrigues parameters, the direction cosine matrix C (see Sec. 1.2) assumes the form

$$C = \frac{\begin{bmatrix} 1 + \rho_1^2 - \rho_2^2 - \rho_3^2 & 2(\rho_1\,\rho_2 - \rho_3) & 2(\rho_3\,\rho_1 + \rho_2) \\ 2(\rho_1\,\rho_2 + \rho_3) & 1 + \rho_2^2 - \rho_3^2 - \rho_1^2 & 2(\rho_2\,\rho_3 - \rho_1) \\ 2(\rho_3\,\rho_1 - \rho_2) & 2(\rho_2\,\rho_3 + \rho_1) & 1 + \rho_3^2 - \rho_1^2 - \rho_2^2 \end{bmatrix}}{1 + \rho_1^2 + \rho_2^2 + \rho_3^2} \tag{4}$$

The Rodrigues vector can be used to establish a simple relationship between the difference and the sum of the vectors $\mathbf{a}$ and $\mathbf{b}$ defined in Sec. 1.1:

$$\mathbf{a} - \mathbf{b} = (\mathbf{a} + \mathbf{b}) \times \boldsymbol{\rho} \tag{5}$$

This relationship will be found useful in connection with a number of derivations, such as the one showing that the following is an expression for a Rodrigues vector that characterizes a simple rotation by means of which a specified change in the relative orientation of A and B can be produced:

$$\boldsymbol{\rho} = \frac{(\boldsymbol{\alpha}_1 - \boldsymbol{\beta}_1) \times (\boldsymbol{\alpha}_2 - \boldsymbol{\beta}_2)}{(\boldsymbol{\alpha}_1 + \boldsymbol{\beta}_1) \cdot (\boldsymbol{\alpha}_2 - \boldsymbol{\beta}_2)} \tag{6}$$

where $\boldsymbol{\alpha}_i$ and $\boldsymbol{\beta}_i$ $(i = 1, 2)$ are vectors fixed in A and B, respectively, and $\boldsymbol{\alpha}_i = \boldsymbol{\beta}_i$ $(i = 1, 2)$ prior to the change in relative orientation. Rodrigues parameters for such a rotation can be expressed as

$$\rho_1 = \frac{C_{32} - C_{23}}{1 + C_{11} + C_{22} + C_{33}} \tag{7}$$

$$\rho_2 = \frac{C_{13} - C_{31}}{1 + C_{11} + C_{22} + C_{33}} \tag{8}$$

$$\rho_3 = \frac{C_{21} - C_{12}}{1 + C_{11} + C_{22} + C_{33}} \tag{9}$$

Derivations The equality of $\boldsymbol{\rho} \cdot \mathbf{a}_i$ and $\boldsymbol{\rho} \cdot \mathbf{b}_i$ [see Eq. (2)] follows from Eqs. (1) and (1.2.22); and Eqs. (3) are obtained by noting that

$$\frac{\boldsymbol{\epsilon}}{\epsilon_4} \underset{(1.3.1,\ 1.3.3)}{=} \boldsymbol{\lambda} \tan\frac{\theta}{2} \underset{(1)}{=} \boldsymbol{\rho} \tag{10}$$

so that

$$\rho_i \underset{(2)}{=} \frac{\boldsymbol{\epsilon} \cdot \mathbf{a}_i}{\epsilon_4} \underset{(1.3.2)}{=} \frac{\epsilon_i}{\epsilon_4} \qquad (i = 1, 2, 3) \tag{11}$$

From Eq. (1.3.4),

$$1 = \epsilon_1^2 + \epsilon_2^2 + \epsilon_3^2 + \epsilon_4^2 \underset{(3)}{=} (\rho_1^2 + \rho_2^2 + \rho_3^2 + 1)\epsilon_4^2 \tag{12}$$

Hence,

$$\epsilon_4^2 = \frac{1}{1 + \rho_1^2 + \rho_2^2 + \rho_3^2} \tag{13}$$

and

$$\begin{aligned} C_{11} &\underset{(1.3.6)}{=} \epsilon_1^2 - \epsilon_2^2 - \epsilon_3^2 + \epsilon_4^2 \underset{(3)}{=} \rho_1^2\,\epsilon_4^2 - \rho_2^2\,\epsilon_4^2 - \rho_3^2\,\epsilon_4^2 + \epsilon_4^2 \\ &= (\rho_1^2 - \rho_2^2 - \rho_3^2 + 1)\epsilon_4^2 = \frac{\rho_1^2 - \rho_2^2 - \rho_3^2 + 1}{1 + \rho_1^2 + \rho_2^2 + \rho_3^2} \end{aligned} \tag{14}$$

in agreement with Eq. (4), and the remaining elements of C are found similarly.

As for Eq. (5), note that cross-multiplication of Eq. (1.3.21) with $\boldsymbol{\epsilon}$ yields

$$\boldsymbol{\epsilon} \times \mathbf{b} = \boldsymbol{\epsilon} \times \mathbf{a} + 2\,\{\epsilon_4\,\boldsymbol{\epsilon} \times (\boldsymbol{\epsilon} \times \mathbf{a}) + \boldsymbol{\epsilon} \times [\boldsymbol{\epsilon} \times (\boldsymbol{\epsilon} \times \mathbf{a})]\} \tag{15}$$

Hence,

$$\begin{aligned} \boldsymbol{\epsilon} \times (\mathbf{a} + \mathbf{b}) &= \boldsymbol{\epsilon} \times \mathbf{a} + \boldsymbol{\epsilon} \times \mathbf{b} \\ &\underset{(15)}{=} 2\{\boldsymbol{\epsilon} \times \mathbf{a} + \epsilon_4\,\boldsymbol{\epsilon} \times (\boldsymbol{\epsilon} \times \mathbf{a}) + \boldsymbol{\epsilon} \times [\boldsymbol{\epsilon} \times (\boldsymbol{\epsilon} \times \mathbf{a})]\} \end{aligned} \tag{16}$$

Now

$$\begin{aligned} \boldsymbol{\epsilon} \times [\boldsymbol{\epsilon} \times (\boldsymbol{\epsilon} \times \mathbf{a})] &= -\boldsymbol{\epsilon}^2\,\boldsymbol{\epsilon} \times \mathbf{a} \underset{(1.3.2)}{=} -(\epsilon_1^2 + \epsilon_2^2 + \epsilon_3^2)\boldsymbol{\epsilon} \times \mathbf{a} \\ &\underset{(1.3.4)}{=} (\epsilon_4^2 - 1)\boldsymbol{\epsilon} \times \mathbf{a} \end{aligned} \tag{17}$$

Consequently,

$$\boldsymbol{\epsilon} \times (\mathbf{a} + \mathbf{b}) \underset{(16,17)}{=} 2\epsilon_4[\epsilon_4\,\boldsymbol{\epsilon} \times \mathbf{a} + \boldsymbol{\epsilon} \times (\boldsymbol{\epsilon} \times \mathbf{a})] \underset{(1.3.21)}{=} \epsilon_4(\mathbf{b} - \mathbf{a}) \tag{18}$$

and

$$\mathbf{a} - \mathbf{b} \underset{(18)}{=} (\mathbf{a} + \mathbf{b}) \times \frac{\boldsymbol{\epsilon}}{\epsilon_4} \underset{(1.3.1,\ 1.3.2)}{=} (\mathbf{a} + \mathbf{b}) \times \boldsymbol{\lambda} \tan\frac{\theta}{2} \underset{(1)}{=} (\mathbf{a} + \mathbf{b}) \times \boldsymbol{\rho} \tag{19}$$

which is Eq. (5).

Equation (6) can now be obtained by observing that

$$\boldsymbol{\alpha}_1 - \boldsymbol{\beta}_1 \underset{(5)}{=} (\boldsymbol{\alpha}_1 + \boldsymbol{\beta}_1) \times \boldsymbol{\rho} \tag{20}$$

and

$$\boldsymbol{\alpha}_2 - \boldsymbol{\beta}_2 \underset{(5)}{=} (\boldsymbol{\alpha}_2 + \boldsymbol{\beta}_2) \times \boldsymbol{\rho} \tag{21}$$

so that

$$
\begin{aligned}
(\boldsymbol{\alpha}_1 - \boldsymbol{\beta}_1) \times (\boldsymbol{\alpha}_2 - \boldsymbol{\beta}_2) &\underset{(20)}{=} [(\boldsymbol{\alpha}_1 + \boldsymbol{\beta}_1) \times \boldsymbol{\rho}] \times (\boldsymbol{\alpha}_2 - \boldsymbol{\beta}_2) \\
&= (\boldsymbol{\alpha}_1 + \boldsymbol{\beta}_1) \cdot (\boldsymbol{\alpha}_2 - \boldsymbol{\beta}_2)\boldsymbol{\rho} - (\boldsymbol{\alpha}_1 + \boldsymbol{\beta}_1)\boldsymbol{\rho} \cdot (\boldsymbol{\alpha}_2 - \boldsymbol{\beta}_2) \\
&\underset{(21)}{=} (\boldsymbol{\alpha}_1 + \boldsymbol{\beta}_1) \cdot (\boldsymbol{\alpha}_2 - \boldsymbol{\beta}_2)\boldsymbol{\rho} - 0
\end{aligned}
\tag{22}
$$

from which Eq. (6) follows immediately.

Finally, to establish the validity of Eqs. (7)–(9), note that

$$
\rho_1 \underset{(3)}{=} \frac{\epsilon_1}{\epsilon_4} \underset{(1.3.15)}{=} \frac{C_{32} - C_{23}}{4\epsilon_4^2} \underset{(1.3.18)}{=} \frac{C_{32} - C_{23}}{1 + C_{11} + C_{22} + C_{33}} \tag{23}
$$

in agreement with Eq. (7); and Eqs. (8) and (9) can be obtained by cyclic permutation of the subscripts in Eq. (7).

Example Referring to Fig. 1.4.1, which depicts the rigid body B previously considered in the example in Sec. 1.1, suppose that B is subjected to a 180° rotation relative to A about an axis parallel to the unit vector $\boldsymbol{\lambda}$; and let $\mathbf{b}_i$ be a unit vector fixed in B and equal to $\mathbf{a}_i$ ($i = 1, 2, 3$) prior to the rotation. The direction cosine matrix C satisfying Eq. (1.2.2) subsequent to the rotation is to be determined.

For $\theta = \pi$ rad, Eq. (1) yields a Rodrigues vector of infinite magnitude, and Eqs. (2) and (4) lead to an indeterminate form of C. To evaluate this

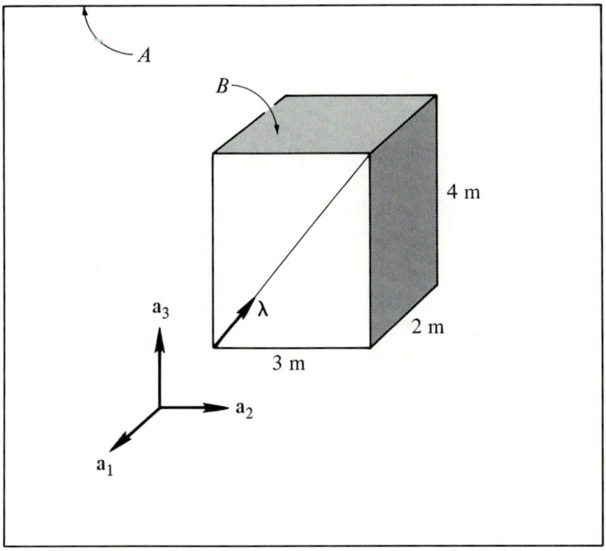

Figure 1.4.1

indeterminate form, one may express each element of C in terms of θ and $\boldsymbol{\lambda} \cdot \mathbf{a}_i$ ($i = 1, 2, 3$) by reference to Eqs. (1) and (2), and then determine the limit approached by the resulting expression as θ approaches π rad. Alternatively, one can use either Eqs. (1.2.23)–(1.2.31) or Eqs. (1.3.6)–(1.3.14) to find the elements of C. The latter equations are particularly convenient, because, for $\theta = \pi$,

$$\boldsymbol{\epsilon} \underset{(1.3.1)}{=} \boldsymbol{\lambda} \tag{24}$$

so that

$$\epsilon_i \underset{(1.3.2)}{=} \boldsymbol{\lambda} \cdot \mathbf{a}_i \qquad (i = 1, 2, 3) \tag{25}$$

and

$$\epsilon_4 \underset{(1.3.3)}{=} 0 \tag{26}$$

Hence, with

$$\boldsymbol{\lambda} = \tfrac{3}{5}\mathbf{a}_2 + \tfrac{4}{5}\mathbf{a}_3 \tag{27}$$

one finds immediately that

$$\epsilon_1 = 0 \qquad \epsilon_2 = \tfrac{3}{5} \qquad \epsilon_3 = \tfrac{4}{5} \tag{28}$$

and substitution into Eqs. (1.3.6)–(1.3.14) then gives

$$C = \frac{1}{25}\begin{bmatrix} -25 & 0 & 0 \\ 0 & -7 & 24 \\ 0 & 24 & 7 \end{bmatrix} \tag{29}$$

1.5 INDIRECT DETERMINATION OF ORIENTATION

If $\mathbf{a}_1$, $\mathbf{a}_2$, $\mathbf{a}_3$, and $\mathbf{b}_1$, $\mathbf{b}_2$, $\mathbf{b}_3$ are dextral sets of orthogonal unit vectors fixed in reference frames A and B respectively, and the orientation of each unit vector in both reference frames is known, then a description of the relative orientation of the two reference frames can be given in terms of direction cosines C_{ij} ($i, j = 1, 2, 3$), for, in accordance with Eq. (1.2.1), these can be found by simply evaluating $\mathbf{a}_i \cdot \mathbf{b}_j$ ($i, j = 1, 2, 3$). But, even when these dot products cannot be evaluated so directly, it may be possible to find C_{ij} ($i, j = 1, 2, 3$). This is the case, for example, when each of two nonparallel vectors, say $\mathbf{p}$ and $\mathbf{q}$, has a known orientation in both A and B, so that the dot products $\mathbf{a}_i \cdot \mathbf{p}$, $\mathbf{a}_i \cdot \mathbf{q}$, $\mathbf{b}_i \cdot \mathbf{p}$, and $\mathbf{b}_i \cdot \mathbf{q}$ ($i = 1, 2, 3$) can be evaluated directly. In that event, one can find C_{ij} as follows: Form a vector $\mathbf{r}$ and a dyadic $\boldsymbol{\sigma}$ by letting

$$\mathbf{r} \triangleq \mathbf{p} \times \mathbf{q} \tag{1}$$

and

$$\boldsymbol{\sigma} \triangleq \frac{\mathbf{p} \times \mathbf{q}\mathbf{r} + \mathbf{q} \times \mathbf{r}\mathbf{p} + \mathbf{r} \times \mathbf{p}\mathbf{q}}{\mathbf{r}^2} \tag{2}$$

Next, express the first member of each dyad in Eq. (2) in terms of $\mathbf{a}_i$, and the second member in terms of $\mathbf{b}_i$ ($i = 1, 2, 3$). Finally, carry out the multiplications indicated in the relationship

$$C_{ij} = \mathbf{a}_i \cdot \boldsymbol{\sigma} \cdot \mathbf{b}_j \qquad (i, j = 1, 2, 3) \tag{3}$$

Derivation It will be shown that the dyadic $\boldsymbol{\sigma}$ defined in Eq. (2) is a unit dyadic, that is, that for every vector $\mathbf{v}$,

$$\mathbf{v} = \mathbf{v} \cdot \boldsymbol{\sigma} \tag{4}$$

The validity of Eq. (3) can then be seen to be an immediate consequence of the definition of C_{ij}, given in Eq. (1.2.1).

If $\mathbf{p}$ and $\mathbf{q}$ are nonparallel, and $\mathbf{r}$ is defined as in Eq. (1), then every vector $\mathbf{v}$ can be expressed as

$$\mathbf{v} = \alpha\mathbf{p} + \beta\mathbf{q} + \gamma\mathbf{r} \tag{5}$$

where α, β, and γ are certain scalars. From Eq. (5),

$$\begin{aligned}\mathbf{v} \cdot (\mathbf{q} \times \mathbf{r}) &= \alpha\mathbf{p} \cdot (\mathbf{q} \times \mathbf{r}) + \beta\mathbf{q} \cdot (\mathbf{q} \times \mathbf{r}) + \gamma\mathbf{r} \cdot (\mathbf{q} \times \mathbf{r}) \\ &= \alpha(\mathbf{p} \times \mathbf{q}) \cdot \mathbf{r} + 0 + 0 \underset{(1)}{=} \alpha\mathbf{r}^2\end{aligned} \tag{6}$$

so that

$$\alpha \underset{(6)}{=} \mathbf{v} \cdot \frac{\mathbf{q} \times \mathbf{r}}{\mathbf{r}^2} \tag{7}$$

Similarly, scalar multiplication of Eq. (5) with $\mathbf{r} \times \mathbf{p}$ and $\mathbf{p} \times \mathbf{q}$ leads to the conclusion that

$$\beta = \mathbf{v} \cdot \frac{\mathbf{r} \times \mathbf{p}}{\mathbf{r}^2} \tag{8}$$

and

$$\gamma = \mathbf{v} \cdot \frac{\mathbf{p} \times \mathbf{q}}{\mathbf{r}^2} \tag{9}$$

Substituting from Eqs. (7)–(9) into Eq. (5), one thus finds that

$$\begin{aligned}\mathbf{v} &= \frac{\mathbf{v} \cdot (\mathbf{q} \times \mathbf{r})\mathbf{p} + \mathbf{v} \cdot (\mathbf{r} \times \mathbf{p})\mathbf{q} + \mathbf{v} \cdot (\mathbf{p} \times \mathbf{q})\mathbf{r}}{\mathbf{r}^2} \\ &= \mathbf{v} \cdot \frac{(\mathbf{q} \times \mathbf{r}\mathbf{p} + \mathbf{r} \times \mathbf{p}\mathbf{q} + \mathbf{p} \times \mathbf{q}\mathbf{r})}{\mathbf{r}^2} \underset{(2)}{=} \mathbf{v} \cdot \boldsymbol{\sigma}\end{aligned} \tag{10}$$

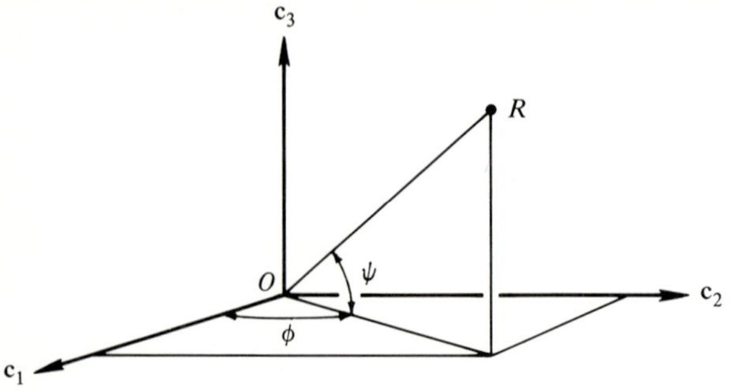

Figure 1.5.1

Example Observations of two stars, P and Q, are made simultaneously from two space vehicles, A and B, in order to generate data to be used in the determination of the relative orientation of A and B. The observations consist of measuring the angles ϕ and ψ shown in Fig. 1.5.1, where O represents either a point fixed in A or a point fixed in B, R is either P or Q, and $\mathbf{c}_1$, $\mathbf{c}_2$, $\mathbf{c}_3$ are orthogonal unit vectors forming a dextral set fixed either in A or B. For the numerical values of these angles given in Table 1.5.1, the direction cosine matrix C is to be determined.

If $\mathbf{R}$ is defined as a unit vector directed from O toward R (see Fig. 1.5.1), then

$$\mathbf{R} = \cos\psi\cos\phi\,\mathbf{c}_1 + \cos\psi\sin\phi\,\mathbf{c}_2 + \sin\psi\,\mathbf{c}_3 \tag{11}$$

Hence, letting $\mathbf{p}$ and $\mathbf{q}$ be unit vectors directed from O toward P and toward Q, respectively, and referring to Table 1.5.1, one can express each of these unit vectors both in terms of $\mathbf{a}_1$, $\mathbf{a}_2$, $\mathbf{a}_3$ and in terms of $\mathbf{b}_1$, $\mathbf{b}_2$, $\mathbf{b}_3$, as indicated in lines 1 and 2 of Table 1.5.2; and these results can then be used to evaluate $\mathbf{r}$ [see Eq. (1)], $\mathbf{q}\times\mathbf{r}$, and $\mathbf{r}\times\mathbf{p}$. Noting that (see line 3 of Table 1.5.2)

$$\mathbf{r}^2 = \frac{2}{16} + \frac{6}{16} + \frac{6}{16} = \frac{7}{8} \tag{12}$$

Table 1.5.1 Angles ϕ and ψ in degrees

	P		Q	
	ϕ	ψ	ϕ	ψ
$c_i = \mathbf{a}_i$	90	45	30	0
$c_i = \mathbf{b}_i$	135	0	90	60

Table 1.5.2 Vectors appearing in $\boldsymbol{\sigma}$

Line	Vector	$\mathbf{a}_1$	$\mathbf{a}_2$	$\mathbf{a}_3$	$\mathbf{b}_1$	$\mathbf{b}_2$	$\mathbf{b}_3$
1	$\mathbf{p}$	0	$\frac{\sqrt{2}}{2}$	$\frac{\sqrt{2}}{2}$	$\frac{-\sqrt{2}}{2}$	$\frac{\sqrt{2}}{2}$	0
2	$\mathbf{q}$	$\frac{\sqrt{3}}{2}$	$\frac{1}{2}$	0	0	$\frac{1}{2}$	$\frac{\sqrt{3}}{2}$
3	$\mathbf{r} = \mathbf{p} \times \mathbf{q}$	$\frac{-\sqrt{2}}{4}$	$\frac{\sqrt{6}}{4}$	$\frac{-\sqrt{6}}{4}$	$\frac{\sqrt{6}}{4}$	$\frac{\sqrt{6}}{4}$	$\frac{-\sqrt{2}}{4}$
4	$\mathbf{q} \times \mathbf{r}$	$\frac{-\sqrt{6}}{8}$	$\frac{3\sqrt{2}}{8}$	$\frac{\sqrt{2}}{2}$			
5	$\mathbf{r} \times \mathbf{p}$	$\frac{\sqrt{3}}{2}$	$\frac{1}{4}$	$-\frac{1}{4}$			

one thus obtains

$$\begin{aligned}\boldsymbol{\sigma} \underset{(2)}{=} \Bigg[&\left(-\frac{\sqrt{2}}{4}\mathbf{a}_1 + \frac{\sqrt{6}}{4}\mathbf{a}_2 - \frac{\sqrt{6}}{4}\mathbf{a}_3\right)\left(\frac{\sqrt{6}}{4}\mathbf{b}_1 + \frac{\sqrt{6}}{4}\mathbf{b}_2 - \frac{\sqrt{2}}{4}\mathbf{b}_3\right) \\ &+ \left(-\frac{\sqrt{6}}{8}\mathbf{a}_1 + \frac{3\sqrt{2}}{8}\mathbf{a}_2 + \frac{\sqrt{2}}{2}\mathbf{a}_3\right)\left(-\frac{\sqrt{2}}{2}\mathbf{b}_1 + \frac{\sqrt{2}}{2}\mathbf{b}_2 + 0\mathbf{b}_3\right) \\ &+ \left(\frac{\sqrt{3}}{2}\mathbf{a}_1 + \frac{1}{4}\mathbf{a}_2 - \frac{1}{4}\mathbf{a}_3\right)\left(0\mathbf{b}_1 + \frac{1}{2}\mathbf{b}_2 + \frac{\sqrt{3}}{2}\mathbf{b}_3\right)\Bigg] \Big/ \frac{7}{8}\end{aligned} \tag{13}$$

and

$$C_{11} \underset{(3)}{=} \mathbf{a}_1 \cdot \boldsymbol{\sigma} \cdot \mathbf{b}_1 = \left(-\frac{\sqrt{2}}{4}\frac{\sqrt{6}}{4} + \frac{\sqrt{6}}{8}\frac{\sqrt{2}}{2}\right) \Big/ \frac{7}{8} = 0 \tag{14}$$

$$C_{12} \underset{(3)}{=} \mathbf{a}_1 \cdot \boldsymbol{\sigma} \cdot \mathbf{b}_2 = \left(-\frac{\sqrt{2}}{4}\frac{\sqrt{6}}{4} - \frac{\sqrt{6}}{8}\frac{\sqrt{2}}{2} + \frac{\sqrt{3}}{2}\frac{1}{2}\right) \Big/ \frac{7}{8} = 0 \tag{15}$$

$$C_{13} \underset{(3)}{=} \mathbf{a}_1 \cdot \boldsymbol{\sigma} \cdot \mathbf{b}_3 = \left(\frac{\sqrt{2}}{4}\frac{\sqrt{2}}{4} + \frac{\sqrt{3}}{2}\frac{\sqrt{3}}{2}\right) \Big/ \frac{7}{8} = 1 \tag{16}$$

and so forth; that is,

$$C = \begin{bmatrix} 0 & 0 & 1 \\ 0 & 1 & 0 \\ -1 & 0 & 0 \end{bmatrix} \tag{17}$$

1.6 SUCCESSIVE ROTATIONS

When a rigid body B is subjected to two successive simple rotations (see Sec. 1.1) in a reference frame A, each of these rotations, as well as a single equivalent rotation (see Sec. 1.3), can be described in terms of direction cosines (see Sec. 1.2), Euler parameters (see Sec. 1.3), and Rodrigues parameters (see Sec. 1.4); and, no matter which method of description is employed, quantities associated with the individual rotations can be related to those characterizing the single equivalent rotation. In discussing such relationships, it is helpful to introduce a fictitious rigid body $\bar{B}$ which moves exactly like B during the first rotation, but remains fixed in A while B performs the second rotation. For analytical purposes, the first rotation can then be regarded as a rotation of $\bar{B}$ relative to A, and the second rotation as one of B relative to $\bar{B}$.

If $\mathbf{a}_i$ ($i = 1, 2, 3$), $\mathbf{b}_i$ ($i = 1, 2, 3$), and $\bar{\mathbf{b}}_i$ ($i = 1, 2, 3$) are three dextral sets of orthogonal unit vectors fixed in A, B, and $\bar{B}$, respectively, such that $\mathbf{a}_i = \mathbf{b}_i = \bar{\mathbf{b}}_i$ ($i = 1, 2, 3$) prior to the first rotation of B in A, and if ${}^AC^{\bar{B}}$, ${}^{\bar{B}}C^B$, and ${}^AC^B$ are the direction cosine matrices characterizing, respectively, the first, the second, and the single equivalent rotation, so that

$$[\bar{\mathbf{b}}_1 \quad \bar{\mathbf{b}}_2 \quad \bar{\mathbf{b}}_3] = [\mathbf{a}_1 \quad \mathbf{a}_2 \quad \mathbf{a}_3]\,{}^AC^{\bar{B}} \tag{1}$$

$$[\mathbf{b}_1 \quad \mathbf{b}_2 \quad \mathbf{b}_3] = [\bar{\mathbf{b}}_1 \quad \bar{\mathbf{b}}_2 \quad \bar{\mathbf{b}}_3]\,{}^{\bar{B}}C^B \tag{2}$$

and

$$[\mathbf{b}_1 \quad \mathbf{b}_2 \quad \mathbf{b}_3] = [\mathbf{a}_1 \quad \mathbf{a}_2 \quad \mathbf{a}_3]\,{}^AC^B \tag{3}$$

then ${}^AC^B$, expressed in terms of ${}^AC^{\bar{B}}$ and ${}^{\bar{B}}C^B$, is given by

$${}^AC^B = {}^AC^{\bar{B}}\,{}^{\bar{B}}C^B \tag{4}$$

Similarly, for Rodrigues vectors,

$${}^A\boldsymbol{\rho}^B = \frac{{}^A\boldsymbol{\rho}^{\bar{B}} + {}^{\bar{B}}\boldsymbol{\rho}^B + {}^{\bar{B}}\boldsymbol{\rho}^B \times {}^A\boldsymbol{\rho}^{\bar{B}}}{1 - {}^A\boldsymbol{\rho}^{\bar{B}} \cdot {}^{\bar{B}}\boldsymbol{\rho}^B} \tag{5}$$

To state the analogous relationship in terms of Euler parameters, we first define three sets of such parameters as follows: With $\bar{\mathbf{b}}_i$ and $\mathbf{b}_i$ ($i = 1, 2, 3$) directed as after the second rotation, and with θ_1, θ_2, and θ denoting respectively the radian measures of the first, the second, and the equivalent rotation,

$${}^A\epsilon_i^{\bar{B}} \triangleq {}^A\boldsymbol{\epsilon}^{\bar{B}} \cdot \mathbf{a}_i = {}^A\boldsymbol{\epsilon}^{\bar{B}} \cdot \bar{\mathbf{b}}_i \qquad (i = 1, 2, 3) \tag{6}$$

$${}^{\bar{B}}\epsilon_i^B \triangleq {}^{\bar{B}}\boldsymbol{\epsilon}^B \cdot \bar{\mathbf{b}}_i = {}^{\bar{B}}\boldsymbol{\epsilon}^B \cdot \mathbf{b}_i \qquad (i = 1, 2, 3) \tag{7}$$

$${}^A\epsilon_i^B \triangleq {}^A\boldsymbol{\epsilon}^B \cdot \mathbf{a}_i = {}^A\boldsymbol{\epsilon}^B \cdot \mathbf{b}_i \qquad (i = 1, 2, 3) \tag{8}$$

$${}^A\epsilon_4^{\bar{B}} \triangleq \cos\frac{\theta_1}{2} \tag{9}$$

$${}^{\bar{B}}\epsilon_4^B \triangleq \cos\frac{\theta_2}{2} \tag{10}$$

$${}^A\epsilon_4^B \triangleq \cos\frac{\theta}{2} \tag{11}$$

It then follows that

$$\begin{bmatrix} {}^A\epsilon_1{}^B \\ {}^A\epsilon_2{}^B \\ {}^A\epsilon_3{}^B \\ {}^A\epsilon_4{}^B \end{bmatrix} = \begin{bmatrix} {}^A\epsilon_4{}^{\bar{B}} & -{}^A\epsilon_3{}^{\bar{B}} & {}^A\epsilon_2{}^{\bar{B}} & {}^A\epsilon_1{}^{\bar{B}} \\ {}^A\epsilon_3{}^{\bar{B}} & {}^A\epsilon_4{}^{\bar{B}} & -{}^A\epsilon_1{}^{\bar{B}} & {}^A\epsilon_2{}^{\bar{B}} \\ -{}^A\epsilon_2{}^{\bar{B}} & {}^A\epsilon_1{}^{\bar{B}} & {}^A\epsilon_4{}^{\bar{B}} & {}^A\epsilon_3{}^{\bar{B}} \\ -{}^A\epsilon_1{}^{\bar{B}} & -{}^A\epsilon_2{}^{\bar{B}} & -{}^A\epsilon_3{}^{\bar{B}} & {}^A\epsilon_4{}^{\bar{B}} \end{bmatrix} \begin{bmatrix} {}^{\bar{B}}\epsilon_1{}^B \\ {}^{\bar{B}}\epsilon_2{}^B \\ {}^{\bar{B}}\epsilon_3{}^B \\ {}^{\bar{B}}\epsilon_4{}^B \end{bmatrix} \tag{12}$$

Furthermore,

$$ {}^A\boldsymbol{\epsilon}^B = {}^A\boldsymbol{\epsilon}^{\bar{B}}\, {}^{\bar{B}}\epsilon_4{}^B + {}^{\bar{B}}\boldsymbol{\epsilon}^B\, {}^A\epsilon_4{}^{\bar{B}} + {}^{\bar{B}}\boldsymbol{\epsilon}^B \times {}^A\boldsymbol{\epsilon}^{\bar{B}} \tag{13}$$

and

$$ {}^A\epsilon_4{}^B = {}^A\epsilon_4{}^{\bar{B}}\, {}^{\bar{B}}\epsilon_4{}^B - {}^A\boldsymbol{\epsilon}^{\bar{B}} \cdot {}^{\bar{B}}\boldsymbol{\epsilon}^B \tag{14}$$

Equations (4), (5), (12), and (13) all reflect the fact that the final orientation of B in A depends upon the order in which the successive rotations are performed. For example, in Eq. (4), ${}^AC^{\bar{B}}$ and ${}^{\bar{B}}C^B$ cannot be interchanged without altering the result, and in Eq. (13) the presence of a cross product shows that order cannot be left out of account.

Repeated use of Eqs. (4)–(14) permits one to construct formulas for quantities characterizing a single rotation that is equivalent to any number of successive rotations. For example, for three successive rotations,

$$ {}^AC^B = {}^AC^{\bar{B}}\, {}^{\bar{B}}C^{\bar{\bar{B}}}\, {}^{\bar{\bar{B}}}C^B \tag{15}$$

where ${}^AC^{\bar{B}}$, ${}^{\bar{B}}C^{\bar{\bar{B}}}$, and ${}^{\bar{\bar{B}}}C^B$ are direction cosine matrices associated with the first, the second, and the third rotation, respectively.

Derivations Substitution from Eq. (1) into Eq. (2) gives

$$[\mathbf{b}_1 \quad \mathbf{b}_2 \quad \mathbf{b}_3] = [\mathbf{a}_1 \quad \mathbf{a}_2 \quad \mathbf{a}_3]\, {}^AC^{\bar{B}}\, {}^{\bar{B}}C^B \tag{16}$$

and comparison of this equation with Eq. (3) shows that Eq. (4) is valid.

To obtain Eq. (5), let $\mathbf{a}$, $\bar{\mathbf{b}}$, and $\mathbf{b}$ be vectors fixed in A, $\bar{B}$, and B, respectively, and choose these in such a way that $\mathbf{a} = \bar{\mathbf{b}} = \mathbf{b}$ prior to the first rotation of B in A. Then, in accordance with Eq. (1.4.5), there exist Rodrigues vectors ${}^A\boldsymbol{\rho}^{\bar{B}}$, ${}^{\bar{B}}\boldsymbol{\rho}^B$, and ${}^A\boldsymbol{\rho}^B$ satisfying the equations

$$\mathbf{a} - \bar{\mathbf{b}} = (\mathbf{a} + \bar{\mathbf{b}}) \times {}^A\boldsymbol{\rho}^{\bar{B}} \tag{17}$$

$$\bar{\mathbf{b}} - \mathbf{b} = (\bar{\mathbf{b}} + \mathbf{b}) \times {}^{\bar{B}}\boldsymbol{\rho}^B \tag{18}$$

and

$$\mathbf{a} - \mathbf{b} = (\mathbf{a} + \mathbf{b}) \times {}^A\boldsymbol{\rho}^B \tag{19}$$

Cross-multiply Eqs. (17) and (18) with ${}^{\bar{B}}\boldsymbol{\rho}^B$ and ${}^A\boldsymbol{\rho}^{\bar{B}}$, respectively; subtract the resulting equations; and use the fact that

$${}^A\boldsymbol{\rho}^{\bar{B}} \cdot \mathbf{a} = {}^A\boldsymbol{\rho}^{\bar{B}} \cdot \bar{\mathbf{b}} \tag{20}$$

and

$${}^{\bar{B}}\boldsymbol{\rho}^B \cdot \bar{\mathbf{b}} = {}^{\bar{B}}\boldsymbol{\rho}^B \cdot \mathbf{b} \tag{21}$$

to eliminate $\bar{\mathbf{b}}$ wherever it appears as a member of a scalar product. This leads to

$$\bar{\mathbf{b}} \times ({}^A\boldsymbol{\rho}^{\bar{B}} + {}^{\bar{B}}\boldsymbol{\rho}^B) = \mathbf{a} \times {}^{\bar{B}}\boldsymbol{\rho}^B + \mathbf{b} \times {}^A\boldsymbol{\rho}^{\bar{B}} + (\mathbf{a} - \mathbf{b}){}^A\boldsymbol{\rho}^{\bar{B}} \cdot {}^{\bar{B}}\boldsymbol{\rho}^B + (\mathbf{a} + \mathbf{b}) \times ({}^{\bar{B}}\boldsymbol{\rho}^B \times {}^A\boldsymbol{\rho}^{\bar{B}}) \tag{22}$$

Next, add Eqs. (17) and (18), eliminate $\bar{\mathbf{b}}$ by using Eq. (22), and solve for $\mathbf{a} - \mathbf{b}$, thus obtaining

$$\mathbf{a} - \mathbf{b} = (\mathbf{a} + \mathbf{b}) \times \frac{{}^A\boldsymbol{\rho}^{\bar{B}} + {}^{\bar{B}}\boldsymbol{\rho}^B + {}^{\bar{B}}\boldsymbol{\rho}^B \times {}^A\boldsymbol{\rho}^{\bar{B}}}{1 - {}^A\boldsymbol{\rho}^{\bar{B}} \cdot {}^{\bar{B}}\boldsymbol{\rho}^B} \tag{23}$$

Together, Eqs. (23) and (19) imply the validity of Eq. (5), for Eqs. (23) and (19) can be satisfied for all choices of the vector $\mathbf{a}$ only if Eq. (5) is satisfied.

As for Eqs. (13) and (14), note that it follows from Eqs. (1.3.1), (1.3.3), and (1.4.1) that

$${}^A\boldsymbol{\rho}^{\bar{B}} = \frac{{}^A\boldsymbol{\epsilon}^{\bar{B}}}{{}^A\epsilon_4^{\bar{B}}} \tag{24}$$

$${}^{\bar{B}}\boldsymbol{\rho}^B = \frac{{}^{\bar{B}}\boldsymbol{\epsilon}^B}{{}^{\bar{B}}\epsilon_4^B} \tag{25}$$

and

$${}^A\boldsymbol{\rho}^B = \frac{{}^A\boldsymbol{\epsilon}^B}{{}^A\epsilon_4^B} \tag{26}$$

Consequently,

$${}^A\boldsymbol{\epsilon}^B \underset{(26)}{=} {}^A\epsilon_4^B\, {}^A\boldsymbol{\rho}^B \underset{(5)}{=} {}^A\epsilon_4^B \frac{{}^A\boldsymbol{\rho}^{\bar{B}} + {}^{\bar{B}}\boldsymbol{\rho}^B + {}^{\bar{B}}\boldsymbol{\rho}^B \times {}^A\boldsymbol{\rho}^{\bar{B}}}{1 - {}^A\boldsymbol{\rho}^{\bar{B}} \cdot {}^{\bar{B}}\boldsymbol{\rho}^B}$$

$$\underset{(24,25)}{=} {}^A\epsilon_4^B \frac{{}^A\boldsymbol{\epsilon}^{\bar{B}}\, {}^{\bar{B}}\epsilon_4^B + {}^{\bar{B}}\boldsymbol{\epsilon}^B\, {}^A\epsilon_4^{\bar{B}} + {}^{\bar{B}}\boldsymbol{\epsilon}^B \times {}^A\boldsymbol{\epsilon}^{\bar{B}}}{{}^A\epsilon_4^{\bar{B}}\, {}^{\bar{B}}\epsilon_4^B - {}^A\boldsymbol{\epsilon}^{\bar{B}} \cdot {}^{\bar{B}}\boldsymbol{\epsilon}^B} \tag{27}$$

and, dot-multiplying each side of this equation with itself and using Eq. (1.3.4), one finds that

$$({}^A\boldsymbol{\epsilon}^B)^2 = ({}^A\epsilon_4^B)^2 \frac{1 - ({}^A\epsilon_4^{\bar{B}}\, {}^{\bar{B}}\epsilon_4^B - {}^A\boldsymbol{\epsilon}^{\bar{B}} \cdot {}^{\bar{B}}\boldsymbol{\epsilon}^B)^2}{({}^A\epsilon_4^{\bar{B}}\, {}^{\bar{B}}\epsilon_4^B - {}^A\boldsymbol{\epsilon}^{\bar{B}} \cdot {}^{\bar{B}}\boldsymbol{\epsilon}^B)^2} \tag{28}$$

But,

$$({}^A\boldsymbol{\epsilon}^B)^2 \underset{(1.3.4)}{=} 1 - ({}^A\epsilon_4^B)^2 \tag{29}$$

Hence,

$$({}^A\epsilon_4^B)^2 \underset{(28,29)}{=} ({}^A\epsilon_4^{\bar{B}}\, {}^{\bar{B}}\epsilon_4^B - {}^A\boldsymbol{\epsilon}^{\bar{B}} \cdot {}^{\bar{B}}\boldsymbol{\epsilon}^B)^2 \tag{30}$$

and

$${}^A\epsilon_4^B \underset{(30)}{=} \pm ({}^A\epsilon_4^{\bar{B}}\, {}^{\bar{B}}\epsilon_4^B - {}^A\boldsymbol{\epsilon}^{\bar{B}} \cdot {}^{\bar{B}}\boldsymbol{\epsilon}^B) \tag{31}$$

so that, using the upper sign, one obtains Eq. (14). Furthermore, substitution into Eq. (27) then yields Eq. (13).

Finally, to establish the validity of Eq. (12), it suffices to show that the four scalar equations implied by this matrix equation can be derived from Eqs. (13) and (14). To this end, one may employ Eqs. (6) and (7) to resolve the right-hand member of Eq. (13) into components parallel to $\bar{\mathbf{b}}_1$, $\bar{\mathbf{b}}_2$, and $\bar{\mathbf{b}}_3$ and then dot-multiply both sides of the resulting equation successively with $\mathbf{b}_1$, $\mathbf{b}_2$, and $\mathbf{b}_3$, using Eq. (8) to evaluate ${}^A\boldsymbol{\epsilon}^B \cdot \mathbf{b}_i$ and Eqs. (1.3.6)–(1.3.14) to form $\bar{\mathbf{b}}_i \cdot \mathbf{b}_j$, which gives, for example,

$$\bar{\mathbf{b}}_2 \cdot \mathbf{b}_3 \underset{(1.3.11)}{=} 2({}^{\bar{B}}\epsilon_2{}^{B}\,{}^{\bar{B}}\epsilon_3{}^{B} - {}^{\bar{B}}\epsilon_1{}^{B}\,{}^{\bar{B}}\epsilon_4{}^{B}) \tag{32}$$

In this way one is led to the first three scalar equations corresponding to Eq. (12), and the fourth is obtained from Eq. (14) by making the substitution

$${}^A\boldsymbol{\epsilon}^{\bar{B}} \cdot {}^{\bar{B}}\boldsymbol{\epsilon}^{B} \underset{(6,7)}{=} {}^A\epsilon_1{}^{\bar{B}}\,{}^{\bar{B}}\epsilon_1{}^{B} + {}^A\epsilon_2{}^{\bar{B}}\,{}^{\bar{B}}\epsilon_2{}^{B} + {}^A\epsilon_3{}^{\bar{B}}\,{}^{\bar{B}}\epsilon_3{}^{B} \tag{33}$$

Example In Fig. 1.6.1, $\mathbf{a}_1$, $\mathbf{a}_2$, and $\mathbf{a}_3$ are mutually perpendicular unit vectors; X and Y are lines perpendicular to $\mathbf{a}_1$ and making fixed angles with $\mathbf{a}_2$ and $\mathbf{a}_3$; and B designates a body that is to be subjected to a 90° rotation about line X and a 180° rotation about line Y, the sense of each of these rotations being that indicated in the sketch.

Suppose that the rotation about X is performed first. Then, if $\mathbf{b}_1$, $\mathbf{b}_2$, and $\mathbf{b}_3$ are unit vectors fixed in B and respectively equal to $\mathbf{a}_1$, $\mathbf{a}_2$, and $\mathbf{a}_3$ prior to the first rotation, there exists a matrix C_x such that, subsequent to the second rotation of B,

$$[\mathbf{b}_1 \quad \mathbf{b}_2 \quad \mathbf{b}_3] = [\mathbf{a}_1 \quad \mathbf{a}_2 \quad \mathbf{a}_3]C_x \tag{34}$$

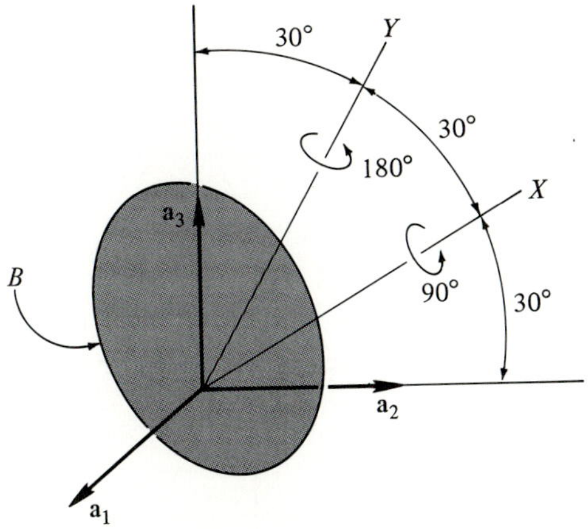

Figure 1.6.1

Similarly, if the rotation about Y is performed first, there exists a matrix C_y such that, subsequent to the second rotation of B,

$$[\mathbf{b}_1 \quad \mathbf{b}_2 \quad \mathbf{b}_3] = [\mathbf{a}_1 \quad \mathbf{a}_2 \quad \mathbf{a}_3]C_y \tag{35}$$

C_x and C_y are to be determined.

In order to find C_x by using Eq. (4), one may first form ${}^A C^{\bar{B}}$ by reference to Eqs. (1.2.23)–(1.2.31) with $\theta = \pi/2$ and

$$\lambda_1 = 0 \qquad \lambda_2 = \frac{\sqrt{3}}{2} \qquad \lambda_3 = \frac{1}{2} \tag{36}$$

which gives

$${}^A C^{\bar{B}} = \begin{bmatrix} 0 & -\frac{1}{2} & \frac{\sqrt{3}}{2} \\ \frac{1}{2} & \frac{3}{4} & \frac{\sqrt{3}}{4} \\ \frac{-\sqrt{3}}{2} & \frac{\sqrt{3}}{4} & \frac{1}{4} \end{bmatrix} \tag{37}$$

Next, to construct the matrix ${}^{\bar{B}}C^B$, one must express a unit vector $\boldsymbol{\lambda}$ which is parallel to Y in terms of suitable unit vectors $\bar{\mathbf{b}}_1$, $\bar{\mathbf{b}}_2$ and $\bar{\mathbf{b}}_3$. This is accomplished by noting that $\boldsymbol{\lambda}$, resolved into components parallel to $\mathbf{a}_1$, $\mathbf{a}_2$, and $\mathbf{a}_3$, is given by

$$\boldsymbol{\lambda} = \frac{1}{2}\mathbf{a}_2 + \frac{\sqrt{3}}{2}\mathbf{a}_3 \tag{38}$$

so that, using Eq. (1.2.9) and ${}^A C^{\bar{B}}$, one obtains

$$\boldsymbol{\lambda} = \left[\frac{1}{2}\left(\frac{1}{2}\right) + \frac{\sqrt{3}}{2}\left(-\frac{\sqrt{3}}{2}\right)\right]\bar{\mathbf{b}}_1 + \left[\frac{1}{2}\left(\frac{3}{4}\right) + \frac{\sqrt{3}}{2}\left(\frac{\sqrt{3}}{4}\right)\right]\bar{\mathbf{b}}_2$$
$$+ \left[\frac{1}{2}\left(\frac{\sqrt{3}}{4}\right) + \frac{\sqrt{3}}{2}\left(\frac{1}{4}\right)\right]\bar{\mathbf{b}}_3$$
$$= -\frac{1}{2}\bar{\mathbf{b}}_1 + \frac{3}{4}\bar{\mathbf{b}}_2 + \frac{\sqrt{3}}{4}\bar{\mathbf{b}}_3 \tag{39}$$

With $\theta = \pi$, Eqs. (1.2.23)–(1.2.31) then provide

$${}^{\bar{B}}C^B = \begin{bmatrix} -\frac{1}{2} & -\frac{3}{4} & \frac{-\sqrt{3}}{4} \\ -\frac{3}{4} & \frac{1}{8} & \frac{3\sqrt{3}}{8} \\ \frac{-\sqrt{3}}{4} & \frac{3\sqrt{3}}{8} & -\frac{5}{8} \end{bmatrix} \tag{40}$$

Consequently,

$$C_x \underset{(4)}{=} {}^AC^{\bar{B}}\,{}^{\bar{B}}C^B = \begin{bmatrix} 0 & \frac{1}{2} & \frac{-\sqrt{3}}{2} \\ -1 & 0 & 0 \\ 0 & \frac{\sqrt{3}}{2} & \frac{1}{2} \end{bmatrix} \tag{41}$$

C_y can be found similarly. Alternatively, one may use Euler parameters, proceeding as follows:

With $\boldsymbol{\lambda}$ expressed in terms of $\mathbf{a}_1$, $\mathbf{a}_2$, and $\mathbf{a}_3$, and with $\theta = \pi$, Eq. (1.3.1) gives

$${}^A\boldsymbol{\epsilon}^{\bar{B}} = \frac{1}{2}\,\mathbf{a}_2 + \frac{\sqrt{3}}{2}\,\mathbf{a}_3 \tag{42}$$

and, from Eq. (1.3.3),

$${}^A\epsilon_4{}^{\bar{B}} = 0 \tag{43}$$

Similarly, for the second rotation

$${}^{\bar{B}}\boldsymbol{\epsilon}^B \underset{(1.3.1)}{=} \left(\frac{\sqrt{3}}{2}\,\mathbf{a}_2 + \frac{1}{2}\,\mathbf{a}_3\right)\frac{\sqrt{2}}{2} \tag{44}$$

and

$${}^{\bar{B}}\epsilon_4{}^B \underset{(1.3.3)}{=} \frac{\sqrt{2}}{2} \tag{45}$$

Hence,

$${}^A\boldsymbol{\epsilon}^B \underset{(13)}{=} \left(\frac{1}{2}\,\mathbf{a}_2 + \frac{\sqrt{3}}{2}\,\mathbf{a}_3\right)\frac{\sqrt{2}}{2} + 0 + \frac{\sqrt{2}}{4}\,\mathbf{a}_1 \tag{46}$$

so that, in accordance with Eq. (8),

$${}^A\epsilon_1{}^B = \frac{\sqrt{2}}{4} \qquad {}^A\epsilon_2{}^B = \frac{\sqrt{2}}{4} \qquad {}^A\epsilon_3{}^B = \frac{\sqrt{6}}{4} \tag{47}$$

while

$${}^A\epsilon_4{}^B \underset{(14)}{=} -\frac{\sqrt{6}}{4} \tag{48}$$

The elements of C_y can now be obtained by using Eqs. (1.3.6)–(1.3.14), which gives

$$C_y = \begin{bmatrix} 0 & 1 & 0 \\ -\frac{1}{2} & 0 & \frac{\sqrt{3}}{2} \\ \frac{\sqrt{3}}{2} & 0 & \frac{1}{2} \end{bmatrix} \tag{49}$$

1.7 ORIENTATION ANGLES

Both for physical and for analytical reasons it is sometimes desirable to describe the orientation of a rigid body B in a reference frame A in terms of three angles. For example, if B is the rotor of a gyroscope whose outer gimbal axis is fixed in a reference frame A, then the angles ϕ, θ, and ψ shown in Fig. 1.7.1 furnish a means for describing the orientation of B in A in a way that is particularly meaningful from a physical point of view.

One scheme for bringing a rigid body B into a desired orientation in a reference frame A is to introduce $\mathbf{a}_1$, $\mathbf{a}_2$, $\mathbf{a}_3$ and $\mathbf{b}_1$, $\mathbf{b}_2$, $\mathbf{b}_3$ as dextral sets of orthogonal unit vectors fixed in A and B, respectively; align $\mathbf{b}_i$ with $\mathbf{a}_i$ ($i = 1, 2, 3$); and subject B successively to an $\mathbf{a}_1$ rotation of amount θ_1, an $\mathbf{a}_2$ rotation of amount θ_2, and an $\mathbf{a}_3$ rotation of amount θ_3. (Recall that, for any unit vector $\boldsymbol{\lambda}$, the phrase "$\boldsymbol{\lambda}$ rotation" means a rotation of B relative to A during which a right-handed screw fixed in B with its axis parallel to $\boldsymbol{\lambda}$ advances in the direction of $\boldsymbol{\lambda}$.) Suitable values of θ_1, θ_2, and θ_3 can be found in terms of elements of the direction cosine matrix C (see Sec. 1.2), which,

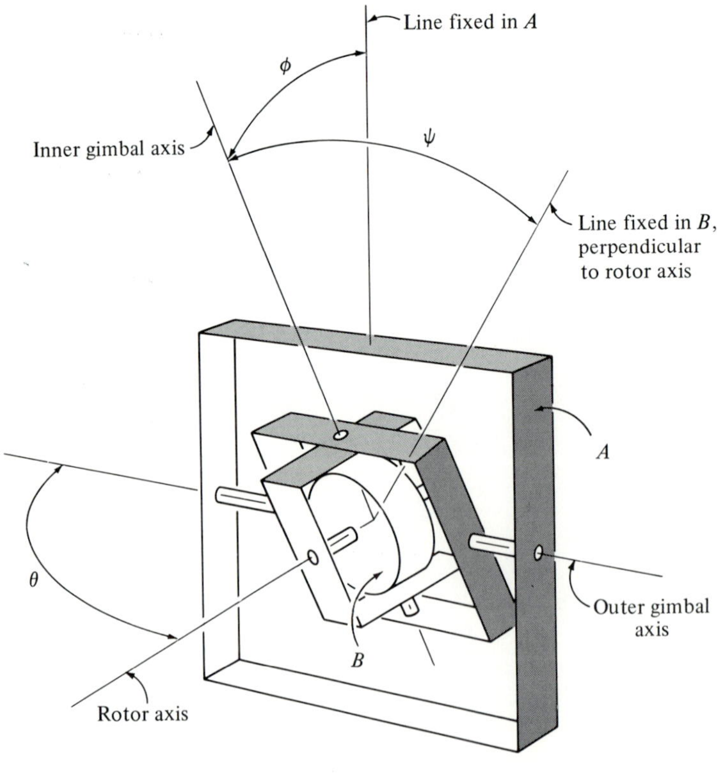

Figure 1.7.1

if s_i and c_i ($i = 1, 2, 3$) denote $\sin\theta_i$ and $\cos\theta_i$ ($i = 1, 2, 3$) respectively, is given by

$$C = \begin{bmatrix} c_2c_3 & s_1s_2c_3 - s_3c_1 & c_1s_2c_3 + s_3s_1 \\ c_2s_3 & s_1s_2s_3 + c_3c_1 & c_1s_2s_3 - c_3s_1 \\ -s_2 & s_1c_2 & c_1c_2 \end{bmatrix} \tag{1}$$

Specifically, if $|C_{31}| \neq 1$, take

$$\theta_2 = \sin^{-1}(-C_{31}) \qquad -\frac{\pi}{2} < \theta_2 < \frac{\pi}{2} \tag{2}$$

Next, after evaluating c_2, define α as

$$\alpha \triangleq \sin^{-1}\frac{C_{32}}{c_2} \qquad -\frac{\pi}{2} \leq \alpha \leq \frac{\pi}{2} \tag{3}$$

and let

$$\theta_1 = \begin{cases} \alpha & \text{if } C_{33} \geq 0 \\ \pi - \alpha & \text{if } C_{33} < 0 \end{cases} \tag{4}$$

Similarly, define β as

$$\beta \triangleq \sin^{-1}\frac{C_{21}}{c_2} \qquad -\frac{\pi}{2} \leq \beta \leq \frac{\pi}{2} \tag{5}$$

and take

$$\theta_3 = \begin{cases} \beta & \text{if } C_{11} \geq 0 \\ \pi - \beta & \text{if } C_{11} < 0 \end{cases} \tag{6}$$

If $|C_{31}| = 1$, take

$$\theta_2 = \begin{cases} -\dfrac{\pi}{2} & \text{if } C_{31} = 1 \\ \dfrac{\pi}{2} & \text{if } C_{31} = -1 \end{cases} \tag{7}$$

and, after defining α as

$$\alpha \triangleq \sin^{-1}(-C_{23}) \qquad -\frac{\pi}{2} \leq \alpha \leq \frac{\pi}{2} \tag{8}$$

let

$$\theta_1 = \begin{cases} \alpha & \text{if } C_{22} \geq 0 \\ \pi - \alpha & \text{if } C_{22} < 0 \end{cases} \tag{9}$$

and

$$\theta_3 = 0 \tag{10}$$

In other words, two rotations suffice in this case.

A second method for accomplishing the same objective is to subject B successively to a $\mathbf{b}_1$ rotation of amount θ_1, a $\mathbf{b}_2$ rotation of amount θ_2, and a $\mathbf{b}_3$ rotation of amount θ_3. The matrix C relating $\mathbf{a}_1$, $\mathbf{a}_2$, $\mathbf{a}_3$ to $\mathbf{b}_1$, $\mathbf{b}_2$, $\mathbf{b}_3$ as in Eq. (1.2.2) subsequent to the last rotation is then given by

$$C = \begin{bmatrix} c_2c_3 & -c_2s_3 & s_2 \\ s_1s_2c_3 + s_3c_1 & -s_1s_2s_3 + c_3c_1 & -s_1c_2 \\ -c_1s_2c_3 + s_3s_1 & c_1s_2s_3 + c_3s_1 & c_1c_2 \end{bmatrix} \tag{11}$$

and, if $|C_{13}| \neq 1$, suitable values of θ_1, θ_2, and θ_3 are obtained by taking

$$\theta_2 = \sin^{-1} C_{13} \qquad -\frac{\pi}{2} < \theta_2 < \frac{\pi}{2} \tag{12}$$

$$\alpha \triangleq \sin^{-1} \frac{-C_{23}}{c_2} \qquad -\frac{\pi}{2} \leq \alpha \leq \frac{\pi}{2} \tag{13}$$

$$\theta_1 = \begin{cases} \alpha & \text{if } C_{33} \geq 0 \\ \pi - \alpha & \text{if } C_{33} < 0 \end{cases} \tag{14}$$

$$\beta \triangleq \sin^{-1} \frac{-C_{12}}{c_2} \qquad -\frac{\pi}{2} \leq \beta \leq \frac{\pi}{2} \tag{15}$$

$$\theta_3 = \begin{cases} \beta & \text{if } C_{11} \geq 0 \\ \pi - \beta & \text{if } C_{11} < 0 \end{cases} \tag{16}$$

whereas, if $|C_{13}| = 1$, one may let

$$\theta_2 = \begin{cases} \dfrac{\pi}{2} & \text{if } C_{13} = 1 \\ -\dfrac{\pi}{2} & \text{if } C_{13} = -1 \end{cases} \tag{17}$$

$$\alpha \triangleq \sin^{-1} C_{32} \qquad -\frac{\pi}{2} \leq \alpha \leq \frac{\pi}{2} \tag{18}$$

$$\theta_1 = \begin{cases} \alpha & \text{if } C_{22} \geq 0 \\ \pi - \alpha & \text{if } C_{22} < 0 \end{cases} \tag{19}$$

$$\theta_3 = 0 \tag{20}$$

so that, once again, only two rotations are required.

The difference between these two procedures for bringing B into a desired orientation in A is that the first involves unit vectors fixed in the reference frame, whereas the second involves unit vectors fixed in the body. What the two methods have in common is that *three distinct* unit vectors are employed in both cases.

It is also possible to bring B into an arbitrary orientation relative to A by performing three successive rotations which involve only *two* distinct unit vectors, and these vectors may be fixed either in the reference frame or in the body. Specifically, if B is subjected successively to an $\mathbf{a}_1$ rotation of amount θ_1, an $\mathbf{a}_2$ rotation of amount θ_2, and again an $\mathbf{a}_1$ rotation, but this time of amount θ_3, then

$$C = \begin{bmatrix} c_2 & s_1s_2 & c_1s_2 \\ s_2s_3 & -s_1c_2s_3 + c_3c_1 & -c_1c_2s_3 - c_3s_1 \\ -s_2c_3 & s_1c_2c_3 + s_3c_1 & c_1c_2c_3 - s_3s_1 \end{bmatrix} \tag{21}$$

and, if $|C_{11}| \neq 1$, one can take

$$\theta_2 = \cos^{-1} C_{11} \qquad 0 < \theta_2 < \pi \tag{22}$$

$$\alpha \triangleq \sin^{-1} \frac{C_{12}}{s_2} \qquad -\frac{\pi}{2} \leq \alpha \leq \frac{\pi}{2} \tag{23}$$

$$\theta_1 = \begin{cases} \alpha & \text{if } C_{13} \geq 0 \\ \pi - \alpha & \text{if } C_{13} < 0 \end{cases} \tag{24}$$

$$\beta \triangleq \sin^{-1} \frac{C_{21}}{s_2} \qquad -\frac{\pi}{2} \leq \beta \leq \frac{\pi}{2} \tag{25}$$

$$\theta_3 = \begin{cases} \beta & \text{if } C_{31} < 0 \\ \pi - \beta & \text{if } C_{31} \geq 0 \end{cases} \tag{26}$$

while, if $|C_{11}| = 1$, one may let

$$\theta_2 = \begin{cases} 0 & \text{if } C_{11} = 1 \\ \pi & \text{if } C_{11} = -1 \end{cases} \tag{27}$$

$$\alpha \triangleq \sin^{-1} (-C_{23}) \qquad -\frac{\pi}{2} \leq \alpha \leq \frac{\pi}{2} \tag{28}$$

$$\theta_1 = \begin{cases} \alpha & \text{if } C_{22} \geq 0 \\ \pi - \alpha & \text{if } C_{22} < 0 \end{cases} \tag{29}$$

$$\theta_3 = 0 \tag{30}$$

Finally, if the successive rotations are a $\mathbf{b}_1$ rotation of amount θ_1, a $\mathbf{b}_2$ rotation of amount θ_2, and again a $\mathbf{b}_1$ rotation, but this time of amount θ_3, then

$$C = \begin{bmatrix} c_2 & s_2s_3 & s_2c_3 \\ s_1s_2 & -s_1c_2s_3 + c_3c_1 & -s_1c_2c_3 - s_3c_1 \\ -c_1s_2 & c_1c_2s_3 + c_3s_1 & c_1c_2c_3 - s_3s_1 \end{bmatrix} \tag{31}$$

and, if $|C_{11}| \neq 1$, θ_1, θ_2, and θ_3 may be found by taking

$$\theta_2 = \cos^{-1} C_{11} \qquad 0 < \theta_2 < \pi \tag{32}$$

$$\alpha \triangleq \sin^{-1} \frac{C_{21}}{s_2} \qquad -\frac{\pi}{2} \leq \alpha \leq \frac{\pi}{2} \tag{33}$$

$$\theta_1 = \begin{cases} \alpha & \text{if } C_{31} < 0 \\ \pi - \alpha & \text{if } C_{31} \geq 0 \end{cases} \tag{34}$$

$$\beta \triangleq \sin^{-1} \frac{C_{12}}{s_2} \qquad -\frac{\pi}{2} \leq \beta \leq \frac{\pi}{2} \tag{35}$$

$$\theta_3 = \begin{cases} \beta & \text{if } C_{13} \geq 0 \\ \pi - \beta & \text{if } C_{13} < 0 \end{cases} \tag{36}$$

while, if $|C_{11}| = 1$, one can use

$$\theta_2 = \begin{cases} 0 & \text{if } C_{11} = 1 \\ \pi & \text{if } C_{11} = -1 \end{cases} \tag{37}$$

$$\alpha \triangleq \sin^{-1} C_{32} \qquad -\frac{\pi}{2} \leq \alpha \leq \frac{\pi}{2} \tag{38}$$

$$\theta_1 = \begin{cases} \alpha & \text{if } C_{22} \geq 0 \\ \pi - \alpha & \text{if } C_{22} < 0 \end{cases} \tag{39}$$

$$\theta_3 = 0 \tag{40}$$

The matrices in Eqs. (1) and (11) are intimately related to each other: either one may be obtained from the other by replacing θ_i with $-\theta_i$ $(i = 1, 2, 3)$ and transposing. The matrices in Eqs. (21) and (31) are related similarly. These facts have the following physical significance, as may be verified by using Eq. (1.6.4): If B is subjected successively to an $\mathbf{a}_1$, an $\mathbf{a}_2$, and an $\mathbf{a}_3$ rotation of amount θ_1, θ_2, and θ_3, respectively, then one can bring B back into its original orientation in A by next subjecting B to successive $-\mathbf{b}_1$, $-\mathbf{b}_2$, and $-\mathbf{b}_3$ rotations of amounts θ_1, θ_2, and θ_3, respectively. Similarly, employing only four unit vectors, one can subject B to successive rotations characterized by $\theta_1 \mathbf{a}_1$, $\theta_2 \mathbf{a}_2$, $\theta_3 \mathbf{a}_1$, $-\theta_1 \mathbf{b}_1$, $-\theta_2 \mathbf{b}_2$, and $-\theta_3 \mathbf{b}_1$ without producing any ultimate change in the orientation of B in A. Furthermore, it does not matter whether the rotations involving unit vectors fixed in A are preceded or followed by those involving unit vectors fixed in B; that is, the sequences of successive rotations represented by $\theta_1 \mathbf{b}_1$, $\theta_2 \mathbf{b}_2$, $\theta_3 \mathbf{b}_3$, $-\theta_1 \mathbf{a}_1$, $-\theta_2 \mathbf{a}_2$, $-\theta_3 \mathbf{a}_3$ and by $\theta_1 \mathbf{b}_1$, $\theta_2 \mathbf{b}_2$, $\theta_3 \mathbf{b}_1$, $-\theta_1 \mathbf{a}_1$, $-\theta_2 \mathbf{a}_2$, $-\theta_3 \mathbf{a}_1$ also have no net effect on the orientation of B in A.

To indicate which set of three angles one is using, one can speak of *space-three angles* in connection with Eqs. (1)–(10), *body-three angles* for Eqs. (11)–(20), *space-two angles* for Eqs. (21)–(30), and *body-two angles* for Eqs. (31)–(40); and this terminology remains meaningful even when the angles and unit vectors employed are denoted by symbols other than those used in Eqs. (1)–(40). Moreover, once one has identified three angles in this way, one can always find appropriate replacements for Eqs. (1), (11), (21), or (31) by direct use of these equations. Suppose, for example, that $\mathbf{x}$, $\mathbf{y}$, $\mathbf{z}$ and $\boldsymbol{\xi}$, $\boldsymbol{\eta}$, $\boldsymbol{\zeta}$ are dextral sets of orthogonal unit vectors fixed in a reference frame A and in a rigid body B, respectively; that $\mathbf{x} = \boldsymbol{\xi}$, $\mathbf{y} = \boldsymbol{\eta}$, and $\mathbf{z} = \boldsymbol{\zeta}$ initially; that B is subjected, successively, to a $\mathbf{z}$ rotation of amount γ, a $\mathbf{y}$ rotation of amount β, and an $\mathbf{x}$ rotation of amount α; and that it is required to find the elements L_{ij} $(i, j = 1, 2, 3)$ of the matrix L such that, subsequent to the last rotation,

$$[\boldsymbol{\xi} \quad \boldsymbol{\eta} \quad \boldsymbol{\zeta}] = [\mathbf{x} \quad \mathbf{y} \quad \mathbf{z}]\, L \tag{41}$$

Then, recognizing α, β, and γ as space-three angles, one can introduce $\mathbf{a}_i$, $\mathbf{b}_i$, and θ_i $(i = 1, 2, 3)$ as

$$\mathbf{a}_1 \triangleq \mathbf{z} \qquad \mathbf{a}_2 \triangleq \mathbf{y} \qquad \mathbf{a}_3 \triangleq -\mathbf{x} \tag{42}$$

$$\mathbf{b}_1 \triangleq \boldsymbol{\zeta} \qquad \mathbf{b}_2 \triangleq \boldsymbol{\eta} \qquad \mathbf{b}_3 \triangleq -\boldsymbol{\xi} \tag{43}$$

and

$$\theta_1 \triangleq \gamma \qquad \theta_2 \triangleq \beta \qquad \theta_3 \triangleq -\alpha \tag{44}$$

in which case the given sequence of rotations is represented by $\theta_1\mathbf{a}_1$, $\theta_2\mathbf{a}_2$, and $\theta_3\mathbf{a}_3$; and L_{ij} can then be found by referring to Eq. (1) to express the scalar products associated with L_{ij} in terms of α, β, and γ. For instance,

$$\begin{aligned} L_{21} &= \mathbf{y} \cdot \boldsymbol{\xi} = \mathbf{a}_2 \cdot (-\mathbf{b}_3) \\ &\underset{(1.2.1)}{=} -C_{23} \underset{(1)}{=} -c_1 s_2 s_3 + c_3 s_1 \\ &= \cos\gamma \sin\beta \sin\alpha + \cos\alpha \sin\gamma \end{aligned} \tag{45}$$

In the spacecraft dynamics literature one encounters a wide variety of direction cosine matrices. Twenty-four such matrices are tabulated in Appendix I.

Derivations To establish the validity of Eq. (1), one may use Eq. (1.6.15), forming ${}^{A}C^{\bar{B}}$, ${}^{\bar{B}}C^{\bar{\bar{B}}}$, and ${}^{\bar{\bar{B}}}C^{B}$ with the aid of Eq. (1.2.35) and Eqs. (1.2.23)–(1.2.31). Specifically, to deal with the $\mathbf{a}_1$ rotation, let

$${}^{A}C^{\bar{B}} \underset{(1.2.35)}{=} \begin{bmatrix} 1 & 0 & 0 \\ 0 & c_1 & -s_1 \\ 0 & s_1 & c_1 \end{bmatrix} \tag{46}$$

Next, to construct a matrix ${}^{\bar{B}}C^{\bar{\bar{B}}}$ that characterizes the $\mathbf{a}_2$ rotation, let ${}^{A}\lambda$ and ${}^{\bar{B}}\lambda$ denote row matrices whose elements are $\mathbf{a}_2 \cdot \mathbf{a}_i$ $(i = 1, 2, 3)$ and $\mathbf{a}_2 \cdot \bar{\mathbf{b}}_i$ $(i = 1, 2, 3)$, respectively. Then

$${}^{A}\lambda = [0 \quad 1 \quad 0] \tag{47}$$

$${}^{\bar{B}}\lambda \underset{(1.2.9)}{=} {}^{A}\lambda\, {}^{A}C^{\bar{B}} \underset{(46,47)}{=} [0 \quad c_1 \quad -s_1] \tag{48}$$

and, from Eqs. (1.2.23)–(1.2.31), with $\lambda_1 = 0$, $\lambda_2 = c_1$, $\lambda_3 = -s_1$, and $\theta = \theta_2$,

$$
{}^{\bar{B}}C^{\bar{\bar{B}}} = \begin{bmatrix} c_2 & s_1 s_2 & c_1 s_2 \\ -s_1 s_2 & 1 + s_1^2(c_2 - 1) & s_1 c_1 (c_2 - 1) \\ -c_1 s_2 & s_1 c_1 (c_2 - 1) & 1 + c_1^2 (c_2 - 1) \end{bmatrix} \tag{49}
$$

A matrix ${}^{A}C^{\bar{\bar{B}}}$ associated with a simple rotation that is equivalent to the first two rotations is now given by

$$
{}^{A}C^{\bar{\bar{B}}} \underset{(1.6.4)}{=} {}^{A}C^{\bar{B}}\,{}^{\bar{B}}C^{\bar{\bar{B}}} \underset{(46,49)}{=} \begin{bmatrix} c_2 & s_1 s_2 & c_1 s_2 \\ 0 & c_1 & -s_1 \\ -s_2 & s_1 c_2 & c_1 c_2 \end{bmatrix} \tag{50}
$$

and, to resolve $\mathbf{a}_3$ into components required for the construction of a matrix ${}^{\bar{\bar{B}}}C^{B}$, one may use Eq. (1.2.9) to obtain

$$
[0 \quad 0 \quad 1]\,{}^{A}C^{\bar{\bar{B}}} \underset{(50)}{=} [-s_2 \quad s_1 c_2 \quad c_1 c_2] \tag{51}
$$

after which Eqs. (1.2.23)–(1.2.31) yield

$$
{}^{\bar{\bar{B}}}C^{B} = \begin{bmatrix} s_2^2 + c_2^2 c_3 & -c_2[c_1 s_3 + s_1 s_2 (1 - c_3)] & c_2[s_1 s_3 - c_1 s_2 (1 - c_3)] \\ c_2[c_1 s_3 + s_1 s_2 (1 - c_3)] & c_3 (1 - s_1^2 c_2^2) + s_1^2 c_2^2 & s_2 s_3 + s_1 c_1 c_2^2 (1 - c_3) \\ -c_2[s_1 s_3 + c_1 s_2 (1 - c_3)] & -s_2 s_3 + s_1 c_1 c_2^2 (1 - c_3) & 1 - (s_2^2 + s_1^2 c_2^2)(1 - c_3) \end{bmatrix} \tag{52}
$$

and substitution from Eqs. (46), (49), and (52) into Eq. (1.6.15) leads directly to Eq. (1).

Equation (11) may be derived by using Eq. (1.6.15) with

$$
{}^{A}C^{\bar{B}} \underset{(1.2.35)}{=} \begin{bmatrix} 1 & 0 & 0 \\ 0 & c_1 & -s_1 \\ 0 & s_1 & c_1 \end{bmatrix} \tag{53}
$$

$$
{}^{\bar{B}}C^{\bar{\bar{B}}} \underset{(1.2.36)}{=} \begin{bmatrix} c_2 & 0 & s_2 \\ 0 & 1 & 0 \\ -s_2 & 0 & c_2 \end{bmatrix} \tag{54}
$$

and

$$
{}^{\bar{\bar{B}}}C^{B} \underset{(1.2.37)}{=} \begin{bmatrix} c_3 & -s_3 & 0 \\ s_3 & c_3 & 0 \\ 0 & 0 & 1 \end{bmatrix} \tag{55}
$$

Equations (2)–(10) and (12)–(20) are immediate consequences of Eq. (1) and Eq. (11), respectively; and Eqs. (21)–(40) can be generated by procedures similar to those employed in the derivation of Eqs. (1)–(20).

Example If unit vectors $\mathbf{a}_1$, $\mathbf{a}_2$, $\mathbf{a}_3$ and $\mathbf{b}_1$, $\mathbf{b}_2$, $\mathbf{b}_3$ are introduced as shown in Fig. 1.7.2, and the angles ϕ, θ, and ψ shown in Fig. 1.7.1 are renamed θ_1, θ_2, and θ_3, respectively, then θ_1, θ_2, and θ_3 are body-two angles such that Eqs (31)–(40) can be used to discuss motions of B in A. However, as will be seen later, it is undesirable to use these angles when dealing with motions during which the rotor axis becomes coincident, or even nearly coincident, with the outer gimbal axis. (Coincidence of these two axes is referred to as *gimbal lock*.) Therefore, it may be convenient to employ in the course of one analysis two modes of description of the orientation of B in A, switching from one to the other whenever θ_2 acquires a value lying in a previously designated range. The following sort of question can then arise: If ϕ_1, ϕ_2, and ϕ_3 are the space-three angles associated with $\mathbf{a}_1$, $\mathbf{a}_2$, $\mathbf{a}_3$ and $\mathbf{b}_1$, $\mathbf{b}_2$, $\mathbf{b}_3$, what are the values of these angles corresponding to $\theta_1 = 30°$, $\theta_2 = 45°$, $\theta_3 = 60°$?

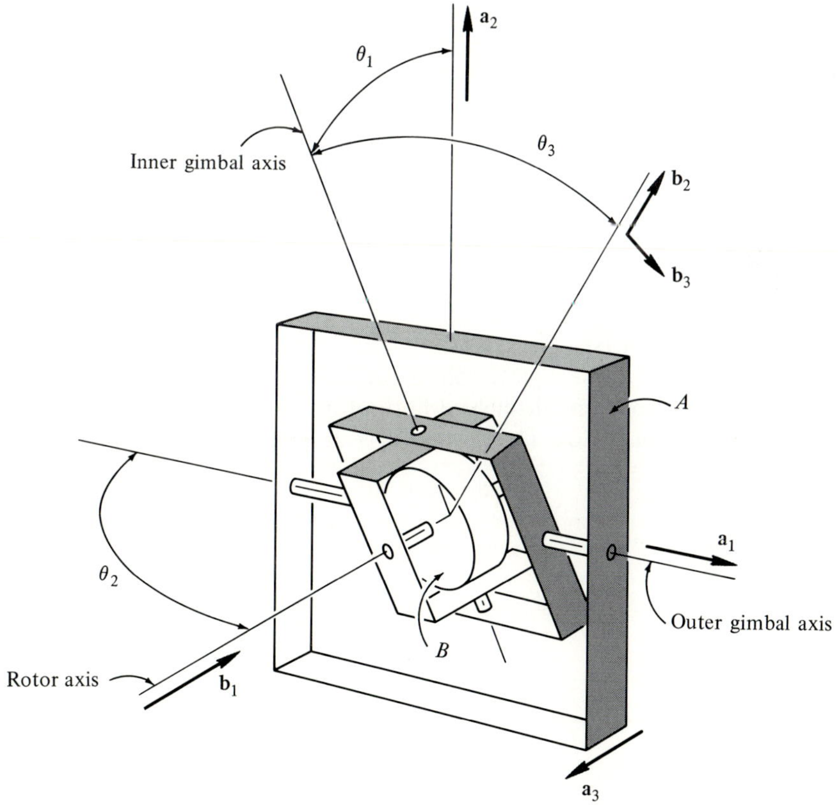

Figure 1.7.2

Inspection of Eqs. (2)–(6) shows that the elements of C required for the evaluation of ϕ_1, ϕ_2, and ϕ_3 are C_{31}, C_{32}, C_{33}, C_{21}, and C_{11}. From Eq. (31),

$$C_{31} = -\frac{\sqrt{3}}{2}\frac{\sqrt{2}}{2} = -0.612 \tag{56}$$

$$C_{32} = \frac{\sqrt{3}}{2}\frac{\sqrt{2}}{2}\frac{\sqrt{3}}{2} + \frac{1}{2}\frac{1}{2} = 0.780 \tag{57}$$

$$C_{33} = \frac{\sqrt{3}}{2}\frac{\sqrt{2}}{2}\frac{1}{2} - \frac{\sqrt{3}}{2}\frac{1}{2} = -0.127 \tag{58}$$

$$C_{21} = \frac{1}{2}\frac{\sqrt{2}}{2} = 0.354 \tag{59}$$

$$C_{11} = \frac{\sqrt{2}}{2} = 0.707 \tag{60}$$

Hence,

$$\phi_2 \underset{(2)}{=} \sin^{-1} 0.612 = 37.7° \tag{61}$$

$$\alpha \underset{(3)}{=} \sin^{-1}\frac{0.780}{0.791} = 80.4° \tag{62}$$

$$\phi_1 \underset{(4)}{=} 99.6° \qquad \beta \underset{(5)}{=} 26.6° \qquad \phi_3 \underset{(6)}{=} 26.6° \tag{63}$$

1.8 SMALL ROTATIONS

When a simple rotation (see Sec. 1.1) is small in the sense that second and higher powers of θ play a negligible role in an analysis involving the rotation, a number of the relationships discussed heretofore can be replaced with simpler ones. For example, Eqs. (1.1.1) and (1.1.2) yield, respectively,

$$\mathbf{b} = \mathbf{a} - \mathbf{a} \times \boldsymbol{\lambda}\theta \tag{1}$$

and

$$\mathbf{C} = \mathbf{U} - \mathbf{U} \times \boldsymbol{\lambda}\theta \tag{2}$$

while Eqs. (1.3.1), (1.3.3), and (1.4.1) give way to

$$\boldsymbol{\epsilon} = \tfrac{1}{2}\boldsymbol{\lambda}\theta \tag{3}$$

$$\epsilon_4 = 1 \tag{4}$$

and

$$\boldsymbol{\rho} = \tfrac{1}{2}\boldsymbol{\lambda}\theta \tag{5}$$

which shows that, to the order of approximation under consideration, the Rodrigues vector is indistinguishable from the Euler vector.

As will be seen presently, analytical descriptions of small rotations frequently involve skew-symmetric matrices. In dealing with these, it is convenient to establish the notational convention that the symbol obtained by placing a tilde over a letter, say q, denotes a skew-symmetric matrix whose off-diagonal elements have values denoted by $\pm q_i$ ($i = 1, 2, 3$), these elements being arranged as follows:

$$\tilde{q} = \begin{bmatrix} 0 & -q_3 & q_2 \\ q_3 & 0 & -q_1 \\ -q_2 & q_1 & 0 \end{bmatrix} \tag{6}$$

Using this convention, one can express the results obtained by neglecting second and higher powers of θ in Eqs. (1.2.23)–(1.2.31) as

$$C = U + \tilde{\lambda}\theta \tag{7}$$

Similarly, Eqs. (1.3.6)–(1.3.14) yield

$$C = U + 2\tilde{\epsilon} \tag{8}$$

Considering two successive small rotations, suppose that ${}^A C^{\bar{B}}$ and ${}^{\bar{B}} C^B$ are direction cosine matrices characterizing the first and second such rotation as in Sec. 1.6. Then, instead of using Eq. (1.6.4), one can express the direction cosine matrix ${}^A C^B$ associated with a single equivalent small rotation as

$${}^A C^B = U + ({}^A C^{\bar{B}} + {}^{\bar{B}} C^B - 2U) \tag{9}$$

Similarly, for three small rotations, Eq. (1.6.15) leads to

$${}^A C^B = U + ({}^A C^{\bar{B}} + {}^{\bar{B}} C^{\bar{\bar{B}}} + {}^{\bar{\bar{B}}} C^B - 3U) \tag{10}$$

Rodrigues vectors ${}^A \boldsymbol{\rho}^{\bar{B}}$, ${}^{\bar{B}} \boldsymbol{\rho}^B$, and ${}^A \boldsymbol{\rho}^B$ associated, respectively, with a first, a second, and an equivalent single small rotation satisfy the equation

$${}^A \boldsymbol{\rho}^B = {}^A \boldsymbol{\rho}^{\bar{B}} + {}^{\bar{B}} \boldsymbol{\rho}^B \tag{11}$$

Both this relationship and Eq. (9) show that the final orientation of B in A is independent of the order in which two successive small rotations are performed.

Finally when θ_1, θ_2, and θ_3 in Eqs. (1.7.1)–(1.7.40) are small in the sense that terms of second or higher degree in these quantities are negligible, then Eqs. (1.7.1) and (1.7.11) each yield [see Eq. (6) for the meaning of $\tilde{\theta}$]

$$C = U + \tilde{\theta} \tag{12}$$

showing that it is immaterial whether one uses space-three angles or body-three angles under these circumstances. The relationship corresponding to Eq. (12) for either space-two or body-two angles, namely

$$C \underset{(1.7.21)}{=} U + \begin{bmatrix} 0 & 0 & \theta_2 \\ 0 & 0 & -(\theta_3 + \theta_1) \\ -\theta_2 & \theta_3 + \theta_1 & 0 \end{bmatrix} \tag{13}$$

is less useful because this equation cannot be solved uniquely for θ_1 and θ_3 as functions of C_{ij} $(i, j = 1, 2, 3)$.

Derivations Equation (1) follows from Eq. (1.1.1) when sin θ is replaced with θ and cos θ with unity. The same substitution in Eq. (1.1.2) leads to Eq. (2). Equations (3) and (4) are obtained by replacing sin $(\theta/2)$ with $\theta/2$ and cos $(\theta/2)$ with unity in Eqs. (1.3.1) and (1.3.3), and Eq. (5) follows from Eq. (1.4.1) when tan $(\theta/2)$ is replaced with $\theta/2$.

Equation (7) follows directly from Eqs. (1.2.23)–(1.2.31), and Eq. (8) results from dropping terms of second degree in ϵ_1, ϵ_2, and/or ϵ_3 when forming C_{ij} in accordance with Eq. (1.3.6)–(1.3.14), which is justified in view of Eqs. (3) and (1.3.2).

To establish the validity of Eq. (9) one may proceed as follows: Using Eq. (7), one can express the matrices ${}^AC^{\bar{B}}$ and ${}^{\bar{B}}C^B$ introduced in Sec. 1.6 as

$$ {}^AC^{\bar{B}} = U + \widetilde{\lambda}\theta \tag{14} $$

and

$$ {}^{\bar{B}}C^B = U + \widetilde{\mu}\phi \tag{15} $$

where θ and ϕ are, respectively, the radian measures of the first and of the second small rotation, and $\widetilde{\lambda}$ and $\widetilde{\mu}$ characterize the associated axes of rotation. Substituting into Eq. (1.6.4), and dropping the product $\widetilde{\lambda}\,\widetilde{\mu}\theta\phi$, one then obtains

$$ \begin{aligned} {}^AC^B &= U + \widetilde{\lambda}\theta + \widetilde{\mu}\phi \\ &= U + (U + \widetilde{\lambda}\theta + U + \widetilde{\mu}\phi - 2U) \\ &= U + ({}^AC^{\bar{B}} + {}^{\bar{B}}C^B - 2U) \end{aligned} \tag{16} $$

A similar procedure leads to Eq. (10).

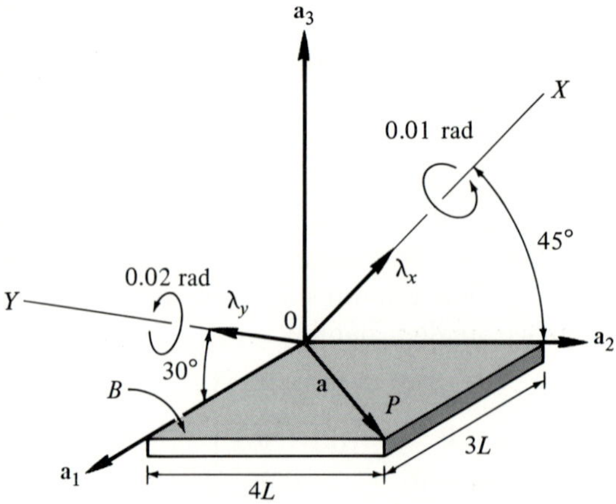

Figure 1.8.1

Finally, Eq. (11) may be obtained by using Eq. (5) in conjunction with Eq. (1.6.5), and Eqs. (12) and (13) result from linearizing in θ_i ($i = 1, 2, 3$) in Eqs. (1.7.11) and (1.7.21), respectively, and using the convention established in Eq. (6).

Example In Fig. 1.8.1, $\mathbf{a}_1$, $\mathbf{a}_2$, and $\mathbf{a}_3$ form a dextral set of orthogonal unit vectors, with $\mathbf{a}_1$ and $\mathbf{a}_2$ parallel to edges of a rectangular plate B; and X and Y designate lines perpendicular to $\mathbf{a}_1$ and $\mathbf{a}_2$, respectively. When B is subjected, successively, to a rotation of amount 0.01 rad about X and a rotation of amount 0.02 rad about Y, the sense of each rotation being that indicated in the sketch, point P traverses a distance d. This distance is to be determined on the assumption that the two rotations can be regarded as small.

If $\mathbf{a}$ designates the position vector of point P relative to point O before the rotations are performed, and $\mathbf{b}$ the position vector of P relative to O subsequent to the second rotation, then

$$d = |\mathbf{b} - \mathbf{a}| \tag{17}$$

with

$$\mathbf{a} = L(3\mathbf{a}_1 + 4\mathbf{a}_2) \tag{18}$$

and

$$\mathbf{b} \underset{(1)}{=} \mathbf{a} - \mathbf{a} \times \boldsymbol{\lambda}\theta \tag{19}$$

where $\boldsymbol{\lambda}$ and θ are, respectively, a unit vector and the radian measure of an angle associated with a single rotation that is equivalent to the two given rotations. To determine the product $\boldsymbol{\lambda}\theta$, let $\boldsymbol{\rho}$ be the Rodrigues vector for this equivalent rotation, in which case

$$\boldsymbol{\lambda}\theta \underset{(5)}{=} 2\boldsymbol{\rho} \tag{20}$$

and refer to Eq. (11) to express $\boldsymbol{\rho}$ as

$$\boldsymbol{\rho} = \boldsymbol{\rho}_x + \boldsymbol{\rho}_y \underset{(5)}{=} \frac{0.01}{2}\boldsymbol{\lambda}_x + \frac{0.02}{2}\boldsymbol{\lambda}_y \tag{21}$$

where $\boldsymbol{\lambda}_x$ and $\boldsymbol{\lambda}_y$ are unit vectors directed as shown in Fig. 1.8.1; that is,

$$\boldsymbol{\lambda}_x = \frac{1}{\sqrt{2}}(\mathbf{a}_2 + \mathbf{a}_3) \tag{22}$$

and

$$\boldsymbol{\lambda}_y = \frac{1}{2}(\sqrt{3}\,\mathbf{a}_1 + \mathbf{a}_3) \tag{23}$$

then

$$\boldsymbol{\lambda}\theta \underset{(20,21)}{=} 0.01\boldsymbol{\lambda}_x + 0.02\boldsymbol{\lambda}_y$$

$$\underset{(22,23)}{=} \frac{0.01}{\sqrt{2}}[\sqrt{6}\,\mathbf{a}_1 + \mathbf{a}_2 + (1 + \sqrt{2})\mathbf{a}_3] \tag{24}$$

$$\mathbf{b} - \mathbf{a} \underset{(19)}{=} -\mathbf{a} \times (\boldsymbol{\lambda}\theta)$$
$$\underset{(18,24)}{=} \frac{0.01}{\sqrt{2}} [-4(1 + \sqrt{2})\mathbf{a}_1 + 3(1 + \sqrt{2})\mathbf{a}_2 + (4\sqrt{6} - 3)\mathbf{a}_3]L \quad (25)$$

and

$$d \underset{(17,25)}{=} L\left[\frac{16(1 + \sqrt{2})^2 + 9(1 + \sqrt{2})^2 + (4\sqrt{6} - 3)^2}{20{,}000}\right]^{1/2} = 0.098L \quad (26)$$

1.9 SCREW MOTION

If P_1 and P_2 are points fixed in a reference frame A, and a point P is moved from P_1 to P_2, then P is said to experience a *displacement* in A, and the position vector of P_2 relative to P_1 is called a *displacement vector* of P in A.

When points of a rigid body B experience displacements in a reference frame A, one speaks of a displacement of B in A; and a displacement of B in A is called a *translation* of B in A if the displacement vectors of all points of B in A are equal to each other.

Every displacement of a rigid body B in a reference frame A can be produced by subjecting B successively to a translation in which a *basepoint* P of B, chosen arbitrarily, is brought from its original to its terminal position, and a simple rotation (see Sec. 1.1) during which P remains fixed in A. The Rodrigues vector (see Sec. 1.4) for the simple rotation is independent of the choice of basepoint, whereas the displacement vector of the basepoint depends on this choice. When the displacement vector of the basepoint is parallel to the Rodrigues vector for the rotation, the displacement under consideration is said to be producible by means of a *screw motion*.

Every displacement of a rigid body B in a reference frame A can be produced by means of a screw motion. In other words, one can always find a basepoint whose displacement vector is parallel to the Rodrigues vector for the simple rotation associated with a displacement of B in A. In fact, there exist infinitely many such basepoints, all lying on a straight line that is parallel to the Rodigues vector and bears the name *screw axis*; and the displacement vectors of all points of B lying on the screw axis are equal to each other and can, therefore, be characterized by a single vector, called the *screw translation vector*. Moreover, the magnitude of the screw translation vector is either smaller than or equal to the magnitude of the displacement vector of any basepoint not lying on the screw axis.

If $\boldsymbol{\delta}$ is the displacement vector of an arbitrary basepoint P, $\boldsymbol{\delta}$ is the Rodrigues vector for the rotation associated with a displacement of B in A, and P^* is a point of B lying on the screw axis (see Fig. 1.9.1), then the position vector $\mathbf{a}^*$ of P^* relative to P prior to the displacement of B in A satisfies the equation

$$\mathbf{a}^* = \frac{\boldsymbol{\rho} \times \boldsymbol{\delta} + (\boldsymbol{\rho} \times \boldsymbol{\delta}) \times \boldsymbol{\rho}}{2\rho^2} + \mu\boldsymbol{\rho} \quad (1)$$

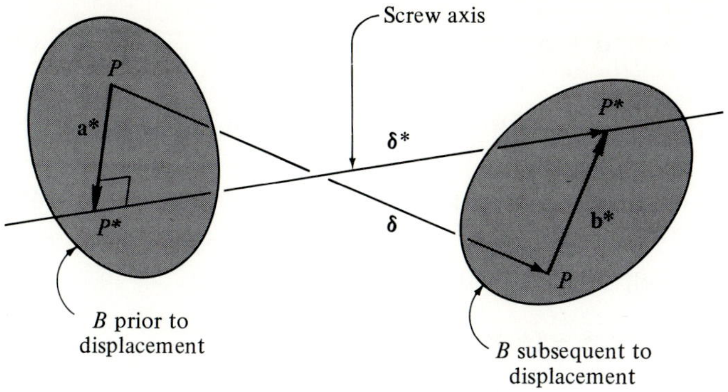

Figure 1.9.1

where μ depends on the choice of P^*; and the screw translation vector $\boldsymbol{\delta}^*$ is given by

$$\boldsymbol{\delta}^* = \frac{\boldsymbol{\rho} \cdot \boldsymbol{\delta}}{\rho^2}\boldsymbol{\rho} \tag{2}$$

Derivation If both P^* and P are points of B selected arbitrarily, and $\mathbf{a}^*$ is the position vector of P^* relative to P prior to the displacement of B in A, while $\mathbf{b}^*$ is the position vector of P^* relative to P subsequent to this displacement, then the displacement vector $\boldsymbol{\delta}^*$ of P^* can be expressed as (see Fig. 1.9.1)

$$\boldsymbol{\delta}^* = \boldsymbol{\delta} + \mathbf{b}^* - \mathbf{a}^* \tag{3}$$

where $\boldsymbol{\delta}$ is the displacement vector of P; and

$$\mathbf{b}^* - \mathbf{a}^* \underset{(1.4.5)}{=} \boldsymbol{\rho} \times (\mathbf{a}^* + \mathbf{b}^*) \tag{4}$$

so that

$$\boldsymbol{\delta}^* \underset{(3,4)}{=} \boldsymbol{\delta} + \boldsymbol{\rho} \times (\mathbf{a}^* + \mathbf{b}^*) \tag{5}$$

Hence, if P^* is to be chosen such that $\boldsymbol{\delta}^*$ is parallel to $\boldsymbol{\rho}$, in which case $\boldsymbol{\rho} \times \boldsymbol{\delta}^*$ is equal to zero, then $\mathbf{a}^*$ must satisfy the equation

$$\boldsymbol{\rho} \times \boldsymbol{\delta} + \boldsymbol{\rho} \times [\boldsymbol{\rho} \times (\mathbf{a}^* + \mathbf{b}^*)] \underset{(5)}{=} 0 \tag{6}$$

or, equivalently,

$$\boldsymbol{\rho} \times \boldsymbol{\delta} + \boldsymbol{\rho} \cdot (\mathbf{a}^* + \mathbf{b}^*)\boldsymbol{\rho} - \rho^2(\mathbf{a}^* + \mathbf{b}^*) = 0 \tag{7}$$

so that

$$\mathbf{a}^* + \mathbf{b}^* \underset{(7)}{=} \frac{\boldsymbol{\rho} \times \boldsymbol{\delta}}{\rho^2} + \frac{\boldsymbol{\rho} \cdot (\mathbf{a}^* + \mathbf{b}^*)}{\rho^2}\boldsymbol{\rho} \tag{8}$$

and
$$\boldsymbol{\delta}^* \underset{(5,8)}{=} \boldsymbol{\delta} + \boldsymbol{\rho} \times \frac{\boldsymbol{\rho} \times \boldsymbol{\delta}}{\boldsymbol{\rho}^2} = \frac{\boldsymbol{\rho} \cdot \boldsymbol{\delta}}{\boldsymbol{\rho}^2}\boldsymbol{\rho} \tag{9}$$

in agreement with Eq. (2). As for Eq. (1), one may solve the equation

$$\mathbf{b}^* - \mathbf{a}^* \underset{(4,8)}{=} \boldsymbol{\rho} \times \frac{\boldsymbol{\rho} \times \boldsymbol{\delta}}{\boldsymbol{\rho}^2} \tag{10}$$

for $\mathbf{b}^*$, substitute the result into Eq. (8), obtaining

$$\mathbf{a}^* = \frac{\boldsymbol{\rho} \times \boldsymbol{\delta} + (\boldsymbol{\rho} \times \boldsymbol{\delta}) \times \boldsymbol{\rho}}{2\boldsymbol{\rho}^2} + \frac{\boldsymbol{\rho} \cdot \mathbf{a}^*}{\boldsymbol{\rho}^2}\boldsymbol{\rho} \tag{11}$$

and then simply define μ as $\boldsymbol{\rho} \cdot \mathbf{a}^*/\boldsymbol{\rho}^2$. Moreover, this equation shows that the locus of basepoints whose displacement vectors are parallel to $\boldsymbol{\rho}$ is a straight line parallel to $\boldsymbol{\rho}$.

The contention that the magnitude of the screw translation vector is either smaller than or equal to the magnitude of the displacement vector of any point not lying on the screw axis is based on the observation that

$$|\boldsymbol{\delta}^*| \underset{(2)}{=} \left| \frac{\boldsymbol{\rho} \cdot \boldsymbol{\delta}}{\boldsymbol{\rho}^2} \boldsymbol{\rho} \right| = \frac{|\boldsymbol{\rho} \cdot \boldsymbol{\delta}|}{|\boldsymbol{\rho}|} \tag{12}$$

or equivalently,

$$|\boldsymbol{\delta}^*| = \left| \frac{\boldsymbol{\rho}}{|\boldsymbol{\rho}|} \cdot \boldsymbol{\delta} \right| \leq |\boldsymbol{\delta}| \tag{13}$$

Example The example in Sec. 1.3 dealt with a displacement of the triangle *ABC* shown in Fig. 1.9.2. The displacement in question was one that could

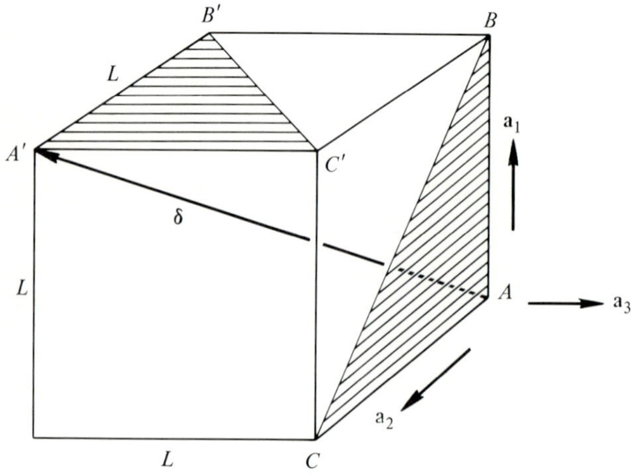

Figure 1.9.2

be produced by performing a translation of the triangle during which point A is brought to A', and following this with a rotation during which point A remains fixed at A'; and the Euler vector $\boldsymbol{\epsilon}$ and Euler parameter ϵ_4 for the rotation were found to be

$$\boldsymbol{\epsilon} = \frac{1}{2}(\mathbf{a}_1 - \mathbf{a}_2 - \mathbf{a}_3) \tag{14}$$

and

$$\epsilon_4 = \frac{1}{2} \tag{15}$$

where $\mathbf{a}_1$, $\mathbf{a}_2$, and $\mathbf{a}_3$ are unit vectors directed as shown in Fig. 1.9.2. To determine how the triangle can be brought into the same ultimate position by means of a screw motion, form $\boldsymbol{\rho}$ by reference to Eqs. (1.4.2) and (1.4.3), obtaining

$$\boldsymbol{\rho} = \mathbf{a}_1 - \mathbf{a}_2 - \mathbf{a}_3 \tag{16}$$

Next, let $\boldsymbol{\delta}$ denote the displacement vector of point A (see Fig. 1.9.2); that is, let

$$\boldsymbol{\delta} = L(\mathbf{a}_1 + \mathbf{a}_2 - \mathbf{a}_3) \tag{17}$$

Then

$$\boldsymbol{\rho} \times \boldsymbol{\delta} = 2L(\mathbf{a}_1 + \mathbf{a}_3) \tag{18}$$

$$(\boldsymbol{\rho} \times \boldsymbol{\delta}) \times \boldsymbol{\rho} = 2L(\mathbf{a}_1 + 2\mathbf{a}_2 - \mathbf{a}_3) \tag{19}$$

and the position vector $\mathbf{a}^*$ of any point P^* on the screw axis relative to point A prior to the displacement of the triangle is given by

$$\mathbf{a}^* \underset{(1)}{=} \frac{2L(\mathbf{a}_1 + \mathbf{a}_3) + 2L(\mathbf{a}_1 + 2\mathbf{a}_2 - \mathbf{a}_3)}{(2)(3)} + \mu\boldsymbol{\rho}$$

$$= \frac{2L}{3}(\mathbf{a}_1 + \mathbf{a}_2) + \mu\boldsymbol{\rho} \tag{20}$$

Hence, if μ is arbitrarily taken equal to zero, then P^* is situated as shown in Fig. 1.9.3 when the triangle is in its original position, and the screw axis, being parallel to $\boldsymbol{\rho}$, appears as indicated. Furthermore, the screw translation vector $\boldsymbol{\delta}^*$ is given by

$$\boldsymbol{\delta}^* \underset{(2)}{=} \frac{L(\mathbf{a}_1 - \mathbf{a}_2 - \mathbf{a}_3) \cdot (\mathbf{a}_1 + \mathbf{a}_2 - \mathbf{a}_3)}{3}(\mathbf{a}_1 - \mathbf{a}_2 - \mathbf{a}_3)$$

$$= \frac{L}{3}(\mathbf{a}_1 - \mathbf{a}_2 - \mathbf{a}_3) \tag{21}$$

so that

$$|\boldsymbol{\delta}^*| = \frac{L}{\sqrt{3}} \tag{22}$$

while the amount θ of the rotation associated with the displacement of the triangle, found in the example in Sec. 1.3, is given by

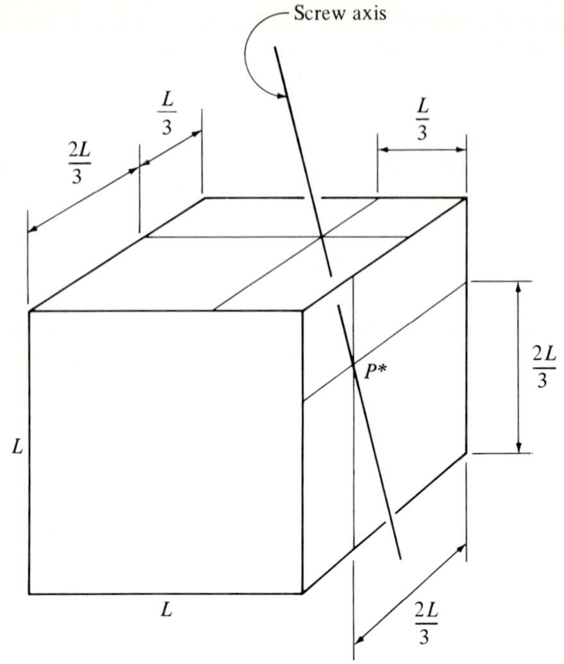

Figure 1.9.3

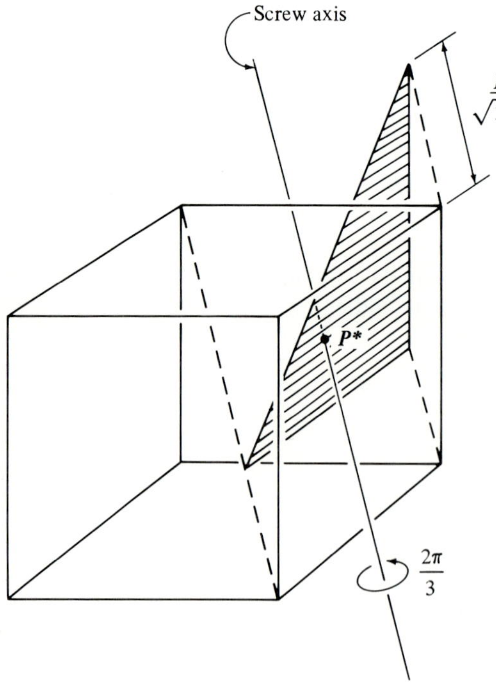

Figure 1.9.4

$$\theta = \frac{2\pi}{3} \text{ rad} \tag{23}$$

Hence, to bring the triangle into the desired position, one may proceed as follows: Perform a translation through a distance $L/\sqrt{3}$, as indicated in Fig. 1.9.4, and follow this with a rotation of amount $2\pi/3$ rad about the screw axis, choosing the sense of the rotation as shown in Fig. 1.9.4.

1.10 ANGULAR VELOCITY MATRIX

If $\mathbf{a}_1$, $\mathbf{a}_2$, $\mathbf{a}_3$, and $\mathbf{b}_1$, $\mathbf{b}_2$, $\mathbf{b}_3$ are two dextral sets of orthogonal unit vectors fixed, respectively, in two reference frames or rigid bodies A and B which are moving relative to each other, then the direction cosine matrix C and its elements C_{ij} $(i, j = 1, 2, 3)$, defined in Sec. 1.2, are functions of time t. The time derivative of C, denoted by $\dot{C}$ and defined in terms of the time derivatives $\dot{C}_{ij}$ of C_{ij} $(i, j = 1, 2, 3)$ as

$$\dot{C} \triangleq \begin{bmatrix} \dot{C}_{11} & \dot{C}_{12} & \dot{C}_{13} \\ \dot{C}_{21} & \dot{C}_{22} & \dot{C}_{23} \\ \dot{C}_{31} & \dot{C}_{32} & \dot{C}_{33} \end{bmatrix} \tag{1}$$

can be expressed as the product of C and a skew-symmetric matrix $\tilde{\omega}$ called an *angular velocity matrix* for B in A and defined as

$$\tilde{\omega} \triangleq C^T \dot{C} \tag{2}$$

In other words, with $\tilde{\omega}$ defined as in Eq. (2),

$$\dot{C} = C\,\tilde{\omega} \tag{3}$$

If functions $\omega_1(t)$, $\omega_2(t)$, and $\omega_3(t)$ are introduced in accordance with the notational convention established in Eq. (1.8.6), that is, by expressing $\tilde{\omega}$ as

$$\tilde{\omega} = \begin{bmatrix} 0 & -\omega_3 & \omega_2 \\ \omega_3 & 0 & -\omega_1 \\ -\omega_2 & \omega_1 & 0 \end{bmatrix} \tag{4}$$

then ω_1, ω_2, and ω_3 are given by

$$\omega_1 = C_{13}\dot{C}_{12} + C_{23}\dot{C}_{22} + C_{33}\dot{C}_{32} \tag{5}$$

$$\omega_2 = C_{21}\dot{C}_{23} + C_{31}\dot{C}_{33} + C_{11}\dot{C}_{13} \tag{6}$$

$$\omega_3 = C_{32}\dot{C}_{31} + C_{12}\dot{C}_{11} + C_{22}\dot{C}_{21} \tag{7}$$

These equations can be expressed more concisely after defining η_{ijk} as

$$\eta_{ijk} \triangleq \tfrac{1}{2}\,\epsilon_{ijk}(\epsilon_{ijk} + 1) \qquad (i, j, k = 1, 2, 3) \tag{8}$$

where ϵ_{ijk} is given by Eq. (1.2.32). (The quantity η_{ijk} is equal to unity when the subscripts appear in cyclic order; otherwise it is equal to zero.) Using the summation convention for repeated subscripts, one can then replace Eqs. (5)–(7) with

$$\omega_i = \eta_{igh}\, \dot{C}_{jg}\, C_{jh} \qquad (i = 1, 2, 3) \tag{9}$$

Similarly, Eqs. (3) can be expressed as

$$\dot{C}_{ij} = C_{ig}\, \omega_h\, \epsilon_{ghj} \qquad (i, j = 1, 2, 3) \tag{10}$$

or, explicitly, as

$$\dot{C}_{11} = C_{12}\, \omega_3 - C_{13}\, \omega_2 \tag{11}$$

$$\dot{C}_{12} = C_{13}\, \omega_1 - C_{11}\, \omega_3 \tag{12}$$

$$\dot{C}_{13} = C_{11}\, \omega_2 - C_{12}\, \omega_1 \tag{13}$$

$$\dot{C}_{21} = C_{22}\, \omega_3 - C_{23}\, \omega_2 \tag{14}$$

$$\dot{C}_{22} = C_{23}\, \omega_1 - C_{21}\, \omega_3 \tag{15}$$

$$\dot{C}_{23} = C_{21}\, \omega_2 - C_{22}\, \omega_1 \tag{16}$$

$$\dot{C}_{31} = C_{32}\, \omega_3 - C_{33}\, \omega_2 \tag{17}$$

$$\dot{C}_{32} = C_{33}\, \omega_1 - C_{31}\, \omega_3 \tag{18}$$

$$\dot{C}_{33} = C_{31}\, \omega_2 - C_{32}\, \omega_1 \tag{19}$$

Equations (10) are known as *Poisson's kinematical equations*.

Derivations Premultiplication of $\widetilde{\omega}$ with C gives

$$C\, \widetilde{\omega} \underset{(2)}{=} C\, C^T\, \dot{C} \underset{(1.2.16)}{=} \dot{C} \tag{20}$$

in agreement with Eq. (3).

To see that $\widetilde{\omega}$ as defined in Eq. (2) is skew-symmetric, note that

$$\widetilde{\omega}^T + \widetilde{\omega} \underset{(2)}{=} (C^T\, \dot{C})^T + C^T\, \dot{C} = \dot{C}^T\, C + C^T\, \dot{C} = \frac{d}{dt}\,(C^T\, C)$$

$$\underset{(1.2.17)}{=} \frac{dU}{dt} = 0 \tag{21}$$

Hence,

$$\widetilde{\omega}^T = -\,\widetilde{\omega} \tag{22}$$

Equations (5)–(7) follow from Eqs. (2) and (4), that is, from

$$\begin{bmatrix} 0 & -\omega_3 & \omega_2 \\ \omega_3 & 0 & -\omega_1 \\ -\omega_2 & \omega_1 & 0 \end{bmatrix} = \begin{bmatrix} C_{11} & C_{21} & C_{31} \\ C_{12} & C_{22} & C_{32} \\ C_{13} & C_{23} & C_{33} \end{bmatrix} \begin{bmatrix} \dot{C}_{11} & \dot{C}_{12} & \dot{C}_{13} \\ \dot{C}_{21} & \dot{C}_{22} & \dot{C}_{23} \\ \dot{C}_{31} & \dot{C}_{32} & \dot{C}_{33} \end{bmatrix} \tag{23}$$

Example The quantities ω_1, ω_2, and ω_3 can be expressed in a simple and revealing form when a body B performs a motion of simple rotation (see Sec. 1.1) in a reference frame A. For, letting θ and λ_i $(i = 1, 2, 3)$ have the same meanings as in Secs. 1.1 and 1.2, and substituting from Eqs. (1.2.23)–(1.2.31) into Eq. (5), one obtains

$$\begin{aligned}\omega_1 = {} & [\lambda_2 \sin\theta + \lambda_3\lambda_1(1 - \cos\theta)](-\lambda_3 \cos\theta + \lambda_1\lambda_2 \sin\theta)\dot{\theta} \\ & - [-\lambda_1 \sin\theta + \lambda_2\lambda_3(1 - \cos\theta)](\lambda_3^2 + \lambda_1^2) \sin\theta\,\dot{\theta} \\ & + [1 - (\lambda_1^2 + \lambda_2^2)(1 - \cos\theta)](\lambda_1 \cos\theta + \lambda_2\lambda_3 \sin\theta)\,\dot{\theta} \\ = {} & [\lambda_1(\lambda_1^2 + \lambda_2^2 + \lambda_3^2) \\ & + (1 - \lambda_1^2 - \lambda_2^2 - \lambda_3^2)(\lambda_1 \cos\theta + \lambda_2\lambda_3 \sin\theta - \lambda_2\lambda_3 \sin\theta \cos\theta)]\dot{\theta}\end{aligned} \tag{24}$$

which, since

$$\lambda_1^2 + \lambda_2^2 + \lambda_3^2 = 1 \tag{25}$$

reduces to

$$\omega_1 = \lambda_1\,\dot{\theta} \tag{26}$$

Similarly,

$$\omega_2 = \lambda_2\,\dot{\theta} \tag{27}$$

and

$$\omega_3 = \lambda_3\,\dot{\theta} \tag{28}$$

1.11 ANGULAR VELOCITY VECTOR

The vector $\boldsymbol{\omega}$ defined as

$$\boldsymbol{\omega} \triangleq \omega_1\,\mathbf{b}_1 + \omega_2\,\mathbf{b}_2 + \omega_3\,\mathbf{b}_3 \tag{1}$$

where ω_i and $\mathbf{b}_i$ $(i = 1, 2, 3)$ have the same meaning as in Sec. 1.10, is called the *angular velocity* of B in (or relative to) A. At times it is convenient to use the more elaborate symbol ${}^A\boldsymbol{\omega}^B$ in place of $\boldsymbol{\omega}$. The symbol ${}^B\boldsymbol{\omega}^A$ then denotes the angular velocity of A in B, and

$${}^A\boldsymbol{\omega}^B = -{}^B\boldsymbol{\omega}^A \tag{2}$$

If the first time-derivative of $\mathbf{b}_i$ in reference frame A is denoted by $\dot{\mathbf{b}}_i$, that is, if $\dot{\mathbf{b}}_i$, is defined as

$$\dot{\mathbf{b}}_i \triangleq \mathbf{a}_j \frac{d}{dt}(\mathbf{a}_j \cdot \mathbf{b}_i) \qquad (i = 1, 2, 3) \tag{3}$$

where the summation convention for repeated subscripts is used and $\mathbf{a}_1$, $\mathbf{a}_2$, $\mathbf{a}_3$ form a dextral set of orthogonal unit vectors fixed in A, then $\boldsymbol{\omega}$ can be expressed as

$$\boldsymbol{\omega} = \mathbf{b}_1\,\dot{\mathbf{b}}_2 \cdot \mathbf{b}_3 + \mathbf{b}_2\,\dot{\mathbf{b}}_3 \cdot \mathbf{b}_1 + \mathbf{b}_3\,\dot{\mathbf{b}}_1 \cdot \mathbf{b}_2 \tag{4}$$

When the motion of B in A is one of simple rotation (see Sec. 1.1), the angular velocity of B in A becomes

$$\boldsymbol{\omega} = \dot{\theta}\,\boldsymbol{\lambda} \tag{5}$$

where θ and $\boldsymbol{\lambda}$ have the same meaning as in Sec. 1.1.

One of the most useful relationships involving angular velocity is that between the first time-derivatives of a vector $\mathbf{v}$ in two reference frames A and B. If these derivatives are denoted by ${}^A d\mathbf{v}/dt$ and ${}^B d\mathbf{v}/dt$, that is, if

$$\frac{{}^A d\mathbf{v}}{dt} \triangleq \mathbf{a}_i \frac{d}{dt}(\mathbf{v} \cdot \mathbf{a}_i) \tag{6}$$

and

$$\frac{{}^B d\mathbf{v}}{dt} \triangleq \mathbf{b}_i \frac{d}{dt}(\mathbf{v} \cdot \mathbf{b}_i) \tag{7}$$

then this relationship assumes the form

$$\frac{{}^A d\mathbf{v}}{dt} = \frac{{}^B d\mathbf{v}}{dt} + {}^A\boldsymbol{\omega}^B \times \mathbf{v} \tag{8}$$

Applied to a vector $\boldsymbol{\beta}$ fixed in B, Eq. (8) gives

$$\frac{{}^A d\boldsymbol{\beta}}{dt} = {}^A\boldsymbol{\omega}^B \times \boldsymbol{\beta} \tag{9}$$

In view of this result one may regard the angular velocity of B in A as an "operator" which, when operating on any vector fixed in B, produces the time-derivative of that vector in A.

Derivations For $i = 2$, the scalar product appearing in Eq. (3) can be expressed as

$$\mathbf{a}_j \cdot \mathbf{b}_2 \underset{(1.2.1)}{=} C_{j2} \tag{10}$$

Consequently,

$$\dot{\mathbf{b}}_2 \underset{(3)}{=} \mathbf{a}_1 \dot{C}_{12} + \mathbf{a}_2 \dot{C}_{22} + \mathbf{a}_3 \dot{C}_{32} \tag{11}$$

and, expressing $\mathbf{b}_3$ as

$$\mathbf{b}_3 \underset{(1.2.1)}{=} \mathbf{a}_1 C_{13} + \mathbf{a}_2 C_{23} + \mathbf{a}_3 C_{33} \tag{12}$$

one finds that

$$\dot{\mathbf{b}}_2 \cdot \mathbf{b}_3 = C_{13}\dot{C}_{12} + C_{23}\dot{C}_{22} + C_{33}\dot{C}_{32} \underset{(1.10.5)}{=} \omega_1 \tag{13}$$

Similarly,

$$\dot{\mathbf{b}}_3 \cdot \mathbf{b}_1 = \omega_2 \tag{14}$$

and

$$\dot{\mathbf{b}}_1 \cdot \mathbf{b}_2 = \omega_3 \tag{15}$$

Substituting into Eq. (1), one thus arrives at Eq. (4).

When the motion of B in A is one of simple rotation, Eqs. (1.10.26)–(1.10.28) may be used to express the angular velocity of B in A as

$$\boldsymbol{\omega} \underset{(1)}{=} (\lambda_1 \, \mathbf{b}_1 + \lambda_2 \, \mathbf{b}_2 + \lambda_3 \, \mathbf{b}_3)\, \dot{\theta} \underset{(1.2.22)}{=} (\boldsymbol{\lambda} \cdot \mathbf{b}_1 \, \mathbf{b}_1 + \boldsymbol{\lambda} \cdot \mathbf{b}_2 \, \mathbf{b}_2 + \boldsymbol{\lambda} \cdot \mathbf{b}_3 \, \mathbf{b}_3)\, \dot{\theta}$$
$$= \boldsymbol{\lambda}\, \dot{\theta} \tag{16}$$

in agreement with Eq. (5).

To establish the validity of Eq. (8), let ${}^A v_i$, ${}^B v_i$, ${}^A v$, and ${}^B v$ have the same meanings as in Sec. 1.2. Then, from Eqs. (6) and (1.2.7),

$$\frac{{}^A d\mathbf{v}}{dt} = {}^A\dot{v}_1 \, \mathbf{a}_1 + {}^A\dot{v}_2 \, \mathbf{a}_2 + {}^A\dot{v}_3 \, \mathbf{a}_3 = {}^A\dot{v}\, [\mathbf{a}_1 \quad \mathbf{a}_2 \quad \mathbf{a}_3]^T \tag{17}$$

Now,

$${}^A\dot{v} \underset{(1.2.9)}{=} \frac{d}{dt}({}^B v \, C^T) = {}^B\dot{v}\, C^T + {}^B v \, \dot{C}^T \tag{18}$$

and

$$[\mathbf{a}_1 \quad \mathbf{a}_2 \quad \mathbf{a}_3]^T \underset{(1.2.2)}{=} C\, [\mathbf{b}_1 \quad \mathbf{b}_2 \quad \mathbf{b}_3]^T \tag{19}$$

Hence,

$$\frac{{}^A d\mathbf{v}}{dt} = {}^B\dot{v}\, C^T C\, [\mathbf{b}_1 \quad \mathbf{b}_2 \quad \mathbf{b}_3]^T + {}^B v \, \dot{C}^T C\, [\mathbf{b}_1 \quad \mathbf{b}_2 \quad \mathbf{b}_3]^T$$
$$\underset{\substack{(1.2.17,\\ 1.10.2)}}{=} {}^B\dot{v}\, [\mathbf{b}_1 \quad \mathbf{b}_2 \quad \mathbf{b}_3]^T + {}^B v \, \tilde{\omega}^T [\mathbf{b}_1 \quad \mathbf{b}_2 \quad \mathbf{b}_3]^T \tag{20}$$

Furthermore, from Eqs. (7) and (1.2.8)

$${}^B\dot{v}\, [\mathbf{b}_1 \quad \mathbf{b}_2 \quad \mathbf{b}_3]^T = \frac{{}^B d\mathbf{v}}{dt} \tag{21}$$

while it follows from Eqs. (1) and (1.10.4) that

$${}^B v \, \tilde{\omega}^T [\mathbf{b}_1 \quad \mathbf{b}_2 \quad \mathbf{b}_3]^T = {}^A\boldsymbol{\omega}^B \times \mathbf{v} \tag{22}$$

Consequently,

$$\frac{{}^A d\mathbf{v}}{dt} = \frac{{}^B d\mathbf{v}}{dt} + {}^A\boldsymbol{\omega}^B \times \mathbf{v} \tag{23}$$

Finally, Eq. (2) follows from the fact that, interchanging A and B in Eq. (8), one obtains

$$\frac{{}^B d\mathbf{v}}{dt} = \frac{{}^A d\mathbf{v}}{dt} + {}^B\boldsymbol{\omega}^A \times \mathbf{v} \tag{24}$$

and, adding corresponding members of this equation and of Eq. (8), one arrives at

$$\frac{{}^A d\mathbf{v}}{dt} + \frac{{}^B d\mathbf{v}}{dt} = \frac{{}^B d\mathbf{v}}{dt} + \frac{{}^A d\mathbf{v}}{dt} + ({}^A\boldsymbol{\omega}^B + {}^B\boldsymbol{\omega}^A) \times \mathbf{v} \tag{25}$$

or

$$({}^A\boldsymbol{\omega}^B + {}^B\boldsymbol{\omega}^A) \times \mathbf{v} = 0 \tag{26}$$

This equation can be satisfied for *all* $\mathbf{v}$ only if

$$ {}^{A}\boldsymbol{\omega}^{B} = -\,{}^{B}\boldsymbol{\omega}^{A} \tag{27}$$

Example When a point P moves on a space curve C fixed in a reference frame A (see Fig. 1.11.1), a dextral set of orthogonal unit vectors $\mathbf{b}_1$, $\mathbf{b}_2$, $\mathbf{b}_3$ can be generated by letting $\mathbf{p}$ be the position vector of P relative to a point P_0 fixed on C and defining $\mathbf{b}_1$, $\mathbf{b}_2$, and $\mathbf{b}_3$ as

$$\mathbf{b}_1 \triangleq \mathbf{p}' \tag{28}$$

$$\mathbf{b}_2 \triangleq \frac{\mathbf{p}''}{|\mathbf{p}''|} \tag{29}$$

$$\mathbf{b}_3 \triangleq \mathbf{p}' \times \frac{\mathbf{p}''}{|\mathbf{p}''|} \tag{30}$$

where primes denote differentiation in A with respect to the arc length displacement s of P relative to P_0. The vector $\mathbf{b}_1$ is called a *vector tangent*, $\mathbf{b}_2$ the *vector principal normal*, and $\mathbf{b}_3$ a *vector binormal* of C at P; and the derivatives of $\mathbf{b}_1$, $\mathbf{b}_2$, and $\mathbf{b}_3$ with respect to s are given by the Serret-Frênet formulas

$$\mathbf{b}_1' = \frac{\mathbf{b}_2}{\rho} \tag{31}$$

$$\mathbf{b}_2' = \frac{-\,\mathbf{b}_1}{\rho} + \lambda \mathbf{b}_3 \tag{32}$$

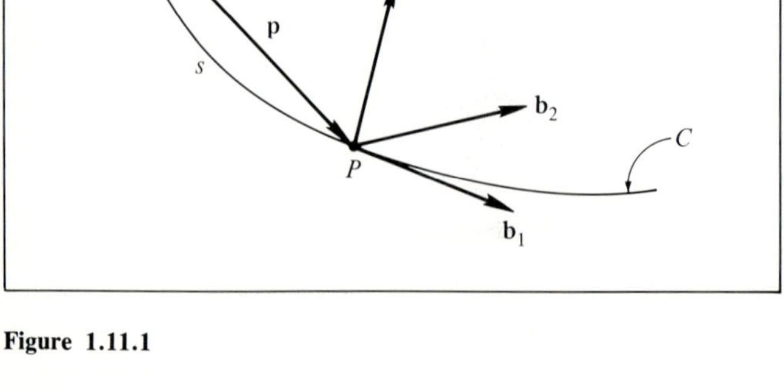

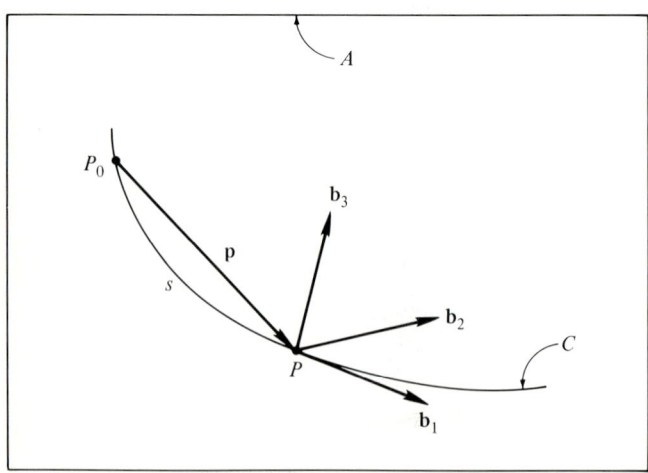

Figure 1.11.1

$$\mathbf{b}_3' = -\lambda \mathbf{b}_2 \tag{33}$$

where ρ and λ, defined as

$$\rho \triangleq \frac{1}{|\mathbf{p}''|} \tag{34}$$

and

$$\lambda \triangleq \rho^2 \mathbf{p}' \cdot \mathbf{p}'' \times \mathbf{p}''' \tag{35}$$

are called the *principal radius of curvature* of C at P and the *torsion* of C at P.

If B designates a reference frame in which $\mathbf{b}_1$, $\mathbf{b}_2$, and $\mathbf{b}_3$ are fixed, the angular velocity $\boldsymbol{\omega}$ of B in A can be expressed in terms of $\mathbf{b}_1$, $\mathbf{b}_2$, $\mathbf{b}_3$, ρ, λ, and $\dot{s}$ by using Eq. (4) together with

$$\dot{\mathbf{b}}_1 = \mathbf{b}_1'\dot{s} \underset{(31)}{=} \frac{\mathbf{b}_2\,\dot{s}}{\rho} \tag{36}$$

$$\dot{\mathbf{b}}_2 \underset{(32)}{=} \left(\frac{-\mathbf{b}_1}{\rho} + \lambda \mathbf{b}_3\right)\dot{s} \tag{37}$$

$$\dot{\mathbf{b}}_3 \underset{(33)}{=} -\lambda \mathbf{b}_2\,\dot{s} \tag{38}$$

$$\dot{\mathbf{b}}_1 \cdot \mathbf{b}_2 \underset{(36)}{=} \frac{\dot{s}}{\rho} \qquad \dot{\mathbf{b}}_2 \cdot \mathbf{b}_3 \underset{(37)}{=} \lambda\,\dot{s} \qquad \dot{\mathbf{b}}_3 \cdot \mathbf{b}_1 \underset{(38)}{=} 0 \tag{39}$$

to obtain

$$\boldsymbol{\omega} \underset{(4)}{=} \left(\lambda \mathbf{b}_1 + \frac{\mathbf{b}_3}{\rho}\right)\dot{s} \tag{40}$$

The term torsion as applied to λ is seen to be particularly appropriate in this context.

1.12 ANGULAR VELOCITY COMPONENTS

The expression for $\boldsymbol{\omega}$ given in Eq. (1.11.1) involves three components, each of which is parallel to a unit vector fixed in B. At times it is necessary to express $\boldsymbol{\omega}$ in other ways, for example, to resolve it into components parallel to unit vectors fixed in A. Whichever resolution is employed, one may wish to know what the physical significance of any one component of $\boldsymbol{\omega}$ is.

In certain situations physical significance can be attributed to angular velocity components by identifying for each component two reference frames such that the angular velocity of one of these relative to the other is equal to the component in question. As will be seen later, this is the case, for example, when the angular velocity of B in A is expressed as in Eq. (1.16.1). In general, however, it is not a simple matter to discover the necessary reference frames. For instance, such reference frames are not readily identifiable for the components $\lambda\dot{s}\mathbf{b}_1$ and $(\dot{s}/\rho)\mathbf{b}_3$ of the angular velocity found in the example in Sec. 1.11.

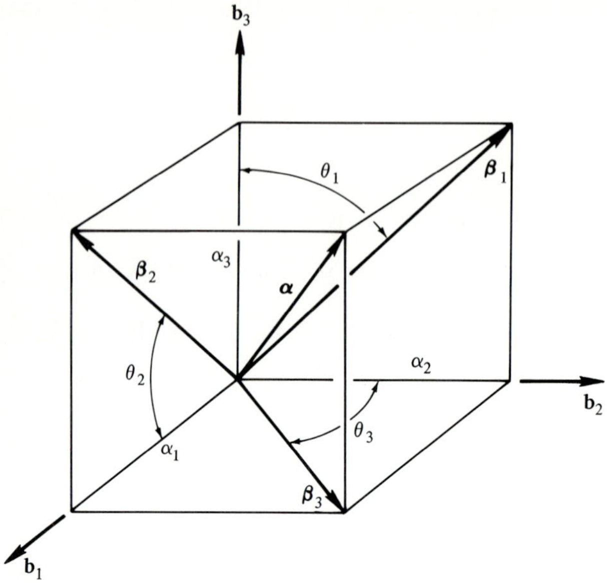

Figure 1.12.1

An essentially geometric interpretation can be given to the quantities ω_1, ω_2, and ω_3 appearing in Eq. (1.11.1), and thus to the components $\boldsymbol{\omega}\cdot\mathbf{b}_i$ ($i = 1, 2, 3$) of $\boldsymbol{\omega}$, by introducing a certain space-average value of the first time-derivative of each of three angles.† Specifically, let $\boldsymbol{\alpha}$ be a generic unit vector fixed in reference frame A, $\boldsymbol{\beta}_i$ the orthogonal projection of $\boldsymbol{\alpha}$ on a plane normal to $\mathbf{b}_i$ ($i = 1, 2, 3$), θ_1 the angle between $\boldsymbol{\beta}_1$ and $\mathbf{b}_3$, θ_2 the angle between $\boldsymbol{\beta}_2$ and $\mathbf{b}_1$, and θ_3 the angle between $\boldsymbol{\beta}_3$ and $\mathbf{b}_2$, as shown in Fig. 1.12.1. Next, letting S be a unit sphere centered at a point O, and designating as P the point of S whose position vector relative to O is parallel to $\boldsymbol{\alpha}$ (see Fig. 1.12.2), associate with P the value of $\dot{\theta}_i$ and define $\overline{\dot{\theta}_i}$ as

$$\overline{\dot{\theta}_i} \triangleq \frac{1}{4\pi}\int \dot{\theta}_i \, d\sigma \qquad (i = 1, 2, 3) \tag{1}$$

where $d\sigma$ is the area of a differential element of S at P. Then

$$\omega_i = \overline{\dot{\theta}_i} \qquad (i = 1, 2, 3) \tag{2}$$

Derivation Defining α_i as

$$\alpha_i \triangleq \boldsymbol{\alpha}\cdot\mathbf{b}_i \qquad (i = 1, 2, 3) \tag{3}$$

†The authors are indebted to Professor R. Skalak of Columbia University for this idea.

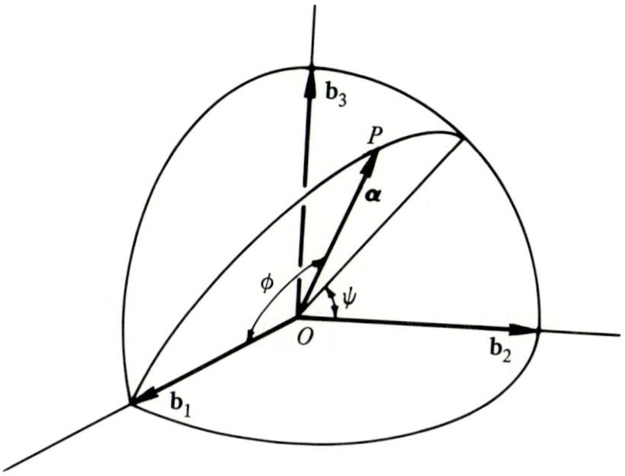

Figure 1.12.2

one can express θ_1 (see Fig. 1.12.1) as

$$\theta_1 = \tan^{-1} \frac{\alpha_2}{\alpha_3} \tag{4}$$

from which it follows that

$$\dot{\theta}_1 = \frac{\dot{\alpha}_2 \alpha_3 - \dot{\alpha}_3 \alpha_2}{\alpha_2^2 + \alpha_3^2} \tag{5}$$

Now, $\frac{{}^B d\boldsymbol{\alpha}}{dt}$ is given both by

$$\frac{{}^B d\boldsymbol{\alpha}}{dt} = \dot{\alpha}_1 \mathbf{b}_1 + \dot{\alpha}_2 \mathbf{b}_2 + \dot{\alpha}_3 \mathbf{b}_3 \tag{6}$$

and by

$$\begin{aligned}\frac{{}^B d\boldsymbol{\alpha}}{dt} &\underset{(1.11.9)}{=} {}^B\boldsymbol{\omega}^A \times \boldsymbol{\alpha} \underset{(1.11.2)}{=} -{}^A\boldsymbol{\omega}^B \times \boldsymbol{\alpha} \\ &\underset{(1.11.1)}{=} (\alpha_2 \omega_3 - \alpha_3 \omega_2)\mathbf{b}_1 \\ &\quad + (\alpha_3 \omega_1 - \alpha_1 \omega_3)\mathbf{b}_2 \\ &\quad + (\alpha_1 \omega_2 - \alpha_2 \omega_1)\mathbf{b}_3\end{aligned} \tag{7}$$

Consequently,

$$\dot{\alpha}_i \underset{(1.10.8)}{=} \eta_{ijk} (\alpha_j \omega_k - \alpha_k \omega_j) \qquad (i = 1, 2, 3) \tag{8}$$

and

$$\begin{aligned}\dot{\theta}_1 &\underset{(5)}{=} \frac{(\alpha_3 \omega_1 - \alpha_1 \omega_3)\alpha_3 - (\alpha_1 \omega_2 - \alpha_2 \omega_1)\alpha_2}{\alpha_2^2 + \alpha_3^2} \\ &= \omega_1 - \frac{\alpha_1 \alpha_2}{\alpha_2^2 + \alpha_3^2}\omega_2 - \frac{\alpha_1 \alpha_3}{\alpha_2^2 + \alpha_3^2}\omega_3\end{aligned} \tag{9}$$

To perform the integration indicated in Eq. (1), introduce the angles ϕ and ψ shown Fig. 1.12.2, noting that $\boldsymbol{\alpha}$ then can be expressed as

$$\boldsymbol{\alpha} = \cos\phi\,\mathbf{b}_1 + \sin\phi\cos\psi\,\mathbf{b}_2 + \sin\phi\sin\psi\,\mathbf{b}_3 \tag{10}$$

so that

$$\alpha_1 = \cos\phi \qquad \alpha_2 = \sin\phi\cos\psi \qquad \alpha_3 = \sin\phi\sin\psi \tag{11}$$

while

$$d\sigma = \sin\phi\, d\phi\, d\psi \tag{12}$$

Consequently,

$$4\pi\overline{\dot{\theta}_1} \underset{(1)}{=} \omega_1 \int_0^{\pi}\left[\int_0^{2\pi} \sin\phi\, d\psi\right]d\phi - \omega_2 \int_0^{\pi}\left[\int_0^{2\pi} \cos\phi\cos\psi\, d\psi\right]d\phi$$
$$- \omega_3 \int_0^{\pi}\left[\int_0^{2\pi} \cos\phi\sin\psi\, d\psi\right]d\phi \tag{13}$$

The first double integral has the value 4π, and the remaining two double integrals are equal to zero. Hence,

$$\overline{\dot{\theta}_1} = \omega_1 \tag{14}$$

Similarly,

$$\overline{\dot{\theta}_2} = \omega_2 \tag{15}$$

and

$$\overline{\dot{\theta}_3} = \omega_3 \tag{16}$$

Example In Fig. 1.12.3, B designates a cylindrical spacecraft whose attitude motion in a reference frame A can be described as a combination of *coning* and *spinning*, the former being characterized by the angle ϕ and involving the motion of the symmetry axis of B on the surface of a cone that is fixed in A and has a constant semivertex angle θ, while the latter is associated with changes in the angle ψ between two lines which intersect on, and are perpendicular to, the symmetry axis of B, one line being fixed in B and the other one intersecting the axis of the cone. Under these circumstances, the direction cosine matrix C such that

$$[\mathbf{b}_1 \quad \mathbf{b}_2 \quad \mathbf{b}_3] = [\mathbf{a}_1 \quad \mathbf{a}_2 \quad \mathbf{a}_3]\, C \tag{17}$$

where $\mathbf{a}_i$ and $\mathbf{b}_i$ $(i = 1, 2, 3)$ are unit vectors directed as shown in Fig. 1.12.3, can be expressed as

$$C \underset{(1.6.15)}{=} C_1(\phi)\, C_3(\theta)\, C_1(\psi) \tag{18}$$

or, after using Eqs. (1.2.35) and (1.2.37), as

$$C = \begin{bmatrix} c\theta & -s\theta\, c\psi & s\theta\, s\psi \\ s\theta\, c\phi & c\theta\, c\phi\, c\psi - s\phi\, s\psi & -c\theta\, c\phi\, s\psi - s\phi\, c\psi \\ s\theta\, s\phi & c\theta\, s\phi\, c\psi + c\phi\, s\psi & -c\theta\, s\phi\, s\psi + c\phi\, c\psi \end{bmatrix} \tag{19}$$

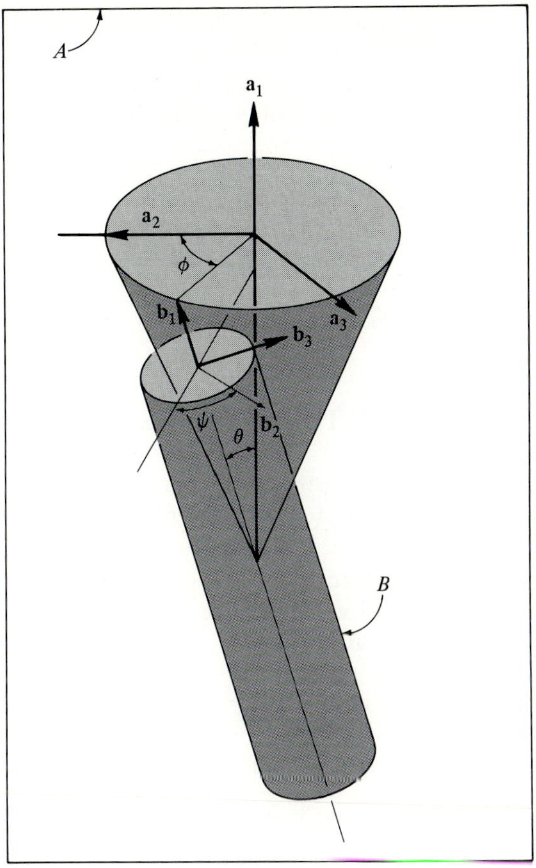

Figure 1.12.3

where sθ and cθ denote sin θ and cos θ, respectively, and similarly for ϕ and ψ. Forming ω_1, ω_2, and ω_3 in accordance with Eqs. (1.10.5)–(1.10.7), one then obtains the following expression for the angular velocity $\boldsymbol{\omega}$ of B in A:

$$\boldsymbol{\omega} = (\dot{\psi} + \dot{\phi}\ \mathrm{c}\theta)\ \mathbf{b}_1 - \dot{\phi}\ \mathrm{s}\theta\ \mathrm{c}\psi\ \mathbf{b}_2 + \dot{\phi}\ \mathrm{s}\theta\ \mathrm{s}\psi\ \mathbf{b}_3 \tag{20}$$

A more efficient method for obtaining this result is described in Sec. 1.16. For present purposes, what is of interest is the fact that the $\mathbf{b}_i$ component ($i = 1, 2, 3$) of $\boldsymbol{\omega}$ does not have a readily apparent physical significance, but that, when $\boldsymbol{\omega}$ is rewritten as

$$\boldsymbol{\omega} = \dot{\psi}\ \mathbf{b}_1 + \dot{\phi}(\mathrm{c}\theta\ \mathbf{b}_1 - \mathrm{s}\theta\ \mathrm{c}\psi\ \mathbf{b}_2 + \mathrm{s}\theta\ \mathrm{s}\psi\ \mathbf{b}_3)$$

$$\underset{(18,19)}{=} \dot{\psi}\ \mathbf{b}_1 + \dot{\phi}\,\mathbf{a}_1 \tag{21}$$

then each component has the same form as the right-hand member of Eq. (1.11.5) and can, therefore, be regarded as the angular velocity of a body performing a motion of simple rotation. Specifically, designating as A_1 a reference frame in which the axis of the cone and the symmetry axis of B are fixed, one can observe that A_1 performs a motion of simple rotation in A; moreover, that B performs such a motion in A_1; and, finally, that the associated angular velocities are

$$^{A}\boldsymbol{\omega}^{A_1} = \dot{\phi}\ \mathbf{a}_1 \tag{22}$$

and

$$^{A_1}\boldsymbol{\omega}^{B} = \dot{\psi}\ \mathbf{b}_1 \tag{23}$$

Thus, it appears that

$$\boldsymbol{\omega} = {}^{A}\boldsymbol{\omega}^{A_1} + {}^{A_1}\boldsymbol{\omega}^{B} \tag{24}$$

1.13 ANGULAR VELOCITY AND EULER PARAMETERS

If $\mathbf{a}_1$, $\mathbf{a}_2$, $\mathbf{a}_3$, and $\mathbf{b}_1$, $\mathbf{b}_2$, $\mathbf{b}_3$ are two dextral sets of orthogonal unit vectors fixed respectively in reference frames or rigid bodies A and B which are moving relative to each other, one can use Eqs. (1.3.15)–(1.3.18) to associate with each instant of time Euler parameters $\epsilon_1, \ldots, \epsilon_4$; and an Euler vector $\boldsymbol{\epsilon}$ can then be formed by reference to Eq. (1.3.2). In terms of $\boldsymbol{\epsilon}$ and ϵ_4, the angular velocity of B in A (see Sec. 1.11) can be expressed as

$$\boldsymbol{\omega} = 2\left(\epsilon_4 \frac{^{B}d\boldsymbol{\epsilon}}{dt} - \dot{\epsilon}_4\,\boldsymbol{\epsilon} - \boldsymbol{\epsilon} \times \frac{^{B}d\boldsymbol{\epsilon}}{dt}\right) \tag{1}$$

Conversely, if $\boldsymbol{\omega}$ is known as a function of time, the Euler parameters can be found by solving the differential equations

$$\frac{^{B}d\boldsymbol{\epsilon}}{dt} = \tfrac{1}{2}(\epsilon_4\,\boldsymbol{\omega} + \boldsymbol{\epsilon} \times \boldsymbol{\omega}) \tag{2}$$

and

$$\dot{\epsilon}_4 = -\tfrac{1}{2}\boldsymbol{\omega} \cdot \boldsymbol{\epsilon} \tag{3}$$

Equations equivalent to Eqs. (1)–(3) can be formulated in terms of matrices ω, ϵ, and E defined as

$$\omega \triangleq [\omega_1 \quad \omega_2 \quad \omega_3 \quad 0] \tag{4}$$

$$\epsilon \triangleq [\epsilon_1 \quad \epsilon_2 \quad \epsilon_3 \quad \epsilon_4] \tag{5}$$

and

$$E = \begin{bmatrix} \epsilon_4 & -\epsilon_3 & \epsilon_2 & \epsilon_1 \\ \epsilon_3 & \epsilon_4 & -\epsilon_1 & \epsilon_2 \\ -\epsilon_2 & \epsilon_1 & \epsilon_4 & \epsilon_3 \\ -\epsilon_1 & -\epsilon_2 & -\epsilon_3 & \epsilon_4 \end{bmatrix} \tag{6}$$

These equations are

$$\omega = 2\,\dot{\epsilon}E \tag{7}$$

and
$$\dot{\epsilon} = \frac{1}{2}\,\omega E^T \tag{8}$$

Derivations Substitution from Eqs. (1.3.6)–(1.3.14) into Eqs. (1.10.5)–(1.10.7) gives

$$\begin{aligned}
\omega_1 &= 4(\epsilon_3\,\epsilon_1 + \epsilon_2\,\epsilon_4)(\dot{\epsilon}_1\,\epsilon_2 + \epsilon_1\,\dot{\epsilon}_2 - \dot{\epsilon}_3\,\epsilon_4 - \epsilon_3\,\dot{\epsilon}_4) \\
&\quad + 4(\epsilon_2\,\epsilon_3 - \epsilon_1\,\epsilon_4)(\epsilon_2\,\dot{\epsilon}_2 - \epsilon_3\,\dot{\epsilon}_3 - \epsilon_1\,\dot{\epsilon}_1 + \epsilon_4\,\dot{\epsilon}_4) \\
&\quad + 2(1 - 2\epsilon_1^2 - 2\epsilon_2^2)(\dot{\epsilon}_2\,\epsilon_3 + \epsilon_2\,\dot{\epsilon}_3 + \dot{\epsilon}_1\,\epsilon_4 + \epsilon_1\,\dot{\epsilon}_4) \\
&= 2(\dot{\epsilon}_1\,\epsilon_4 + \dot{\epsilon}_2\,\epsilon_3 - \dot{\epsilon}_3\,\epsilon_2 - \dot{\epsilon}_4\,\epsilon_1)
\end{aligned} \tag{9}$$

$$\omega_2 = 2(\dot{\epsilon}_2\,\epsilon_4 + \dot{\epsilon}_3\,\epsilon_1 - \dot{\epsilon}_1\,\epsilon_3 - \dot{\epsilon}_4\,\epsilon_2) \tag{10}$$

$$\omega_3 = 2(\dot{\epsilon}_3\,\epsilon_4 + \dot{\epsilon}_1\,\epsilon_2 - \dot{\epsilon}_2\,\epsilon_1 - \dot{\epsilon}_4\,\epsilon_3) \tag{11}$$

and these are three of the four scalar equations corresponding to Eq. (7). The fourth is

$$0 \underset{(7)}{=} 2(\dot{\epsilon}_1\,\epsilon_1 + \dot{\epsilon}_2\,\epsilon_2 + \dot{\epsilon}_3\,\epsilon_3 + \dot{\epsilon}_4\,\epsilon_4) \tag{12}$$

and this equation is satisfied because

$$\dot{\epsilon}_1\,\epsilon_1 + \dot{\epsilon}_2\,\epsilon_2 + \dot{\epsilon}_3\,\epsilon_3 + \dot{\epsilon}_4\,\epsilon_4 = \frac{1}{2}\frac{d}{dt}(\epsilon_1^2 + \epsilon_2^2 + \epsilon_3^2 + \epsilon_4^2) \underset{(1.3.4)}{=} 0 \tag{13}$$

Thus the validity of Eq. (7) is established; and Eq. (1) can be obtained by noting that

$$\begin{aligned}
\boldsymbol{\omega} &\underset{(1.11.1)}{=} \omega_1\,\mathbf{b}_1 + \omega_2\,\mathbf{b}_2 + \omega_3\,\mathbf{b}_3 \\
&\underset{(7)}{=} 2[(\dot{\epsilon}_1\,\epsilon_4 + \dot{\epsilon}_2\,\epsilon_3 - \dot{\epsilon}_3\,\epsilon_2 - \dot{\epsilon}_4\,\epsilon_1)\,\mathbf{b}_1 \\
&\quad + (\dot{\epsilon}_2\,\epsilon_4 + \dot{\epsilon}_3\,\epsilon_1 - \dot{\epsilon}_1\,\epsilon_3 - \dot{\epsilon}_4\,\epsilon_2)\,\mathbf{b}_2 \\
&\quad + (\dot{\epsilon}_3\,\epsilon_4 + \dot{\epsilon}_1\,\epsilon_2 - \dot{\epsilon}_2\,\epsilon_1 - \dot{\epsilon}_4\,\epsilon_3)\,\mathbf{b}_3] \\
&= 2[\epsilon_4(\dot{\epsilon}_1\,\mathbf{b}_1 + \dot{\epsilon}_2\,\mathbf{b}_2 + \dot{\epsilon}_3\,\mathbf{b}_3) - \dot{\epsilon}_4(\epsilon_1\,\mathbf{b}_1 + \epsilon_2\,\mathbf{b}_2 + \epsilon_3\,\mathbf{b}_3) \\
&\quad + (\dot{\epsilon}_2\,\epsilon_3 - \dot{\epsilon}_3\,\epsilon_2)\,\mathbf{b}_1 + (\dot{\epsilon}_3\,\epsilon_1 - \dot{\epsilon}_1\,\epsilon_3)\,\mathbf{b}_2 + (\dot{\epsilon}_1\,\epsilon_2 - \dot{\epsilon}_2\,\epsilon_1)\,\mathbf{b}_3] \\
&\underset{(1.3.2)}{=} 2\left(\epsilon_4\,\frac{{}^B d\boldsymbol{\epsilon}}{dt} - \dot{\epsilon}_4\,\boldsymbol{\epsilon} - \boldsymbol{\epsilon} \times \frac{{}^B d\boldsymbol{\epsilon}}{dt}\right)
\end{aligned} \tag{14}$$

Postmultiplication of both sides of Eq. (7) with E^T gives

$$\omega E^T = 2\,\dot{\epsilon} E E^T \tag{15}$$

Now, using Eq. (1.3.4) and referring to Eq. (6), one finds that

$$E E^T = \begin{bmatrix} 1 & 0 & 0 & 0 \\ 0 & 1 & 0 & 0 \\ 0 & 0 & 1 & 0 \\ 0 & 0 & 0 & 1 \end{bmatrix} \tag{16}$$

Consequently,

$$\omega E^T = 2\,\dot{\boldsymbol{\epsilon}} \tag{17}$$

in agreement with Eq. (8).

Finally,

$$\dot{\epsilon}_4 \underset{(8)}{=} -\tfrac{1}{2}(\omega_1\,\epsilon_1 + \omega_2\,\epsilon_2 + \omega_3\,\epsilon_3) \underset{(1.11.1,\ 1.3.2)}{=} -\tfrac{1}{2}\,\boldsymbol{\omega}\cdot\boldsymbol{\epsilon} \tag{18}$$

as in Eq. (3); and

$$\frac{{}^B d\boldsymbol{\epsilon}}{dt} \underset{(1.3.2)}{=} \dot{\epsilon}_1\,\mathbf{b}_1 + \dot{\epsilon}_2\,\mathbf{b}_2 + \dot{\epsilon}_3\,\mathbf{b}_3$$

$$\underset{(8)}{=} \tfrac{1}{2}\,[(\omega_1\,\epsilon_4 - \omega_2\,\epsilon_3 + \omega_3\,\epsilon_2)\,\mathbf{b}_1 + (\omega_1\,\epsilon_3 + \omega_2\,\epsilon_4 - \omega_3\,\epsilon_1)\,\mathbf{b}_2 - (\omega_1\,\epsilon_2 - \omega_2\,\epsilon_1 - \omega_3\,\epsilon_4)\,\mathbf{b}_3] \tag{19}$$

The right-hand member of this equation is equal to that of Eq. (2).

Example Suppose that the inertia ellipsoid of B for the mass center B^* of B is an ellipsoid of revolution whose axis of revolution is parallel to $\mathbf{b}_3$. Then, if $\mathbf{I}$ denotes the inertia dyadic of B for B^*, and if I and J are defined as

$$I \triangleq \mathbf{b}_1 \cdot \mathbf{I} \cdot \mathbf{b}_1 = \mathbf{b}_2 \cdot \mathbf{I} \cdot \mathbf{b}_2 \tag{20}$$

and

$$J \triangleq \mathbf{b}_3 \cdot \mathbf{I} \cdot \mathbf{b}_3 \tag{21}$$

the angular momentum $\mathbf{H}$ of B in A with respect to B^* is given by

$$\mathbf{H} = I\omega_1\,\mathbf{b}_1 + I\omega_2\,\mathbf{b}_2 + J\omega_3\,\mathbf{b}_3 \tag{22}$$

and the first time-derivative of $\mathbf{H}$ in A can be expressed as

$$\frac{{}^A d\mathbf{H}}{dt} \underset{(1.11.8)}{=} \frac{{}^B d\mathbf{H}}{dt} + \boldsymbol{\omega}\times\mathbf{H}$$

$$\underset{(22)}{=} [I\,\dot{\omega}_1 + (J - I)\omega_2\,\omega_3]\mathbf{b}_1 + [I\,\dot{\omega}_2 - (J - I)\omega_3\,\omega_1]\mathbf{b}_2 + J\,\dot{\omega}_3\,\mathbf{b}_3 \tag{23}$$

Hence, if B moves under the action of forces the sum of whose moments about B^* is equal to zero, and if A is an inertial reference frame, so that, in accordance with the angular momentum principle, ${}^A d\mathbf{H}/dt$ is equal to zero, then ω_1, ω_2, and ω_3 are governed by the differential equations

$$\dot{\omega}_1 - \frac{I - J}{I}\,\omega_2\,\omega_3 = 0 \tag{24}$$

$$\dot{\omega}_2 + \frac{I - J}{I}\,\omega_3\,\omega_1 = 0 \tag{25}$$

$$\dot{\omega}_3 = 0 \tag{26}$$

Letting $\overline{\omega}_i$ denote the value of ω_i ($i = 1, 2, 3$) at $t = 0$, and defining a constant s as

$$s \triangleq \frac{I - J}{I} \bar{\omega}_3 \tag{27}$$

one can express the general solution of Eqs. (24)–(26) as

$$\omega_1 = \bar{\omega}_1 \cos st + \bar{\omega}_2 \sin st \tag{28}$$

$$\omega_2 = -\bar{\omega}_1 \sin st + \bar{\omega}_2 \cos st \tag{29}$$

$$\omega_3 = \bar{\omega}_3 \tag{30}$$

and, to determine the orientation of B in A, one then can seek the solution of the differential equations

$$\dot{\epsilon}_1 \underset{(8)}{=} \frac{1}{2}(\omega_1 \epsilon_4 - \omega_2 \epsilon_3 + \omega_3 \epsilon_2)$$

$$\underset{(28\text{–}30)}{=} \frac{1}{2}[(\bar{\omega}_1 \cos st + \bar{\omega}_2 \sin st)\epsilon_4 + (\bar{\omega}_1 \sin st - \bar{\omega}_2 \cos st)\epsilon_3 + \bar{\omega}_3 \epsilon_2] \tag{31}$$

$$\dot{\epsilon}_2 \underset{(8)}{=} \frac{1}{2}(\omega_1 \epsilon_3 + \omega_2 \epsilon_4 - \omega_3 \epsilon_1)$$

$$\underset{(28\text{–}30)}{=} \frac{1}{2}[(\bar{\omega}_1 \cos st + \bar{\omega}_2 \sin st)\epsilon_3 - (\bar{\omega}_1 \sin st - \bar{\omega}_2 \cos st)\epsilon_4 - \bar{\omega}_3 \epsilon_1] \tag{32}$$

$$\dot{\epsilon}_3 \underset{(8)}{=} \frac{1}{2}(-\omega_1 \epsilon_2 + \omega_2 \epsilon_1 + \omega_3 \epsilon_4)$$

$$\underset{(28\text{–}30)}{=} \frac{1}{2}[-(\bar{\omega}_1 \cos st + \bar{\omega}_2 \sin st)\epsilon_2 + (-\bar{\omega}_1 \sin st + \bar{\omega}_2 \cos st)\epsilon_1 + \bar{\omega}_3 \epsilon_4] \tag{33}$$

$$\dot{\epsilon}_4 \underset{(8)}{=} -\frac{1}{2}(\omega_1 \epsilon_1 + \omega_2 \epsilon_2 + \omega_3 \epsilon_3)$$

$$\underset{(28\text{–}30)}{=} -\frac{1}{2}[(\bar{\omega}_1 \cos st + \bar{\omega}_2 \sin st)\epsilon_1 + (-\bar{\omega}_1 \sin st + \bar{\omega}_2 \cos st)\epsilon_2 + \bar{\omega}_3 \epsilon_3] \tag{34}$$

using as initial conditions

$$\epsilon_1 = \epsilon_2 = \epsilon_3 = 0 \qquad \epsilon_4 = 1 \qquad \text{at } t = 0 \tag{35}$$

which means that the unit vectors $\mathbf{a}_1$, $\mathbf{a}_2$, and $\mathbf{a}_3$ have been chosen such that $\mathbf{a}_i = \mathbf{b}_i$ $(i = 1, 2, 3)$ at $t = 0$.

Since Eqs. (31)–(34) have time-dependent coefficients, they cannot be solved for $\epsilon_1, \ldots, \epsilon_4$ by simple analytical procedures. However, attacking the physical problem at hand by a different method (see Sec. 3.1), and defining a quantity p as

$$p \triangleq \left[\bar{\omega}_1^2 + \bar{\omega}_2^2 + \left(\bar{\omega}_3 \frac{J}{I}\right)^2\right]^{1/2} \tag{36}$$

one can show that ϵ_1, ϵ_2, ϵ_3, and ϵ_4 are given by

$$\epsilon_1 = \frac{\sin(pt/2)}{p}\left(\bar{\omega}_1 \cos\frac{st}{2} + \bar{\omega}_2 \sin\frac{st}{2}\right) \tag{37}$$

$$\epsilon_2 = \frac{\sin (pt/2)}{p}\left(-\bar{\omega}_1 \sin \frac{st}{2} + \bar{\omega}_2 \cos \frac{st}{2}\right) \tag{38}$$

$$\epsilon_3 = \bar{\omega}_3 \frac{J}{Ip} \sin \frac{pt}{2} \cos \frac{st}{2} + \cos \frac{pt}{2} \sin \frac{st}{2} \tag{39}$$

$$\epsilon_4 = -\bar{\omega}_3 \frac{J}{Ip} \sin \frac{pt}{2} \sin \frac{st}{2} + \cos \frac{pt}{2} \cos \frac{st}{2} \tag{40}$$

and it may be verified that these expressions do, indeed, satisfy Eqs. (31)–(35).

1.14 ANGULAR VELOCITY AND RODRIGUES PARAMETERS

If $\mathbf{a}_1$, $\mathbf{a}_2$, $\mathbf{a}_3$ and $\mathbf{b}_1$, $\mathbf{b}_2$, $\mathbf{b}_3$ are two dextral sets of orthogonal unit vectors fixed respectively in reference frames or rigid bodies A and B which are moving relative to each other, one can use Eqs. (1.4.7)–(1.4.9) to associate with each instant of time Rodrigues parameters ρ_1, ρ_2, and ρ_3; and a Rodrigues vector $\boldsymbol{\rho}$ can then be formed by reference to Eq. (1.4.2). The angular velocity of B in A (see Sec. 1.11), expressed in terms of $\boldsymbol{\rho}$, is given by

$$\boldsymbol{\omega} = \frac{2}{1 + \boldsymbol{\rho}^2}\left(\frac{^B d\boldsymbol{\rho}}{dt} - \boldsymbol{\rho} \times \frac{^B d\boldsymbol{\rho}}{dt}\right) \tag{1}$$

Conversely, if $\boldsymbol{\omega}$ is known as a function of time, the Rodrigues vector can be found by solving the differential equation

$$\frac{^B d\boldsymbol{\rho}}{dt} = \frac{1}{2}(\boldsymbol{\omega} + \boldsymbol{\rho} \times \boldsymbol{\omega} + \boldsymbol{\rho}\boldsymbol{\rho} \cdot \boldsymbol{\omega}) \tag{2}$$

Equations equivalent to Eqs. (1) and (2) can be formulated in terms of matrices ω, ρ, and $\widetilde{\rho}$ defined as

$$\omega \triangleq [\omega_1 \quad \omega_2 \quad \omega_3] \tag{3}$$

$$\rho \triangleq [\rho_1 \quad \rho_2 \quad \rho_3] \tag{4}$$

and

$$\widetilde{\rho} \triangleq \begin{bmatrix} 0 & -\rho_3 & \rho_2 \\ \rho_3 & 0 & -\rho_1 \\ -\rho_2 & \rho_1 & 0 \end{bmatrix} \tag{5}$$

These equations are

$$\omega = \frac{2\dot{\rho}(U + \widetilde{\rho})}{1 + \rho\rho^T} \tag{6}$$

and

$$\dot{\rho} = \frac{1}{2}\omega(U - \widetilde{\rho} + \rho^T\rho) \tag{7}$$

Like its counterparts for the direction cosine matrix and for Euler parameters [see Eqs. (1.10.3) and (1.13.8)], Eq. (7) is, in general, an equation with variable coefficients. Since it is, moreover, nonlinear, one must usually resort to numerical methods to obtain solutions.

Derivations Using Eqs. (1.3.1), (1.3.3), and (1.4.1), one can express $\boldsymbol{\epsilon}$ and ϵ_4 as

$$\boldsymbol{\epsilon} = \boldsymbol{\rho}(1 + \boldsymbol{\rho}^2)^{-1/2} \tag{8}$$

and

$$\epsilon_4 = (1 + \boldsymbol{\rho}^2)^{-1/2} \tag{9}$$

respectively. Consequently,

$$\frac{{}^B d\boldsymbol{\epsilon}}{dt} = \frac{{}^B d\boldsymbol{\rho}}{dt}(1 + \boldsymbol{\rho}^2)^{-1/2} - \boldsymbol{\rho}(1 + \boldsymbol{\rho}^2)^{-3/2}\, \boldsymbol{\rho} \cdot \frac{{}^B d\boldsymbol{\rho}}{dt} \tag{10}$$

$$\dot{\epsilon}_4 = -(1 + \boldsymbol{\rho}^2)^{-3/2}\, \boldsymbol{\rho} \cdot \frac{{}^B d\boldsymbol{\rho}}{dt} \tag{11}$$

and

$$\boldsymbol{\omega} \underset{(1.13.1)}{=} 2\left[\frac{{}^B d\boldsymbol{\rho}}{dt}(1 + \boldsymbol{\rho}^2)^{-1} - \boldsymbol{\rho}(1 + \boldsymbol{\rho}^2)^{-2}\, \boldsymbol{\rho} \cdot \frac{{}^B d\boldsymbol{\rho}}{dt}\right.$$
$$\left. + \boldsymbol{\rho}(1 + \boldsymbol{\rho}^2)^{-2}\, \boldsymbol{\rho} \cdot \frac{{}^B d\boldsymbol{\rho}}{dt} - \boldsymbol{\rho} \times \frac{{}^B d\boldsymbol{\rho}}{dt}(1 + \boldsymbol{\rho}^2)^{-1}\right] \tag{12}$$

which is equivalent to Eq. (1).

Cross-multiplication of Eq. (1) with $\boldsymbol{\rho}$ yields

$$\boldsymbol{\omega} \times \boldsymbol{\rho} = \frac{2}{1 + \boldsymbol{\rho}^2}\left(\frac{{}^B d\boldsymbol{\rho}}{dt} \times \boldsymbol{\rho} - \boldsymbol{\rho}^2 \frac{{}^B d\boldsymbol{\rho}}{dt} + \boldsymbol{\rho}\, \frac{{}^B d\boldsymbol{\rho}}{dt} \cdot \boldsymbol{\rho}\right)$$
$$\underset{(1)}{=} \boldsymbol{\omega} - 2\frac{{}^B d\boldsymbol{\rho}}{dt} + \frac{2}{1 + \boldsymbol{\rho}^2}\frac{{}^B d\boldsymbol{\rho}}{dt} \cdot \boldsymbol{\rho}\boldsymbol{\rho} \tag{13}$$

while dot-multiplication produces

$$\boldsymbol{\omega} \cdot \boldsymbol{\rho} = \frac{2}{1 + \boldsymbol{\rho}^2}\frac{{}^B d\boldsymbol{\rho}}{dt} \cdot \boldsymbol{\rho} \tag{14}$$

Consequently,

$$\boldsymbol{\omega} \times \boldsymbol{\rho} = \boldsymbol{\omega} - 2\frac{{}^B d\boldsymbol{\rho}}{dt} + \boldsymbol{\omega} \cdot \boldsymbol{\rho}\boldsymbol{\rho} \tag{15}$$

in agreement with Eq. (2).

The validity of Eqs. (6) and (7) may be verified by carrying out the indicated matrix multiplications and then comparing the associated scalar equations with the scalar equations corresponding to Eqs. (1) and (2).

Example The "spin-up" problem for an axially symmetric spacecraft B can be formulated most simply as follows: taking the axis of revolution of the inertia ellipsoid of B for the mass center B^* of B to be parallel to $\mathbf{b}_3$, assuming that B is subjected to the action of a system of forces whose resultant moment about B^* is equal to $M\mathbf{b}_3$, where M is a constant, and letting ω_1, ω_2, and ω_3 have the values

$$\omega_1 = \overline{\omega}_1 \qquad \omega_2 = \omega_3 = 0 \tag{16}$$

at time $t = 0$, determine the orientation of B in an inertial reference frame A for $t > 0$. (The reason for taking ω_2 equal to zero at $t = 0$ is that the unit vectors $\mathbf{b}_1$ and $\mathbf{b}_2$ can always be chosen such that $\mathbf{b}_2$ is perpendicular to $\boldsymbol{\omega}$ at $t = 0$, in which case $\omega_2 = \boldsymbol{\omega} \cdot \mathbf{b}_2 = 0$. As for ω_3, this is taken equal to zero at $t = 0$ because the satellite is presumed to have either no rotational motion or to be tumbling initially, tumbling here referring to a motion such that the angular velocity is perpendicular to the symmetry axis.)

Letting $\mathbf{I}$ denote the inertia dyadic of B for B^*, and defining I and J as

$$I \triangleq \mathbf{b}_1 \cdot \mathbf{I} \cdot \mathbf{b}_1 = \mathbf{b}_2 \cdot \mathbf{I} \cdot \mathbf{b}_2 \tag{17}$$

and

$$J \triangleq \mathbf{b}_3 \cdot \mathbf{I} \cdot \mathbf{b}_3 \tag{18}$$

one can use the angular momentum principle to obtain the following differential equations governing ω_1, ω_2, and ω_3:

$$\dot{\omega}_1 = \frac{I - J}{I} \omega_2 \omega_3 \tag{19}$$

$$\dot{\omega}_2 = -\frac{I - J}{I} \omega_3 \omega_1 \tag{20}$$

$$\dot{\omega}_3 = \frac{M}{J} \tag{21}$$

Since M and J are constants,

$$\omega_3 \underset{(21,16)}{=} \frac{M}{J} t \tag{22}$$

and

$$\dot{\omega}_1 \underset{(19,22)}{=} \frac{I - J}{I} \frac{M}{J} t \, \omega_2 \tag{23}$$

$$\dot{\omega}_2 \underset{(20,22)}{=} -\frac{I - J}{I} \frac{M}{J} t \, \omega_1 \tag{24}$$

The solution of these equations is facilitated by introducing a function ϕ as

$$\phi \triangleq \frac{I-J}{I}\frac{M}{J}\frac{t^2}{2} \tag{25}$$

Then

$$\dot{\omega}_1 \underset{(23,25)}{=} \dot{\phi}\,\omega_2 \tag{26}$$

$$\dot{\omega}_2 \underset{(24,25)}{=} -\dot{\phi}\,\omega_1 \tag{27}$$

or

$$\frac{d\omega_1}{d\phi} \underset{(26)}{=} \omega_2 \tag{28}$$

$$\frac{d\omega_2}{d\phi} \underset{(27)}{=} -\omega_1 \tag{29}$$

so that

$$\frac{d^2\omega_1}{d\phi^2} + \omega_1 \underset{(28,29)}{=} 0 \tag{30}$$

$$\omega_1 = C_1 \sin\phi + C_2 \cos\phi \tag{31}$$

and

$$\omega_2 \underset{(28,31)}{=} C_1 \cos\phi - C_2 \sin\phi \tag{32}$$

where C_1 and C_2 are constants which can be evaluated by noting that ϕ [see Eq. (25)] vanishes at $t = 0$. That is,

$$\overline{\omega}_1 \underset{(16,31)}{=} C_2 \tag{33}$$

and

$$0 \underset{(16,32)}{=} C_1 \tag{34}$$

Consequently,

$$\omega_1 \underset{(31)}{=} \overline{\omega}_1 \cos\phi \tag{35}$$

and

$$\omega_2 \underset{(32)}{=} -\overline{\omega}_1 \sin\phi \tag{36}$$

Equations governing the Rodrigues parameters ρ_1, ρ_2, and ρ_3 can now be formulated by referring to Eqs. (3), (4), (5), and (7) to obtain

$$\begin{aligned} 2\dot{\rho}_1 &= \omega_1(1+\rho_1^2) + \omega_2(\rho_1\rho_2 - \rho_3) + \omega_3(\rho_3\rho_1 + \rho_2) \\ &= \overline{\omega}_1 \cos\phi(1+\rho_1^2) - \overline{\omega}_1 \sin\phi(\rho_1\rho_2 - \rho_3) + \frac{M}{J}t\,(\rho_3\rho_1 + \rho_2) \end{aligned} \tag{37}$$

$$\begin{aligned} 2\dot{\rho}_2 &= \omega_1(\rho_1\rho_2 + \rho_3) + \omega_2(1+\rho_2^2) + \omega_3(\rho_2\rho_3 - \rho_1) \\ &= \overline{\omega}_1 \cos\phi(\rho_1\rho_2 + \rho_3) - \overline{\omega}_1 \sin\phi(1+\rho_2^2) + \frac{M}{J}t\,(\rho_2\rho_3 - \rho_1) \end{aligned} \tag{38}$$

$$\begin{aligned} 2\dot{\rho}_3 &= \omega_1(\rho_3\rho_1 - \rho_2) + \omega_2(\rho_2\rho_3 + \rho_1) + \omega_3(1+\rho_3^2) \\ &= \overline{\omega}_1 \cos\phi(\rho_3\rho_1 - \rho_2) - \overline{\omega}_1 \sin\phi(\rho_2\rho_3 + \rho_1) + \frac{M}{J}t\,(1+\rho_3^2) \end{aligned} \tag{39}$$

and, if $\mathbf{a}_1$, $\mathbf{a}_2$, and $\mathbf{a}_3$ are chosen such that $\mathbf{a}_i = \mathbf{b}_i$ ($i = 1, 2, 3$) at $t = 0$, then ρ_1, ρ_2, and ρ_3 must satisfy the initial conditions

$$\rho_i(0) = 0 \qquad (i = 1, 2, 3) \tag{40}$$

Suppose now that one wished to study the behavior of the symmetry axis of B, say for $0 \le \bar{\omega}_1 t \le 10.0$, by plotting the angle θ between this axis and the line fixed in A with which the symmetry axis coincides initially. Once the dimensionless parameters J/I and $M/(J\bar{\omega}_1^2)$ have been specified, ρ_1, ρ_2, and ρ_3 can be evaluated by integrating Eqs. (37)–(39) numerically, and θ is then given by

$$\theta = \cos^{-1}(\mathbf{a}_3 \cdot \mathbf{b}_3) \underset{(1.2.2,\ 1.2.3)}{=} \cos^{-1} C_{33} \underset{(1.4.4)}{=} \cos^{-1} \frac{1 - \rho_1^2 - \rho_2^2 + \rho_3^2}{1 + \rho_1^2 + \rho_2^2 + \rho_3^2} \tag{41}$$

Table 1.14.1 shows values of ρ_1, ρ_2, ρ_3, and θ obtained in this way for $J/I = 0.5$ and $M/(J\bar{\omega}_1^2) = 0.1$. The largest value of $\bar{\omega}_1 t$ appearing in the table is 3.0, rather than 10.0, because during integration from 3.0 to 3.5 the values of ρ_1, ρ_2, and ρ_3 became so large that the integration could not be continued. To overcome this obstacle, Eqs. (37)–(41) were replaced with [see Eq. (1.13.8)]

$$2\dot{\epsilon}_1 = \bar{\omega}_1 \cos\phi\ \epsilon_4 + \bar{\omega}_1 \sin\phi\ \epsilon_3 + \frac{M}{J} t\ \epsilon_2 \tag{42}$$

$$2\dot{\epsilon}_2 = \bar{\omega}_1 \cos\phi\ \epsilon_3 - \bar{\omega}_1 \sin\phi\ \epsilon_4 - \frac{M}{J} t\ \epsilon_1 \tag{43}$$

$$2\dot{\epsilon}_3 = -\bar{\omega}_1 \cos\phi\ \epsilon_2 - \bar{\omega}_1 \sin\phi\ \epsilon_1 + \frac{M}{J} t\ \epsilon_4 \tag{44}$$

Table 1.14.1

$\bar{\omega}_1 t$	ρ_1	ρ_2	ρ_3	θ (deg)
0.0	0.00	0.00	0.00	0
0.5	0.26	−0.00	0.00	29
1.0	0.55	−0.01	0.03	57
1.5	0.93	−0.04	0.06	86
2.0	1.56	−0.11	0.13	115
2.5	3.06	−0.33	0.27	143
3.0	16.94	−2.69	1.41	169

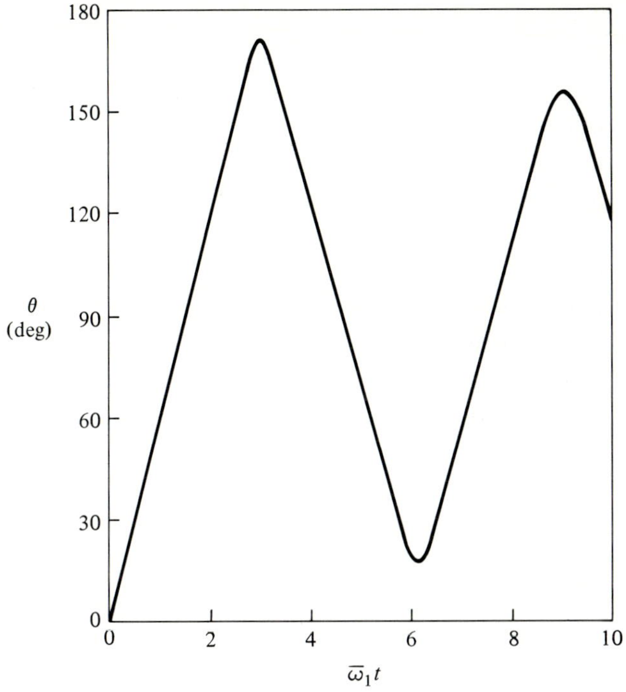

Figure 1.14.1

$$2\dot{\epsilon}_4 = -\bar{\omega}_1 \cos\phi\ \epsilon_1 + \bar{\omega}_1 \sin\phi\ \epsilon_2 - \frac{M}{J} t\ \epsilon_3 \tag{45}$$

$$\epsilon_1(0) = \epsilon_2(0) = \epsilon_3(0) = 0 \qquad \epsilon_4(0) = 1 \tag{46}$$

and

$$\theta \underset{(1.3.14)}{=} \cos^{-1}(1 - 2\epsilon_1^2 - 2\epsilon_2^2) \tag{47}$$

respectively, and a numerical integration of these equations, performed without difficulty because $-1 \le \epsilon_i \le 1$ $(i = 1, \ldots, 4)$, produced the values listed in Table 1.14.2. These results not only permit one to plot θ versus $\bar{\omega}_1 t$, as has been done in Fig. 1.14.1, but they indicate quite clearly why the numerical solution of Eqs. (37)–(39) could not proceed smoothly: ϵ_4 changes sign between $\bar{\omega}_1 t = 3.0$ and $\bar{\omega}_1 t = 3.5$, and again between $\bar{\omega}_1 t = 8.5$ and $\bar{\omega}_1 t = 9.0$, whereas ϵ_i $(i = 1, 2, 3)$ do not change sign in these intervals. Hence ϵ_4 vanishes at two points at which ϵ_i $(i = 1, 2, 3)$ do not vanish, and since

$$\rho_i \underset{(1.4.3)}{=} \frac{\epsilon_i}{\epsilon_4} \qquad (i = 1, 2, 3) \tag{48}$$

the Rodrigues parameters become infinite at these two points.

Table 1.14.2

$\bar{\omega}_1 t$	ϵ_1	ϵ_2	ϵ_3	ϵ_4	θ (deg)
0.0	0.00	0.00	0.00	1.00	0
0.5	0.25	0.00	0.01	0.97	29
1.0	0.48	−0.01	0.02	0.88	57
1.5	0.68	−0.03	0.05	0.73	86
2.0	0.84	−0.06	0.07	0.54	115
2.5	0.94	−0.10	0.08	0.31	143
3.0	0.98	−0.16	0.08	0.06	169
3.5	0.96	−0.21	0.06	−0.20	157
4.0	0.86	−0.26	0.00	−0.43	129
4.5	0.71	−0.30	−0.01	−0.64	100
5.0	0.50	−0.30	−0.18	−0.79	71
5.5	0.26	−0.26	−0.30	−0.88	44
6.0	0.02	−0.17	−0.41	−0.89	20
6.5	−0.22	−0.04	−0.51	−0.83	26
7.0	−0.42	0.13	−0.57	−0.70	52
7.5	−0.55	0.33	−0.58	−0.50	80
8.0	−0.61	0.53	−0.51	−0.28	108
8.5	−0.59	0.71	−0.37	−0.06	136
9.0	−0.49	0.84	−0.17	0.14	155
9.5	−0.33	0.90	0.09	0.28	146
10.0	−0.13	0.86	0.36	0.34	120

1.15 INDIRECT DETERMINATION OF ANGULAR VELOCITY

When a rigid body B can be observed from a vantage point fixed in a reference frame A, the angular velocity $\boldsymbol{\omega}$ of B in A can be determined by using Eq. (1.11.4). If observations permitting such a direct evaluation of $\boldsymbol{\omega}$ cannot be made, it may, nevertheless, be possible to find $\boldsymbol{\omega}$. This is the case, for example, when two vectors, say $\mathbf{p}$ and $\mathbf{q}$, can each be observed from a vantage point fixed in A as well as from one fixed in B, for $\boldsymbol{\omega}$ can then be found by using the relationship

$$\boldsymbol{\omega} = \frac{({}^A d\mathbf{p}/dt - {}^B d\mathbf{p}/dt) \times ({}^A d\mathbf{q}/dt - {}^B d\mathbf{q}/dt)}{({}^A d\mathbf{p}/dt - {}^B d\mathbf{p}/dt) \cdot \mathbf{q}} \tag{1}$$

To carry out the algebraic operations indicated in this equation, one must be able to express all vectors in a common basis. This can be accomplished by using Eq. (1.2.9) after forming a direction cosine matrix by reference to Eq. (1.5.3).

Derivation From Eq. (1.11.8),

$$\frac{{}^{A}d\mathbf{p}}{dt} - \frac{{}^{B}d\mathbf{p}}{dt} = \boldsymbol{\omega} \times \mathbf{p} \tag{2}$$

and

$$\frac{{}^{A}d\mathbf{q}}{dt} - \frac{{}^{B}d\mathbf{q}}{dt} = \boldsymbol{\omega} \times \mathbf{q} \tag{3}$$

Hence

$$\begin{aligned}\left(\frac{{}^{A}d\mathbf{p}}{dt} - \frac{{}^{B}d\mathbf{p}}{dt}\right) \times \left(\frac{{}^{A}d\mathbf{q}}{dt} - \frac{{}^{B}d\mathbf{q}}{dt}\right) &= \left(\frac{{}^{A}d\mathbf{p}}{dt} - \frac{{}^{B}d\mathbf{p}}{dt}\right) \times (\boldsymbol{\omega} \underset{(3)}{\times} \mathbf{q}) \\ &= \left(\frac{{}^{A}d\mathbf{p}}{dt} - \frac{{}^{B}d\mathbf{p}}{dt}\right) \cdot \mathbf{q}\boldsymbol{\omega} - \left(\frac{{}^{A}d\mathbf{p}}{dt} - \frac{{}^{B}d\mathbf{p}}{dt}\right) \cdot \boldsymbol{\omega}\mathbf{q} \\ &= \left(\frac{{}^{A}d\mathbf{p}}{dt} - \frac{{}^{B}d\mathbf{p}}{dt}\right) \cdot \mathbf{q}\boldsymbol{\omega} - (\boldsymbol{\omega} \underset{(2)}{\times} \mathbf{p}) \cdot \boldsymbol{\omega}\mathbf{q} \\ &= \left(\frac{{}^{A}d\mathbf{p}}{dt} - \frac{{}^{B}d\mathbf{p}}{dt}\right) \cdot \mathbf{q}\boldsymbol{\omega} + 0 \end{aligned} \tag{4}$$

and, solving for $\boldsymbol{\omega}$, one arrives at Eq. (1).

Example Observations of two stars, P and Q, are made simultaneously from two space vehicles, A and B, these observations consisting of meas-

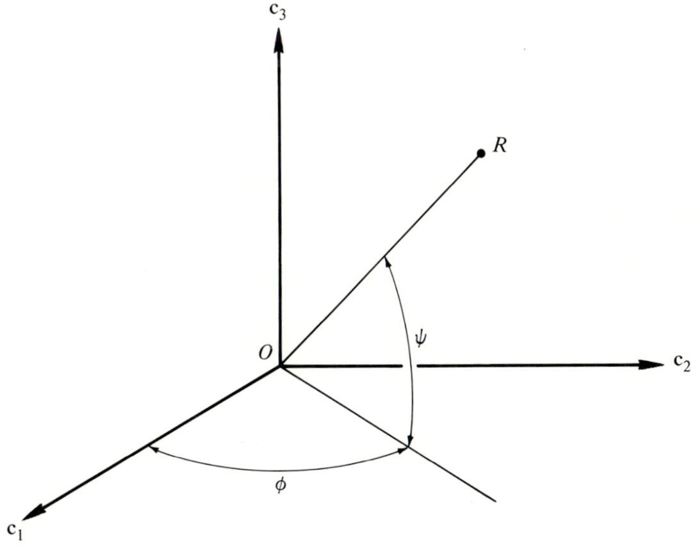

Figure 1.15.1

Table 1.15.1

	P				Q			
	ϕ (deg)	ψ (deg)	$\dot{\phi}$ (rad/sec)	$\dot{\psi}$ (rad/sec)	ϕ (deg)	ψ (deg)	$\dot{\phi}$ (rad/sec)	$\dot{\psi}$ (rad/sec)
$\mathbf{c}_i = \mathbf{a}_i$	90	45	−2	1	30	0	0	$\frac{1}{2} - \sqrt{3}$
$\mathbf{c}_i = \mathbf{b}_i$	135	0	0	$-3\sqrt{2}/2$	90	60	$3\sqrt{3}$	0

uring the angles ϕ and ψ shown in Fig. 1.15.1, where O represents either a point fixed in A or a point fixed in B, R is either P or Q, and $\mathbf{c}_1$, $\mathbf{c}_2$, $\mathbf{c}_3$ are orthogonal unit vectors forming a dextral set fixed either in A or in B.

For a certain instant, the angles and their first time-derivatives are found to have the values shown in Table 1.15.1. The angular velocity $\boldsymbol{\omega}$ of B in A at that instant is to be determined.

The situation under consideration is the same as that discussed in the Example in Sec. 1.5. Hence, if $\mathbf{v}$ is any vector and ${}^A v$ and ${}^B v$ are row matrices having $\mathbf{v} \cdot \mathbf{a}_i$ and $\mathbf{v} \cdot \mathbf{b}_i$ $(i = 1, 2, 3)$ as elements, then

$$ {}^B v \underset{(1.2.9)}{=} {}^A v \begin{bmatrix} 0 & 0 & 1 \\ 0 & 1 & 0 \\ -1 & 0 & 0 \end{bmatrix} \tag{5} $$

Furthermore, if $\mathbf{R}$ is again defined as a unit vector directed from O toward R (see Fig. 1.15.1), then

$$ \mathbf{R} = \cos\psi \cos\phi \, \mathbf{c}_1 + \cos\psi \sin\phi \, \mathbf{c}_2 + \sin\psi \, \mathbf{c}_3 \tag{6} $$

and the first time-derivative of $\mathbf{R}$ in a reference frame C in which $\mathbf{c}_1$, $\mathbf{c}_2$, and $\mathbf{c}_3$ are fixed is given by

$$ \begin{aligned} \frac{{}^C d\mathbf{R}}{dt} \underset{(6)}{=} & -(\sin\psi \cos\phi \, \dot{\psi} + \cos\psi \sin\phi \, \dot{\phi})\mathbf{c}_1 \\ & -(\sin\psi \sin\phi \, \dot{\psi} - \cos\psi \cos\phi \, \dot{\phi})\mathbf{c}_2 \\ & + \cos\psi \, \dot{\psi} \, \mathbf{c}_3 \end{aligned} \tag{7} $$

Consequently, letting $\mathbf{p}$ and $\mathbf{q}$ be unit vectors directed from O toward P and Q respectively, and referring to Table 1.15.1, one can express the time-derivatives of $\mathbf{p}$ and $\mathbf{q}$ in A as

$$ \frac{{}^A d\mathbf{p}}{dt} \underset{(7)}{=} \sqrt{2} \, \mathbf{a}_1 - \frac{\sqrt{2}}{2} \mathbf{a}_2 + \frac{\sqrt{2}}{2} \mathbf{a}_3 \tag{8} $$

Table 1.15.2 Vectors appearing in Eq. (1)

Line	Vector	$\mathbf{b}_1$	$\mathbf{b}_2$	$\mathbf{b}_3$
1	$\frac{^A d\mathbf{p}}{dt}$	$-\frac{\sqrt{2}}{2}$	$-\frac{\sqrt{2}}{2}$	$\sqrt{2}$
2	$\frac{^A d\mathbf{q}}{dt}$	$\sqrt{3}-\frac{1}{2}$	0	0
3	$\frac{^B d\mathbf{p}}{dt}$	0	0	$-\frac{3\sqrt{2}}{2}$
4	$\frac{^B d\mathbf{q}}{dt}$	$-\frac{3\sqrt{3}}{2}$	0	0
5	$\frac{^A d\mathbf{p}}{dt}-\frac{^B d\mathbf{p}}{dt}$	$-\frac{\sqrt{2}}{2}$	$-\frac{\sqrt{2}}{2}$	$\frac{5\sqrt{2}}{2}$
6	$\frac{^A d\mathbf{q}}{dt}-\frac{^B d\mathbf{q}}{dt}$	$\frac{5\sqrt{3}}{2}-\frac{1}{2}$	0	0
7	$\left(\frac{^A d\mathbf{p}}{dt}-\frac{^B d\mathbf{p}}{dt}\right)\times\left(\frac{^A d\mathbf{q}}{dt}-\frac{^B d\mathbf{q}}{dt}\right)$	0	$\frac{5\sqrt{2}}{2}\left(\frac{5\sqrt{3}}{2}-\frac{1}{2}\right)$	$\frac{\sqrt{2}}{2}\left(\frac{5\sqrt{3}}{2}-\frac{1}{2}\right)$

and

$$\frac{^A d\mathbf{q}}{dt} \underset{(7)}{=} \left(\frac{1}{2}-\sqrt{3}\right)\mathbf{a}_3 \tag{9}$$

Next, one can express these derivatives in terms of $\mathbf{b}_1$, $\mathbf{b}_2$, and $\mathbf{b}_3$, as indicated in lines 1 and 2 of Table 1.15.2, and lines 3 and 4 can be constructed similarly. Lines 5, 6, and 7 then can be formed by purely algebraic operations, and the scalar product appearing in the denominator of the right-hand member of Eq. (1) is given by (see line 2 of Table 1.5.2)

$$\left(\frac{^A d\mathbf{p}}{dt}-\frac{^B d\mathbf{p}}{dt}\right)\cdot\mathbf{q} = -\frac{\sqrt{2}}{2}\left(\frac{1}{2}\right)+\frac{5\sqrt{2}}{2}\left(\frac{\sqrt{3}}{2}\right) = \left(\frac{5\sqrt{3}}{2}-\frac{1}{2}\right)\frac{\sqrt{2}}{2} \tag{10}$$

Consequently,

$$\boldsymbol{\omega} \underset{(1)}{=} \frac{\frac{5\sqrt{2}}{2}\left(\frac{5\sqrt{3}}{2}-\frac{1}{2}\right)\mathbf{b}_2+\frac{\sqrt{2}}{2}\left(\frac{5\sqrt{3}}{2}-\frac{1}{2}\right)\mathbf{b}_3}{\left(\frac{5\sqrt{3}}{2}-\frac{1}{2}\right)\frac{\sqrt{2}}{2}}$$

$$= 5\mathbf{b}_2+\mathbf{b}_3 \qquad \text{rad/sec} \tag{11}$$

1.16 AUXILIARY REFERENCE FRAMES

The angular velocity of a rigid body B in a reference frame A (see Sec. 1.11) can be expressed in the following form involving n auxiliary reference frames $A_1, \ldots, A_n$:

$$ {}^{A}\boldsymbol{\omega}^{B} = {}^{A}\boldsymbol{\omega}^{A_1} + {}^{A_1}\boldsymbol{\omega}^{A_2} + \cdots + {}^{A_{n-1}}\boldsymbol{\omega}^{A_n} + {}^{A_n}\boldsymbol{\omega}^{B} \tag{1} $$

This relationship, the *addition theorem for angular velocities*, is particularly useful when each term in the right-hand member represents the angular velocity of a body performing a motion of simple rotation (see Sec. 1.1) and can, therefore, be expressed as in Eq. (1.11.5).

Derivation For any vector **c** fixed in B,

$$ \frac{{}^{A}d\mathbf{c}}{dt} \underset{(1.11.9)}{=} {}^{A}\boldsymbol{\omega}^{B} \times \mathbf{c} \tag{2} $$

$$ \frac{{}^{A_1}d\mathbf{c}}{dt} \underset{(1.11.9)}{=} {}^{A_1}\boldsymbol{\omega}^{B} \times \mathbf{c} \tag{3} $$

and

$$ \frac{{}^{A}d\mathbf{c}}{dt} \underset{(1.11.8)}{=} \frac{{}^{A_1}d\mathbf{c}}{dt} + {}^{A}\boldsymbol{\omega}^{A_1} \times \mathbf{c} \tag{4} $$

so that

$$ {}^{A}\boldsymbol{\omega}^{B} \times \mathbf{c} \underset{(2\text{-}4)}{=} {}^{A}\boldsymbol{\omega}^{A_1} \times \mathbf{c} + {}^{A_1}\boldsymbol{\omega}^{B} \times \mathbf{c} \tag{5} $$

or, since this equation is satisfied for *every* **c** fixed in B,

$$ {}^{A}\boldsymbol{\omega}^{B} = {}^{A}\boldsymbol{\omega}^{A_1} + {}^{A_1}\boldsymbol{\omega}^{B} \tag{6} $$

which shows that Eq. (1) is valid for $n = 1$. Proceeding similarly, one can verify that

$$ {}^{A_1}\boldsymbol{\omega}^{B} = {}^{A_1}\boldsymbol{\omega}^{A_2} + {}^{A_2}\boldsymbol{\omega}^{B} \tag{7} $$

and substitution into Eq. (6) then yields

$$ {}^{A}\boldsymbol{\omega}^{B} = {}^{A}\boldsymbol{\omega}^{A_1} + {}^{A_1}\boldsymbol{\omega}^{A_2} + {}^{A_2}\boldsymbol{\omega}^{B} \tag{8} $$

which is Eq. (1) for $n = 2$. The validity of Eq. (1) for any value of n can thus be established by applying this procedure a sufficient number of times.

Example In Fig. 1.16.1, θ, ϕ, and ψ designate angles used to describe the orientation of a rigid cone B in a reference frame A. These angles are formed by lines described as follows: L_1 and L_2 are perpendicular to each other and fixed in A; L_3 is the axis of symmetry of B; L_4 is perpendicular to L_2 and intersects L_2 and L_3; L_5 is perpendicular to L_3 and intersects L_2 and L_3; L_6 is perpendicular to L_3 and is fixed in B; and L_7 is perpendicular to L_2 and L_4. To find the angular velocity of B in A, one can designate as A_1 a reference frame in which L_2, L_4, and L_7 are fixed, and as A_2 a reference frame in which L_3, L_5, and L_7 are fixed, observing that L_2 is then fixed both in A and A_1, L_7 is fixed both in A_1 and A_2, and L_3 is fixed both in A_2 and B, so that, in accordance with Eq. (1.11.5),

$$ {}^{A}\boldsymbol{\omega}^{A_1} = \dot{\phi}\boldsymbol{\lambda}_2 \qquad {}^{A_1}\boldsymbol{\omega}^{A_2} = \dot{\theta}\boldsymbol{\lambda}_7 \qquad {}^{A_2}\boldsymbol{\omega}^{B} = \dot{\psi}\boldsymbol{\lambda}_3 \tag{9} $$

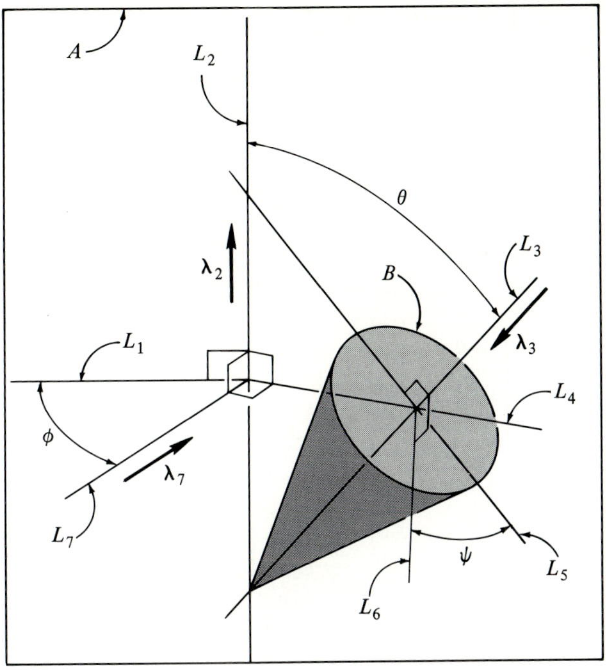

Figure 1.16.1

where $\boldsymbol{\lambda}_2$, $\boldsymbol{\lambda}_3$, and $\boldsymbol{\lambda}_7$ are unit vectors directed as shown in Fig. 1.16.1. It then follows immediately that

$$^A\boldsymbol{\omega}^B \underset{(1)}{=} \dot{\phi}\boldsymbol{\lambda}_2 + \dot{\theta}\boldsymbol{\lambda}_7 + \dot{\psi}\boldsymbol{\lambda}_3 \tag{10}$$

1.17 ANGULAR VELOCITY AND ORIENTATION ANGLES

When the orientation of a rigid body B in a reference frame A is described by specifying the time dependence of orientation angles θ_1, θ_2, and θ_3 (see Sec. 1.7), the angular velocity of B in A (see Sec. 1.11) can be found by using the relationship

$$[\omega_1 \quad \omega_2 \quad \omega_3] = [\dot{\theta}_1 \quad \dot{\theta}_2 \quad \dot{\theta}_3]\, M \tag{1}$$

where M is a 3×3 matrix whose elements are functions of θ_1, θ_2, and θ_3. Conversely, if ω_1, ω_2, and ω_3 are known as functions of time, then θ_1, θ_2, and θ_3 can be evaluated by solving the differential equations

$$[\dot{\theta}_1 \quad \dot{\theta}_2 \quad \dot{\theta}_3] = [\omega_1 \quad \omega_2 \quad \omega_3]\, M^{-1} \tag{2}$$

For space-three angles, the matrices M and M^{-1} are

$$M = \begin{bmatrix} 1 & 0 & 0 \\ 0 & c_1 & -s_1 \\ -s_2 & s_1c_2 & c_1c_2 \end{bmatrix} \tag{3}$$

and

$$M^{-1} = \frac{1}{c_2}\begin{bmatrix} c_2 & 0 & 0 \\ s_1s_2 & c_1c_2 & s_1 \\ c_1s_2 & -s_1c_2 & c_1 \end{bmatrix} \tag{4}$$

For body-three angles,

$$M = \begin{bmatrix} c_2c_3 & -c_2s_3 & s_2 \\ s_3 & c_3 & 0 \\ 0 & 0 & 1 \end{bmatrix} \tag{5}$$

and

$$M^{-1} = \frac{1}{c_2}\begin{bmatrix} c_3 & c_2s_3 & -s_2c_3 \\ -s_3 & c_2c_3 & s_2s_3 \\ 0 & 0 & c_2 \end{bmatrix} \tag{6}$$

For space-two angles,

$$M = \begin{bmatrix} 1 & 0 & 0 \\ 0 & c_1 & -s_1 \\ c_2 & s_1s_2 & c_1s_2 \end{bmatrix} \tag{7}$$

and

$$M^{-1} = \frac{1}{s_2}\begin{bmatrix} s_2 & 0 & 0 \\ -s_1c_2 & c_1s_2 & s_1 \\ -c_1c_2 & -s_1s_2 & c_1 \end{bmatrix} \tag{8}$$

Finally, for body-two angles,

$$M = \begin{bmatrix} c_2 & s_2s_3 & s_2c_3 \\ 0 & c_3 & -s_3 \\ 1 & 0 & 0 \end{bmatrix} \tag{9}$$

and

$$M^{-1} = \frac{1}{s_2}\begin{bmatrix} 0 & 0 & s_2 \\ s_3 & s_2c_3 & -c_2s_3 \\ c_3 & -s_2s_3 & -c_2c_3 \end{bmatrix} \tag{10}$$

When c_2 vanishes, M as given by Eq. (3) or by Eq. (5) is a singular matrix, and M^{-1} is thus undefined. Hence, given ω_1, ω_2, and ω_3 one cannot use Eq. (2) to determine $\dot{\theta}_1$, $\dot{\theta}_2$, and $\dot{\theta}_3$ if θ_1, θ_2, and θ_3 are space-three or body-three

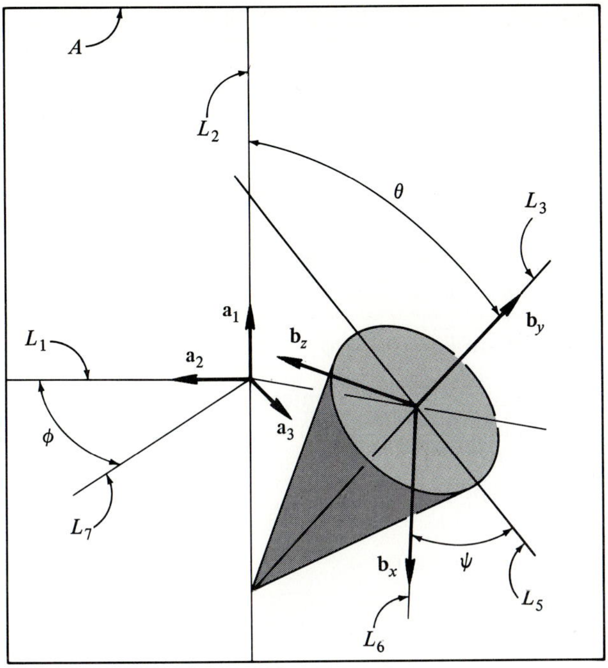

Figure 1.17.1

angles and $c_2 = 0$. Similarly, if θ_1, θ_2, and θ_3 are space-two or body-two angles, Eq. (2) involves an undefined matrix when s_2 is equal to zero.

When the angles and unit vectors employed in an analysis are denoted by symbols other than those used in connection with Eqs. (1)–(10), appropriate replacements for these equations can be obtained directly from Eqs. (1)–(10) whenever the angles have been identified as regards type, that is, as being space-three angles, body-three angles, etc. Suppose, for example, that in the course of an analysis involving the cone shown in Fig. 1.16.1, and previously considered in the example in Sec. 1.16, unit vectors $\mathbf{b}_x$, $\mathbf{b}_y$, and $\mathbf{b}_z$, fixed in B as shown in Fig. 1.17.1, have been introduced, and it is now desired to find ω_x, ω_y, and ω_z, defined as

$$\omega_x \triangleq \boldsymbol{\omega} \cdot \mathbf{b}_x \qquad \omega_y \triangleq \boldsymbol{\omega} \cdot \mathbf{b}_y \qquad \omega_z \triangleq \boldsymbol{\omega} \cdot \mathbf{b}_z \tag{11}$$

where $\boldsymbol{\omega}$ denotes the angular velocity of B in A. This can be done easily by regarding ϕ, θ, and ψ as body-two angles, that is, by introducing unit vectors $\mathbf{a}_1$, $\mathbf{a}_2$, $\mathbf{a}_3$ as shown in Fig. 1.17.1, defining $\mathbf{b}_1$, $\mathbf{b}_2$, and $\mathbf{b}_3$ as

$$\mathbf{b}_1 \triangleq \mathbf{b}_y \qquad \mathbf{b}_2 \triangleq \mathbf{b}_z \qquad \mathbf{b}_3 \triangleq \mathbf{b}_x \tag{12}$$

and taking

$$\theta_1 = \phi \qquad \theta_2 = -\theta \qquad \theta_3 = -\psi \tag{13}$$

For it then follows immediately from Eqs. (1) and (9) that

$$[\omega_y \ \omega_z \ \omega_x] = [\dot{\phi} \ -\dot{\theta} \ -\dot{\psi}] \begin{bmatrix} \cos\theta & \sin\theta \sin\psi & -\sin\theta\cos\psi \\ 0 & \cos\psi & \sin\psi \\ 1 & 0 & 0 \end{bmatrix} \tag{14}$$

so that

$$\omega_x = -\dot{\phi} \sin\theta \cos\psi - \dot{\theta} \sin\psi \tag{15}$$

$$\omega_y = \dot{\phi} \cos\theta - \dot{\psi} \tag{16}$$

$$\omega_z = \dot{\phi} \sin\theta \sin\psi - \dot{\theta} \cos\psi \tag{17}$$

In the spacecraft dynamics literature one encounters a wide variety of differential equations relating angular velocity measure numbers to orientation angles and their time-derivatives. Twenty-four such sets of kinematical differential equations are tabulated in Appendix II.

Derivations From Eqs. (1.10.5) and (1.7.1),

$$\begin{aligned} \omega_1 &= (c_1 s_2 c_3 + s_3 s_1) \frac{d}{dt} (s_1 s_2 c_3 - s_3 c_1) \\ &\quad + (c_1 s_2 s_3 - c_3 s_1) \frac{d}{dt} (s_1 s_2 s_3 + c_3 c_1) + c_1 c_2 \frac{d}{dt} (s_1 c_2) \\ &= \dot{\theta}_1 - \dot{\theta}_3 s_2 \end{aligned} \tag{18}$$

Similarly, from Eqs. (1.10.6) and (1.7.1),

$$\omega_2 = \dot{\theta}_2 c_1 + \dot{\theta}_3 s_1 c_2 \tag{19}$$

and from Eqs. (1.10.7) and (1.7.1)

$$\omega_3 = -\dot{\theta}_2 s_1 + \dot{\theta}_3 c_1 c_2 \tag{20}$$

These three equations are the three scalar equations corresponding to Eq. (1) when M is given by Eq. (3).

Equation (2) follows from Eq. (1) and from the definition of the inverse of a matrix; and the validity of Eq. (4) may be established by noting that the product of the right-hand members of Eqs. (3) and (4) is equal to U, the unit matrix. Proceeding similarly, but using Eq. (1.7.11), (1.7.21), or (1.7.31) in place of Eq. (1.7.1), and Eqs. (5) and (6), Eqs. (7) and (8), or Eqs. (9) and (10) in place of Eqs. (3) and (4), one can demonstrate the validity of Eqs. (5)–(10).

Example Figure 1.17.2 shows the gyroscopic system previously discussed in Sec. 1.7, where it was mentioned that one may wish to employ space-three angles ϕ_1, ϕ_2, and ϕ_3, as well as the body-two angles θ_1, θ_2, and θ_3 shown in Fig. 1.17.2, when analyzing motions during which θ_2 becomes

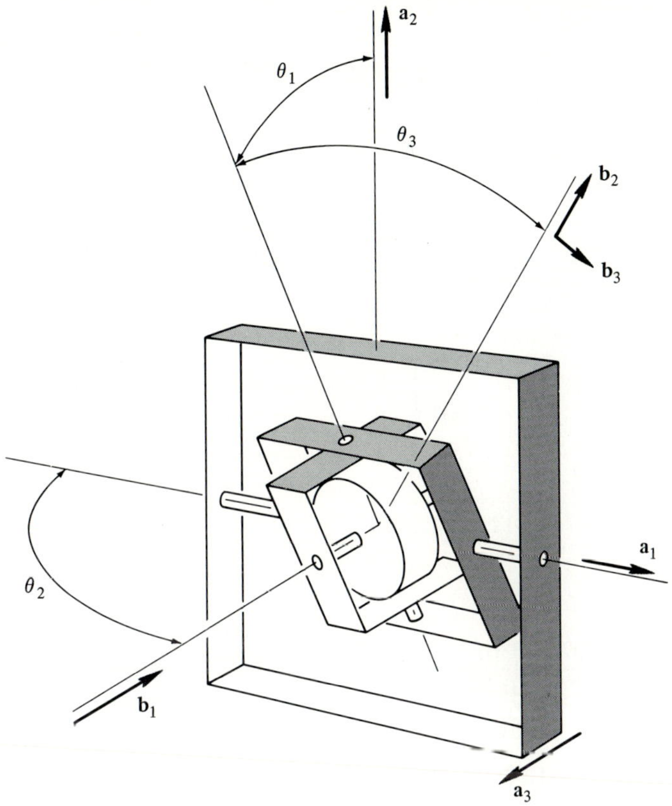

Figure 1.17.2

small or equal to zero. Given θ_i and $\dot{\theta}_i$ $(i = 1, 2, 3)$, one must then be able to evaluate ϕ_i and $\dot{\phi}_i$ $(i = 1, 2, 3)$.

Suppose that, as in the example in Sec. 1.7, $\theta_1 = 30°$, $\theta_2 = 45°$, and $\theta_3 = 60°$ at a certain instant and that, furthermore, $\dot{\theta}_1 = 1.00$, $\dot{\theta}_2 = 2.00$, $\dot{\theta}_3 = 3.00$ rad/sec. What are the values of $\dot{\phi}_1$, $\dot{\phi}_2$, and $\dot{\phi}_3$ at this instant?

From Eqs. (2) and (4),

$$[\dot{\phi}_1 \quad \dot{\phi}_2 \quad \dot{\phi}_3] = \frac{[\omega_1 \quad \omega_2 \quad \omega_3]}{\cos\phi_2}\begin{bmatrix} \cos\phi_2 & 0 & 0 \\ \sin\phi_1 \sin\phi_2 & \cos\phi_1 \cos\phi_2 & \sin\phi_1 \\ \cos\phi_1 \sin\phi_2 & -\sin\phi_1 \cos\phi_2 & \cos\phi_1 \end{bmatrix} \tag{21}$$

or, using the values of ϕ_1 and ϕ_2 found previously,

$$[\dot{\phi}_1 \quad \dot{\phi}_2 \quad \dot{\phi}_3] = \frac{[\omega_1 \quad \omega_2 \quad \omega_3]}{0.791} \begin{bmatrix} 0.791 & 0 & 0 \\ 0.603 & -0.138 & 0.985 \\ -0.107 & -0.780 & -0.174 \end{bmatrix} \tag{22}$$

Now, from Eqs. (1) and (9),

$$[\omega_1 \quad \omega_2 \quad \omega_3] = [\dot{\theta}_1 \quad \dot{\theta}_2 \quad \dot{\theta}_3] \begin{bmatrix} \cos\theta_2 & \sin\theta_2 \sin\theta_3 & \sin\theta_2 \cos\theta_3 \\ 0 & \cos\theta_2 & -\sin\theta_3 \\ 1 & 0 & 0 \end{bmatrix}$$

$$= [1.00 \quad 2.00 \quad 3.00] \begin{bmatrix} 0.707 & 0.612 & 0.354 \\ 0 & 0.500 & -0.866 \\ 1 & 0 & 0 \end{bmatrix}$$

$$= [3.707 \quad 1.612 \quad -1.378] \tag{23}$$

Hence

$$[\dot{\phi}_1 \quad \dot{\phi}_2 \quad \dot{\phi}_3] = \frac{[3.707 \quad 1.612 \quad -1.378]}{0.791} \begin{bmatrix} 0.791 & 0 & 0 \\ 0.603 & -0.138 & 0.985 \\ -0.107 & -0.780 & -0.174 \end{bmatrix}$$

$$= [5.12 \quad 1.08 \quad 2.31] \tag{24}$$

and

$$\dot{\phi}_1 = 5.12 \qquad \dot{\phi}_2 = 1.08 \qquad \dot{\phi}_3 = 2.31 \text{ rad/sec} \tag{25}$$

1.18 SLOW, SMALL ROTATIONAL MOTIONS

If $\mathbf{a}_1$, $\mathbf{a}_2$, $\mathbf{a}_3$, and $\mathbf{b}_1$, $\mathbf{b}_2$, $\mathbf{b}_3$ are two dextral sets of orthogonal unit vectors fixed respectively in reference frames or rigid bodies A and B which are moving relative to each other, one can use Eqs. (1.3.18) and (1.3.20) to associate with each instant of time an angle θ, and the motion is called a slow, small rotational motion when all terms of second or higher degree in θ and $\dot{\theta}$ play a negligible role in an analysis of the motion. Under these circumstances, a number of the relationships discussed previously can be replaced with simpler ones. Specifically, in place of Eqs. (1.13.1) and (1.13.3) one may then use

$$\boldsymbol{\omega} = 2\frac{{}^B d\boldsymbol{\epsilon}}{dt} \tag{1}$$

and

$$\epsilon_4 = 1 \tag{2}$$

Equations (1.14.1) and (1.14.2) can be replaced with

$$\boldsymbol{\omega} = 2\,\frac{{}^{B}d\boldsymbol{\rho}}{dt} \tag{3}$$

and, if θ_1, θ_2, and θ_3 are chosen such that terms of second or higher degree in θ_i and/or $\dot{\theta}_j$ $(i, j = 1, 2, 3)$ are negligible, then Eq. (1.17.1) together with Eq. (1.17.3) or Eq. (1.17.5) leads to

$$[\omega_1 \quad \omega_2 \quad \omega_3] = [\dot{\theta}_1 \quad \dot{\theta}_2 \quad \dot{\theta}_3] \tag{4}$$

which shows that it does not matter whether one uses space-three angles or body-three angles when dealing with slow, small rotational motions.

Derivations From Eqs. (1.8.3) and (1.8.4),

$$\boldsymbol{\epsilon} = \frac{1}{2}\boldsymbol{\lambda}\theta \tag{5}$$

and

$$\epsilon_4 = 1 \tag{6}$$

Hence,

$$\frac{{}^{B}d\boldsymbol{\epsilon}}{dt} = \frac{1}{2}\left(\frac{{}^{B}d\boldsymbol{\lambda}}{dt}\,\theta + \boldsymbol{\lambda}\dot{\theta}\right) \tag{7}$$

and, substituting into Eq. (1.13.1) and retaining only terms of first degree in θ and $\dot{\theta}$, one obtains

$$\boldsymbol{\omega} = 2\left[\frac{1}{2}\left(\frac{{}^{B}d\boldsymbol{\lambda}}{dt}\,\theta + \boldsymbol{\lambda}\dot{\theta}\right)\right] = 2\,\frac{{}^{B}d\boldsymbol{\epsilon}}{dt} \tag{8}$$

in agreement with Eq. (1). Equation (2) is the same as Eq. (1.8.4).

Equation (3) follows immediately from Eq. (1), since $\boldsymbol{\rho}$ and $\boldsymbol{\epsilon}$ are equal to each other to the order of approximation under consideration, as is apparent from Eqs. (1.8.3) and (1.8.5). Finally, Eq. (4) results from substituting M as given in Eq. (1.17.3) or (1.17.5) into Eq. (1.17.1) and then dropping all nonlinear terms.

Example In Fig. 1.18.1, B designates a rigid body that is attached by means of elastic supports to a space vehicle A which is moving in such a way that the angular velocity ${}^{N}\boldsymbol{\omega}^{A}$ of A in a Newtonian reference frame N is given by

$${}^{N}\boldsymbol{\omega}^{A} = \omega_1\mathbf{a}_1 + \omega_2\mathbf{a}_2 + \omega_3\mathbf{a}_3 \tag{9}$$

where ω_1, ω_2, ω_3 are constants and $\mathbf{a}_1$, $\mathbf{a}_2$, $\mathbf{a}_3$ form a dextral set of orthogonal unit vectors fixed in A. Point B^* is the mass center of B, and $\mathbf{b}_1$, $\mathbf{b}_2$, $\mathbf{b}_3$ are unit vectors parallel to principal axes of inertia of B for B^*, the associated moments of inertia having the values I_1, I_2, and I_3.

In preparation for the formulation of equations of motion of B, the

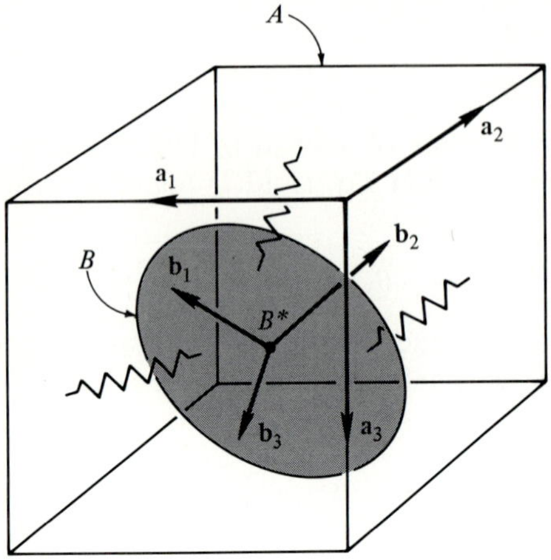

Figure 1.18.1

first time-derivative, in N, of the angular momentum $\mathbf{H}$ of B relative to B^* in N is to be determined, assuming that all rotational motions of B in A are slow, small motions. The orientation of B in A is to be described in terms of body-three angles θ_1, θ_2, and θ_3, all of which vanish when $\mathbf{a}_i = \mathbf{b}_i$ $(i = 1, 2, 3)$.

The angular velocity ${}^N\boldsymbol{\omega}^B$ of B in N can be expressed as

$$ {}^N\boldsymbol{\omega}^B \underset{(1.16.1)}{=} {}^N\boldsymbol{\omega}^A + {}^A\boldsymbol{\omega}^B \tag{10} $$

Referring to Eqs. (1.2.5) and (1.8.12), one can write

$$ \begin{aligned} {}^N\boldsymbol{\omega}^A \underset{(8)}{=} & (\omega_1 + \omega_2\theta_3 - \omega_3\theta_2)\mathbf{b}_1 \\ & + (\omega_2 + \omega_3\theta_1 - \omega_1\theta_3)\mathbf{b}_2 \\ & + (\omega_3 + \omega_1\theta_2 - \omega_2\theta_1)\mathbf{b}_3 \end{aligned} \tag{11} $$

and, from Eq. (4),

$$ {}^A\boldsymbol{\omega}^B = \dot{\theta}_1\mathbf{b}_1 + \dot{\theta}_2\mathbf{b}_2 + \dot{\theta}_3\mathbf{b}_3 \tag{12} $$

Hence,

$$ {}^N\boldsymbol{\omega}^B \underset{(10-12)}{=} (\omega_1 + \omega_2\theta_3 - \omega_3\theta_2 + \dot{\theta}_1)\mathbf{b}_1 + \cdots \tag{13} $$

and

$$ \mathbf{H} \underset{(13)}{=} I_1(\omega_1 + \omega_2\theta_3 - \omega_3\theta_2 + \dot{\theta}_1)\mathbf{b}_1 + \cdots \tag{14} $$

To evaluate the first time-derivative of $\mathbf{H}$ in N, it is convenient to use the relationship

$$ \frac{{}^N d\mathbf{H}}{dt} \underset{(1.11.8)}{=} \frac{{}^B d\mathbf{H}}{dt} + {}^N\boldsymbol{\omega}^B \times \mathbf{H} \tag{15} $$

with

$$ \frac{{}^B d\mathbf{H}}{dt} \underset{(14)}{=} I_1(\omega_2\dot{\theta}_3 - \omega_3\dot{\theta}_2 + \ddot{\theta}_1)\mathbf{b}_1 + \cdots \tag{16} $$

and

$$^N\boldsymbol{\omega}^B \times \mathbf{H} \underset{(13,14)}{=}$$

$$(I_3 - I_2)(\omega_2 + \omega_3\theta_1 - \omega_1\theta_3 + \dot{\theta}_2)(\omega_3 + \omega_1\theta_2 - \omega_2\theta_1 + \dot{\theta}_3)\mathbf{b}_1 + \cdots \tag{17}$$

where, however, all nonlinear terms are to be dropped. Thus, one finds that

$$\frac{^N d\mathbf{H}}{dt} = \Big\{ I_1\ddot{\theta}_1 + (-I_1 - I_2 + I_3)\omega_3\dot{\theta}_2 - (-I_1 + I_2 - I_3)\omega_2\dot{\theta}_3$$

$$+ (I_3 - I_2)[\omega_2\omega_3 - (\omega_2^2 - \omega_3^2)\theta_1 + \omega_1\omega_2\theta_2 - \omega_1\omega_3\theta_3]\Big\}\mathbf{b}_1 + \cdots \tag{18}$$

1.19 INSTANTANEOUS AXIS

At an instant at which the angular velocity $\boldsymbol{\omega}$ of a rigid body B in a reference frame A is equal to zero, the velocities of all points of B in A are equal to each other. Whenever $\boldsymbol{\omega}$ is not equal to zero, there exist infinitely many points of B whose velocities in A are parallel to $\boldsymbol{\omega}$ or equal to zero. These points all have the same velocity $\mathbf{v}^*$ in A and they form a straight line parallel to $\boldsymbol{\omega}$ and called the *instantaneous axis* of B in A. The magnitude of $\mathbf{v}^*$ is smaller than the magnitude of the velocity in A of any point of B not lying on the instantaneous axis.

If $\mathbf{v}^Q$ is the velocity in A of an arbitrarily selected *basepoint* Q of B, and P^* is a point of the instantaneous axis, then the position vector $\mathbf{r}^*$ of P^* relative to Q can be expressed as

$$\mathbf{r}^* = \frac{\boldsymbol{\omega} \times \mathbf{v}^Q}{\omega^2} + \mu^*\boldsymbol{\omega} \tag{1}$$

where μ^* depends on the choice of P^*; and $\mathbf{v}^*$ is given by

$$\mathbf{v}^* = \frac{\boldsymbol{\omega} \cdot \mathbf{v}^Q}{\omega^2}\,\boldsymbol{\omega} \tag{2}$$

Derivation In Fig. 1.19.1 both P and Q are arbitrarily selected points of B, $\mathbf{p}$ and $\mathbf{q}$ are their respective position vectors relative to a point O that is fixed in A, and $\mathbf{r}$ is the position vector of P relative to Q. Hence,

$$\mathbf{p} = \mathbf{q} + \mathbf{r} \tag{3}$$

and

$$\frac{^A d\mathbf{p}}{dt} \underset{(3)}{=} \frac{^A d\mathbf{q}}{dt} + \frac{^A d\mathbf{r}}{dt} \underset{(1.11.9)}{=} \frac{^A d\mathbf{q}}{dt} + \boldsymbol{\omega} \times \mathbf{r}$$

or, since the velocities $\mathbf{v}^P$ and $\mathbf{v}^Q$ of P and Q in A are equal to $^A d\mathbf{p}/dt$ and $^A d\mathbf{q}/dt$, respectively,

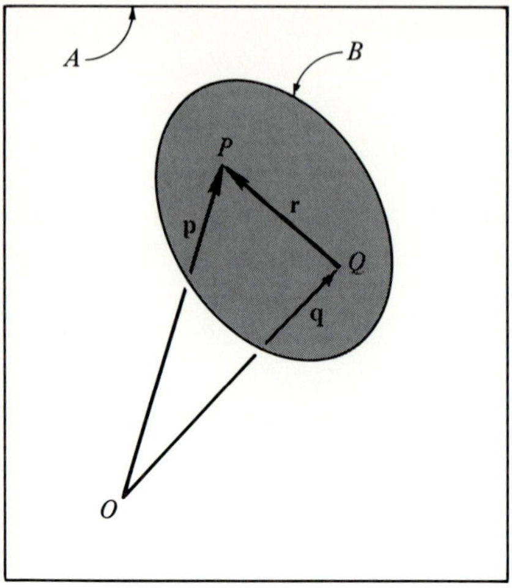

Figure 1.19.1

$$\mathbf{v}^P = \mathbf{v}^Q + \boldsymbol{\omega} \times \mathbf{r} \tag{4}$$

If $\boldsymbol{\omega} \neq 0$, the vector $\mathbf{r}$ can always be expressed as the sum of a vector, say $\mathbf{s}$, that is perpendicular to $\boldsymbol{\omega}$, and the vector $\mu\boldsymbol{\omega}$, where μ is a certain scalar; that is,

$$\mathbf{r} = \mathbf{s} + \mu\boldsymbol{\omega} \tag{5}$$

with

$$\boldsymbol{\omega} \cdot \mathbf{s} = 0 \tag{6}$$

Consequently, $\mathbf{v}^P$ can be expressed as

$$\mathbf{v}^P \underset{(4,5)}{=} \mathbf{v}^Q + \boldsymbol{\omega} \times \mathbf{s} \tag{7}$$

and

$$\boldsymbol{\omega} \times \mathbf{v}^P \underset{(7)}{=} \boldsymbol{\omega} \times \mathbf{v}^Q + \boldsymbol{\omega} \cdot \mathbf{s}\boldsymbol{\omega} - \omega^2\mathbf{s}$$

$$\underset{(6)}{=} \boldsymbol{\omega} \times \mathbf{v}^Q - \omega^2\mathbf{s} \tag{8}$$

If P is now taken to be a point P^* whose velocity $\mathbf{v}^*$ in A is parallel to $\boldsymbol{\omega}$, and the associated values of $\mathbf{r}$, $\mathbf{s}$, and μ are called $\mathbf{r}^*$, $\mathbf{s}^*$, and μ^*, then

$$0 \underset{(8)}{=} \boldsymbol{\omega} \times \mathbf{v}^Q - \omega^2\mathbf{s}^* \tag{9}$$

$$\mathbf{r}^* \underset{(5)}{=} \mathbf{s}^* + \mu^*\boldsymbol{\omega} \underset{(9)}{=} \frac{\boldsymbol{\omega} \times \mathbf{v}^Q}{\omega^2} + \mu^*\boldsymbol{\omega} \tag{10}$$

in agreement with Eq. (1), and

$$\mathbf{v}^* \underset{(7)}{=} \mathbf{v}^Q + \boldsymbol{\omega} \times \mathbf{s}^* \underset{(9)}{=} \mathbf{v}^Q + \frac{\boldsymbol{\omega} \times (\boldsymbol{\omega} \times \mathbf{v}^Q)}{\omega^2}$$

$$= \frac{\omega^2\mathbf{v}^Q + \boldsymbol{\omega} \cdot \mathbf{v}^Q\boldsymbol{\omega} - \omega^2\mathbf{v}^Q}{\omega^2} = \frac{\boldsymbol{\omega} \cdot \mathbf{v}^Q}{\omega^2}\,\boldsymbol{\omega} \tag{11}$$

in agreement with Eq. (2).

Example In Fig. 1.19.2, B represents a slowly spinning cylindrical satellite whose mass center B^* moves on a circular orbit of radius R fixed in a reference frame A. Throughout this motion the symmetry axis of B is constrained to remain tangent to the circle while B rotates about this axis at a constant rate such that a plane fixed in B and passing through the axis becomes parallel to the orbit plane twice during each orbital revolution of B^*. The instantaneous axis of B in A is to be located for a typical instant during the motion.

Letting A^* be the center of the circle on which B^* moves, and designating as C a reference frame in which the normal to the circle at A^* and the line joining A^* to B^* are both fixed, one can express the angular velocity of B in A as

$$\boldsymbol{\omega} \underset{(1.16.1)}{=} {}^A\boldsymbol{\omega}^C + {}^C\boldsymbol{\omega}^B \tag{12}$$

Furthermore, if Ω denotes the rate at which the line joining A^* to B^* rotates in A, then

$$ {}^A\boldsymbol{\omega}^C \underset{(1.11.5)}{=} \Omega \mathbf{c}_2 \tag{13}$$

and

$$ {}^C\boldsymbol{\omega}^B \underset{(1.11.5)}{=} \Omega \mathbf{c}_1 \tag{14}$$

where $\mathbf{c}_1$ and $\mathbf{c}_2$ are unit vectors directed as in Fig. 1.19.2. Hence

$$\boldsymbol{\omega} = \Omega(\mathbf{c}_1 + \mathbf{c}_2) \tag{15}$$

The velocity $\mathbf{v}^{B^*}$ of B^* in A is given by

$$\mathbf{v}^{B^*} = R\Omega\mathbf{c}_1 \tag{16}$$

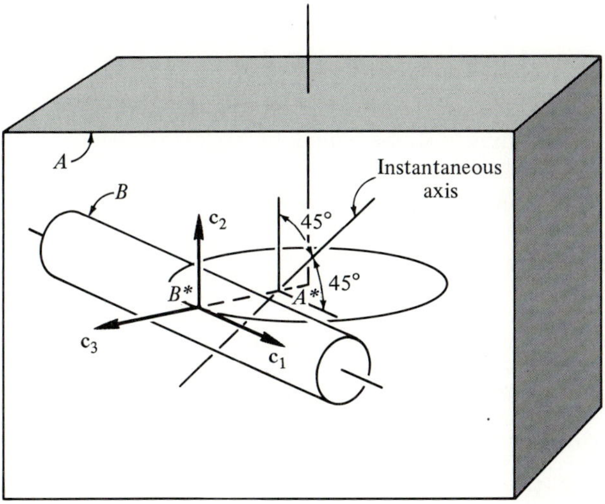

Figure 1.19.2

Consequently, if P^* is a point on the instantaneous axis of B in A, then the position vector $\mathbf{r}^*$ of P^* relative to B^* is given by

$$\mathbf{r}^* \underset{(1)}{=} \frac{\Omega(\mathbf{c}_1 + \mathbf{c}_2) \times (R\,\Omega\mathbf{c}_1)}{2\Omega^2} + \mu^*\Omega(\mathbf{c}_1 + \mathbf{c}_2)$$

$$= -\frac{R}{2}\mathbf{c}_3 + \mu^*\Omega(\mathbf{c}_1 + \mathbf{c}_2) \tag{17}$$

Hence, the instantaneous axis of B in A passes through the midpoint of the line joining A^* to B^*, is perpendicular to $\mathbf{c}_3$, and makes a 45° angle with each of $\mathbf{c}_1$ and $\mathbf{c}_2$, as indicated in Fig. 1.19.2.

1.20 ANGULAR ACCELERATION

The angular acceleration $\boldsymbol{\alpha}$ of a rigid body B in a reference frame A is defined as the first time-derivative in A of the angular velocity $\boldsymbol{\omega}$ of B in A (see Sec. 1.11):

$$\boldsymbol{\alpha} \triangleq \frac{{}^A d\boldsymbol{\omega}}{dt} \tag{1}$$

Frequently, it is convenient to resolve both $\boldsymbol{\omega}$ and $\boldsymbol{\alpha}$ into components parallel to unit vectors fixed in a reference frame C, that is, to express $\boldsymbol{\omega}$ and $\boldsymbol{\alpha}$ as

$$\boldsymbol{\omega} = {}^C\omega_1\mathbf{c}_1 + {}^C\omega_2\mathbf{c}_2 + {}^C\omega_3\mathbf{c}_3 \tag{2}$$

and

$$\boldsymbol{\alpha} = {}^C\alpha_1\mathbf{c}_1 + {}^C\alpha_2\mathbf{c}_2 + {}^C\alpha_3\mathbf{c}_3 \tag{3}$$

where $\mathbf{c}_1$, $\mathbf{c}_2$, $\mathbf{c}_3$ form a dextral set of orthogonal unit vectors. When this is done,

$${}^C\alpha_i = {}^C\dot{\omega}_i + \boldsymbol{\Omega} \times \boldsymbol{\omega} \cdot \mathbf{c}_i \qquad (i = 1, 2, 3) \tag{4}$$

where $\boldsymbol{\Omega}$ is the angular velocity of C in A. In other words, depending on the motion of C in A, ${}^C\alpha_i$ may, or may not, be equal to ${}^C\dot{\omega}_i$.

Derivations Using Eq. (1.11.8), one can express $\boldsymbol{\alpha}$ as

$$\boldsymbol{\alpha} \underset{(1)}{=} \frac{{}^C d\boldsymbol{\omega}}{dt} + \boldsymbol{\Omega} \times \boldsymbol{\omega}$$

$$\underset{(2)}{=} {}^C\dot{\omega}_1\mathbf{c}_1 + {}^C\dot{\omega}_2\mathbf{c}_2 + {}^C\dot{\omega}_3\mathbf{c}_3 + \boldsymbol{\Omega} \times \boldsymbol{\omega} \tag{5}$$

Consequently, when $\boldsymbol{\alpha}$ is expressed as in Eq. (3), then

$${}^C\alpha_1\mathbf{c}_1 + {}^C\alpha_2\mathbf{c}_2 + {}^C\alpha_3\mathbf{c}_3 = {}^C\dot{\omega}_1\mathbf{c}_1 + {}^C\dot{\omega}_2\mathbf{c}_2 + {}^C\dot{\omega}_3\mathbf{c}_3 + \boldsymbol{\Omega} \times \boldsymbol{\omega} \tag{6}$$

and dot-multiplication with $\mathbf{c}_i$ $(i = 1, 2, 3)$ gives

$$ {}^C\alpha_i = {}^C\dot{\omega}_i + \mathbf{\Omega} \times \boldsymbol{\omega} \cdot \mathbf{c}_i \qquad (i = 1, 2, 3) \tag{7} $$

Example Figure 1.20.1 depicts the system previously considered in the example in Sec. 1.12. In addition to the unit vectors used previously, orthogonal unit vectors $\mathbf{c}_1$, $\mathbf{c}_2$, and $\mathbf{c}_3$ are shown, and a reference frame C, in which these are fixed, is indicated. Considering only motions such that $\dot{\phi}$ and $\dot{\psi}$, as well as θ, remain constant, the quantities ${}^A\alpha_i$, ${}^B\alpha_i$, and ${}^C\alpha_i$ $(i = 1, 2, 3)$ are to be determined, these being defined as

$$ {}^A\alpha_i \triangleq \boldsymbol{\alpha} \cdot \mathbf{a}_i \qquad {}^B\alpha_i \triangleq \boldsymbol{\alpha} \cdot \mathbf{b}_i \qquad {}^C\alpha_i \triangleq \boldsymbol{\alpha} \cdot \mathbf{c}_i \qquad (i = 1, 2, 3) \tag{8} $$

where $\boldsymbol{\alpha}$ is the angular acceleration of B in A.

The angular velocity $\boldsymbol{\omega}$ of B in A can be expressed as

$$ \boldsymbol{\omega} = (\dot{\psi}\ c\theta + \dot{\phi})\mathbf{a}_1 + \dot{\psi}\ s\theta\ c\phi\ \mathbf{a}_2 + \dot{\psi}\ s\theta\ s\phi\ \mathbf{a}_3 \tag{9} $$

or as

$$ \boldsymbol{\omega} = (\dot{\psi} + \dot{\phi}\ c\theta)\mathbf{b}_1 - \dot{\phi}\ s\theta\ c\psi\ \mathbf{b}_2 + \dot{\phi}\ s\theta\ s\psi\ \mathbf{b}_3 \tag{10} $$

or as

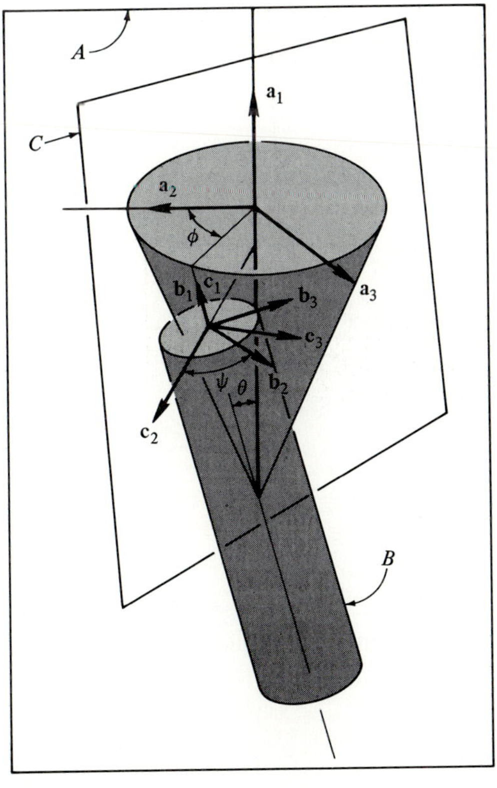

Figure 1.20.1

$$\boldsymbol{\omega} = (\dot{\psi} + \dot{\phi}\ c\theta)\mathbf{c}_1 - \dot{\phi}\ s\theta\ \mathbf{c}_2 \tag{11}$$

Using Eq. (4), with C replaced by A, and hence $\boldsymbol{\Omega} = 0$, one obtains by reference to Eq. (9)

$$^A\alpha_1 = \frac{d}{dt}(\dot{\psi}\ c\theta + \dot{\phi}) = 0 \tag{12}$$

$$^A\alpha_2 = \frac{d}{dt}(\dot{\psi}\ s\theta\ c\phi) = -\dot{\psi}\dot{\phi}\ s\theta\ s\phi \tag{13}$$

$$^A\alpha_3 = \frac{d}{dt}(\dot{\psi}\ s\theta\ s\phi) = \dot{\psi}\dot{\phi}\ s\theta\ c\phi \tag{14}$$

Similarly, with C replaced by B in Eq. (4), so that $\boldsymbol{\Omega} = \boldsymbol{\omega}$, Eq. (10) permits one to write

$$^B\alpha_1 = \frac{d}{dt}(\dot{\psi} + \dot{\phi}\ c\theta) = 0 \tag{15}$$

$$^B\alpha_2 = \frac{d}{dt}(-\dot{\phi}\ s\theta\ c\psi) = \dot{\phi}\dot{\psi}\ s\theta\ s\psi \tag{16}$$

$$^B\alpha_3 = \frac{d}{dt}(\dot{\phi}\ s\theta\ s\psi) = \dot{\phi}\dot{\psi}\ s\theta\ c\psi \tag{17}$$

Finally, with

$$\boldsymbol{\Omega} = {}^A\boldsymbol{\omega}^C = \dot{\phi}(c\theta\ \mathbf{c}_1 - s\theta\ \mathbf{c}_2) \tag{18}$$

so that

$$\boldsymbol{\Omega} \times \boldsymbol{\omega} \underset{(11,18)}{=} \dot{\phi}\dot{\psi}\ s\theta\ \mathbf{c}_3 \tag{19}$$

it follows from Eq. (4) together with Eqs. (11) and (19) that

$$^C\alpha_1 = \frac{d}{dt}(\dot{\psi} + \dot{\phi}\ c\theta) + \dot{\phi}\dot{\psi}\ s\theta\ \mathbf{c}_3 \cdot \mathbf{c}_1 = 0 \tag{20}$$

$$^C\alpha_2 = \frac{d}{dt}(-\dot{\phi}\ s\theta) + \dot{\phi}\dot{\psi}\ s\theta\ \mathbf{c}_3 \cdot \mathbf{c}_2 = 0 \tag{21}$$

and

$$^C\alpha_3 = \dot{\phi}\dot{\psi}\ s\theta \tag{22}$$

Thus, with the exception of $^C\alpha_3$, every one of the quantities defined in Eqs. (8) is equal to the time-derivative of the corresponding angular velocity measure number.

1.21 PARTIAL ANGULAR VELOCITIES AND PARTIAL VELOCITIES

For reasons that will become apparent later, it is often convenient, when dealing with a system S whose configuration in a reference frame A is characterized by n generalized coordinates $q_1, \ldots, q_n$, to introduce n quantities $u_1, \ldots, u_n$, called *generalized speeds* for S in A, as linear combinations of $\dot{q}_1, \ldots, \dot{q}_n$ by means of equations of the form

$$u_r \triangleq \sum_{s=1}^{n} Y_{rs}\, \dot{q}_s + Z_r \qquad (r = 1, \ldots, n) \tag{1}$$

where Y_{rs} and Z_r are functions of $q_1, \ldots, q_n$, and t, and Y_{rs} $(r, s = 1, \ldots, n)$ are chosen such that Eq. (1) can be solved uniquely for $\dot{q}_1, \ldots, \dot{q}_n$. When this is done, $\boldsymbol{\omega}$, the angular velocity in A of a rigid body B belonging to S, and $\mathbf{v}$, the velocity in A of a particle P belonging to S, can be expressed *uniquely* as

$$\boldsymbol{\omega} = \sum_{r=1}^{n} \boldsymbol{\omega}_r\, u_r + \boldsymbol{\omega}_t \tag{2}$$

and

$$\mathbf{v} = \sum_{r=1}^{n} \mathbf{v}_r\, u_r + \mathbf{v}_t \tag{3}$$

where $\boldsymbol{\omega}_r$, $\mathbf{v}_r$ $(r - 1, \ldots, n)$, $\boldsymbol{\omega}_t$, and $\mathbf{v}_t$ are functions of $q_1, \ldots, q_n$, and t. The vector $\boldsymbol{\omega}_r$, called the rth *partial angular velocity* of B in A, and the vector $\mathbf{v}_r$, called the rth *partial velocity* of P in A, are formed by inspection of expressions having the forms of the right-hand members of Eqs. (2) and (3), respectively; $u_1, \ldots, u_n$ need not be introduced in the formal manner indicated in Eq. (1). When dealing with a specific problem, one selects the generalized speeds so as to arrive at especially simple expressions for angular velocities of rigid bodies and velocities of points of the system under consideration, frequently doing this without first explicitly designating generalized coordinates. At times it is convenient to take $u_r = \dot{q}_r$ for some or all values of r.

As will become apparent in Chap. 4, the use of generalized speeds, partial angular velocities, and partial velocities permits one to formulate expressions for generalized forces in a particularly effective way and enables one to construct, with a minimum amount of labor, equations of motion having the simplest form possible.

Derivations Solution of Eq. (1) for $\dot{q}_1, \ldots, \dot{q}_n$ leads to

$$\dot{q}_s = \sum_{r=1}^{n} W_{sr}\, u_r + X_s \qquad (s = 1, \ldots, n) \tag{4}$$

where W_{sr} and X_s are certain functions of $q_1, \ldots, q_n$, and t. Now, if $\mathbf{b}_1$, $\mathbf{b}_2$, $\mathbf{b}_3$ form a dextral set of mutually perpendicular unit vectors fixed in B, and if $\dot{\mathbf{b}}_i$ denotes the first time-derivative of $\mathbf{b}_i$ in A, then

$$\dot{\mathbf{b}}_i = \sum_{s=1}^{n} \frac{\partial \mathbf{b}_i}{\partial q_s} \dot{q}_s + \frac{\partial \mathbf{b}_i}{\partial t}$$

$$\underset{(4)}{=} \sum_{s=1}^{n} \frac{\partial \mathbf{b}_i}{\partial q_s} \left(\sum_{r=1}^{n} W_{sr} u_r + X_s \right) + \frac{\partial \mathbf{b}_i}{\partial t}$$

$$= \sum_{r=1}^{n} \sum_{s=1}^{n} \frac{\partial \mathbf{b}_i}{\partial q_s} W_{sr} u_r + \sum_{s=1}^{n} \frac{\partial \mathbf{b}_i}{\partial q_s} X_s + \frac{\partial \mathbf{b}_i}{\partial t} \qquad (i = 1, 2, 3) \tag{5}$$

where all partial differentiations of $\mathbf{b}_i$ are performed in A. Consequently,

$$\boldsymbol{\omega} \underset{(1.11.4)}{=} \mathbf{b}_1 \dot{\mathbf{b}}_2 \cdot \mathbf{b}_3 + \mathbf{b}_2 \dot{\mathbf{b}}_3 \cdot \mathbf{b}_1 + \mathbf{b}_3 \dot{\mathbf{b}}_1 \cdot \mathbf{b}_2$$

$$\underset{(5)}{=} \mathbf{b}_1 \left(\sum_{r=1}^{n} \sum_{s=1}^{n} \frac{\partial \mathbf{b}_2}{\partial q_s} \cdot \mathbf{b}_3 W_{sr} u_r + \sum_{s=1}^{n} \frac{\partial \mathbf{b}_2}{\partial q_s} \cdot \mathbf{b}_3 X_s + \frac{\partial \mathbf{b}_2}{\partial t} \cdot \mathbf{b}_3 \right)$$

$$+ \mathbf{b}_2 \left(\sum_{r=1}^{n} \sum_{s=1}^{n} \frac{\partial \mathbf{b}_3}{\partial q_s} \cdot \mathbf{b}_1 W_{sr} u_r + \sum_{s=1}^{n} \frac{\partial \mathbf{b}_3}{\partial q_s} \cdot \mathbf{b}_1 X_s + \frac{\partial \mathbf{b}_3}{\partial t} \cdot \mathbf{b}_1 \right)$$

$$+ \mathbf{b}_3 \left(\sum_{r=1}^{n} \sum_{s=1}^{n} \frac{\partial \mathbf{b}_1}{\partial q_s} \cdot \mathbf{b}_2 W_{sr} u_r + \sum_{s=1}^{n} \frac{\partial \mathbf{b}_1}{\partial q_s} \cdot \mathbf{b}_2 X_s + \frac{\partial \mathbf{b}_1}{\partial t} \cdot \mathbf{b}_2 \right) \tag{6}$$

and, if $\boldsymbol{\omega}_r$ and $\boldsymbol{\omega}_t$ are defined as

$$\boldsymbol{\omega}_r \triangleq \sum_{s=1}^{n} \left(\mathbf{b}_1 \frac{\partial \mathbf{b}_2}{\partial q_s} \cdot \mathbf{b}_3 + \mathbf{b}_2 \frac{\partial \mathbf{b}_3}{\partial q_s} \cdot \mathbf{b}_1 + \mathbf{b}_3 \frac{\partial \mathbf{b}_1}{\partial q_s} \cdot \mathbf{b}_2 \right) W_{sr} \qquad (r = 1, \ldots, n) \tag{7}$$

and

$$\boldsymbol{\omega}_t \triangleq \sum_{s=1}^{n} \left(\mathbf{b}_1 \frac{\partial \mathbf{b}_2}{\partial q_s} \cdot \mathbf{b}_3 + \mathbf{b}_2 \frac{\partial \mathbf{b}_3}{\partial q_s} \cdot \mathbf{b}_1 + \mathbf{b}_3 \frac{\partial \mathbf{b}_1}{\partial q_s} \cdot \mathbf{b}_2 \right) X_s$$

$$+ \mathbf{b}_1 \frac{\partial \mathbf{b}_2}{\partial t} \cdot \mathbf{b}_3 + \mathbf{b}_2 \frac{\partial \mathbf{b}_3}{\partial t} \cdot \mathbf{b}_1 + \mathbf{b}_3 \frac{\partial \mathbf{b}_1}{\partial t} \cdot \mathbf{b}_2 \tag{8}$$

respectively, then substitution from Eqs. (7) and (8) into Eq. (6) leads directly to Eq. (2).

If $\mathbf{p}$ is the position vector from a point O fixed in A to a generic particle P of S, and if $\dot{\mathbf{p}}$ denotes the first time-derivative of $\mathbf{p}$ in A, then, by definition,

$$\mathbf{v} \triangleq \dot{\mathbf{p}} = \sum_{s=1}^{n} \frac{\partial \mathbf{p}}{\partial q_s} \dot{q}_s + \frac{\partial \mathbf{p}}{\partial t}$$

$$\underset{(4)}{=} \sum_{s=1}^{n} \frac{\partial \mathbf{p}}{\partial q_s} \left(\sum_{r=1}^{n} W_{sr} u_r + X_s \right) + \frac{\partial \mathbf{p}}{\partial t}$$

$$= \sum_{r=1}^{n} \sum_{s=1}^{n} \frac{\partial \mathbf{p}}{\partial q_s} W_{sr} u_r + \sum_{s=1}^{n} \frac{\partial \mathbf{p}}{\partial q_s} X_s + \frac{\partial \mathbf{p}}{\partial t} \tag{9}$$

Hence, after defining $\mathbf{v}_r$ and $\mathbf{v}_t$ as

$$\mathbf{v}_r \triangleq \sum_{s=1}^{n} \frac{\partial \mathbf{p}}{\partial q_s} W_{sr} \qquad (r = 1, \ldots, n) \tag{10}$$

and

$$\mathbf{v}_t \triangleq \sum_{s=1}^{n} \frac{\partial \mathbf{p}}{\partial q_s} X_s + \frac{\partial \mathbf{p}}{\partial t} \tag{11}$$

one arrives at Eq. (3).

Example In Fig. 1.21.1, B represents a rigid body whose mass center, B^*, is made to move with a constant speed V on a circular orbit C that is fixed in a reference frame A and has a radius R. $\mathbf{a}_1$, $\mathbf{a}_2$, $\mathbf{a}_3$ and $\mathbf{b}_1$, $\mathbf{b}_2$, $\mathbf{b}_3$ form dextral sets of mutually perpendicular unit vectors fixed in A and B, respectively, and $\boldsymbol{\tau}$ is a unit vector tangent to C at B^*.

Because the motion of B^* in A is prescribed, three generalized coordinates suffice to characterize the configuration of B in A. Without introducing coordinates explicitly, one can define generalized speeds in terms of $\mathbf{b}_r$ and the angular velocity $\boldsymbol{\omega}$ of B in A as

$$u_r \triangleq \boldsymbol{\omega} \cdot \mathbf{b}_r \qquad (r = 1, 2, 3) \tag{12}$$

from which it follows immediately that

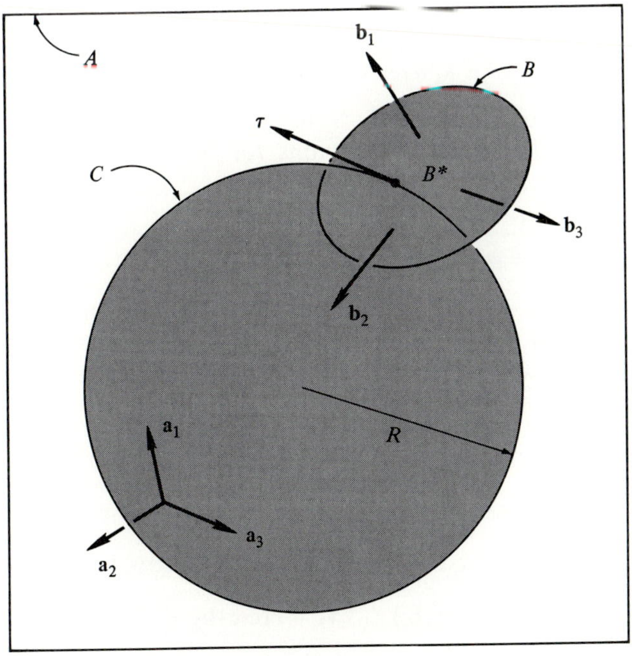

Figure 1.21.1

$$\boldsymbol{\omega} = u_1\mathbf{b}_1 + u_2\mathbf{b}_2 + u_3\mathbf{b}_3 \tag{13}$$

Comparing Eqs. (2) and (13), one thus finds that the partial angular velocities of B in A are given by

$$\boldsymbol{\omega}_r = \mathbf{b}_r \qquad (r = 1, 2, 3) \tag{14}$$

If P is a point of B whose position vector relative to B^* is the vector $r\mathbf{b}_1$, then the velocity $\mathbf{v}$ of P in A can be expressed as

$$\mathbf{v} = V\boldsymbol{\tau} + \boldsymbol{\omega} \times (r\mathbf{b}_1)$$

$$\underset{(13)}{=} V\boldsymbol{\tau} + r(u_3\mathbf{b}_2 - u_2\mathbf{b}_3) \tag{15}$$

The partial velocities of P in A are thus [compare Eqs. (3) and (15)]

$$\mathbf{v}_1 = 0 \qquad \mathbf{v}_2 = -r\mathbf{b}_3 \qquad \mathbf{v}_3 = r\mathbf{b}_2 \tag{16}$$

Suppose one decided to use for generalized coordinates space-three angles q_1, q_2, q_3 relating the unit vectors $\mathbf{a}_1, \mathbf{a}_2, \mathbf{a}_3$ to unit vectors $\mathbf{b}_1, \mathbf{b}_2, \mathbf{b}_3$ as in Sec. 1.7. Then $\boldsymbol{\omega}$ would be given by [see Eqs. (1.17.1) and (1.17.3)]

$$\boldsymbol{\omega} = (\dot{q}_1 - s_2\dot{q}_3)\mathbf{b}_1 + (c_1\dot{q}_2 + s_1c_2\dot{q}_3)\mathbf{b}_2 + (-s_1\dot{q}_2 + c_1c_2\dot{q}_3)\mathbf{b}_3 \tag{17}$$

and u_1, u_2, u_3 would be related to q_1, q_2, q_3 as follows [see Eqs. (13) and (17)]:

$$u_1 = \dot{q}_1 - s_2\dot{q}_3 \tag{18}$$

$$u_2 = c_1\dot{q}_2 + s_1c_2\dot{q}_3 \tag{19}$$

$$u_3 = -s_1\dot{q}_2 + c_1c_2\dot{q}_3 \tag{20}$$

These equations have the form of Eq. (1). Note that it was not necessary to use them in order to formulate expressions for the partial angular velocities $\boldsymbol{\omega}_r$ and partial velocities $\mathbf{v}_r$ $(r = 1, 2, 3)$.

As an alternative to Eq. (12), one could define u_r as

$$u_r \triangleq \dot{q}_r \qquad (r = 1, 2, 3) \tag{21}$$

Clearly, these equations are simpler than Eqs. (18)–(20); but expressions for $\boldsymbol{\omega}$, $\boldsymbol{\omega}_r$, $\mathbf{v}$, and $\mathbf{v}_r$ $(r = 1, 2, 3)$ tend to be more complicated when u_r is defined in this way than when Eqs. (12) are employed. Specifically, the relationships

$$\boldsymbol{\omega} = (u_1 - s_2u_3)\mathbf{b}_1 + (c_1u_2 + s_1c_2u_3)\mathbf{b}_2 + (-s_1u_2 + c_1c_2u_3)\mathbf{b}_3 \tag{22}$$

$$\boldsymbol{\omega}_1 = \mathbf{b}_1 \qquad \boldsymbol{\omega}_2 = c_1\mathbf{b}_2 - s_1\mathbf{b}_3 \qquad \boldsymbol{\omega}_3 = -s_2\mathbf{b}_1 + s_1c_2\mathbf{b}_2 + c_1c_2\mathbf{b}_3 \tag{23}$$

$$\mathbf{v} = V\boldsymbol{\tau} + r[(-s_1u_2 + c_1c_2u_3)\mathbf{b}_2 - (c_1u_2 + s_1c_2u_3)\mathbf{b}_3] \tag{24}$$

$$\mathbf{v}_1 = 0 \qquad \mathbf{v}_2 = r(-s_1\mathbf{b}_2 - c_1\mathbf{b}_3) \qquad \mathbf{v}_3 = r(c_1c_2\mathbf{b}_2 - s_1c_2\mathbf{b}_3) \tag{25}$$

take the place of Eqs. (13)–(16).

CHAPTER

TWO

GRAVITATIONAL FORCES

The solution of most spacecraft dynamics problems requires consideration of gravitational forces and moments. In this chapter, Newton's law of gravitation is used as the point of departure for the development of force and moment expressions of particular interest in this context.

In the first eight sections of the chapter, only vector-algebraic methods are employed; that is, no use is made of partial differential calculus. One section is then devoted to the development of partial differentiation techniques that come into play in the remaining sections, which deal with force functions. Since in Chaps. 3 and 4 only sparing use is made of material from Secs. 2.9–2.18, these sections (and Probs. 2.18–2.38) may be omitted by readers wishing to move on to Chaps. 3 and 4 as rapidly as possible.

2.1 GRAVITATIONAL INTERACTION OF TWO PARTICLES

A particle P of mass m experiences in the presence of a particle $\bar{P}$ of mass $\bar{m}$ a force $\mathbf{F}$ acting along the line joining P to $\bar{P}$, directed from P toward $\bar{P}$, and having a magnitude proportional to the product of m and $\bar{m}$ and inversely proportional to the square of the distance between P and $\bar{P}$. Hence, if $\mathbf{p}$ is the position vector from $\bar{P}$ to P (see Fig. 2.1.1), the force $\mathbf{F}$ can be expressed as†

†The superscript 2 on a vector indicates scalar multiplication of the vector with itself, i.e., squaring the magnitude of the vector.

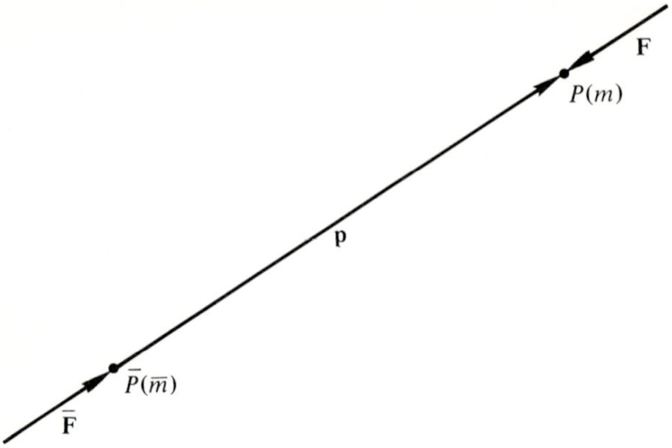

Figure 2.1.1

$$\mathbf{F} = -G\bar{m}m\,\mathbf{p}(\mathbf{p}^2)^{-3/2} \tag{1}$$

where G is the universal gravitational constant, given numerically by† $G \approx 6.6732 \times 10^{-11}\ \mathrm{N \cdot m^2\ kg^{-2}}$.

The force $\bar{\mathbf{F}}$ experienced by $\bar{P}$ in the presence of P is

$$\bar{\mathbf{F}} = -\mathbf{F} \tag{2}$$

in conformity with Newton's third law.

Example If $\bar{m}$ is the mass of the Earth, m the mass of the Moon, and $\mathbf{p}$ the vector from the mass center of the Earth to the mass center of the Moon, and if the numerical values of $\bar{m}$, m and $|\mathbf{p}|$ are given approximately by $\bar{m} \approx 5.97 \times 10^{24}$ kg, $m \approx 7.34 \times 10^{22}$ kg, and $|\mathbf{p}| \approx 3.844 \times 10^{8}$ m, what is the magnitude of the force $\mathbf{F}$ exerted on the Moon by the Earth?

As Eq. (1) applies solely to particles, it can be used for the purpose at hand only with the supposition that the Earth and Moon can be replaced with particles situated at their mass centers and having masses $\bar{m}$ and m, respectively. In that case,

$$|\mathbf{F}| \underset{(1)}{=} G\bar{m}m\,|\mathbf{p}|(\mathbf{p}^2)^{-3/2} \approx 1.98 \times 10^{20}\ \mathrm{N} \tag{3}$$

2.2 FORCE EXERTED ON A BODY BY A PARTICLE

The system of gravitational forces exerted by a particle $\bar{P}$ of mass $\bar{m}$ on the particles of a (not necessarily rigid) body B is equivalent to a single force $\bar{\mathbf{F}}$

†E. A. Mechtly, "The International System of Units: Physical Constants and Conversion Factors," NASA SP-7012, revised, 1969.

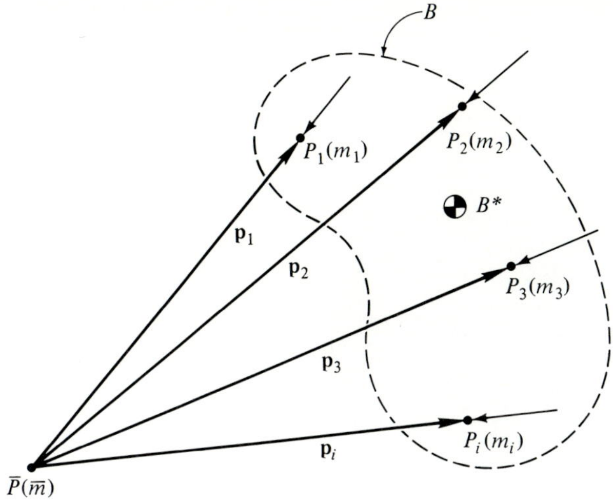

Figure 2.2.1

whose line of action passes through $\bar{P}$, but not necessarily through the mass center B^* of B. If B consists of particles $P_1, \ldots, P_N$ of masses $m_1, \ldots, m_N$ and if $\mathbf{p}_1, \ldots, \mathbf{p}_N$ are the position vectors of $P_1, \ldots, P_N$ relative to $\bar{P}$ (see Fig. 2.2.1), then $\mathbf{F}$ is given by

$$\mathbf{F} = -G\bar{m} \sum_{i=1}^{N} m_i \mathbf{p}_i (\mathbf{p}_i^2)^{-3/2} \tag{1}$$

whereas, if B is a continuous distribution of matter, ρ is the mass density of B at a generic point P of B, $\mathbf{p}$ is the position vector of P relative to $\bar{P}$ (see Fig. 2.2.2), and $d\tau$ is the length, area, or volume of a differential element of the figure (curve, surface, or solid) occupied by B, then $\mathbf{F}$ can be expressed as

$$\mathbf{F} = -G\bar{m} \int \mathbf{p}(\mathbf{p}^2)^{-3/2} \rho \, d\tau \tag{2}$$

Once the line of action of the force $\mathbf{F}$ in Eq. (1) or Eq. (2) has been established, it is always possible to locate a point B' on this line such that the force exerted by $\bar{P}$ on a particle placed at B' and having a mass equal to that of B is equal to the force $\mathbf{F}$ exerted by $\bar{P}$ on B. The point B', called the *center of gravity* of B for the attracting particle $\bar{P}$, does not, in general, coincide with the mass center of B. If R' is the distance from $\bar{P}$ to the center of gravity B', then

$$R' = \left(\frac{G\bar{m}m}{|\mathbf{F}|}\right)^{1/2} \tag{3}$$

where $\mathbf{F}$ is obtained from Eq. (1) or Eq. (2), and m is the mass of B.

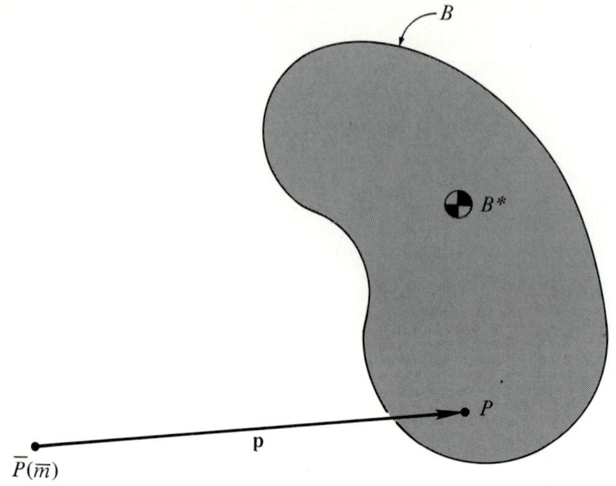

Figure 2.2.2

Derivations The force $d\mathbf{F}$ exerted by $\bar{P}$ on a differential element of B at point P acts along the line joining P to $\bar{P}$ and is given by

$$d\mathbf{F} \underset{(2.1.1)}{=} -G\bar{m}\mathbf{p}(\mathbf{p}^2)^{-3/2}\rho\, d\tau \tag{4}$$

By definition, two systems of forces are equivalent if they have equal resultants and equal moments about one point. Now, on the one hand, the resultant of the system of forces exerted by $\bar{P}$ on all differential elements of B is given by

$$\int d\mathbf{F} = -G\bar{m}\int \mathbf{p}(\mathbf{p}^2)^{-3/2}\rho\, d\tau \tag{5}$$

and the moment of the system of forces about point $\bar{P}$ is equal to zero, because $d\mathbf{F}$, acting along the line joining P to $\bar{P}$, has zero moment about $\bar{P}$. On the other hand, the resultant of a system of forces containing but one force $\mathbf{F}$ is $\mathbf{F}$ itself, and the moment of this system about point $\bar{P}$ is zero if the line of action of $\mathbf{F}$ passes through $\bar{P}$. Hence, if $\mathbf{F}$ is given by Eq. (2) and acts along a line passing through $\bar{P}$, then $\mathbf{F}$ is equivalent to the system of forces exerted by $\bar{P}$ on all differential elements of B.

A parallel proof may be constructed for Eq. (1), which replaces Eq. (2) when B consists of a finite number of particles.

The truth of the assertion that the line of action of the force $\mathbf{F}$ does not necessarily pass through B^* is most easily demonstrated by an example. Finally, Eq. (3) follows directly from Eq. (2.1.1) and the definition of R'.

Example A uniform thin rod B of length L and mass m is subjected to the gravitational attraction of a particle $\bar{P}$ of mass $\bar{m}$, as shown in Fig.

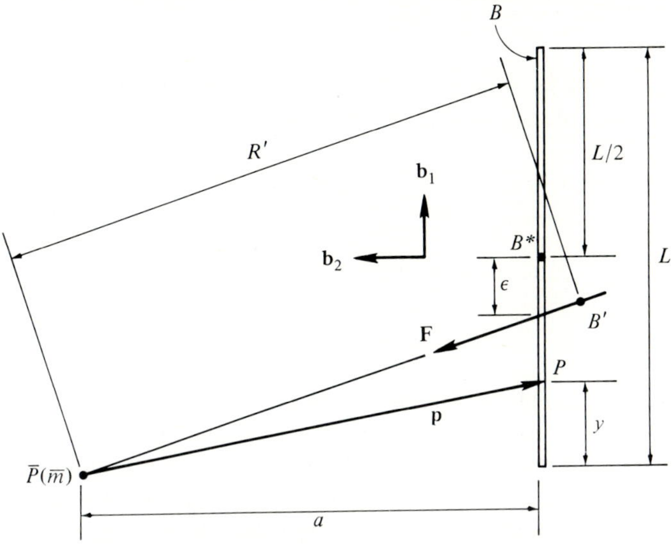

Figure 2.2.3

2.2.3. The system of forces exerted by $\bar{P}$ on the particles comprising B is to be replaced with a force **F** whose line of action passes through $\bar{P}$, and an expression is to be found for the distance R' between $\bar{P}$ and the center of gravity B' of B for $\bar{P}$.

If the rod is regarded as matter distributed along a straight line segment, then the position vector **p** of a generic point P of the rod relative to $\bar{P}$ can be expressed as

$$\mathbf{p} = y\mathbf{b}_1 - a\mathbf{b}_2 \tag{6}$$

where $\mathbf{b}_1$ and $\mathbf{b}_2$ are unit vectors directed as shown in Fig. 2.2.3 and where y varies from 0 to L. The mass density ρ of B at P is then equal to m/L, and a differential element of B has a length dy. Hence, the force **F**, resolved into components parallel to $\mathbf{b}_1$ and $\mathbf{b}_2$, can be written

$$\mathbf{F} = F_1\mathbf{b}_1 + F_2\mathbf{b}_2 \underset{(2)}{=} -G\bar{m}\int_0^L (y\mathbf{b}_1 - a\mathbf{b}_2)(a^2 + y^2)^{-3/2}\frac{m}{L}\,dy \tag{7}$$

and integration yields

$$F_1 = -\frac{G\bar{m}m[(1 + a^2/L^2)^{1/2} - a/L]}{a(a^2 + L^2)^{1/2}} \tag{8}$$

$$F_2 = \frac{G\bar{m}m}{a(a^2 + L^2)^{1/2}} \tag{9}$$

From Fig. 2.2.3 it can be seen that the distance ϵ between the mass center B^* of B and the intersection of B with the line of action of $\mathbf{F}$ is related to F_1 and F_2 by

$$-\frac{F_1}{F_2} = \frac{L/2 - \epsilon}{a} \tag{10}$$

so that

$$\frac{\epsilon}{L} \underset{(10)}{=} \frac{1}{2} + \frac{aF_1}{LF_2} \underset{(8,9)}{=} \frac{1}{2} - \frac{a}{L}\left[\left(1 + \frac{a^2}{L^2}\right)^{1/2} - \frac{a}{L}\right] \tag{11}$$

The ratio ϵ/L from Eq. (11) is plotted versus a/L in Fig. 2.2.4. Since two bodies cannot occupy the same point in space, the limiting case $a/L = 0$ must be excluded from consideration. However, as this limit is approached, the ratio ϵ/L approaches the value 1/2 and the line of action of $\mathbf{F}$ approaches coincidence with B. For any finite value of a/L, Fig. 2.2.4 indicates that the line of action of $\mathbf{F}$ cannot pass through both $\bar{P}$ and B^*.

The distance R' from $\bar{P}$ to the center of gravity B' of B for $\bar{P}$ is given by

$$R' \underset{(3)}{=} \left[\frac{G\bar{m}m}{(F_1^2 + F_2^2)^{1/2}}\right]^{1/2} \underset{(8,9)}{=} \frac{(aL)^{1/2}}{\{2[1 - (a/L)(1 + a^2/L^2)^{-1/2}]\}^{1/4}} \tag{12}$$

It is clear from this example that the location of the center of gravity B' is not, in general, a property of body B alone, but depends on the position of the attracting particle $\bar{P}$.

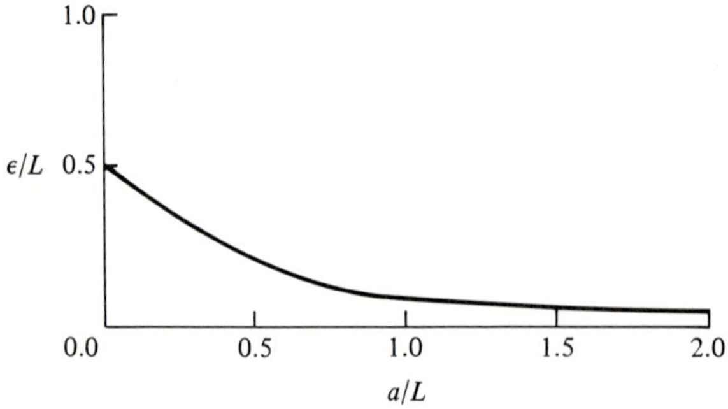

Figure 2.2.4

2.3 FORCE EXERTED ON A SMALL BODY BY A PARTICLE

When a body B is subjected to the gravitational attraction of a particle $\bar{P}$ removed so far from the mass center B^* of B that the largest distance from B^* to any point P of B is smaller than the distance R between $\bar{P}$ and B^*, a useful form of the expression for the force $\mathbf{F}$ given in Eq. (2.2.2) can be found as follows: replace $\mathbf{p}$ with the sum of $\mathbf{R}$, the position vector of B^* relative to $\bar{P}$, and $\mathbf{r}$, the position vector of P relative to B^* (see Fig. 2.3.1), and then expand the integrand in ascending powers of $|\mathbf{r}|/R$ to obtain

$$\mathbf{F} = -\frac{G\bar{m}m}{R^2}\left[\mathbf{a}_1 + \sum_{i=2}^{\infty} \mathbf{f}^{(i)}\right] \tag{1}$$

where $\mathbf{f}^{(i)}$ is a collection of terms of ith degree in $|\mathbf{r}|/R$, m and $\bar{m}$ are the masses of B and $\bar{P}$, G is the universal gravitational constant, and $\mathbf{a}_1$ is a unit vector directed from $\bar{P}$ toward B^*, so that

$$\mathbf{R} = R\mathbf{a}_1 \tag{2}$$

In particular, $\mathbf{f}^{(2)}$ is given by

$$\mathbf{f}^{(2)} = \frac{1}{mR^2}\left\{\frac{3}{2}\,[\text{tr}(\mathbf{I}) - 5\mathbf{a}_1 \cdot \mathbf{I} \cdot \mathbf{a}_1]\mathbf{a}_1 + 3\mathbf{I} \cdot \mathbf{a}_1\right\} \tag{3}$$

where $\mathbf{I}$ is the central inertia dyadic of B, and $\text{tr}(\mathbf{I})$ denotes a scalar invariant of $\mathbf{I}$, called the trace of $\mathbf{I}$ and defined in terms of any mutually orthogonal unit vectors $\mathbf{n}_1$, $\mathbf{n}_2$, and $\mathbf{n}_3$ as

$$\text{tr}(\mathbf{I}) \triangleq \mathbf{n}_1 \cdot \mathbf{I} \cdot \mathbf{n}_1 + \mathbf{n}_2 \cdot \mathbf{I} \cdot \mathbf{n}_2 + \mathbf{n}_3 \cdot \mathbf{I} \cdot \mathbf{n}_3 \tag{4}$$

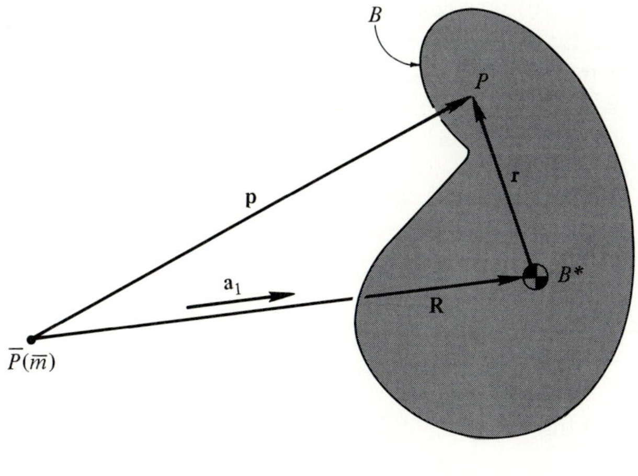

Figure 2.3.1

Equation (1) suggests two useful approximations, namely,

$$\mathbf{F} \approx \hat{\mathbf{F}} \triangleq -\frac{G\overline{m}m}{R^2}\mathbf{a}_1 \tag{5}$$

and

$$\mathbf{F} \approx \widetilde{\mathbf{F}} \triangleq -\frac{G\overline{m}m}{R^2}[\mathbf{a}_1 + \mathbf{f}^{(2)}] \tag{6}$$

Two expressions for $\mathbf{f}^{(2)}$ which sometimes furnish convenient alternatives to the one given in Eq. (3) are obtained as follows: introduce unit vectors $\mathbf{a}_2$ and $\mathbf{a}_3$ such that $\mathbf{a}_1$, $\mathbf{a}_2$, and $\mathbf{a}_3$ form a dextral, orthogonal set, and let $\mathbf{b}_1$, $\mathbf{b}_2$, and $\mathbf{b}_3$ be unit vectors respectively parallel to principal axes of inertia of B for B^* and also forming a dextral orthogonal set. Next, define I_{jk} and I_j as

$$I_{jk} \triangleq \mathbf{a}_j \cdot \mathbf{I} \cdot \mathbf{a}_k \qquad (j, k = 1, 2, 3) \tag{7}$$

and

$$I_j \triangleq \mathbf{b}_j \cdot \mathbf{I} \cdot \mathbf{b}_j \qquad (j = 1, 2, 3) \tag{8}$$

Finally, let

$$C_{ij} \triangleq \mathbf{a}_i \cdot \mathbf{b}_j \qquad (i, j = 1, 2, 3) \tag{9}$$

Then $\mathbf{f}^{(2)}$ may be written either as

$$\mathbf{f}^{(2)} = \frac{3}{mR^2}\left[\frac{1}{2}(I_{22} + I_{33} - 2I_{11})\mathbf{a}_1 + I_{21}\,\mathbf{a}_2 + I_{31}\,\mathbf{a}_3\right] \tag{10}$$

or as

$$\begin{aligned}\mathbf{f}^{(2)} = \frac{3}{mR^2}\Big\{&\frac{1}{2}[I_1(1 - 3C_{11}{}^2) + I_2(1 - 3C_{12}{}^2) + I_3(1 - 3C_{13}{}^2)]\mathbf{a}_1\\ &+ (I_1C_{21}C_{11} + I_2C_{22}C_{12} + I_3C_{23}C_{13})\mathbf{a}_2\\ &+ (I_1C_{31}C_{11} + I_2C_{32}C_{12} + I_3C_{33}C_{13})\mathbf{a}_3\Big\}\end{aligned} \tag{11}$$

The relative simplicity of Eq. (10) is a result of the use of variable moments and products of inertia [see Eq. (7)]. By way of contrast, the principal moments of inertia appearing in Eq. (11) are constants, and the orientation of B relative to $\mathbf{a}_1$, $\mathbf{a}_2$, and $\mathbf{a}_3$ now comes into evidence through the direction cosines relating the two sets of unit vectors $\mathbf{a}_1$, $\mathbf{a}_2$, $\mathbf{a}_3$ and $\mathbf{b}_1$, $\mathbf{b}_2$, $\mathbf{b}_3$.

Derivations Replacing $\mathbf{p}$ in Eq. (2.2.2) by $\mathbf{R} + \mathbf{r}$ (see Fig. 2.3.1) provides

$$\mathbf{F} = -G\overline{m}\int(\mathbf{R} + \mathbf{r})(\mathbf{R}^2 + 2\mathbf{R}\cdot\mathbf{r} + \mathbf{r}^2)^{-3/2}\,\rho\,d\tau \tag{12}$$

In terms of the vector $\mathbf{q}$ defined as

$$\mathbf{q} \triangleq \frac{\mathbf{r}}{R} \tag{13}$$

and the unit vector $\mathbf{a}_1$ satisfying

$$\mathbf{a}_1 \underset{(2)}{=} \frac{\mathbf{R}}{R} \tag{14}$$

the force becomes

$$\mathbf{F} \underset{(12-14)}{=} -\frac{G\overline{m}}{R^2}\int(\mathbf{a}_1 + \mathbf{q})(1 + 2\mathbf{a}_1 \cdot \mathbf{q} + \mathbf{q}^2)^{-3/2}\,\rho\,d\tau \tag{15}$$

and application of the binomial series expansion

$$(1 + x)^n = 1 + nx + \frac{1}{2!}n(n-1)x^2 + \frac{1}{3!}n(n-1)(n-2)x^3 + \cdots \tag{16}$$

(valid for $x < 1$) to the exponentiated quantity then yields, for $2\mathbf{a}_1 \cdot \mathbf{q} + \mathbf{q}^2 < 1$,

$$\mathbf{F} = -\frac{G\overline{m}}{R^2}\int\left\{\mathbf{a}_1\left[1 - \frac{3}{2}(2\mathbf{a}_1 \cdot \mathbf{q} + \mathbf{q}^2) + \frac{15}{2}(\mathbf{a}_1 \cdot \mathbf{q})^2 + \cdots\right]\right.$$
$$\left. + \mathbf{q}(1 - 3\mathbf{a}_1 \cdot \mathbf{q} + \cdots)\right\}\rho\,d\tau \tag{17}$$

where three dots represent terms of degree three and higher in $|\mathbf{q}|$. This expression can be simplified by taking advantage of the fact that B^* is the mass center of B, for this means that

$$\int \mathbf{q}\rho\,d\tau = 0 \tag{18}$$

so that

$$\mathbf{F} \underset{(17,18)}{=} -\frac{G\overline{m}}{R^2}\int\left[\mathbf{a}_1 - \frac{3}{2}\mathbf{a}_1\mathbf{q}^2 + \frac{15}{2}\mathbf{a}_1(\mathbf{a}_1 \cdot \mathbf{q})^2 - 3\mathbf{q}\mathbf{a}_1 \cdot \mathbf{q}\right]\rho\,d\tau + \cdots \tag{19}$$

Furthermore, replacing $\mathbf{q}$ with $\mathbf{r}/R$ and observing that

$$\int \rho\,d\tau = m \tag{20}$$

one arrives at

$$\mathbf{F} \underset{(13,19,20)}{=} -\frac{G\overline{m}m}{R^2}\mathbf{a}_1 + \frac{3G\overline{m}}{2R^4}\left(\mathbf{a}_1\int \mathbf{r}^2\rho\,d\tau - 5\mathbf{a}_1\mathbf{a}_1 \cdot \int \mathbf{r}\mathbf{r}\rho\,d\tau \cdot \mathbf{a}_1\right.$$
$$\left. + 2\int \mathbf{r}\mathbf{r}\rho\,d\tau \cdot \mathbf{a}_1\right) + \cdots \tag{21}$$

One can relate the integrals appearing in Eq. (21) to inertia properties of B by introducing two quantities, namely, the central inertia dyadic $\mathbf{I}$ of B, defined as

$$\mathbf{I} \triangleq \int(\mathbf{U}\mathbf{r}^2 - \mathbf{r}\mathbf{r})\rho\,d\tau \tag{22}$$

where $\mathbf{U}$ denotes the unit dyadic, and the trace of $\mathbf{I}$, defined in Eq. (4); for it follows from Eqs. (22) and (4) that

$$\text{tr}(\mathbf{I}) = \int[3\mathbf{r}^2 - (\mathbf{n}_1 \cdot \mathbf{r}\mathbf{r} \cdot \mathbf{n}_1 + \mathbf{n}_2 \cdot \mathbf{r}\mathbf{r} \cdot \mathbf{n}_2 + \mathbf{n}_3 \cdot \mathbf{r}\mathbf{r} \cdot \mathbf{n}_3)]\rho\,d\tau$$
$$= \int(3\mathbf{r}^2 - \mathbf{r}^2)\rho\,d\tau = 2\int \mathbf{r}^2\rho\,d\tau \tag{23}$$

so that

$$\int \mathbf{r}^2 \rho \, d\tau \underset{(23)}{=} \frac{1}{2} \operatorname{tr}(\mathbf{I}) \tag{24}$$

and

$$\int \mathbf{r}\mathbf{r}\rho \, d\tau \underset{(22)}{=} \mathbf{U} \int \mathbf{r}^2 \rho \, d\tau - \mathbf{I} \underset{(24)}{=} \mathbf{U} \frac{\operatorname{tr}(\mathbf{I})}{2} - \mathbf{I} \tag{25}$$

Consequently,

$$\mathbf{F} \underset{(21,24,25)}{=} -\frac{G\overline{m}m}{R^2}\mathbf{a}_1 - \frac{3G\overline{m}}{2R^4}[\operatorname{tr}(\mathbf{I}) - 5\mathbf{a}_1 \cdot \mathbf{I} \cdot \mathbf{a}_1]\mathbf{a}_1 - \frac{3G\overline{m}}{R^4}\mathbf{I} \cdot \mathbf{a}_1 + \cdots \tag{26}$$

and the equivalence of Eqs. (1) and (26) becomes apparent if one uses Eq. (3) and recognizes that the three dots in Eq. (26) stand for terms of degree three and higher in $|\mathbf{r}|/R$, these terms being represented in Eq. (1) by $\mathbf{f}^{(i)}$ for $i \geq 3$.

Referring to Eq. (7), one can express $\mathbf{I}$ as

$$\begin{aligned} \mathbf{I} = {} & I_{11}\mathbf{a}_1\mathbf{a}_1 + I_{12}\mathbf{a}_1\mathbf{a}_2 + I_{13}\mathbf{a}_1\mathbf{a}_3 \\ & + I_{21}\mathbf{a}_2\mathbf{a}_1 + I_{22}\mathbf{a}_2\mathbf{a}_2 + I_{23}\mathbf{a}_2\mathbf{a}_3 \\ & + I_{31}\mathbf{a}_3\mathbf{a}_1 + I_{32}\mathbf{a}_3\mathbf{a}_2 + I_{33}\mathbf{a}_3\mathbf{a}_3 \end{aligned} \tag{27}$$

from which it follows that

$$\mathbf{I} \cdot \mathbf{a}_1 = I_{11}\mathbf{a}_1 + I_{21}\mathbf{a}_2 + I_{31}\mathbf{a}_3 \tag{28}$$

Substituting from this equation into Eq. (3) and noting that

$$\operatorname{tr}(\mathbf{I}) \underset{(4,7)}{=} I_{11} + I_{22} + I_{33} \tag{29}$$

while

$$\mathbf{a}_1 \cdot \mathbf{I} \cdot \mathbf{a}_1 \underset{(7)}{=} I_{11} \tag{30}$$

one arrives at Eq. (10).

Finally, one can obtain Eq. (11) from Eq. (10) after observing that, in view of Eq. (8) and of the assumption that $\mathbf{b}_1$, $\mathbf{b}_2$, and $\mathbf{b}_3$ are parallel to principal axes of inertia of B for B^*, $\mathbf{I}$ can be expressed as

$$\mathbf{I} = I_1\mathbf{b}_1\mathbf{b}_1 + I_2\mathbf{b}_2\mathbf{b}_2 + I_3\mathbf{b}_3\mathbf{b}_3 \tag{31}$$

so that

$$\begin{aligned} I_{11} & \underset{(7)}{=} I_1\mathbf{a}_1 \cdot \mathbf{b}_1\mathbf{b}_1 \cdot \mathbf{a}_1 + I_2\mathbf{a}_1 \cdot \mathbf{b}_2\mathbf{b}_2 \cdot \mathbf{a}_1 + I_3\mathbf{a}_1 \cdot \mathbf{b}_3\mathbf{b}_3 \cdot \mathbf{a}_1 \\ & \underset{(9)}{=} I_1 C_{11}^2 + I_2 C_{12}^2 + I_3 C_{13}^2 \end{aligned} \tag{32}$$

and, similarly,

$$I_{21} = I_1 C_{21} C_{11} + I_2 C_{22} C_{12} + I_3 C_{23} C_{13} \tag{33}$$

and so forth. Substitution into Eq. (10) then yields Eq. (11) if one makes use of the fact that $C_{i1}^2 + C_{i2}^2 + C_{i3}^2 = 1$ for $i = 1, 2, 3$.

Example An approximate expression is required for the force $\mathbf{F}$ exerted on a uniform thin rod B of mass m and length L by a particle $\overline{P}$ of mass

$\bar{m}$ located relative to B as shown in Fig. 2.3.2, with $R >> L$. Body B is to be idealized as matter distributed uniformly along a straight line segment.

When $\mathbf{f}^{(3)}$, $\mathbf{f}^{(4)}$, etc., are omitted, Eq. (6) provides the required approximate force expression. If $\mathbf{b}_1$, $\mathbf{b}_2$, and $\mathbf{b}_3$ are introduced as a set of dextral, orthogonal unit vectors, with $\mathbf{b}_1$ parallel to the rod axis, then the inertia dyadic $\mathbf{I}$ of B for B^* is given by

$$\mathbf{I} = \frac{mL^2}{12}(\mathbf{b}_2\mathbf{b}_2 + \mathbf{b}_3\mathbf{b}_3) \tag{34}$$

which, when substituted into Eq. (3), introduces into the force expression the dot products $\mathbf{b}_2 \cdot \mathbf{a}_1$ and $\mathbf{b}_3 \cdot \mathbf{a}_1$. To evaluate these, let ψ be the angle between $\mathbf{b}_1$ and $\mathbf{a}_1$, and note that

$$\mathbf{b}_2 \cdot \mathbf{a}_1 = -\sin\psi \qquad \mathbf{b}_3 \cdot \mathbf{a}_1 = 0 \tag{35}$$

so that

$$\mathbf{f}^{(2)} \underset{(3)}{=} \frac{1}{mR^2}\left[\frac{3}{2}\left(\frac{mL^2}{6} - \frac{5mL^2}{12}\sin^2\psi\right)\mathbf{a}_1 - \frac{3mL^2}{12}\sin\psi\,\mathbf{b}_2\right] \tag{36}$$

Hence, after using

$$\mathbf{b}_2 = \mathbf{a}_2\cos\psi - \mathbf{a}_1\sin\psi \tag{37}$$

and omitting $\mathbf{f}^{(3)}$, $\mathbf{f}^{(4)}$, etc., one arrives at

$$\mathbf{F} \underset{(6)}{\approx} \tilde{\mathbf{F}} \underset{(6,36,37)}{=} \frac{G\bar{m}m}{R^2}\left\{\mathbf{a}_1\left[1 + \frac{L^2}{8R^2}(2 - 3\sin^2\psi)\right] - \mathbf{a}_2\frac{L^2}{8R^2}\sin 2\psi\right\} \tag{38}$$

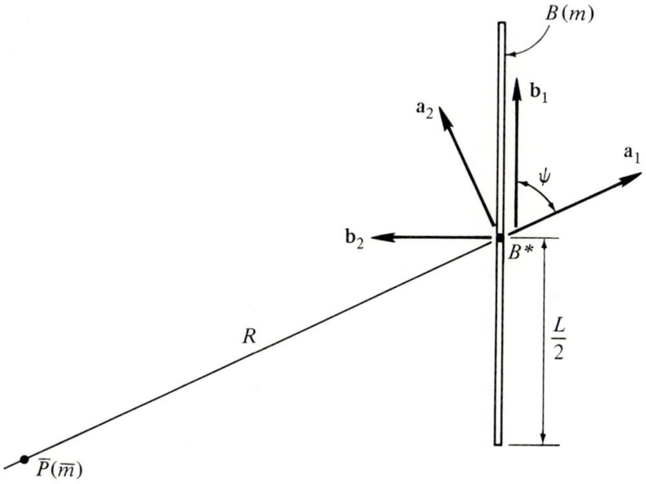

Figure 2.3.2

This result could have been obtained as easily from Eqs. (1), (6), and (11), with the substitution of direction cosines available from Fig. 2.3.2 as

$$\begin{bmatrix} C_{11} & C_{12} & C_{13} \\ C_{21} & C_{22} & C_{23} \\ C_{31} & C_{32} & C_{33} \end{bmatrix} = \begin{bmatrix} \cos\psi & -\sin\psi & 0 \\ \sin\psi & \cos\psi & 0 \\ 0 & 0 & 1 \end{bmatrix} \tag{39}$$

and with $I_1 = 0$, $I_2 = I_3 = mL^2/12$.

When the rod is aligned with the line joining its mass center B^* to $\bar{P}$, so that $\psi = 0$, the force becomes

$$\tilde{\mathbf{F}} \underset{(38)}{=} -\frac{G\bar{m}m}{R^2}\left(1 + \frac{L^2}{4R^2}\right)\mathbf{a}_1 \tag{40}$$

and when the rod is perpendicular to this line, so that $\psi = \pi/2$ rad,

$$\tilde{\mathbf{F}} \underset{(38)}{=} -\frac{G\bar{m}m}{R^2}\left(1 - \frac{L^2}{8R^2}\right)\mathbf{a}_1 \tag{41}$$

Since in these special cases lines passing through B^* and parallel to $\mathbf{a}_1$, $\mathbf{a}_2$, and $\mathbf{a}_3$, are central principal axes of inertia of B, Eqs. (40) and 41) could have been obtained most directly by using Eq. (10), with $I_{21} = I_{31} = 0$ and, for $\psi = 0$,

$$I_{11} = 0 \qquad I_{22} = I_{33} = \frac{mL^2}{12} \tag{42}$$

whereas, for $\psi = \pi/2$,

$$I_{11} = \frac{mL^2}{12} \qquad I_{22} = 0 \qquad I_{33} = \frac{mL^2}{12} \tag{43}$$

Note that the term $(G\bar{m}mL^2/8R^4)\ \sin 2\psi\ \mathbf{a}_2$ in Eq. (38) represents a force directed normal to the line joining $\bar{P}$ to B^*. Such force components cause bodies to move in orbits differing from the Keplerian orbits associated with particles, but this effect is so small that it can generally be ignored, even in high-precision orbital calculations.

Certain differences between the example in Sec. 2.2 and the present example should be noted carefully. For instance, Eq. (38) is valid only when $(L/R)\ \mathbf{a}_1 \cdot \mathbf{b}_1 + L^2/4R^3 < 1$, as required for the underlying series expansion in Eq. (17), whereas Eqs. (2.2.7)–(2.2.9) in the example in Sec. 2.2 are not subject to any such restriction; and the line through $\bar{P}$ and normal to B passes through the end of B in Fig. 2.2.3, but need not do so in Fig. 2.3.2. For the special case in which $\bar{P}$ in Fig. 2.3.2 lies on the normal to B passing through the end of B, as in Fig. 2.2.3, cos ψ and sin ψ are given by

$$\cos\psi = \frac{L}{2R} \tag{44}$$

and

$$\sin\psi = \left[1 - \left(\frac{L}{2R^2}\right)^2\right]^{1/2} \tag{45}$$

but the form of Eq. (38) applicable to this case would not be obtained simply by substituting these expressions into Eq. (38); rather, it is necessary to expand sin ψ in ascending powers of L/R and then to drop terms of degree three or higher after substituting into Eq. (38), which gives

$$\widetilde{\mathbf{F}} = -\frac{G\bar{m}m}{R^2}\left(1 - \frac{L^2}{8R^2}\right)\mathbf{a}_1 \tag{46}$$

2.4 FORCE EXERTED ON A SMALL BODY BY A SMALL BODY

When the distance R between the mass centers B^* and $\bar{B}^*$ of two (not necessarily rigid) bodies B and $\bar{B}$ exceeds in the case of each body the greatest distance from the mass center to any point of the body, the system of gravitational forces exerted on B by $\bar{B}$ has a resultant $\mathbf{F}$ which can be expressed as

$$\mathbf{F} = -\frac{G\bar{m}m}{R^2}\left[\mathbf{a}_1 + \sum_{i=2}^{\infty}\mathbf{f}^{(i)} + \sum_{i=2}^{\infty}\bar{\mathbf{f}}^{(i)} + \sum_{i=2}^{\infty}\sum_{j=2}^{\infty}\mathbf{f}^{(ij)}\right] \tag{1}$$

where $\mathbf{a}_1$ is a unit vector directed from $\bar{B}^*$ toward B^*, G is the universal gravitational constant, and m and $\bar{m}$ are the masses of B and $\bar{B}$, respectively, and where $\mathbf{f}^{(i)}$ is a collection of terms of ith degree in $|\mathbf{r}|/R$, $\bar{\mathbf{f}}^{(i)}$ is a collection of terms of ith degree in $|\bar{\mathbf{r}}|/R$, and $\mathbf{f}^{(ij)}$ is a collection of terms in the product $(|\mathbf{r}|/R)^i(|\bar{\mathbf{r}}|/R)^j$, with $\mathbf{r}$ and $\bar{\mathbf{r}}$ position vectors of generic points of B and $\bar{B}$ relative to B^* and $\bar{B}^*$, respectively. In particular,

$$\mathbf{f}^{(2)} \triangleq \frac{1}{mR^2}\left\{\frac{3}{2}[\text{tr}(\mathbf{I}) - 5\mathbf{a}_1 \cdot \mathbf{I} \cdot \mathbf{a}_1]\mathbf{a}_1 + 3\mathbf{I} \cdot \mathbf{a}_1\right\} \tag{2}$$

and

$$\bar{\mathbf{f}}^{(2)} \triangleq \frac{1}{\bar{m}R^2}\left\{\frac{3}{2}[\text{tr}(\bar{\mathbf{I}}) - 5\mathbf{a}_1 \cdot \bar{\mathbf{I}} \cdot \mathbf{a}_1]\mathbf{a}_1 + 3\bar{\mathbf{I}} \cdot \mathbf{a}_1\right\} \tag{3}$$

where $\mathbf{I}$ and $\bar{\mathbf{I}}$ are the inertia dyadics of B for B^* and of $\bar{B}$ for $\bar{B}^*$, respectively.

If $\mathbf{a}_2$ and $\mathbf{a}_3$ are defined so as to establish a dextral, orthogonal set of unit vectors $\mathbf{a}_1$, $\mathbf{a}_2$, $\mathbf{a}_3$, then $\mathbf{f}^{(2)}$ and $\bar{\mathbf{f}}^{(2)}$ can be expressed in terms of these unit vectors and the moments and products of inertia of B and $\bar{B}$ for axes parallel to $\mathbf{a}_1$, $\mathbf{a}_2$, and $\mathbf{a}_3$ and passing through the mass centers of the individual bodies. To this end, I_{jk} and $\bar{I}_{jk}$ are defined as

$$I_{jk} \triangleq \mathbf{a}_j \cdot \mathbf{I} \cdot \mathbf{a}_k \qquad (j, k = 1, 2, 3) \tag{4}$$

and

$$\bar{I}_{jk} \triangleq \mathbf{a}_j \cdot \bar{\mathbf{I}} \cdot \mathbf{a}_k \qquad (j, k = 1, 2, 3) \tag{5}$$

after which $\mathbf{f}^{(2)}$ and $\bar{\mathbf{f}}^{(2)}$ may be written

$$\mathbf{f}^{(2)} = \frac{3}{mR^2}\left[\frac{1}{2}(I_{22} + I_{33} - 2I_{11})\mathbf{a}_1 + I_{21}\mathbf{a}_2 + I_{31}\mathbf{a}_3\right] \tag{6}$$

and
$$\bar{\mathbf{f}}^{(2)} = \frac{3}{\bar{m}R^2}\left[\frac{1}{2}(\bar{I}_{22} + \bar{I}_{33} - 2\bar{I}_{11})\mathbf{a}_1 + \bar{I}_{21}\mathbf{a}_2 + \bar{I}_{31}\mathbf{a}_3\right] \tag{7}$$

Alternatively, $\mathbf{f}^{(2)}$ and $\bar{\mathbf{f}}^{(2)}$ can be expressed in terms of principal moments of inertia of B for B^* and of $\bar{B}$ for $\bar{B}^*$. To accomplish this, two sets of dextral, orthogonal unit vectors, $\mathbf{b}_1$, $\mathbf{b}_2$, $\mathbf{b}_3$, and $\bar{\mathbf{b}}_1$, $\bar{\mathbf{b}}_2$, $\bar{\mathbf{b}}_3$, parallel to principal axes of inertia of B for B^* and of $\bar{B}$ for $\bar{B}^*$, respectively, are introduced, and I_j, $\bar{I}_j$, C_{ij}, and $\bar{C}_{ij}$ are defined as

$$I_j = \mathbf{b}_j \cdot \mathbf{I} \cdot \mathbf{b}_j \qquad (j = 1, 2, 3) \tag{8}$$

$$\bar{I}_j = \bar{\mathbf{b}}_j \cdot \bar{\mathbf{I}} \cdot \bar{\mathbf{b}}_j \qquad (j = 1, 2, 3) \tag{9}$$

and

$$C_{ij} \triangleq \mathbf{a}_i \cdot \mathbf{b}_j \qquad (i, j = 1, 2, 3) \tag{10}$$

$$\bar{C}_{ij} \triangleq \mathbf{a}_i \cdot \bar{\mathbf{b}}_j \qquad (i, j = 1, 2, 3) \tag{11}$$

Thus one obtains

$$\begin{aligned}\mathbf{f}^{(2)} = \frac{3}{mR^2}\Big\{&\frac{1}{2}[I_1(1 - 3C_{11}^2) + I_2(1 - 3C_{12}^2) + I_3(1 - 3C_{13}^2)]\mathbf{a}_1\\ &+ (I_1 C_{21} C_{11} + I_2 C_{22} C_{12} + I_3 C_{23} C_{13})\mathbf{a}_2\\ &+ (I_1 C_{31} C_{11} + I_2 C_{32} C_{12} + I_3 C_{33} C_{13})\mathbf{a}_3\Big\}\end{aligned} \tag{12}$$

and

$$\begin{aligned}\bar{\mathbf{f}}^{(2)} = \frac{3}{\bar{m}R^2}\Big\{&\frac{1}{2}[\bar{I}_1(1 - 3\bar{C}_{11}^2) + \bar{I}_2(1 - 3\bar{C}_{12}^2) + \bar{I}_3(1 - 3\bar{C}_{13}^2)]\mathbf{a}_1\\ &+ (\bar{I}_1 \bar{C}_{21} \bar{C}_{11} + \bar{I}_2 \bar{C}_{22} \bar{C}_{12} + \bar{I}_3 \bar{C}_{23} \bar{C}_{13})\mathbf{a}_2\\ &+ (\bar{I}_1 \bar{C}_{31} \bar{C}_{11} + \bar{I}_2 \bar{C}_{32} \bar{C}_{12} + \bar{I}_3 \bar{C}_{33} \bar{C}_{13})\mathbf{a}_3\Big\}\end{aligned} \tag{13}$$

A useful approximation to $\mathbf{F}$ in Eq. (1) may be obtained by defining $\tilde{\mathbf{F}}$ such that

$$\mathbf{F} \approx \tilde{\mathbf{F}} \triangleq -\frac{G\bar{m}m}{R^2}[\mathbf{a}_1 + \mathbf{f}^{(2)} + \bar{\mathbf{f}}^{(2)}] \tag{14}$$

with $\mathbf{f}^{(2)}$ and $\bar{\mathbf{f}}^{(2)}$ given by either Eqs. (2) and (3), or Eqs. (6) and (7), or Eqs. (12) and (13).

Derivations To establish the validity of Eqs. (1), (2), (3), (6), (7), (12), and (13), expressions given in Sec. 2.3 may be used to represent the force exerted on B by a differential element of $\bar{B}$ located at a generic point $\bar{P}$ of $\bar{B}$ (see Fig. 2.4.1), and the total force $\mathbf{F}$ applied to B by $\bar{B}$ then can be obtained by integrating over the figure of $\bar{B}$. Specifically, if $\bar{R}$ is the distance between $\bar{P}$

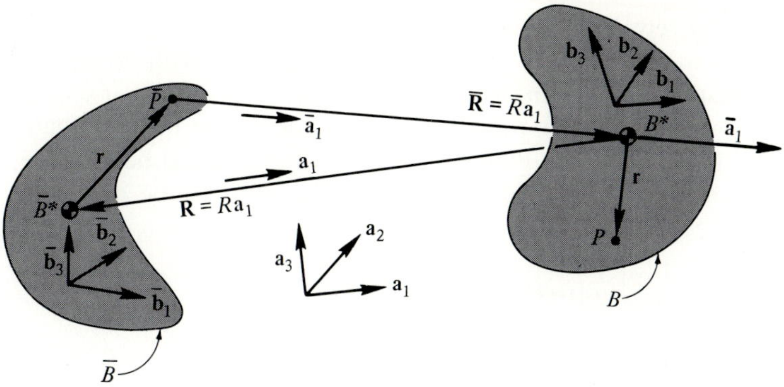

Figure 2.4.1

and B^*, $\bar{\mathbf{a}}_1$ is a unit vector directed from $\bar{P}$ toward B^*, $\bar{\rho}$ is the mass density of $\bar{B}$ at $\bar{P}$, and $d\bar{\tau}$ is the volume of a differential element of $\bar{B}$ at $\bar{P}$, then, from Eqs. (2.3.1)–(2.3.3) with R, $\mathbf{a}_1$, and m replaced by $\bar{R}$, $\bar{\mathbf{a}}_1$, and $\bar{\rho}\, d\bar{\tau}$, respectively,

$$\mathbf{F} = -\int \frac{Gm}{\bar{R}^2}\left(\bar{\mathbf{a}}_1 + \frac{1}{m\bar{R}^2}\left\{\frac{3}{2}[\text{tr}(\mathbf{I}) - 5\bar{\mathbf{a}}_1 \cdot \mathbf{I} \cdot \bar{\mathbf{a}}_1]\bar{\mathbf{a}}_1 + 3\mathbf{I} \cdot \bar{\mathbf{a}}_1\right\} + \cdots\right)\bar{\rho}\, d\bar{\tau} \tag{15}$$

where the three dots represent terms of third or higher degree in $|\mathbf{r}|/R$. Now, if $\bar{\mathbf{R}}$ denotes the position vector of B^* relative to $\bar{P}$, and $\mathbf{R}$ the position vector from B^* to $\bar{B}^*$, then (see Fig. 2.4.1)

$$\frac{\bar{\mathbf{a}}_1}{\bar{R}^2} = \bar{\mathbf{R}}(\bar{\mathbf{R}}^2)^{-3/2} = -(\mathbf{R} + \bar{\mathbf{r}})(\mathbf{R}^2 + 2\mathbf{R} \cdot \bar{\mathbf{r}} + \bar{\mathbf{r}}^2)^{-3/2} \tag{16}$$

Hence,

$$\mathbf{F} \underset{(15,16)}{=} Gm\int(\mathbf{R} + \bar{\mathbf{r}})(\mathbf{R}^2 + 2\mathbf{R} \cdot \bar{\mathbf{r}} + \bar{\mathbf{r}}^2)^{-3/2}\bar{\rho}\, d\bar{\tau}$$
$$- G\int\left(\frac{1}{\bar{R}^4}\left\{\frac{3}{2}[\text{tr}(\mathbf{I}) - 5\bar{\mathbf{a}}_1 \cdot \mathbf{I} \cdot \bar{\mathbf{a}}_1]\bar{\mathbf{a}}_1 + 3\mathbf{I} \cdot \bar{\mathbf{a}}_1\right\} + \cdots\right)\bar{\rho}\, d\bar{\tau} \tag{17}$$

The first integral in this equation has precisely the same form as the integral in Eq. (2.3.12). Consequently, proceeding as before, one obtains results analogous to Eqs. (2.3.26), namely,

$$Gm\int(\mathbf{R} + \bar{\mathbf{r}})(\mathbf{R}^2 + 2\mathbf{R} \cdot \bar{\mathbf{r}} + \bar{\mathbf{r}}^2)^{-3/2}\bar{\rho}\, d\bar{\tau}$$
$$= -\frac{G\bar{m}m}{R^2}\left(\mathbf{a}_1 + \frac{1}{\bar{m}R^2}\left\{\frac{3}{2}[\text{tr}(\bar{\mathbf{I}}) - 5\mathbf{a}_1 \cdot \bar{\mathbf{I}} \cdot \mathbf{a}_1]\mathbf{a}_1 + 3\bar{\mathbf{I}} \cdot \mathbf{a}_1 + \cdots\right\}\right) \tag{18}$$

where the three dots represent terms of third or higher degree in $|\bar{\mathbf{r}}|/R$. In the second integral in Eq. (17), $\bar{R}$ and $\bar{\mathbf{a}}_1$ may be replaced with R and $\mathbf{a}_1$, respectively, because every term in the integrand involves quantities of second or higher degree in $|\mathbf{r}|$, so that no terms of interest for the purposes at hand are lost through this replacement; and, once the replacement has been made, the portion of the integral displayed explicitly in Eq. (17) can be evaluated readily. Thus, one obtains

$$\mathbf{F} \underset{(17,18)}{=} -\frac{G\bar{m}m}{R^2}\left(\mathbf{a}_1 + \frac{1}{\bar{m}R^2}\left\{\frac{3}{2}[\mathrm{tr}(\bar{\mathbf{I}}) - 5\mathbf{a}_1 \cdot \bar{\mathbf{I}} \cdot \mathbf{a}_1]\mathbf{a}_1 + 3\bar{\mathbf{I}} \cdot \mathbf{a}_1 + \cdots\right\}\right.$$
$$\left. + \frac{1}{mR^2}\left\{\frac{3}{2}[\mathrm{tr}(\mathbf{I}) - 5\mathbf{a}_1 \cdot \mathbf{I} \cdot \mathbf{a}_1]\mathbf{a}_1 + 3\mathbf{I} \cdot \mathbf{a}_1 + \cdots\right\}\right) \tag{19}$$

where the three dots now represent terms of third or higher degree either in $|\mathbf{r}|/R$ or in $|\bar{\mathbf{r}}|/R$ or terms involving the product $(|\mathbf{r}|/R)^i(|\bar{\mathbf{r}}|/R)^j$, with neither i nor j equal to unity, because $\mathbf{r}$ and $\bar{\mathbf{r}}$ are drawn from the mass centers of B and $\bar{B}$, respectively. Equation (1) now follows directly from Eq. (19) together with the definitions in Eqs. (2) and (3).

The relationship between Eqs. (2), (6), and (12) is completely analogous to that between Eqs. (2.3.3), (2.3.10), and (2.3.11); similarly, for Eqs. (3), (7), and (13). Hence, to establish the validity of Eqs. (6), (7), and (13), one can proceed exactly as in the corresponding derivations in Sec. 2.3.

Example An approximate expression is required for the force exerted by a homogeneous, oblate spheroid $\bar{B}$ on a homogeneous rectangular parallelepiped B having the dimensions shown in Fig. 2.4.2. The approximation to $\mathbf{F}$ denoted by $\tilde{\mathbf{F}}$ in Eq. (14) is to be employed, and the three additive terms in this expression are to be compared with each other. For purposes of numerical comparison, the values $\bar{\alpha} = 6.38 \times 10^6$ m, $\bar{\beta} = 6.36 \times 10^6$ m, $\alpha = 16$ m, $\beta < \alpha$, and $R = 6.6\bar{\alpha} = 4.2108 \times 10^7$ m are to be used. The system then roughly approximates that of the Earth ($\bar{B}$) and a large, synchronous-altitude† artificial satellite (B).

Equations (12) and (13) provide a convenient point of departure for the required comparisons. The principal moments of inertia appearing in these equations are given by

$$I_1 = \frac{m}{12}(\alpha^2 + \alpha^2) \tag{20}$$

$$I_2 = I_3 = \frac{m}{12}(\alpha^2 + \beta^2) \tag{21}$$

† A satellite in a circular, equatorial orbit at synchronous altitude remains above a point fixed on the Earth.

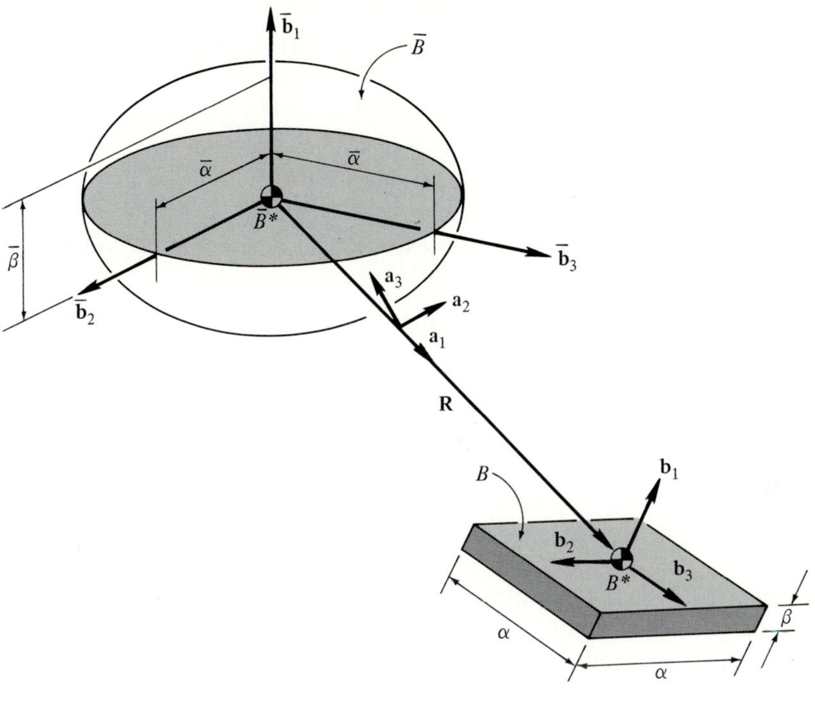

Figure 2.4.2

for the rectangular parallelepiped B, and by

$$\bar{I}_1 = \frac{\bar{m}}{5}(\bar{\alpha}^2 + \bar{\alpha}^2) \tag{22}$$

$$\bar{I}_2 = \bar{I}_3 = \frac{\bar{m}}{5}(\bar{\alpha}^2 + \bar{\beta}^2) \tag{23}$$

for the spheroid $\bar{B}$. Later work is facilitated by eliminating β and $\bar{\beta}$ through the introduction of the eccentricity $\bar{\epsilon}$ of the spheroid $\bar{B}$ with the substitution

$$\bar{\beta}^2 = \bar{\alpha}^2(1 - \bar{\epsilon}^2) \tag{24}$$

and a similar quantity ϵ for body B with the substitution

$$\beta^2 = \alpha^2(1 - \epsilon^2) \tag{25}$$

The vector $\mathbf{f}^{(2)}$ then becomes

$$\mathbf{f}^{(2)} \underset{(12)}{=} \frac{\alpha^2}{4R^2}\left\{\left[3 - 3(C_{11}{}^2 + C_{12}{}^2 + C_{13}{}^2) - \frac{\epsilon^2}{2}(2 - 3C_{12}{}^2 - 3C_{13}{}^2)\right]\mathbf{a}_1\right.$$

$$+ 2\left[C_{21}C_{11} + C_{22}C_{12} + C_{23}C_{13} - \frac{\epsilon^2}{2}(C_{22}C_{12} + C_{23}C_{13})\right]\mathbf{a}_2$$

(*continued on next page*)

$$+ 2\left[C_{31}C_{11} + C_{32}C_{12} + C_{33}C_{13} - \frac{\epsilon^2}{2}(C_{32}C_{12} + C_{33}C_{13})\right]\mathbf{a}_3\Big\} \tag{26}$$

or, after simplification by reference to Eq. (1.2.14),

$$\mathbf{f}^{(2)} \underset{(26)}{=} \frac{\alpha^2\epsilon^2}{4R^2}\left[\frac{1}{2}(1 - 3C_{11}^{\ 2})\mathbf{a}_1 + C_{21}C_{11}\mathbf{a}_2 + C_{31}C_{11}\mathbf{a}_3\right] \tag{27}$$

By using Eq. (1.2.15), one can now express the magnitude $|\mathbf{f}^{(2)}|$ of $\mathbf{f}^{(2)}$ as

$$|\mathbf{f}^{(2)}| \underset{(27)}{=} \frac{\alpha^2\epsilon^2}{8R^2}(1 - 2C_{11}^{\ 2} + 5C_{11}^{\ 4})^{1/2} \tag{28}$$

Similarly,

$$\bar{\mathbf{f}}^{(2)} = \frac{3\bar{\alpha}^2\bar{\epsilon}^2}{5R^2}\left[\frac{1}{2}(1 - 3\bar{C}_{11}^{\ 2})\mathbf{a}_1 + \bar{C}_{21}\bar{C}_{11}\mathbf{a}_2 + \bar{C}_{31}\bar{C}_{11}\mathbf{a}_3\right] \tag{29}$$

and,

$$|\bar{\mathbf{f}}^{(2)}| = \frac{3\bar{\alpha}^2\bar{\epsilon}^2}{10R^2}(1 - 2\bar{C}_{11}^{\ 2} + 5\bar{C}_{11}^{\ 4})^{1/2} \tag{30}$$

Figure 2.4.3, showing a plot of the function $(1 - 2x^2 + 5x^4)^{1/2}$ versus x in the range $-1 \le x \le 1$, can be used to find extremals of $|\mathbf{f}^{(2)}|$ and $|\bar{\mathbf{f}}^{(2)}|$ by substituting C_{11} and $\bar{C}_{11}$ for x. By reference to Eqs. (28) and (30) it can thus be concluded that

$$\frac{\alpha^2\epsilon^2}{4\sqrt{5}R^2} \le |\mathbf{f}^{(2)}| \le \frac{\alpha^2\epsilon^2}{4R^2} \tag{31}$$

and

$$\frac{3\bar{\alpha}^2\bar{\epsilon}^2}{5\sqrt{5}R^2} \le |\bar{\mathbf{f}}^{(2)}| \le \frac{3\bar{\alpha}^2\bar{\epsilon}^2}{5R^2} \tag{32}$$

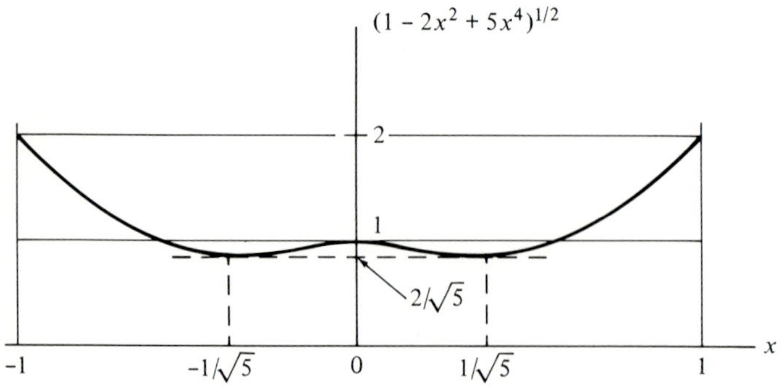

Figure 2.4.3

and substitution of the given numerical values then yields

$$\frac{|\mathbf{f}^{(2)}|_{\max}}{|\bar{\mathbf{f}}^{(2)}|_{\min}} = \frac{5\sqrt{5}}{12}\frac{\alpha^2\epsilon^2}{\bar{\alpha}^2\bar{\epsilon}^2} \approx \frac{5\sqrt{5}(16)^2\epsilon^2}{12(6.38)^2 \times 10^{12}(0.006)} \approx 10^{-9}\epsilon^2 < 10^{-9} \tag{33}$$

and

$$\frac{|\bar{\mathbf{f}}^{(2)}|_{\max}}{|\mathbf{a}_1|} = \frac{3\bar{\alpha}^2\bar{\epsilon}^2}{5R^2} \approx \frac{3(0.006)}{5(6.6)^2} \approx 10^{-4} \tag{34}$$

Equation (34) shows that, for a large, synchronous-altitude satellite of the Earth, the leading term in the series for $\mathbf{F}$ in Eq. (1) has a magnitude far exceeding that of $\bar{\mathbf{f}}^{(2)}$, a vector which reflects the oblateness of the Earth through the presence of $\bar{\epsilon}$ in Eq. (30); and [see Eq. (33)] $|\mathbf{f}^{(2)}|$ is much smaller than even $|\bar{\mathbf{f}}^{(2)}|$.

It is noteworthy that $\mathbf{f}^{(2)}$ [see Eq. (27)] depends on C_{11}, C_{21}, and C_{31}, and hence on the orientation of the satellite relative to $\mathbf{a}_1$, $\mathbf{a}_2$, and $\mathbf{a}_3$, whereas $\bar{\mathbf{f}}^{(2)}$ [see Eq. (29)] involving $\bar{C}_{11}$, $\bar{C}_{21}$, and $\bar{C}_{31}$ depends on the orientation of the satellite's orbital plane relative to the spheroid. For an equatorial orbit, $\bar{\mathbf{f}}^{(2)}$ reduces to

$$\bar{\mathbf{f}}^{(2)} \underset{(29)}{=} -\frac{3\bar{\alpha}^2\bar{\epsilon}^2}{5R^2}\mathbf{a}_1 \tag{35}$$

and the addition of this vector to the term $\mathbf{a}_1$ in Eq. (14) is of little consequence [see Eq. (34)]. By way of contrast, if B^* moves in a non-equatorial orbit, oblateness effects may be significant despite the fact that $\bar{\mathbf{f}}^{(2)}$ is small in comparison with $\mathbf{a}_1$, because $\bar{\mathbf{f}}^{(2)}$ then has components perpendicular to $\mathbf{a}_1$.

2.5 CENTROBARIC BODIES

As noted in Sec. 2.2, the center of gravity B' of a body B for an attracting particle $\bar{P}$ does not in general coincide with the mass center B^* of B. However, there exist bodies for which the center of gravity and center of mass necessarily coincide. Such bodies are called *centrobaric*.† Thus, a body B of mass m is centrobaric if the force $\mathbf{F}$ exerted on B by every particle $\bar{P}$ of mass $\bar{m}$ is given by

$$\mathbf{F} = -\frac{G\bar{m}m}{R^2}\mathbf{a}_1 \tag{1}$$

where G is the universal gravitational constant, R is the distance between $\bar{P}$ and B^*, and $\mathbf{a}_1$ is a unit vector directed from $\bar{P}$ toward B^*.

† In the classical literature, the term *center of gravity* is sometimes so defined that *only* a centrobaric body has a center of gravity; then the center of mass coincides with the center of gravity whenever the latter exists.

Centrobaric bodies, which possess a variety of shapes and mass distributions, have the following property: a centrobaric body has the same moment of inertia about every line passing through its mass center. In other words, the central inertia ellipsoid of a centrobaric body is a sphere. However, not every body possessing this property is centrobaric.

Derivations The existence of centrobaric bodies is most easily established by citing a specific case (see the example, in which a solid sphere is shown to be centrobaric if its mass density at any point is a function only of the distance from the point to the center of the sphere).

To prove that a centrobaric body B has the same moment of inertia about every line passing through its mass center B^*, it is sufficient to show that the quantity $\mathbf{a}_1 \cdot \mathbf{I} \cdot \mathbf{a}_1$ has a value independent of $\mathbf{a}_1$, where $\mathbf{I}$ is the inertia dyadic of B for B^*. Now, taken in conjunction, Eq. (1) and the series expansion for $\mathbf{F}$ given in Eq. (2.3.1) imply

$$\sum_{i=2}^{\infty} \mathbf{f}^{(i)} = 0 \tag{2}$$

The terms in this summation are independent of each other in the sense that $\mathbf{f}^{(2)}$ is proportional to R^{-2}, $\mathbf{f}^{(3)}$ is proportional to R^{-3}, and so forth. It follows that they must vanish separately, that is, that

$$\mathbf{f}^{(i)} = 0 \qquad (i = 2, \ldots, \infty) \tag{3}$$

Hence,

$$\mathbf{a}_1 \cdot \mathbf{f}^{(2)} = 0 \tag{4}$$

and, using the expression for $\mathbf{f}^{(2)}$ given in Eq. (2.3.3), one finds that

$$\mathbf{a}_1 \cdot \mathbf{I} \cdot \mathbf{a}_1 = \tfrac{1}{3}\,\mathrm{tr}(\mathbf{I}) \tag{5}$$

Since $\mathrm{tr}(\mathbf{I})$ is an invariant, it follows that the moment of inertia of B about the line that passes through B^* and is parallel to $\mathbf{a}_1$ has the same value for all orientations of $\mathbf{a}_1$ relative to B.

Example To show that a solid sphere S of mass m is centrobaric if the mass density ρ at a point P depends only on the distance r between P and the center S^* of S, the force $\mathbf{F}$ exerted on S by a particle $\bar{P}$ of mass $\bar{m}$ will be calculated.

The position vector $\mathbf{p}$ of P relative to $\bar{P}$, expressed in terms of the spherical polar coordinates r, θ, ψ, the distance R between $\bar{P}$ and S^*, and the unit vectors $\mathbf{a}_1$, $\mathbf{a}_2$, $\mathbf{a}_3$ shown in Fig. 2.5.1, is given by

$$\mathbf{p} = (R + r\cos\psi)\mathbf{a}_1 + r\sin\psi\sin\theta\,\mathbf{a}_2 + r\sin\psi\cos\theta\,\mathbf{a}_3 \tag{6}$$

The volume $d\tau$ of a differential element of S is

$$d\tau = r^2 \sin\psi\, dr\, d\theta\, d\psi \tag{7}$$

and, with $\rho = \rho(r)$, the mass m of S can be expressed as

$$m = \int \rho \, d\tau = \int_0^a r^2 \rho \int_0^\pi \sin\psi \int_0^{2\pi} d\theta \, d\psi \, dr \tag{8}$$

or

$$m = 4\pi \int_0^a r^2 \rho \, dr \tag{9}$$

where a is the radius of S.

The force $\mathbf{F}$ exerted by $\overline{P}$ on S is given by

$$\mathbf{F} \underset{(2.2.2)}{=} -G\overline{m} \int \mathbf{p}(\mathbf{p}^2)^{-3/2} \rho \, d\tau \tag{10}$$

or, in somewhat more explicit terms, by

$$\mathbf{F} \underset{(10,6,7)}{=} -G\overline{m} \int_0^a r^2 \rho \int_0^\pi \sin\psi (R^2 + 2Rr\cos\psi + r^2)^{-3/2} \int_0^{2\pi} \mathbf{p} \, d\theta \, d\psi \, dr \tag{11}$$

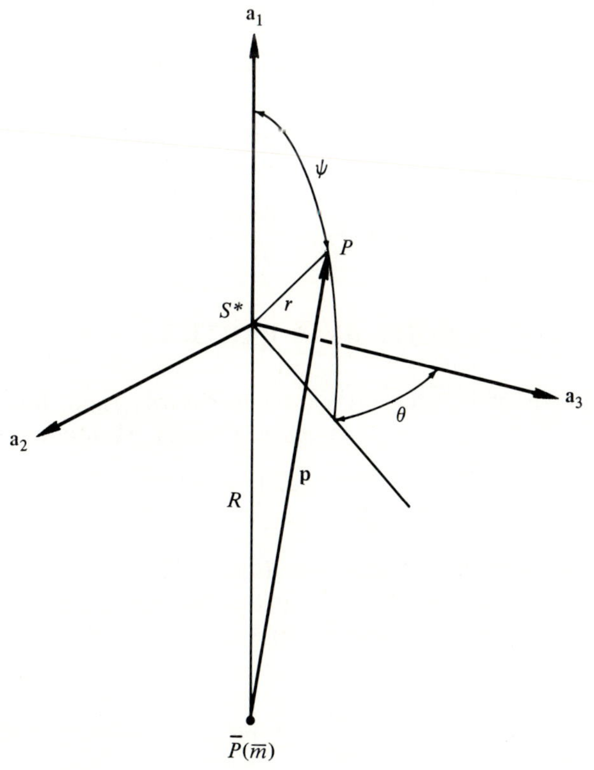

Figure 2.5.1

Using Eq. (6), one obtains the innermost integral in Eq. (11),

$$\int_0^{2\pi} [(R + r\cos\psi)\mathbf{a}_1 + r\sin\psi\sin\theta\,\mathbf{a}_2 + r\sin\psi\cos\theta\,\mathbf{a}_3]\,d\theta = 2\pi(R + r\cos\psi)\mathbf{a}_1 \qquad (12)$$

so that

$$\mathbf{F} \underset{(11,12)}{=} -2\pi G\bar{m}\int_0^a r^2\rho\int_0^\pi \sin\psi(R^2 + 2Rr\cos\psi + r^2)^{-3/2}(R + r\cos\psi)\,d\psi\,dr\,\mathbf{a}_1 \qquad (13)$$

Integration is now facilitated by the introduction of a new variable v defined such that

$$v^2 \triangleq R^2 + 2Rr\cos\psi + r^2 \qquad (14)$$

which implies

$$\sin\psi\,d\psi \underset{(14)}{=} -\frac{v\,dv}{Rr} \qquad (15)$$

and

$$R + r\cos\psi \underset{(14)}{=} \frac{v^2 + R^2 - r^2}{2R} \qquad (16)$$

so that substitution into Eq. (13) leads to

$$\mathbf{F} = \frac{\pi G\bar{m}}{R^2}\int_0^a r\rho\int_{R+r}^{R-r}\frac{v^2 + R^2 - r^2}{v^2}\,dv\,dr\,\mathbf{a}_1$$

$$= -\frac{4\pi G\bar{m}}{R^2}\int_0^a r^2\rho\,dr\,\mathbf{a}_1 \underset{(9)}{=} -\frac{G\bar{m}m}{R^2}\mathbf{a}_1 \qquad (17)$$

2.6 MOMENT EXERTED ON A BODY BY A PARTICLE

The system of gravitational forces exerted on a (not necessarily rigid) body B of mass m by a particle $\bar{P}$ of mass $\bar{m}$ produces a moment $\mathbf{M}$ about the mass center B^* of B. $\mathbf{M}$ is given by

$$\mathbf{M} = -\mathbf{R} \times \mathbf{F} \qquad (1)$$

where $\mathbf{F}$ is the resultant of the system of forces [see Eqs. (2.2.1), (2.2.2), and (2.3.1)] and $\mathbf{R}$ is the position vector from $\bar{P}$ to B^*. If the distance R between $\bar{P}$ and B^* exceeds the greatest distance from B^* to any point P of B, this moment can be expressed as

$$\mathbf{M} = \frac{3G\bar{m}}{R^3}\mathbf{a}_1 \times \mathbf{I}\cdot\mathbf{a}_1 + \frac{G\bar{m}m}{R}\sum_{i=3}^{\infty}\mathbf{m}^{(i)} \qquad (2)$$

where $\mathbf{a}_1$ is a unit vector directed from $\bar{P}$ toward B^*, G is the universal gravitational constant, $\mathbf{I}$ is the inertia dyadic of B for B^*, and the dimensionless vector $\mathbf{m}^{(i)}$ is a collection of terms of ith degree in $|\mathbf{r}|/R$, with $\mathbf{r}$ the position vector of a typical point P of B relative to B^*.

Equation (2) suggests the approximation

$$\mathbf{M} \approx \widetilde{\mathbf{M}} \triangleq \frac{3G\overline{m}}{R^3}\, \mathbf{a}_1 \times \mathbf{I} \cdot \mathbf{a}_1 \tag{3}$$

Expressed in terms of scalars $I_{jk}(j, k = 1, 2, 3)$ defined as

$$I_{jk} \triangleq \mathbf{a}_j \cdot \mathbf{I} \cdot \mathbf{a}_k \qquad (j, k = 1, 2, 3) \tag{4}$$

the vector $\widetilde{\mathbf{M}}$ is given by

$$\widetilde{\mathbf{M}} = \frac{3G\overline{m}}{R^3}\,(I_{21}\,\mathbf{a}_3 - I_{31}\,\mathbf{a}_2) \tag{5}$$

Alternatively, one can introduce a dextral set of orthogonal unit vectors $\mathbf{b}_1$, $\mathbf{b}_2$, $\mathbf{b}_3$ parallel to the principal axes of inertia of B for B^*, and express $\widetilde{\mathbf{M}}$ in terms of the principal moments of inertia I_1, I_2, I_3 of B for B^* and the direction cosines C_{ij} $(i, j = 1, 2, 3)$ defined, respectively, as

$$I_j \triangleq \mathbf{b}_j \cdot \mathbf{I} \cdot \mathbf{b}_j \qquad (j = 1, 2, 3) \tag{6}$$

and

$$C_{ij} \triangleq \mathbf{a}_i \cdot \mathbf{b}_j \qquad (i, j = 1, 2, 3) \tag{7}$$

thus obtaining

$$\widetilde{\mathbf{M}} = \frac{3G\overline{m}}{R^3}\,[\mathbf{b}_1(I_3 - I_2)C_{12}C_{13} + \mathbf{b}_2(I_1 - I_3)C_{13}C_{11} + \mathbf{b}_3(I_2 - I_1)C_{11}C_{12}] \tag{8}$$

Equation (5), despite its apparent simplicity, is less useful than Eq. (8), both because the products of inertia I_{21} and I_{31} vary with the orientation of $\mathbf{a}_1$ relative to B and because the rotational equations of dynamics are generally most easily formulated in terms of the vector basis $\mathbf{b}_1$, $\mathbf{b}_2$, $\mathbf{b}_3$.

It should be noted that $\widetilde{\mathbf{M}}$, which by Eq. (3) is the first term in Eq. (2), can vanish when $\mathbf{M}$ does not vanish. Hence, when using $\widetilde{\mathbf{M}}$ as an approximation to $\mathbf{M}$ one is *not necessarily* retaining the largest term in the series expansion for $\mathbf{M}$. Specifically, $\widetilde{\mathbf{M}}$ vanishes whenever $\mathbf{a}_1$ is parallel to a central principal axis of inertia of B, but this state of affairs need not produce a zero value for $\mathbf{M}$ (see Prob. 2.14). Moreover, $\widetilde{\mathbf{M}}$ is identically zero for any body with a spherical central inertia ellipsoid, whereas $\mathbf{M}$ is identically zero only if the body is also centrobaric (see Sec. 2.5 and Prob. 2.13).

Derivations If B is a continuous distribution of matter, then the resultant $\mathbf{F}$ of the gravitational forces exerted on B by $\overline{P}$ can be expressed in terms of the force $d\mathbf{F}$ exerted by $\overline{P}$ on a differential element of B at a generic point P of B as

$$\mathbf{F} = \int d\mathbf{F} \tag{9}$$

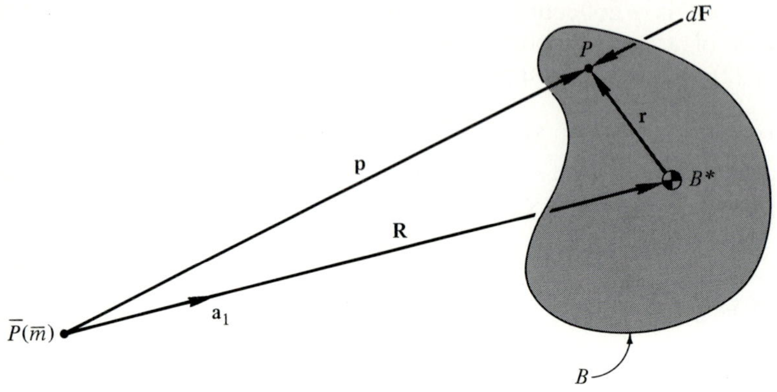

Figure 2.6.1

and, if **r** is the position vector of P relative to the mass center B^* of B, then the moment **M** of the system of gravitational forces about B^* is

$$\mathbf{M} = \int \mathbf{r} \times d\mathbf{F} \tag{10}$$

Substituting

$$\mathbf{r} = -\mathbf{R} + \mathbf{p} \tag{11}$$

where **R** is the position vector of B^* relative to $\bar{P}$ and **p** is the position vector of P relative to $\bar{P}$ (see Fig. 2.6.1), yields

$$\mathbf{M} = \int (-\mathbf{R} + \mathbf{p}) \times d\mathbf{F} \underset{(9)}{=} -\mathbf{R} \times \mathbf{F} + \int \mathbf{p} \times d\mathbf{F} \tag{12}$$

But $d\mathbf{F}$ is parallel to **p**, so that $\mathbf{p} \times d\mathbf{F} = 0$ and

$$\mathbf{M} \underset{(12)}{=} -\mathbf{R} \times \mathbf{F} \tag{13}$$

in agreement with Eq. (1). Alternatively, this result can be inferred from Sec. 2.2 and from the fact that two equivalent systems of forces have equal moments about every point. Hence it applies also when B consists of a finite number of particles, $P_1, \ldots, P_N$.

The validity of Eq. (2) is to be established subject to the same restrictions on R and on the dimensions of B as apply to the expression for **F** in Eq. (2.3.1). Under these circumstances, substitution from Eqs. (2.3.1) and (2.3.2) into Eq. (13) yields

$$\mathbf{M} = \frac{G\bar{m}m}{R}\,\mathbf{a}_1 \times \sum_{i=2}^{\infty} \mathbf{f}^{(i)} = \frac{G\bar{m}m}{R}\left[\mathbf{a}_1 \times \mathbf{f}^{(2)} + \sum_{i=3}^{\infty} \mathbf{a}_1 \times \mathbf{f}^{(i)}\right] \tag{14}$$

and use of Eqs. (2.3.3) and (2.3.4) gives

$$\mathbf{M} = \frac{3G\overline{m}}{R^3}\,\mathbf{a}_1 \times \mathbf{I} \cdot \mathbf{a}_1 + \frac{G\overline{m}m}{R}\sum_{i=3}^{\infty} \mathbf{a}_1 \times \mathbf{f}^{(i)} \tag{15}$$

which is equivalent to Eq. (2) if $\mathbf{m}^{(i)}$ is defined as

$$\mathbf{m}^{(i)} \triangleq \mathbf{a}_1 \times \mathbf{f}^{(i)} \qquad (i = 3, \ldots, \infty) \tag{16}$$

Substitution into Eq. (3) of

$$\mathbf{a}_1 \underset{(1.2.2)}{=} C_{11}\mathbf{b}_1 + C_{12}\mathbf{b}_2 + C_{13}\mathbf{b}_3 \tag{17}$$

and

$$\mathbf{I} \underset{(4)}{=} \sum_{j=1}^{3}\sum_{k=1}^{3} I_{jk}\,\mathbf{a}_j\mathbf{a}_k \tag{18}$$

leads to Eq. (5), and use of

$$\mathbf{I} \underset{(6)}{=} I_1\mathbf{b}_1\mathbf{b}_1 + I_2\mathbf{b}_2\mathbf{b}_2 + I_3\mathbf{b}_3\mathbf{b}_3 \tag{19}$$

in place of Eq. (18) produces Eq. (8).

Example The vector $\widetilde{\mathbf{M}}$ defined in Eq. (3) is to be used to approximate the moment $\mathbf{M}$ exerted by a particle $\overline{P}$ of mass $\overline{m}$ about the mass center B^* of a homogeneous, right circular cylinder B having moments of inertia J and I, respectively, about the symmetry axis and about any line through B^* and normal to the symmetry axis. The result is to be expressed in terms of two angles, ϕ and θ, used to specify the orientation of the symmetry axis of B relative to a dextral, orthogonal set of unit vectors $\mathbf{a}_1$, $\mathbf{a}_2$, $\mathbf{a}_3$. Specifically, as shown in Fig. 2.6.2, θ is the angle between $\mathbf{a}_3$ and the symmetry axis of B, and ϕ is the angle between $\mathbf{a}_2$ and the intersection of the plane P passing through B^* and normal to $\mathbf{a}_3$ with the plane Q determined by the symmetry axis of B and a line passing through B^* and parallel to $\mathbf{a}_3$.

Of the alternative expressions for $\widetilde{\mathbf{M}}$ given in Eqs. (3), (5) and (8), the last is the most convenient, since suitable unit vectors $\mathbf{b}_1$, $\mathbf{b}_2$, and $\mathbf{b}_3$ can be introduced readily, for example, as shown in Fig. 2.6.2, where $\mathbf{b}_1$ is normal to plane Q. Using Eq. (7), one thus finds

$$C_{11} = \cos\phi \qquad C_{12} = -\cos\theta\sin\phi \qquad C_{13} = \sin\theta\sin\phi \tag{20}$$

and, in accordance with Eq. (6), $I_1 = I_2 = I$, $I_3 = J$. Hence,

$$\widetilde{\mathbf{M}} \underset{(8)}{=} \frac{3G\overline{m}}{R^3}(I - J)\sin\theta\sin\phi(\mathbf{b}_1\cos\theta\sin\phi + \mathbf{b}_2\cos\phi) \tag{21}$$

If $I \neq J$, $\widetilde{\mathbf{M}}$ vanishes only when at least one of the following conditions is fulfilled: $\sin\phi = 0$, $\sin\theta = 0$, or $\cos\theta = \cos\phi = 0$. In all of these cases, the symmetry axis of B is either normal or parallel to the line joining $\overline{P}$

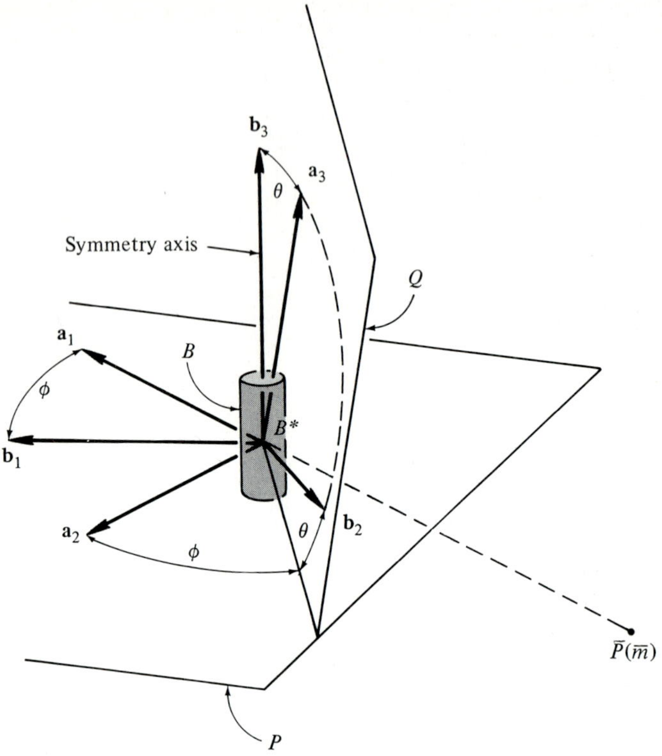

Figure 2.6.2

and B^*, and considerations of symmetry indicate that **M** also vanishes under these circumstances. Thus it appears that $\tilde{\mathbf{M}} = 0$ can imply $\mathbf{M} = 0$ for a particular body, although this implication is not valid in general.

2.7 MOMENT EXERTED ON A SMALL BODY BY A SMALL BODY

When the distance R between the mass centers B^* and $\bar{B}^*$ of two (not necessarily rigid) bodies B and $\bar{B}$ exceeds the greatest distance from the mass center of either body to any point of that body, the system of gravitational forces exerted on B by $\bar{B}$ produces a moment **M** about B^* which can be expressed as

$$\mathbf{M} = \frac{3G\bar{m}}{R^3}\,\mathbf{a}_1 \times \mathbf{I} \cdot \mathbf{a}_1 + \frac{G\bar{m}m}{R} \sum_{i=3}^{\infty} \mathbf{m}^{(i)} + \frac{G\bar{m}m}{R} \sum_{i=2}^{\infty} \sum_{j=2}^{\infty} \mathbf{m}^{(ij)} \qquad (1)$$

where $\mathbf{a}_1$ is a unit vector directed from $\bar{B}^*$ toward B^*, **I** is the central inertia dyadic of B, m and $\bar{m}$ are, respectively, the masses of B and $\bar{B}$, G is the universal gravitational constant, and the dimensionless vector $\mathbf{m}^{(i)}$ is a collection of terms of ith degree in $|\mathbf{r}|/R$ while the dimensionless vector $\mathbf{m}^{(ij)}$ is a collection of terms in the product $(|\mathbf{r}|/R)^i(|\bar{\mathbf{r}}|/R)^j$, with $\mathbf{r}$ and $\bar{\mathbf{r}}$ the position vectors of generic points of B and $\bar{B}$ relative to B^* and $\bar{B}^*$, respectively.

The similarity of Eqs. (1) and (2.6.2) suggests that an approximate relationship similar to Eq. (2.6.3), namely,

$$\mathbf{M} \approx \widetilde{\mathbf{M}} \triangleq \frac{3G\bar{m}}{R^3}\, \mathbf{a}_1 \times \mathbf{I} \cdot \mathbf{a}_1 \tag{2}$$

may prove useful. The vector $\widetilde{\mathbf{M}}$ thus defined can be expressed as

$$\widetilde{\mathbf{M}} = \frac{3G\bar{m}}{R^3}\,(I_{21}\,\mathbf{a}_3 - I_{31}\,\mathbf{a}_2) \tag{3}$$

or as

$$\widetilde{\mathbf{M}} = \frac{3G\bar{m}}{R^3}\,[\mathbf{b}_1(I_3 - I_2)C_{12}C_{13} + \mathbf{b}_2(I_1 - I_3)C_{13}C_{11} + \mathbf{b}_3(I_2 - I_1)C_{11}C_{12}] \tag{4}$$

where $\mathbf{a}_i$, $\mathbf{b}_i$, I_i, I_{ij}, and C_{ij} have the same meaning as in Sec. 2.6, and $\widetilde{\mathbf{M}}$, that is, the first term in Eq. (1), can be dominated by other terms in the series. A significant difference between Eq. (2.6.2) and Eq. (1) is that the former can be replaced with Eq. (2.6.1), whereas the latter cannot be so replaced, even if $\mathbf{R}$ and $\mathbf{F}$ are redefined, respectively, as the position vector of B^* relative to $\bar{B}^*$ and as the resultant force exerted on B by $\bar{B}$.

Derivations The validity of Eq. (1) can be established by using Eq. (2.6.2) to represent the moment exerted on B about B^* by a differential element of $\bar{B}$ located at a generic point $\bar{P}$ of $\bar{B}$ (see Fig. 2.7.1) and then integrating over

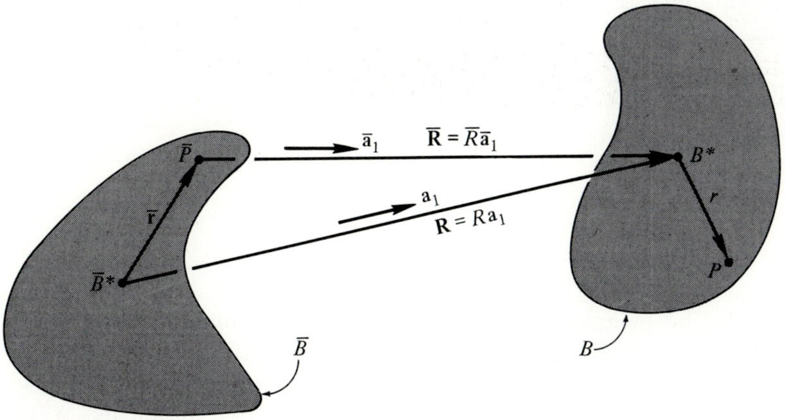

Figure 2.7.1

the figure of $\bar{B}$. Specifically, if $\bar{R}$ is the distance between $\bar{P}$ and B^*, $\bar{\mathbf{a}}_1$ is a unit vector directed from $\bar{P}$ toward B^*, $\bar{\rho}$ is the mass density of $\bar{B}$ at $\bar{P}$, and $d\bar{\tau}$ is the volume of a differential element of $\bar{B}$ at $\bar{P}$, then, from Eq. (2.6.2), with R, $\mathbf{a}_1$, and $\bar{m}$ replaced by $\bar{R}$, $\bar{\mathbf{a}}_1$, and $\bar{\rho}\, d\bar{\tau}$, respectively,

$$\mathbf{M} = \int \left(\frac{3G}{\bar{R}^3} \bar{\mathbf{a}}_1 \times \mathbf{I} \cdot \bar{\mathbf{a}}_1 + \cdots \right) \bar{\rho}\, d\bar{\tau} \tag{5}$$

where the three dots represent terms of third or higher degree in $|\mathbf{r}|/R$. Because $\mathbf{I}$ consists of terms of second degree in $|\mathbf{r}|/R$, the integration in Eq. (5) can produce only terms of second and higher degree in $|\mathbf{r}|/R$, and the substitution of $\mathbf{a}_1$ for $\bar{\mathbf{a}}_1$ and R for $\bar{R}$ cannot result in the loss of any terms of interest. These substitutions permit the first term in parentheses to be removed from the integrand, leaving as a factor an integral equal to the mass $\bar{m}$ of $\bar{B}$, in agreement with the first term in Eq. (1); and they yield the first series in Eq. (1). The second series in Eq. (1) reflects the deviations of $\mathbf{a}_1$ from $\bar{\mathbf{a}}_1$ and R from $\bar{R}$. Every term in this series involves $|\bar{\mathbf{r}}|/R$, and the series does not contain terms linear in $|\bar{\mathbf{r}}|/R$ because $\bar{\mathbf{r}}$ is drawn from the mass center of $\bar{B}$.

Equations (3) and (4) can be obtained from Eq. (2) by a procedure analogous to that used to derive Eqs. (2.6.5) and (2.6.8) from Eq. (2.6.3).

Example In the example of Sec. 2.4 (see Fig. 2.4.2) an approximate expression was developed for the force exerted by a homogeneous oblate spheroid $\bar{B}$ on a homogeneous rectangular parallelepiped B. Now an approximate expression is to be obtained for $\mathbf{M}$, the moment exerted by $\bar{B}$ on B about the mass center B^* of B. To this end, $\tilde{\mathbf{M}}$ as given by Eq. (4) is to be formed, and the magnitude of $\tilde{\mathbf{M}}$ is to be determined for $\bar{m} = 6 \times 10^{24}$ kg, $m = 160$ kg, $\alpha = 16$ m, $\beta = 4$ m, $R = 4.2108 \times 10^7$ m, and $G = 6.6732 \times 10^{-11}\ \text{N} \cdot \text{m}^2\ \text{kg}^{-2}$.

As in the example of Sec. 2.4, the system roughly approximates the Earth ($\bar{B}$) and a large, synchronous-altitude artificial satellite (B).

The symmetry of B permits the substitution of I_2 for I_3 in Eq. (4), furnishing

$$\tilde{\mathbf{M}} = \frac{3G\bar{m}}{R^3} (I_1 - I_2) C_{11} (C_{13} \mathbf{b}_2 - C_{12} \mathbf{b}_3) \tag{6}$$

from which

$$|\tilde{\mathbf{M}}| = \frac{3G\bar{m}}{R^3} |(I_1 - I_2) C_{11} (C_{13}{}^2 + C_{12}{}^2)^{1/2}|$$

$$\underset{(1.2.14)}{=} \frac{3G\bar{m}}{R^3} |(I_1 - I_2) C_{11} (1 - C_{11}{}^2)^{1/2}| \tag{7}$$

Variations in the magnitude of $\tilde{\mathbf{M}}$ resulting from changes in the relative orientation of B and $\bar{B}$ thus depend only on changes in the angle between $\mathbf{a}_1$ and $\mathbf{b}_1$, and $|\tilde{\mathbf{M}}|$ attains its maximum value when this angle is equal to $\pi/4$ rad, since the derivative of $|\tilde{\mathbf{M}}|$ with respect to C_{11} vanishes when C_{11} is $1/\sqrt{2}$. Thus, for any body B with $I_3 = I_2$,

$$|\tilde{\mathbf{M}}|_{\max} = \frac{3G\bar{m}}{2R^3}\,|I_1 - I_2| \tag{8}$$

and for the given rectangular parallelepiped, for which

$$I_1 = \frac{m\alpha^2}{6} \tag{9}$$

and

$$I_2 = I_3 = \frac{m}{12}\,(\alpha^2 + \beta^2) \tag{10}$$

$|\tilde{\mathbf{M}}|_{\max}$ becomes

$$|\tilde{\mathbf{M}}|_{\max} = \frac{G\bar{m}m(\alpha^2 - \beta^2)}{8R^3} \tag{11}$$

or, when the given numerical values are used,

$$|\tilde{\mathbf{M}}| \approx 2.6 \times 10^{-5}\ \text{N}\cdot\text{m} \tag{12}$$

2.8 PROXIMATE BODIES

When dealing with the gravitational interaction of two bodies in close proximity, one can at times make effective use of results which, at first glance, may appear to be inapplicable because they were encountered initially in connection with the analysis of the gravitational interaction of two widely separated bodies. For instance, consider Eq. (2.7.2), which furnishes an approximation that becomes ever better as the distance R between B^* and $\bar{B}^*$ (see Fig. 2.8.1) grows in comparison with the largest dimension of either body. Equation (2.7.2) can be useful also when B and $\bar{B}$ are near each other (see Fig. 2.8.2), provided $\bar{B}$ does not differ too much from a centrobaric body (see Sec. 2.5) and R is sufficiently large in comparison with the largest dimension of B, for $\bar{B}$ then acts nearly like a particle of mass $\bar{m}$ situated at the mass center $\bar{B}^*$ of $\bar{B}$; the resultant moment about B^* of the forces exerted on B

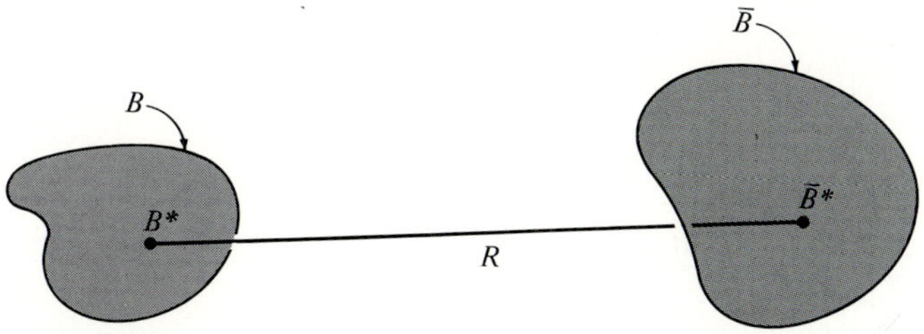

Figure 2.8.1

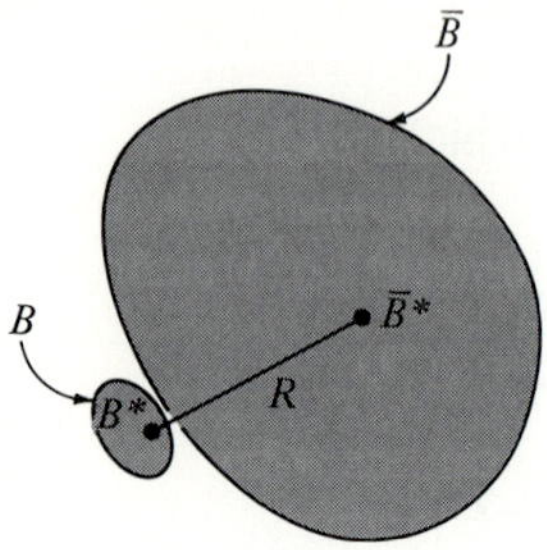

Figure 2.8.2

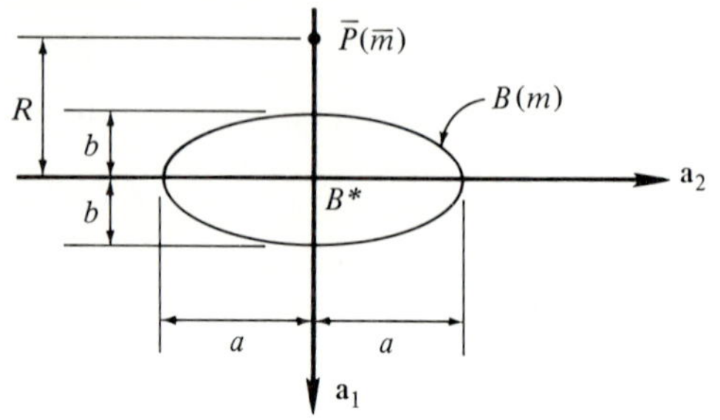

Figure 2.8.3

by such a particle is given to a good approximation by Eq. (2.6.3); and this equation is identical with Eq. (2.7.2).

Example Figure 2.8.3 shows a particle $\bar{P}$ of mass $\bar{m}$ situated on the axis of revolution of a uniform oblate spheroid B of mass m, at a distance R from the mass center B^* of B. To explore the utility of Eq. (2.3.6) in situations involving proximity of a particle and a body, the ratio $\tilde{F}/F$ is to be plotted versus β for various values of ϵ, where $\tilde{F}$ is the magnitude of $\tilde{\mathbf{F}}$ as given in Eq. (2.3.6), F is the magnitude of $\mathbf{F}$ as given in Eq. (2.2.2), and β and ϵ are defined as (see Fig. 2.8.3 for a and b)

$$\beta \triangleq \frac{b}{R} \tag{1}$$

and

$$\epsilon \triangleq \left[1 - \left(\frac{b}{a}\right)^2\right]^{1/2} \tag{2}$$

If $\mathbf{a}_1$ and $\mathbf{a}_2$ are unit vectors as shown in Fig. 2.8.3, and $\mathbf{a}_3 = \mathbf{a}_1 \times \mathbf{a}_2$, then the associated moments and products of inertia of B for B^* are given by

$$I_{11} = \frac{2ma^2}{5} \qquad I_{22} = I_{33} = \frac{m}{5}(a^2 + b^2) \qquad I_{12} = I_{23} = I_{31} = 0 \tag{3}$$

Hence,

$$\mathbf{f}^{(2)} \underset{(2.3.10)}{=} -\frac{3}{5R^2}(a^2 - b^2)\mathbf{a}_1 \tag{4}$$

$$\tilde{\mathbf{F}} \underset{(2.3.6)}{=} -\frac{G\bar{m}m}{R^2}\left[1 - \frac{3}{5R^2}(a^2 - b^2)\right]\mathbf{a}_1$$

$$\underset{(1,2)}{=} -\frac{G\bar{m}m}{R^2}\left(1 - \frac{3}{5}\frac{\beta^2\epsilon^2}{1 - \epsilon^2}\right)\mathbf{a}_1 \tag{5}$$

and
$$\widetilde{F} \triangleq |\widetilde{\mathbf{F}}| = \frac{G\overline{m}m}{R^2}\left|1 - \frac{3}{5}\frac{\beta^2\epsilon^2}{1-\epsilon^2}\right| \tag{6}$$

For the evaluation of the integral in Eq. (2.2.2), it is convenient to introduce the coordinates r, θ, and z shown in Fig. 2.8.4 and to note that $\mathbf{p}$, the position vector from $\overline{P}$ to P, then can be expressed as

$$\mathbf{p} = (R - z)\mathbf{a}_1 + r \sin\theta\, \mathbf{a}_2 + r \cos\theta\, \mathbf{a}_3 \tag{7}$$

while

$$d\tau = r\, d\theta\, dr\, dz \tag{8}$$

As for ρ, the mass density of B, this is given by $\rho = 3m/(4\pi ba^2)$. Hence,

$$\mathbf{F} \underset{(2.2.2)}{=} -\frac{3G\overline{m}m}{4\pi ba^2}\int_{\theta_1}^{\theta_2}\int_{z_1}^{z_2}\int_{r_1}^{r_2}\frac{(R-z)\mathbf{a}_1 + r\sin\theta\,\mathbf{a}_2 + r\cos\theta\,\mathbf{a}_3}{[(R-z)^2 + r^2]^{3/2}}\, r\, dr\, dz\, d\theta \tag{9}$$

where

$$\theta_1 = 0 \qquad \theta_2 = 2\pi \qquad z_1 = -b \qquad z_2 = b \qquad r_1 = 0$$

$$r_2 = a\left[1 - \left(\frac{z}{b}\right)^2\right]^{1/2} \tag{10}$$

and, after carrying out the indicated integrations and using Eqs. (1) and (2) to eliminate b/R and a/b, one arrives at

$$F \triangleq |\mathbf{F}| \underset{(9)}{=} \frac{3G\overline{m}m(1-\epsilon^2)}{R^2\beta^2\epsilon^2}\left|1 - \frac{(1-\epsilon^2)^{1/2}}{2\beta\epsilon}[H(1) - H(-1)]\right| \tag{11}$$

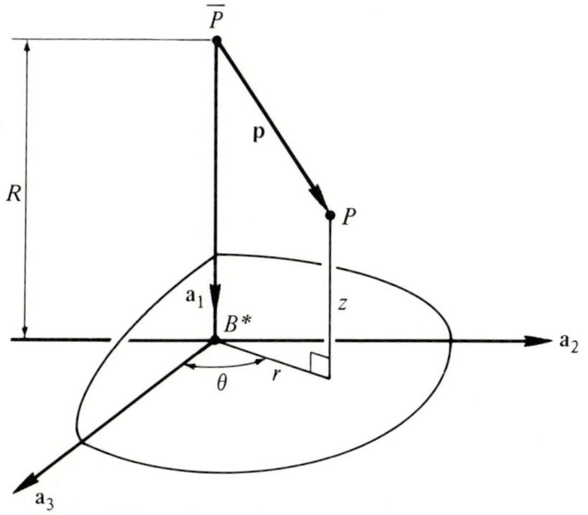

Figure 2.8.4

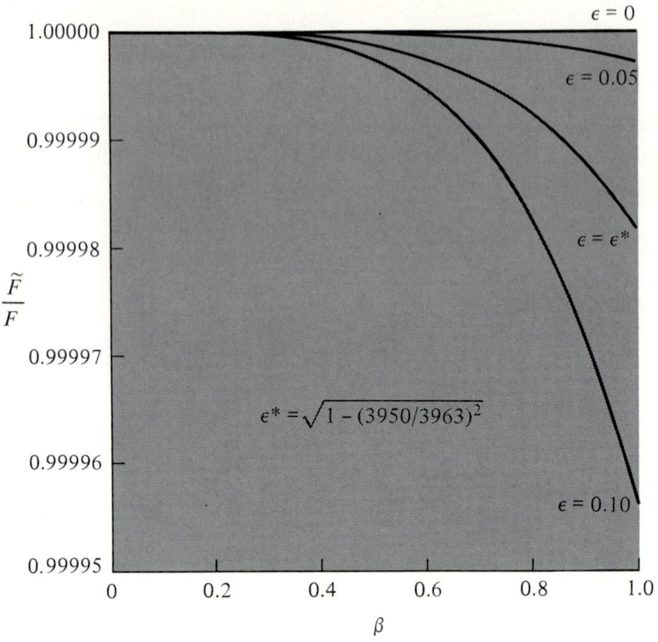

Figure 2.8.5

where H is defined as

$$H(\xi) \triangleq \sin^{-1}\frac{1-\epsilon^2(1-\beta\xi)}{[1-\epsilon^2(1-\beta^2)]^{1/2}} \tag{12}$$

Consequently,

$$\frac{\tilde{F}}{F} \underset{(6,11)}{=} \frac{\beta^2\epsilon^2}{3(1-\epsilon^2)}\,\frac{|1-\frac{3}{5}[\beta^2\epsilon^2/(1-\epsilon^2)]|}{|1-[(1-\epsilon^2)^{1/2}/(2\beta\epsilon)][H(1)-H(-1)]|} \tag{13}$$

In Fig. 2.8.5, $\tilde{F}/F$ as given by Eq. (13) is plotted versus β for four values of ϵ. Note that small values of β correspond to placing $\bar{P}$ at a great distance from B^* [see Eq. (1)]. Hence it is not surprising that, regardless of the values of ϵ, $\tilde{F}/F$ approaches unity as β approaches zero. Conversely, values of β near unity represent situations in which the particle comes relatively close to the spheroid, and the error one then makes when using $\tilde{F}$ in place of F can be seen to depend on ϵ, larger departures of $\tilde{F}/F$ from unity being associated with larger values of ϵ, that is, with a more pronounced flattening of the ellipsoid [see Eq. (2)]. However, it appears that even for $\beta \approx 1.0$, that is, for a particle that is nearly in contact with the spheroid, $\tilde{F}/F$ departs only very slightly from unity so long as $\epsilon < 0.10$, which is the case, for example, when $\epsilon = \epsilon^*$, where ϵ^* is the eccentricity of a spheroid whose major semidiameters are equal to the

Earth's polar and equatorial radii. Consequently, Eq. (2.3.6) may be expected to yield highly accurate results when used in an analysis concerned with a near-Earth satellite idealized as a particle.

2.9 DIFFERENTIATION WITH RESPECT TO A VECTOR

It is sometimes convenient to employ vector operations on scalar functions in order to express gravitational forces and moments. These operations may be interpreted as ordinary and/or partial differentiations of scalar functions with respect to vector variables. In some cases the vector in question is a position vector, and the ordinary derivative is then a spatial gradient. When the differentiation of a scalar function produces a force, the scalar function is called a *force function*. In other cases, differentiations with respect to other vectors are employed, as in the representation of gravitational moments in terms of derivatives with respect to unit vectors. To unify the presentation of this subject in Secs. 2.10–2.18, the mathematical tools there employed are first discussed briefly in the present section.

If a scalar quantity F depends on a vector $\mathbf{v}$, then it is useful to define a vector denoted by $\nabla_v F$ as follows: introduce an arbitrary set of mutually perpendicular unit vectors $\mathbf{a}_1$, $\mathbf{a}_2$, and $\mathbf{a}_3$; let $v_i \triangleq \mathbf{v} \cdot \mathbf{a}_i$ $(i = 1, 2, 3)$; regard F as a function of v_1, v_2, and v_3; and let

$$\nabla_v F \triangleq \frac{\partial F}{\partial v_1}\mathbf{a}_1 + \frac{\partial F}{\partial v_2}\mathbf{a}_2 + \frac{\partial F}{\partial v_3}\mathbf{a}_3 \tag{1}$$

The vector $\nabla_v F$ constructed according to Eq. (1) is invariant with respect to the choice of vector basis $\mathbf{a}_1$, $\mathbf{a}_2$, $\mathbf{a}_3$. The operation denoted by ∇_v may be termed differentiation with respect to $\mathbf{v}$.

Similarly, if a vector quantity $\mathbf{F}$ depends on a vector $\mathbf{v}$, and for some arbitrary orthogonal vector basis $\mathbf{a}_1$, $\mathbf{a}_2$, $\mathbf{a}_3$ one lets $F_i \triangleq \mathbf{F} \cdot \mathbf{a}_i$ and $v_i \triangleq \mathbf{v} \cdot \mathbf{a}_i$ $(i = 1, 2, 3)$, then a scalar and a dyadic may be defined as

$$\nabla_v \cdot \mathbf{F} \triangleq \frac{\partial F_1}{\partial v_1} + \frac{\partial F_2}{\partial v_2} + \frac{\partial F_3}{\partial v_3} \tag{2}$$

and

$$\nabla_v \mathbf{F} \triangleq (\nabla_v F_1)\mathbf{a}_1 + (\nabla_v F_2)\mathbf{a}_2 + (\nabla_v F_3)\mathbf{a}_3 \tag{3}$$

The quantities defined by Eqs. (2) and (3) do not depend upon the choice of vector basis $\mathbf{a}_1$, $\mathbf{a}_2$, $\mathbf{a}_3$.

Further useful quantities are at times obtained by cascading some of these definitions. For example, when the operation defined in Eq. (2) is performed on the vector defined in Eq. (1), the result is a scalar denoted by $\nabla_v^2 F$:

$$\nabla_v^2 F \triangleq \nabla_v \cdot (\nabla_v F) \tag{4}$$

When $\mathbf{v}$ is a position vector $\mathbf{p}$, and when $\mathbf{p}$ can be inferred from the context, the subscript $\mathbf{v} = \mathbf{p}$ is often omitted; the quantities ∇F, $\nabla \cdot \mathbf{F}$, $\nabla \mathbf{F}$, and $\nabla^2 F$ are then called the *spatial gradient* of F, the *spatial divergence* of $\mathbf{F}$, the *spatial gradient* of $\mathbf{F}$, and the *spatial Laplacian* of F, respectively.

As a consequence of the definitions given in Eqs. (1)–(4) and of various theorems of the differential calculus of functions of one or more scalar variables, the quantities $\nabla_{\mathbf{v}}F$, $\nabla_{\mathbf{v}} \cdot \mathbf{F}$, and $\nabla_{\mathbf{v}}{}^2 F$ satisfy many relationships having counterparts in this calculus. For example, if $\mathbf{v} = \mathbf{w} + \mathbf{c}$, where $\mathbf{c}$ is independent of $\mathbf{v}$, and if† $F(\mathbf{v})$ and $G(\mathbf{w})$ are functions such that $G(\mathbf{w}) = F(\mathbf{w} + \mathbf{c})$, then

$$\nabla_{\mathbf{v}}F = \nabla_{\mathbf{w}}G \tag{5}$$

Or, suppose that $\mathbf{v}(g_1, \ldots, g_n)$ denotes a vector function in a reference frame A of the scalar independent variables $g_1, \ldots, g_n$. Suppose further, that $F(\mathbf{v})$ and $G(g_1, \ldots, g_n)$ are functions such that $G(g_1, \ldots, g_n) = F[\mathbf{v}(g_1, \ldots, g_n)]$. Finally, let $\mathbf{a}_1$, $\mathbf{a}_2$, $\mathbf{a}_3$ be mutually perpendicular unit vectors fixed in A, express $\nabla_{\mathbf{v}}F$ as in Eq. (1), and denote the partial derivative of $\mathbf{v}$ with respect to g_r in A by $\partial \mathbf{v}/\partial g_r$. Then

$$\frac{\partial G}{\partial g_r} = \nabla_{\mathbf{v}}F \cdot \frac{\partial \mathbf{v}}{\partial g_r} \qquad (r = 1, \ldots, n) \tag{6}$$

Three differentiation formulas involving a unit dyadic $\mathbf{U}$ and a unit vector $\mathbf{u}$ having the same direction as a vector $\mathbf{v}$ will prove useful in the sequel. These are

$$\nabla_{\mathbf{v}}\mathbf{v} = \mathbf{U} \tag{7}$$

$$\nabla_{\mathbf{v}}v = \mathbf{u} \tag{8}$$

and

$$\nabla_{\mathbf{v}}\mathbf{u} = v^{-1}(\mathbf{U} - \mathbf{uu}) \tag{9}$$

where $v \triangleq \mathbf{v} \cdot \mathbf{u}$, so that $\mathbf{v} = v\mathbf{u}$.

In dealing with gravitational moments, it is at times convenient to invoke the idea of *partial* differentiation with respect to a vector in a reference frame, in the following sense: suppose that $\mathbf{a}_1$, $\mathbf{a}_2$, $\mathbf{a}_3$, v, $\mathbf{v}$, $\mathbf{u}$, and F have the same meanings as heretofore; let $u_i \triangleq \mathbf{u} \cdot \mathbf{a}_i$ $(i = 1, 2, 3)$; and let $\mathbf{v}(\mathbf{u}, v)$ and $G(\mathbf{u}, v)$ denote, respectively, a vector function and a scalar function of u_1, u_2, u_3, and v, choosing $G(\mathbf{u}, v)$ such that $G(\mathbf{u}, v) \triangleq F[\mathbf{v}(\mathbf{u}, v)]$. Then, defining $\partial G/\partial \mathbf{u}$ as

$$\frac{\partial G}{\partial \mathbf{u}} \triangleq \frac{\partial G}{\partial u_1}\mathbf{a}_1 + \frac{\partial G}{\partial u_2}\mathbf{a}_2 + \frac{\partial G}{\partial u_3}\mathbf{a}_3 \tag{10}$$

one can write

$$\nabla_{\mathbf{v}}F = \frac{\partial G}{\partial \mathbf{u}} \cdot \nabla_{\mathbf{v}}\mathbf{u} + \frac{\partial G}{\partial v}\mathbf{u} \tag{11}$$

† As a notational convenience, a scalar quantity such as F which depends upon the vector $\mathbf{v}$ is designated $F(\mathbf{v})$; the functional representation of this quantity [as required by Eqs. (1)–(4)] is written $F(v_1, v_2, v_3)$ or $F(v_1', v_2', v_3')$, where $v_i \triangleq \mathbf{v} \cdot \mathbf{a}_i$ and $v_i' \triangleq \mathbf{v} \cdot \mathbf{a}_i'$ $(i = 1, 2, 3)$ for arbitrary vector bases $\mathbf{a}_1$, $\mathbf{a}_2$, $\mathbf{a}_3$ and $\mathbf{a}_1'$, $\mathbf{a}_2'$, $\mathbf{a}_3'$.

or, in view of Eq. (9),

$$\nabla_{\mathbf{v}} F = \frac{1}{v}\frac{\partial G}{\partial \mathbf{u}} + \left(\frac{\partial G}{\partial v} - \frac{1}{v}\frac{\partial G}{\partial \mathbf{u}} \cdot \mathbf{u}\right)\mathbf{u} \tag{12}$$

It follows from this equation and from $\mathbf{v} = v\mathbf{u}$ that

$$\mathbf{v} \times \nabla_{\mathbf{v}} F = \mathbf{u} \times \frac{\partial G}{\partial \mathbf{u}} \tag{13}$$

Derivations In characterizing a quantity as a function of a vector, one implies that the quantity is represented in terms of the scalar components of the vector by a functional relationship that is independent of the vector basis employed. Thus, for example, if the scalar quantity F is a function of the vector $\mathbf{v}$, and $v_i \triangleq \mathbf{v} \cdot \mathbf{a}_i$ and $v_i' \triangleq \mathbf{v} \cdot \mathbf{a}_i'$ $(i = 1, 2, 3)$ for arbitrary orthogonal vector bases $\mathbf{a}_1, \mathbf{a}_2, \mathbf{a}_3$ and $\mathbf{a}_1', \mathbf{a}_2', \mathbf{a}_3'$, then, in functional notation,

$$F(v_1, v_2, v_3) = F(v_1', v_2', v_3') \tag{14}$$

Using this relationship, one can prove the invariance of $\nabla_{\mathbf{v}} F$, $\nabla_{\mathbf{v}} \cdot \mathbf{F}$, $\nabla_{\mathbf{v}} \mathbf{F}$, and $\nabla_{\mathbf{v}}^2 F$ with respect to the choice of vector basis by establishing the equality of alternative representations of each of these quantities in terms of the arbitrary orthogonal vector bases $\mathbf{a}_1, \mathbf{a}_2, \mathbf{a}_3$ and $\mathbf{a}_1', \mathbf{a}_2', \mathbf{a}_3'$. This objective can be accomplished as follows:

Letting $C_{ij} \triangleq \mathbf{a}_i \cdot \mathbf{a}_j'$ $(i, j = 1, 2, 3)$, one can write (using the summation convention)

$$\mathbf{a}_i = \mathbf{a}_j' C_{ij} \tag{15}$$

$$F_i - F_j' C_{ij} \tag{16}$$

and

$$v_i' = v_j C_{ji} \tag{17}$$

from the last of which it follows that

$$\frac{\partial v_i'}{\partial v_j} = C_{ji} \tag{18}$$

Consequently, using familiar differentiation theorems, one can write

$$\frac{\partial F}{\partial v_i} \underset{(14)}{=} \frac{\partial F}{\partial v_j'}\frac{\partial v_j'}{\partial v_i} \underset{(18)}{=} \frac{\partial F}{\partial v_j'} C_{ij} \tag{19}$$

and

$$\frac{\partial F}{\partial v_i}\mathbf{a}_i \underset{(19,15)}{=} \frac{\partial F}{\partial v_j'} C_{ij}\, \mathbf{a}_k' C_{ik} \tag{20}$$

But,

$$C_{ij} C_{ik} \underset{(1.2.15)}{=} \delta_{jk} \tag{21}$$

Hence,

$$\frac{\partial F}{\partial v_i}\mathbf{a}_i \underset{(20,21)}{=} \frac{\partial F}{\partial v_j'}\mathbf{a}_k' \delta_{jk} = \frac{\partial F}{\partial v_i'}\mathbf{a}_i' \tag{22}$$

which establishes the invariance of $\nabla_v F$. For $\nabla_v \cdot \mathbf{F}$, one has [see Eq. (2)]

$$\frac{\partial F_i}{\partial v_i} \underset{(16)}{=} \frac{\partial F_j'}{\partial v_i} C_{ij} = \frac{\partial F_j'}{\partial v_k'} \frac{\partial v_k'}{\partial v_i} C_{ij}$$

$$\underset{(18)}{=} \frac{\partial F_j'}{\partial v_k'} C_{ik} C_{ij} \underset{(1.2.15)}{=} \frac{\partial F_i'}{\partial v_i'} \tag{23}$$

and the proofs for $\nabla_v \mathbf{F}$ and $\nabla_v \cdot \nabla_v F$ proceed similarly.

When $\mathbf{F}$ in Eq. (3) is replaced with $\mathbf{v}$, one has

$$\nabla_v \mathbf{v} = (\nabla_v v_i) \mathbf{a}_i \tag{24}$$

Now,

$$\nabla_v v_i \underset{(1)}{=} \frac{\partial v_i}{\partial v_j} \mathbf{a}_j = \delta_{ij} \mathbf{a}_j = \mathbf{a}_i \tag{25}$$

Hence,

$$\nabla_v \mathbf{v} = \mathbf{a}_i \mathbf{a}_i = \mathbf{U} \tag{26}$$

in agreement with Eq. (7).

If $\mathbf{v} = v\mathbf{u}$, so that $\mathbf{u}$ has the same direction as $\mathbf{v}$, then

$$v = (\mathbf{v} \cdot \mathbf{v})^{1/2} \tag{27}$$

and

$$\nabla_v v = \nabla_v (\mathbf{v} \cdot \mathbf{v})^{1/2} = \tfrac{1}{2}(\mathbf{v} \cdot \mathbf{v})^{-1/2} [(\nabla_v \mathbf{v}) \cdot \mathbf{v} + \mathbf{v} \cdot (\nabla_v \mathbf{v})]$$

$$\underset{(26,27)}{=} \tfrac{1}{2} v^{-1}(\mathbf{v} + \mathbf{v}) = v^{-1}\mathbf{v} = \mathbf{u} \tag{28}$$

which proves the validity of Eq. (8).

As for Eq. (9),

$$\nabla_v \mathbf{u} = \nabla_v (v^{-1} \mathbf{v}) = (\nabla_v v^{-1})\mathbf{v} + v^{-1} \nabla_v \mathbf{v}$$

$$= (-v^{-2} \nabla_v v)\mathbf{v} + v^{-1} \nabla_v \mathbf{v}$$

$$\underset{(7,8)}{=} (-v^{-2} \mathbf{u}) v\mathbf{u} + v^{-1} \mathbf{U}$$

$$= v^{-1}(\mathbf{U} - \mathbf{uu}) \tag{29}$$

Finally, in connection with Eq. (11), note that both u_i and v can be expressed as functions of v_1, v_2, v_3 (where $v_i \triangleq \mathbf{v} \cdot \mathbf{a}_i$); that is,

$$u_i = v_i(v_1^2 + v_2^2 + v_3^2)^{-1/2} \qquad v = (v_1^2 + v_2^2 + v_3^2)^{1/2} \tag{30}$$

It follows that

$$\frac{\partial F}{\partial v_i} = \frac{\partial G}{\partial u_j} \frac{\partial u_j}{\partial v_i} + \frac{\partial G}{\partial v} \frac{\partial v}{\partial v_i} \tag{31}$$

so that

$$\nabla_v F \underset{(1)}{=} \frac{\partial F}{\partial v_i} \mathbf{a}_i \underset{(31)}{=} \frac{\partial G}{\partial u_j} \frac{\partial u_j}{\partial v_i} \mathbf{a}_i + \frac{\partial G}{\partial v} \frac{\partial v}{\partial v_i} \mathbf{a}_i$$

(continued on next page)

$$\underset{(1)}{=} \frac{\partial G}{\partial u_j} \nabla_v u_j + \frac{\partial G}{\partial v} \nabla_v v$$

$$\underset{(10,8)}{=} (\nabla_v u_j)\mathbf{a}_j \cdot \frac{\partial G}{\partial u} + \frac{\partial G}{\partial v}\mathbf{u}$$

$$\underset{(3)}{=} (\nabla_v \mathbf{u}) \cdot \frac{\partial G}{\partial \mathbf{u}} + \frac{\partial G}{\partial v}\mathbf{u} \tag{32}$$

and this is equivalent to Eq. (11) since $\nabla_v \mathbf{u}$ is a symmetric dyadic, as is apparent from Eq. (9).

Example In Fig. 2.9.1, r, λ, β are spherical coordinates of a point P, and $\mathbf{b}_1$, $\mathbf{b}_2$, $\mathbf{b}_3$ are unit vectors pointing, respectively, in the directions in which P moves when r, λ, β are made to vary, one at a time. The position vector $\mathbf{p}$ of P relative to O may be expressed as

$$\mathbf{p}(r, \lambda, \beta) = r(\mathrm{c}\lambda\ \mathrm{c}\beta\ \mathbf{a}_1 + \mathrm{s}\lambda\ \mathrm{c}\beta\ \mathbf{a}_2 + \mathrm{s}\beta\ \mathbf{a}_3) \tag{33}$$

where $\mathbf{a}_1$, $\mathbf{a}_2$, $\mathbf{a}_3$, are mutually perpendicular unit vectors fixed in a reference frame A. If $F(\mathbf{p})$ denotes a function of $\mathbf{p}$, and if $G(r, \lambda, \beta)$ is defined as

$$G(r, \lambda, \beta) \overset{\triangle}{=} F[\mathbf{p}(r, \lambda, \beta)] \tag{34}$$

then Eq. (6) can be used to express the spatial gradient of F in the following frequently convenient form:

$$\nabla F = \frac{\partial G}{\partial r}\mathbf{b}_1 + \frac{1}{r\,\mathrm{c}\beta}\frac{\partial G}{\partial \lambda}\mathbf{b}_2 + \frac{1}{r}\frac{\partial G}{\partial \beta}\mathbf{b}_3 \tag{35}$$

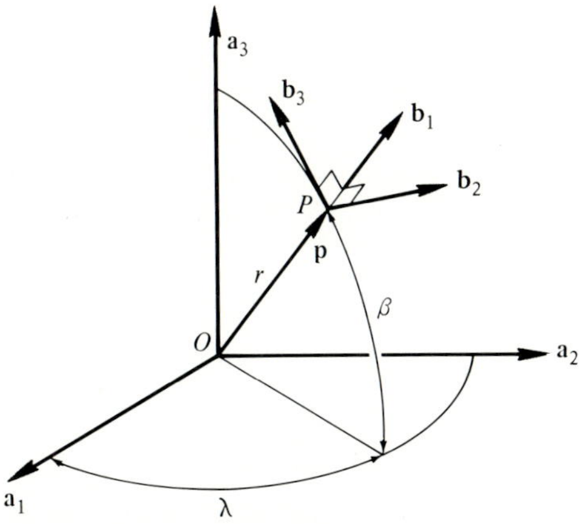

Figure 2.9.1

To show this, one may begin by evaluating the partial derivatives of **p** with respect to r, λ, and β in A:

$$\frac{\partial \mathbf{p}}{\partial r} \underset{(33)}{=} \mathrm{c}\lambda\, \mathrm{c}\beta\, \mathbf{a}_1 + \mathrm{s}\lambda\, \mathrm{c}\beta\, \mathbf{a}_2 + \mathrm{s}\beta\, \mathbf{a}_3 = \mathbf{b}_1 \tag{36}$$

$$\frac{\partial \mathbf{p}}{\partial \lambda} \underset{(33)}{=} r(-\mathrm{s}\lambda\, \mathrm{c}\beta\, \mathbf{a}_1 + \mathrm{c}\lambda\, \mathrm{c}\beta\, \mathbf{a}_2) = r\, \mathrm{c}\beta\, \mathbf{b}_2 \tag{37}$$

$$\frac{\partial \mathbf{p}}{\partial \beta} \underset{(33)}{=} r(-\mathrm{c}\lambda\, \mathrm{s}\beta\, \mathbf{a}_1 + \mathrm{s}\lambda\, \mathrm{s}\beta\, \mathbf{a}_2 + \mathrm{c}\beta\, \mathbf{a}_3) = r\mathbf{b}_3 \tag{38}$$

Next, from Eq. (6), and with $\boldsymbol{\nabla}$ written in place of $\boldsymbol{\nabla}_{\mathbf{p}}$,

$$\frac{\partial G}{\partial r} = \boldsymbol{\nabla} F \cdot \frac{\partial \mathbf{p}}{\partial r} \underset{(36)}{=} \boldsymbol{\nabla} F \cdot \mathbf{b}_1 \tag{39}$$

$$\frac{\partial G}{\partial \lambda} = \boldsymbol{\nabla} F \cdot \frac{\partial \mathbf{p}}{\partial \lambda} \underset{(37)}{=} r\, \mathrm{c}\beta\, \boldsymbol{\nabla} F \cdot \mathbf{b}_2 \tag{40}$$

$$\frac{\partial G}{\partial \beta} = \boldsymbol{\nabla} F \cdot \frac{\partial \mathbf{p}}{\partial \beta} \underset{(38)}{=} r\, \boldsymbol{\nabla} F \cdot \mathbf{b}_3 \tag{41}$$

Now, since $\mathbf{b}_1\mathbf{b}_1 + \mathbf{b}_2\mathbf{b}_2 + \mathbf{b}_3\mathbf{b}_3$ is a unit dyadic, one may write

$$\begin{aligned} \boldsymbol{\nabla} F &= \boldsymbol{\nabla} F \cdot (\mathbf{b}_1\mathbf{b}_1 + \mathbf{b}_2\mathbf{b}_2 + \mathbf{b}_3\mathbf{b}_3) \\ &= \boldsymbol{\nabla} F \cdot \mathbf{b}_1\mathbf{b}_1 + \boldsymbol{\nabla} F \cdot \mathbf{b}_2\mathbf{b}_2 + \boldsymbol{\nabla} F \cdot \mathbf{b}_3\mathbf{b}_3 \end{aligned} \tag{42}$$

Solving Eqs. (39)–(41) for $\boldsymbol{\nabla} F \cdot \mathbf{b}_1$, $\boldsymbol{\nabla} F \cdot \mathbf{b}_2$, and $\boldsymbol{\nabla} F \cdot \mathbf{b}_3$, and substituting into Eq. (42), one arrives at Eq. (35).

The spatial Laplacian of F may be expressed as

$$\nabla^2 F = \frac{1}{r^2}\left[\frac{\partial}{\partial r}\left(r^2 \frac{\partial G}{\partial r}\right) + \sec^2 \beta \frac{\partial^2 G}{\partial \lambda^2} + \sec \beta \frac{\partial}{\partial \beta}\left(\mathrm{c}\beta \frac{\partial G}{\partial \beta}\right)\right] \tag{43}$$

by using Eq. (4) in conjunction with Eq. (35), as follows.

$$\begin{aligned} \nabla^2 F \underset{(4)}{=} \boldsymbol{\nabla} \cdot \boldsymbol{\nabla} F &\underset{(35)}{=} \left(\frac{\partial}{\partial r} \mathbf{b}_1 + \frac{1}{r\, \mathrm{c}\beta} \frac{\partial}{\partial \lambda} \mathbf{b}_2 + \frac{1}{r} \frac{\partial}{\partial \beta} \mathbf{b}_3\right) \cdot \boldsymbol{\nabla} F \\ &= \mathbf{b}_1 \cdot \frac{\partial}{\partial r}(\boldsymbol{\nabla} F) + \frac{1}{r\, \mathrm{c}\beta} \mathbf{b}_2 \cdot \frac{\partial}{\partial \lambda}(\boldsymbol{\nabla} F) + \frac{1}{r} \mathbf{b}_3 \cdot \frac{\partial}{\partial \beta}(\boldsymbol{\nabla} F) \end{aligned} \tag{44}$$

The partial differentiations indicated in this equation must be performed in reference frame A. Displaying only terms that will not be eliminated by subsequent dot-multiplication, one has

$$\frac{\partial}{\partial r}(\boldsymbol{\nabla} F) \underset{(35)}{=} \frac{\partial^2 G}{\partial r^2} \mathbf{b}_1 + \cdots \tag{45}$$

$$\frac{\partial}{\partial \lambda}(\boldsymbol{\nabla} F) \underset{(35)}{=} \frac{\partial G}{\partial r} \frac{\partial \mathbf{b}_1}{\partial \lambda} + \frac{1}{r\, \mathrm{c}\beta} \frac{\partial^2 G}{\partial \lambda^2} \mathbf{b}_2 + \frac{1}{r} \frac{\partial G}{\partial \beta} \frac{\partial \mathbf{b}_3}{\partial \lambda} + \cdots$$

(continued on next page)

$$= \frac{\partial G}{\partial r} \text{c}\beta\, \mathbf{b}_2 + \frac{1}{r\,\text{c}\beta} \frac{\partial^2 G}{\partial \lambda^2} \mathbf{b}_2 - \frac{1}{r} \frac{\partial G}{\partial \beta} \text{s}\beta\, \mathbf{b}_2 + \cdots \tag{46}$$

$$\frac{\partial}{\partial \beta}(\nabla F) \underset{(35)}{=} \frac{\partial G}{\partial r} \frac{\partial \mathbf{b}_1}{\partial \beta} + \frac{1}{r} \frac{\partial^2 G}{\partial \beta^2} \mathbf{b}_3 + \cdots$$

$$= \frac{\partial G}{\partial r} \mathbf{b}_3 + \frac{1}{r} \frac{\partial^2 G}{\partial \beta^2} \mathbf{b}_3 + \cdots \tag{47}$$

and substitution into Eq. (44) then yields

$$\nabla^2 F = \frac{\partial^2 G}{\partial r^2} + \frac{2}{r} \frac{\partial G}{\partial r} + \frac{1}{r^2\, \text{c}^2\beta} \frac{\partial^2 G}{\partial \lambda^2} - \frac{\text{s}\beta}{r^2\, \text{c}\beta} \frac{\partial G}{\partial \beta} + \frac{1}{r^2} \frac{\partial^2 G}{\partial \beta^2} \tag{48}$$

which is equivalent to Eq. (43).

2.10 FORCE FUNCTION FOR TWO PARTICLES

The gravitational force **F** exerted on a particle P of mass m by a particle $\bar{P}$ of mass $\bar{m}$ (see Sec. 2.1) can be expressed as

$$\mathbf{F} = \nabla_{\mathbf{p}} V \triangleq \nabla V \tag{1}$$

where **p** is the position vector of P relative to $\bar{P}$ and V is given by

$$V = G\bar{m}mp^{-1} + C \tag{2}$$

with p defined as

$$p \triangleq (\mathbf{p}^2)^{1/2} \tag{3}$$

and C an arbitrary constant, while G is the universal gravitational constant.

A scalar function of a vector variable is called a *force function* if the derivative of the function with respect to the variable is equal to a force. Thus V is a force function associated with the gravitational interaction of two particles.

The spatial Laplacian of V (see Sec. 2.9) is zero:

$$\nabla^2 V = 0 \tag{4}$$

Equation (4) is known as *Laplace's equation*. Any solution of this equation is called a *spherical harmonic*; in the context of gravitational problems it is called a *gravitational potential*. For the purpose of characterizing the interaction force **F** between two particles, however, only that special spherical harmonic given by Eq. (2) provides the force function appropriate for substitution into Eq. (1).

Derivations Differentiation of V with respect to **p** gives

$$\nabla V \underset{(2)}{=} G\bar{m}m\,\nabla p^{-1} \underset{(2.9.8)}{=} -G\bar{m}mp^{-2}\,\mathbf{u} \tag{5}$$

where **u** is the unit vector in the direction of **p**, that is,

$$\mathbf{u} \triangleq p^{-1}\mathbf{p} \tag{6}$$

Consequently,

$$\nabla V = -G\bar{m}mp^{-3}\mathbf{p} \underset{(2.1.1)}{=} \mathbf{F} \tag{7}$$

The Laplacian of V is

$$\nabla^2 V \underset{(2.9.4)}{=} \nabla \cdot \nabla V \underset{(7)}{=} -G\bar{m}m\nabla \cdot (p^{-3}\mathbf{p})$$

$$\underset{\substack{(2.9.8,\\2.9.2)}}{=} -G\bar{m}m(3p^{-4}\mathbf{u} \cdot \mathbf{p} - 3p^{-3}) = 0 \tag{8}$$

Example A particle P of mass m and two particles P_1 and P_2, each of mass $\bar{m}$, are situated as shown in Fig. 2.10.1, where r, λ, β are spherical coordinates of P, and $\mathbf{b}_1$, $\mathbf{b}_2$, $\mathbf{b}_3$ are unit vectors pointing in the directions in which P moves when r, λ, β are made to increase, one at a time. A force function V for the resultant gravitational force **F** acting on P is to be constructed, and this function is to be used to express **F** in terms of components parallel to $\mathbf{b}_1$, $\mathbf{b}_2$, $\mathbf{b}_3$.

The forces $\mathbf{F}_1$ and $\mathbf{F}_2$ exerted on P by P_1 and P_2, respectively, can be expressed as

$$\mathbf{F}_i = \nabla_{\mathbf{p}_i} V_i \qquad (i = 1, 2) \tag{9}$$

where $\mathbf{p}_i$ is the position vector of P relative to P_i and where $V_1(\mathbf{p}_1)$ and $V_2(\mathbf{p}_2)$ are given by

$$V_i \underset{(2)}{=} G\bar{m}mp_i^{-1} + C_i \qquad (i = 1, 2) \tag{10}$$

with p_i equal to the magnitude of $\mathbf{p}_i$. Hence

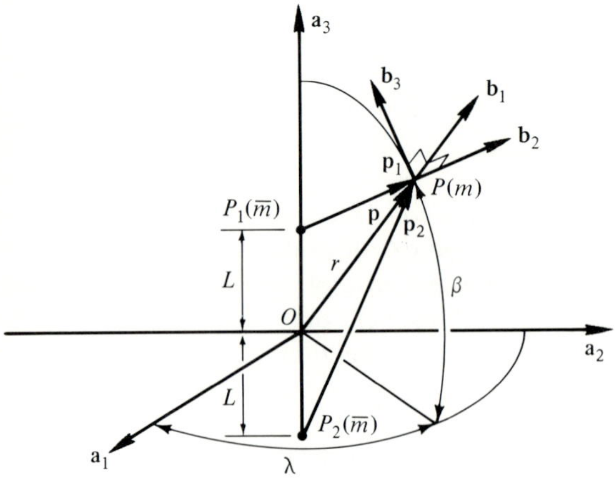

Figure 2.10.1

$$\mathbf{F} = \mathbf{F}_1 + \mathbf{F}_2 \underset{(9)}{=} \nabla_{\mathbf{p}_1} V_1 + \nabla_{\mathbf{p}_2} V_2 \tag{11}$$

If $\mathbf{p}$ is now introduced as the position vector of P relative to O, then $\mathbf{p}_i = \mathbf{p} + \mathbf{c}_i$ $(i = 1, 2)$, where $\mathbf{c}_i$ is independent of $\mathbf{p}_i$ (see Fig. 2.10.1); and, if $W_i(\mathbf{p})$ is defined as

$$W_i(\mathbf{p}) \triangleq V_i(\mathbf{p} + \mathbf{c}_i) \qquad (i = 1, 2) \tag{12}$$

then

$$\nabla_{\mathbf{p}_i} V_i \underset{(2.9.5)}{=} \nabla_{\mathbf{p}} W_i \qquad (i = 1, 2) \tag{13}$$

Hence,

$$\mathbf{F} \underset{(11,13)}{=} \nabla_{\mathbf{p}} W_1 + \nabla_{\mathbf{p}} W_2 = \nabla_{\mathbf{p}}(W_1 + W_2) = \nabla_{\mathbf{p}} V \tag{14}$$

if V is defined as

$$V \triangleq W_1 + W_2 \underset{(12)}{=} V_1 + V_2 \underset{(10)}{=} G\overline{m}m(p_1^{-1} + p_2^{-1}) + C \tag{15}$$

V is the desired force function.

To express $\mathbf{F}$ in terms of components parallel to $\mathbf{b}_1$, $\mathbf{b}_2$, $\mathbf{b}_3$, note that

$$p_1 = (r^2 + L^2 - 2rL \sin \beta)^{1/2} \tag{16}$$

and

$$p_2 = (r^2 + L^2 + 2rL \sin \beta)^{1/2} \tag{17}$$

and define a function W of r, λ, β as

$$W \triangleq G\overline{m}m[(r^2 + L^2 - 2rL \sin \beta)^{-1/2} + (r^2 + L^2 + 2rL \sin \beta)^{-1/2}] + C \tag{18}$$

Then

$$V \underset{(15\text{–}18)}{=} W \tag{19}$$

and, in accordance with Eq. (2.9.35),

$$\begin{aligned}
\nabla_{\mathbf{p}} V &= \frac{\partial W}{\partial r}\mathbf{b}_1 + \frac{1}{r \cos \beta}\frac{\partial W}{\partial \lambda}\mathbf{b}_2 + \frac{1}{r}\frac{\partial W}{\partial \beta}\mathbf{b}_3 \\
&\underset{(18)}{=} -G\overline{m}m\{[(r^2 + L^2 - 2rL \sin \beta)^{-3/2}(r - L \sin \beta) \\
&\quad + (r^2 + L^2 + 2rL \sin \beta)^{-3/2}(r + L \sin \beta)]\mathbf{b}_1 \\
&\quad - [(r^2 + L^2 - 2rL \sin \beta)^{-3/2} L \cos \beta \\
&\quad - (r^2 + L^2 + 2rL \sin \beta)^{-3/2} L \cos \beta]\mathbf{b}_3\}
\end{aligned} \tag{20}$$

Substituting from Eq. (20) into Eq. (14), one arrives at the desired expression for $\mathbf{F}$.

2.11 FORCE FUNCTION FOR A BODY AND A PARTICLE

The resultant gravitational force **F** exerted by a particle $\overline{P}$ of mass $\overline{m}$ on the particles of a (not necessarily rigid) body B (see Sec. 2.2) can be expressed as

$$\mathbf{F} = \nabla_{\mathbf{R}} V \tag{1}$$

where **R** is the position vector of the mass center B^* of B relative to $\overline{P}$ and V is a force function given by

$$V = G\overline{m} \sum_{i=1}^{N} m_i p_i^{-1} + C \tag{2}$$

with

$$p_i \triangleq (\mathbf{p}_i^2)^{1/2} \qquad (i = 1, \ldots, N) \tag{3}$$

and

$$\mathbf{p}_i \triangleq \mathbf{R} + \mathbf{r}_i \qquad (i = 1, \ldots, N) \tag{4}$$

Here B is presumed to consist of particles $P_1, \ldots, P_N$ of masses $m_1, \ldots, m_N$, $\mathbf{p}_i$ is the position vector of P_i relative to $\overline{P}$, assumed nonzero, $\mathbf{r}_i$ is the position vector of P_i relative to B^*, C is an arbitrary constant, and G is the universal gravitational constant. If B is a continuous distribution of matter not including the point occupied by $\overline{P}$, then V is given by

$$V = G\overline{m} \int p^{-1} \rho \, d\tau + C \tag{5}$$

with

$$p \triangleq (\mathbf{p}^2)^{1/2} \tag{6}$$

and $\mathbf{p} = \mathbf{R} + \mathbf{r}$, where ρ is the mass density of B at a generic point P of B, $\mathbf{p}$ is the position vector of P relative to $\overline{P}$, $\mathbf{r}$ is the position vector of P relative to B^*, and $d\tau$ is the length, area, or volume of a differential element of the figure (curve, surface, or solid) occupied by B.

The force functions in Eqs. (2) and (5) both satisfy Laplace's equation:

$$\nabla^2 V = 0 \tag{7}$$

Derivations Differentiation of V with respect to **R** gives

$$\nabla_{\mathbf{R}} V \underset{(2)}{=} -G\overline{m} \sum_{i=1}^{N} m_i p_i^{-2} \nabla_{\mathbf{R}} p_i \tag{8}$$

Now,

$$\nabla_{\mathbf{R}} p_i \underset{(3,4)}{=} \nabla_{\mathbf{R}} [(\mathbf{R} + \mathbf{r}_i)^2]^{1/2} = [(\mathbf{R} + \mathbf{r}_i)^2]^{-1/2} (\mathbf{R} + \mathbf{r}_i) \cdot \nabla_{\mathbf{R}} (\mathbf{R} + \mathbf{r}_i)$$

$$\underset{(3,4)}{=} p_i^{-1} \mathbf{p}_i \cdot \nabla_{\mathbf{R}} (\mathbf{R} + \mathbf{r}_i) \underset{(2.9.7)}{=} p_i^{-1} \mathbf{p}_i \cdot (\mathbf{U} + 0) = p_i^{-1} \mathbf{p}_i \tag{9}$$

Hence,

$$\nabla_{\mathbf{R}} V \underset{(8,9)}{=} -G\overline{m} \sum_{i=1}^{N} m_i p_i^{-3} \mathbf{p}_i \underset{(2.2.1)}{=} \mathbf{F} \tag{10}$$

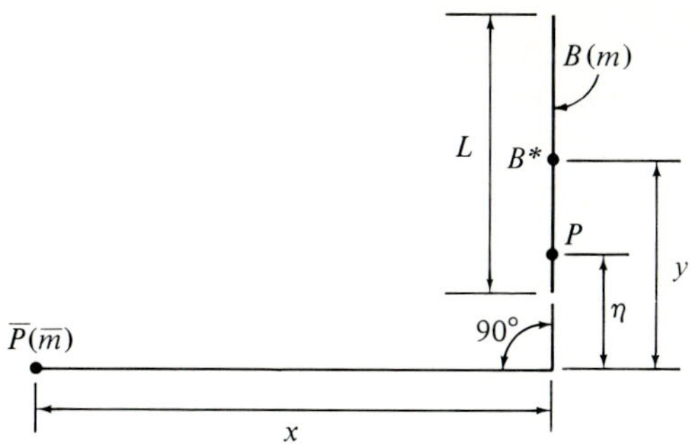

Figure 2.11.1

A parallel derivation shows that Eq. (1) remains valid when Eq. (2) is replaced by Eq. (5).

Equation (7) follows directly from Eq. (2.10.8), since the order of the operations of evaluating the gradient and integrating over B is immaterial.

Example A uniform thin rod B of length L and mass m is subjected to the gravitational attraction of a particle $\bar{P}$ of mass $\bar{m}$. The value of a force function V associated with the resultant gravitational force exerted by $\bar{P}$ on B is to be determined in terms of the Cartesian coordinates x, y of $\bar{P}$ for a coordinate system with origin at the mass center B^* of B, as shown in Fig. 2.11.1. The constant C in Eq. (3) is to be chosen such that V approaches zero when $y = L/2$ and L/x approaches zero.

If the rod is regarded as matter distributed along a straight line segment, then the distance p from $\bar{P}$ to a generic point P of the rod can be expressed as $p = (x^2 + \eta^2)^{1/2}$, where η varies from $y - L/2$ to $y + L/2$. The mass density ρ of B at P is equal to m/L, and a differential element of B has a length $d\eta$. Hence,

$$V \underset{(3)}{=} \frac{G\bar{m}m}{L} \int_{y-L/2}^{y+L/2} (x^2 + \eta^2)^{-1/2}\, d\eta + C$$

$$= \frac{G\bar{m}m}{L} \ln\left[\frac{y + L/2 + \sqrt{x^2 + (y + L/2)^2}}{y - L/2 + \sqrt{x^2 + (y - L/2)^2}}\right] + C \qquad (11)$$

and, when $y = L/2$, the limit approached by V as L/x approaches zero is C. Hence $C = 0$.

2.12 FORCE FUNCTION FOR A SMALL BODY AND A PARTICLE

When a body B is subjected to the gravitational attraction of a particle $\bar{P}$ removed so far from the mass center B^* of B that the largest distance from B^* to any point P of B is smaller than the distance R between $\bar{P}$ and B^*, a useful form of the expressions for the force function V given in Eqs. (2.11.2) and (2.11.5) can be found as follows: replace p_i and p with $[(\mathbf{R} + \mathbf{r}_i)^2]^{1/2}$ and with $[(\mathbf{R} + \mathbf{r})^2]^{1/2}$, respectively, and expand the integrand in ascending powers of $|\mathbf{r}_i|/R$ or $|\mathbf{r}|/R$ to obtain

$$V = \frac{G\overline{m}m}{R}\left[1 + \sum_{i=2}^{\infty} v^{(i)}\right] + C \tag{1}$$

where $v^{(i)}$ is a collection of terms of the ith degree in $|\mathbf{r}_i|/R$ or $|\mathbf{r}|/R$, m and $\overline{m}$ are the masses of B and $\overline{P}$, C is an arbitrary constant, and G is the universal gravitational constant. In particular, $v^{(2)}$ is given by

$$v^{(2)} = \frac{1}{2mR^2}[\text{tr}(\mathbf{I}) - 3I_{11}] \tag{2}$$

where tr(**I**) is the trace of the central inertia dyadic **I** of B, and I_{11} is the moment of inertia of B about a line connecting $\overline{P}$ and B^*, so that (see Fig. 2.3.1)

$$I_{11} \triangleq \mathbf{a}_1 \cdot \mathbf{I} \cdot \mathbf{a}_1 \tag{3}$$

Because I_{11} depends upon the orientation of $\mathbf{a}_1$ relative to B, Eq. (2) is sometimes less convenient than an alternative form involving central moments and products of inertia of B for an arbitrary vector basis $\mathbf{b}_1'$, $\mathbf{b}_2'$, $\mathbf{b}_3'$. With

$$I'_{jk} \triangleq \mathbf{b}_j' \cdot \mathbf{I} \cdot \mathbf{b}_k' \qquad (j, k = 1, 2, 3) \tag{4}$$

the required expression is

$$v^{(2)} = \frac{1}{2mR^2}[I'_{11}(1 - 3C'^2_{11}) + I'_{22}(1 - 3C'^2_{12}) + I'_{33}(1 - 3C'^2_{13}) - 6(I'_{12}C'_{11}C'_{12} + I'_{13}C'_{11}C'_{13} + I'_{23}C'_{12}C'_{13})] \tag{5}$$

where $C'_{1j} \triangleq \mathbf{a}_1 \cdot \mathbf{b}_j'$ $(j = 1, 2, 3)$. When B is a rigid body and $\mathbf{b}_1'$, $\mathbf{b}_2'$, and $\mathbf{b}_3'$ are fixed in B, the scalars I'_{jk} $(j, k = 1, 2, 3)$ in Eq. (5) become constants, whereas I_{11} in Eq. (2) remains a variable.

For the special case in which the body-fixed unit vectors are parallel to central principal axes of inertia of B,

$$v^{(2)} = \frac{1}{2mR^2}[I_1(1 - 3C_{11}^2) + I_2(1 - 3C_{12}^2) + I_3(1 - 3C_{13}^2)] \tag{6}$$

where I_j and C_{1j} $(j = 1, 2, 3)$ are given, respectively, by Eqs. (2.3.8) and (2.3.9).

Equations (5) and (6) assume particularly convenient forms when the indicated direction cosines are written in terms of the spherical coordinates shown in Fig. 2.12.1, namely,

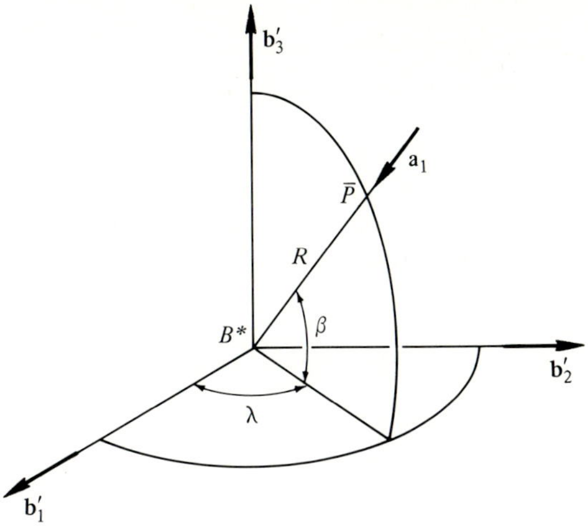

Figure 2.12.1

$$v^{(2)} = \frac{1}{4mR^2}\,[(I'_{11} + I'_{22} - 2I'_{33})(3\sin^2\beta - 1) - 3(I'_{11} - I'_{22})\cos^2\beta\cos 2\lambda$$
$$- 6(I'_{12}\cos^2\beta\sin 2\lambda + I'_{13}\sin 2\beta\cos\lambda + I'_{23}\sin 2\beta\sin\lambda)] \tag{7}$$

and

$$v^{(2)} = \frac{1}{4mR^2}\,[(I_1 + I_2 - 2I_3)(3\sin^2\beta - 1) - 3(I_1 - I_2)\cos^2\beta\cos 2\lambda] \tag{8}$$

Often it is desirable to introduce some measure of the dimensions of B into $v^{(2)}$ in order to obtain an expression involving dimensionless constants and dimensionless variables. If the symbol R_B is assigned to the selected normalizing dimension, and the symbols

$$J \triangleq -\frac{1}{2mR_B{}^2}\,(I_1 + I_2 - 2I_3) \tag{9}$$

and

$$\epsilon \triangleq -\frac{3}{2mR_B{}^2}\,(I_1 - I_2) \tag{10}$$

are introduced, then Eq. (8) may be rewritten

$$v^{(2)} = -\frac{J}{2}\left(\frac{R_B}{R}\right)^2 (3\sin^2\beta - 1) + \frac{\epsilon}{2}\left(\frac{R_B}{R}\right)^2 \cos^2\beta\cos 2\lambda \tag{11}$$

If $\widetilde{\mathbf{F}}$ is the approximation to $\mathbf{F}$ defined in Eq. (2.3.6), then $\widetilde{\mathbf{F}}$ can be expressed as

$$\widetilde{\mathbf{F}} = \nabla_{\mathbf{R}} \widetilde{V} \tag{12}$$

where $\widetilde{V}$ is given by

$$\widetilde{V} = \frac{G\bar{m}m}{R}[1 + v^{(2)}] + C \tag{13}$$

and $\mathbf{R}$ is the position vector of B^* relative to $\bar{P}$. In other words, $\widetilde{V}$ is a force function suitable for dealing with the gravitational interaction of a body and a remote particle when $\widetilde{\mathbf{F}}$ is used to approximate $\mathbf{F}$.

Derivations Replacing p in Eq. (2.11.5) with $[(\mathbf{R} + \mathbf{r})^2]^{1/2}$ gives

$$V = G\bar{m} \int (\mathbf{R}^2 + 2\mathbf{R} \cdot \mathbf{r} + \mathbf{r}^2)^{-1/2} \rho \, d\tau + C \tag{14}$$

Introducing $\mathbf{q}$ as $\mathbf{q} \triangleq \mathbf{r}/R$, noting that $\mathbf{R}/R = \mathbf{a}_1$, and applying the binomial series, convergent for $|2\mathbf{R} \cdot \mathbf{r} + \mathbf{r}^2| < 1$, leads to

$$\begin{aligned} V &= \frac{G\bar{m}}{R} \int (1 + 2\mathbf{a}_1 \cdot \mathbf{q} + \mathbf{q}^2)^{-1/2} \rho \, d\tau + C \\ &= \frac{G\bar{m}}{R} \int \left[1 - \mathbf{a}_1 \cdot \mathbf{q} - \frac{1}{2}\mathbf{q}^2 + \frac{3}{2}(\mathbf{a}_1 \cdot \mathbf{q})^2 + \cdots \right] \rho \, d\tau + C \\ &= \frac{G\bar{m}}{R} \left\{ \int \rho \, d\tau - \frac{\mathbf{a}_1}{R} \cdot \int \mathbf{r} \rho \, d\tau \right. \\ &\quad \left. - \frac{1}{2R^2} \int [\mathbf{r}^2 - 3(\mathbf{a}_1 \cdot \mathbf{r})^2] \rho \, d\tau + \cdots \right\} + C \end{aligned} \tag{15}$$

Now,

$$\int \rho \, d\tau = m \tag{16}$$

$$\int \mathbf{r} \rho \, d\tau = 0 \tag{17}$$

and

$$\int [\mathbf{r}^2 - 3(\mathbf{a}_1 \cdot \mathbf{r})^2] \rho \, d\tau = 3 \int [\mathbf{r}^2 - (\mathbf{a}_1 \cdot \mathbf{r})^2] \rho \, d\tau - 2 \int \mathbf{r}^2 \rho \, d\tau \tag{18}$$

Furthermore,

$$\int [\mathbf{r}^2 - (\mathbf{a}_1 \cdot \mathbf{r})^2] \rho \, d\tau = I_{11} \tag{19}$$

and

$$2 \int \mathbf{r}^2 \rho \, d\tau = \text{tr}(\mathbf{I}) \tag{20}$$

Hence,

$$\int [\mathbf{r}^2 - 3(\mathbf{a}_1 \cdot \mathbf{r})^2]\rho \, d\tau \underset{(18\text{-}20)}{=} 3I_{11} - \text{tr}(\mathbf{I}) \tag{21}$$

and

$$V \underset{(15\text{-}17,21)}{=} \frac{G\bar{m}m}{R}\left\{1 + \frac{1}{2mR^2}[\text{tr}(\mathbf{I}) - 3I_{11}] + \cdots\right\} + C \tag{22}$$

in agreement with Eqs. (1) and (2).

Equation (5) follows directly from Eq. (2) when I_{11} is formed in accordance with Eq. (3) and **I** is expressed as

$$\mathbf{I} \underset{(4)}{=} \sum_{j=1}^{3}\sum_{k=1}^{3} \mathbf{b}_j' I_{jk}' \mathbf{b}_k' \tag{23}$$

With the aid of these relationships, one finds that

$$\begin{aligned} v^{(2)} &= \frac{1}{2mR^2}\left(\sum_{j=1}^{3} I_{jj}' - 3\sum_{j=1}^{3}\sum_{k=1}^{3} C_{1j}' I_{jk}' C_{1k}'\right) \\ &= \frac{1}{2mR^2}[I_{11}'(1 - 3C_{11}'^2) + I_{22}'(1 - 3C_{12}'^2) + I_{33}'(1 - 3C_{13}'^2) \\ &\quad - 3(I_{12}'C_{11}'C_{12}' + I_{13}'C_{11}'C_{13}' + I_{21}'C_{12}'C_{11}' + I_{23}'C_{12}'C_{13}' \\ &\quad + I_{31}'C_{13}'C_{11}' + I_{32}'C_{13}'C_{12}')] \end{aligned} \tag{24}$$

which, since $I_{jk}' = I_{kj}'$, reduces to Eq. (5). Equation (6) is merely the special case of Eq. (5) with vanishing products of inertia and new notation, and Eqs. (7) and (8) are the special cases of Eqs. (5) and (6) in which the relationship

$$\mathbf{a}_1 = -\cos\beta\cos\lambda\,\mathbf{b}_1' - \cos\beta\sin\lambda\,\mathbf{b}_2' - \sin\beta\,\mathbf{b}_3' \tag{25}$$

(see Fig. 2.12.1) has been used to obtain for the required direction cosines the expressions

$$C_{11}' = -\cos\beta\cos\lambda \qquad C_{12}' = -\cos\beta\sin\lambda \qquad C_{13}' = -\sin\beta \tag{26}$$

Substitution from Eqs. (26) into Eq. (5) produces

$$\begin{aligned} v^{(2)} &= \frac{1}{2mR^2}[I_{11}'(1 - 3\cos^2\beta\cos^2\lambda) + I_{22}'(1 - 3\cos^2\beta\sin^2\lambda) \\ &\quad + I_{33}'(1 - 3\sin^2\beta) - 6(I_{12}'\cos^2\beta\cos\lambda\sin\lambda \\ &\quad + I_{13}'\cos\beta\cos\lambda\sin\beta + I_{23}'\sin\beta\cos\beta\sin\lambda)] \\ &= \frac{1}{2mR^2}\left\{I_{11}'\left[1 - 3\cos^2\beta\left(\frac{1}{2} + \frac{1}{2}\cos 2\lambda\right)\right]\right. \end{aligned}$$

(continued on next page)

$$+ I'_{22}\left[1 - 3\cos^2\beta\left(\frac{1}{2} - \frac{1}{2}\cos 2\lambda\right)\right]$$

$$- I'_{33}(3\sin^2\beta - 1) - 6\left[I'_{12}\cos^2\beta\left(\frac{1}{2}\sin 2\lambda\right)\right.$$

$$\left.\left. + I'_{13}\cos\lambda\left(\frac{1}{2}\sin 2\beta\right) + I'_{23}\sin\lambda\left(\frac{1}{2}\sin 2\beta\right)\right]\right\}$$

$$= \frac{1}{4mR^2}\{I'_{11}[2 - 3(1 - \sin^2\beta)] + I'_{22}[2 - 3(1 - \sin^2\beta)]$$

$$- 2I'_{33}(3\sin^2\beta - 1) - 3(I'_{11} - I'_{22})\cos^2\beta\cos 2\lambda$$

$$- 6(I'_{12}\cos^2\beta\sin 2\lambda + I'_{13}\cos\lambda\sin 2\beta + I'_{23}\sin\lambda\sin 2\beta)\}$$

$$= \frac{1}{4mR^2}[(I'_{11} + I'_{22} - 2I'_{33})(3\sin^2\beta - 1)$$

$$- 3(I'_{11} - I'_{22})\cos^2\beta\cos 2\lambda - 6(I'_{12}\cos^2\beta\sin 2\lambda$$

$$+ I'_{13}\sin 2\beta\cos\lambda + I'_{23}\sin 2\beta\sin\lambda)] \tag{27}$$

in agreement with Eq. (7). Equation (8) is the special case of Eq. (7) involving principal moments of inertia, and Eq. (11) is a restatement of Eq. (8) in terms of the symbols defined in Eqs. (9) and (10).

Differentiation of Eq. (13) gives

$$\nabla_{\mathbf{R}}\widetilde{V} = -\frac{G\overline{m}m}{R^2}\nabla_{\mathbf{R}}R[1 + \nu^{(2)}] + \frac{G\overline{m}m}{R}\nabla_{\mathbf{R}}\nu^{(2)} \tag{28}$$

From Eqs. (2.9.8) and (2.9.9),

$$\nabla_{\mathbf{R}}R = \mathbf{a}_1 \qquad \nabla_{\mathbf{R}}\mathbf{a}_1 = R^{-1}(\mathbf{U} - \mathbf{a}_1\mathbf{a}_1) \tag{29}$$

and from Eq. (2), with I_{11} replaced by $\mathbf{a}_1 \cdot \mathbf{I} \cdot \mathbf{a}_1$,

$$\nabla_{\mathbf{R}}\nu^{(2)} = -\frac{1}{mR^3}\nabla_{\mathbf{R}}R[\mathrm{tr}(\mathbf{I}) - 3\mathbf{a}_1 \cdot \mathbf{I} \cdot \mathbf{a}_1] - \frac{3}{mR^2}\nabla_{\mathbf{R}}\mathbf{a}_1 \cdot \mathbf{I} \cdot \mathbf{a}_1$$

$$\underset{(29)}{=} -\frac{1}{mR^3}[\mathbf{a}_1\mathrm{tr}(\mathbf{I}) - 6\mathbf{a}_1\mathbf{a}_1 \cdot \mathbf{I} \cdot \mathbf{a}_1 + 3\mathbf{I} \cdot \mathbf{a}_1] \tag{30}$$

Substituting from Eqs. (29) and (30) into Eq. (28) and using Eq. (2), one can thus write

$$\nabla_{\mathbf{R}}\widetilde{V} = -\frac{G\overline{m}m}{R^2}\left\{\mathbf{a}_1 + \frac{1}{2mR^2}[\mathbf{a}_1\mathrm{tr}(\mathbf{I}) - 3\mathbf{a}_1\mathbf{a}_1 \cdot \mathbf{I} \cdot \mathbf{a}_1]\right.$$

(continued on next page)

$$+\frac{1}{mR^2}[\mathbf{a}_1\,\mathrm{tr}(\mathbf{I}) - 6\mathbf{a}_1\mathbf{a}_1\cdot\mathbf{I}\cdot\mathbf{a}_1 + 3\mathbf{I}\cdot\mathbf{a}_1]\Big\}$$

$$= -\frac{G\overline{m}m}{R^2}\left(\mathbf{a}_1 + \frac{1}{mR^2}\left\{\frac{3}{2}[\mathrm{tr}(\mathbf{I}) - 5\mathbf{a}_1\cdot\mathbf{I}\cdot\mathbf{a}_1]\,\mathbf{a}_1 + 3\mathbf{I}\cdot\mathbf{a}_1\right\}\right)$$

$$\underset{(2.3.3)}{=} -\frac{G\overline{m}m}{R^2}[\mathbf{a}_1 + \mathbf{f}^{(2)}] \underset{(2.3.6)}{=} \widetilde{\mathbf{F}} \tag{31}$$

which establishes the validity of Eq. (12).

Example Figure. 2.12.2 shows a thin rod B of mass m and a particle $\overline{P}$ of mass $\overline{m}$. If the distance R between $\overline{P}$ and the mass center B^* of B is larger than the length L of B, and if B is idealized as matter distributed along a straight line segment, then, in accordance with the results of the example in Sec. 2.3, the gravitational force $\mathbf{F}$ exerted on B by $\overline{P}$ is approximately equal to a force $\widetilde{\mathbf{F}}$ given by (see Fig. 2.12.2 for ψ and $\mathbf{a}_1$)

$$\widetilde{\mathbf{F}} = -\frac{G\overline{m}m}{R^2}\left[\mathbf{a}_1 + \frac{L^2}{8R^2}(2 - 3\sin^2\psi)\mathbf{a}_1 - \frac{L^2}{8R^2}\sin 2\psi\,\mathbf{a}_2\right] \tag{32}$$

A force function V such that Eq. (12) is satisfied can be found by using

$$\mathbf{I} = \frac{mL^2}{12}(\mathbf{b}_2\mathbf{b}_2 + \mathbf{b}_3\mathbf{b}_3) \tag{33}$$

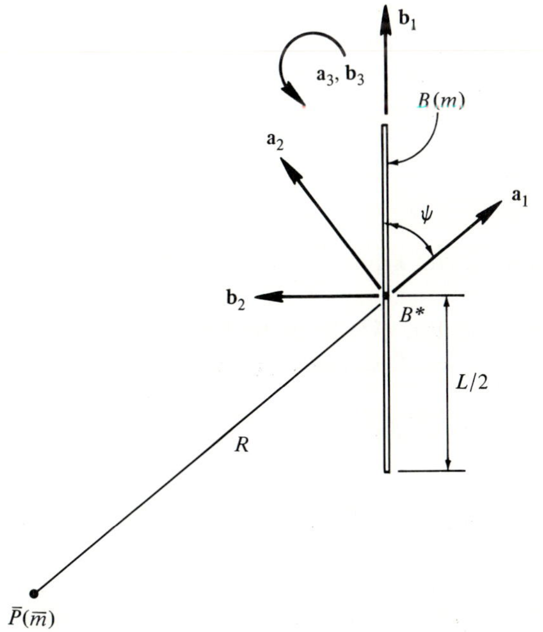

Figure 2.12.2

to evaluate tr(**I**) and I_{11} as

$$\mathrm{tr}(\mathbf{I}) = \frac{mL^2}{6} \qquad I_{11} = \mathbf{a}_1 \cdot \mathbf{I} \cdot \mathbf{a}_1 = \frac{mL^2}{12} \sin^2 \psi \tag{34}$$

after which substitution into Eqs. (2) and (13) gives

$$\tilde{V} = \frac{G\bar{m}m}{R}\left[1 + \frac{L^2}{24R^2}(2 - 3\sin^2\psi)\right] + C \tag{35}$$

In order to obtain this result by means of Eq. (11) rather than Eq. (2), one must recognize by comparing Figs. 2.12.1 and 2.12.2 that now $\beta = 0$ and $\lambda = -\psi$. With

$$I_1 = 0 \qquad I_2 = I_3 = \frac{mL^2}{12} \tag{36}$$

Eq. (11) thus provides

$$\nu^{(2)} \underset{(11)}{=} -\frac{J}{2}\left(\frac{R_B}{R}\right)^2(-1) + \frac{\epsilon}{2}\left(\frac{R_B}{R}\right)^2(1)\cos 2\psi \tag{37}$$

where

$$J \underset{(9)}{=} -\frac{1}{2mR_B{}^2}\left(0 + \frac{mL^2}{12} - \frac{2mL^2}{12}\right) = \frac{L^2}{24R_B{}^2} \tag{38}$$

and

$$\epsilon \underset{(10)}{=} -\frac{3}{2mR_B{}^2}\left(0 - \frac{mL^2}{12}\right) = \frac{L^2}{8R_B{}^2} \tag{39}$$

Although it is customary to make a specific choice for R_B (such as $R_B = L/2$) in order to obtain numerical values for J and ϵ, it is apparent that in combining Eqs. (37)–(39) one can cancel R_B from $\nu^{(2)}$, so that the choice of R_B does not affect $\nu^{(2)}$; that is,

$$\nu^{(2)} \underset{(37\text{-}39)}{=} \frac{L^2}{R^2}\left(\frac{1}{48} + \frac{1}{16}\cos 2\psi\right)$$

$$= \frac{L^2}{48R^2}[1 + 3(1 - 2\sin^2\psi)] = \frac{L^2}{24R^2}(2 - 3\sin^2\psi) \tag{40}$$

and substitution from Eq. (40) into Eq. (13) leads to Eq. (35).

To verify that Eq. (12) is indeed satisfied, one may proceed as follows:

$$\nabla_{\mathbf{R}}\tilde{V} \underset{(35)}{=} -\frac{G\bar{m}m}{R^2}\nabla_{\mathbf{R}}R\left[1 + \frac{L^2}{24R^2}(2 - 3\sin^2\psi)\right]$$

$$+ \frac{G\bar{m}m}{R}\left[-\frac{L^2}{12R^3}(2 - 3\sin^2\psi)\nabla_{\mathbf{R}}R - \frac{L^2}{8R^2}\sin 2\psi\, \nabla_{\mathbf{R}}\psi\right] \tag{41}$$

Now,

$$\mathbf{R} = R(\cos\psi\, \mathbf{b}_1 - \sin\psi\, \mathbf{b}_2) \tag{42}$$

Hence,

$$\nabla_{\mathbf{R}} \mathbf{R} = \nabla_{\mathbf{R}} R(\cos\psi\, \mathbf{b}_1 - \sin\psi\, \mathbf{b}_2)$$
$$+ R(-\sin\psi \nabla_{\mathbf{R}}\psi\, \mathbf{b}_1 - \cos\psi\, \nabla_{\mathbf{R}}\psi\, \mathbf{b}_2) \tag{43}$$

or, in view of Eqs. (2.9.7) and (2.9.8),

$$\mathbf{U} = \mathbf{a}_1(\cos\psi\, \mathbf{b}_1 - \sin\psi\, \mathbf{b}_2) - R\, \nabla_{\mathbf{R}}\psi(\sin\psi\, \mathbf{b}_1 + \cos\psi\, \mathbf{b}_2)$$
$$= \mathbf{a}_1 \mathbf{a}_1 - R\nabla_{\mathbf{R}}\psi\, \mathbf{a}_2 \tag{44}$$

Dot-multiplication with $\mathbf{a}_2$ thus gives

$$\mathbf{a}_2 = -R\nabla_{\mathbf{R}}\psi \tag{45}$$

and using $\nabla_{\mathbf{R}} R = \mathbf{a}_1$ and $\nabla_{\mathbf{R}}\psi = -\mathbf{a}_2/R$, one obtains

$$\nabla_{\mathbf{R}} \widetilde{V} \underset{(41)}{=} -\frac{G\bar{m}m}{R^2}\left[\mathbf{a}_1 + \frac{L^2}{8R^2}(2 - 3\sin^2\psi)\,\mathbf{a}_1 - \frac{L^2}{8R^2}\sin^2\psi\, \mathbf{a}_2\right] \underset{(32)}{=} \widetilde{\mathbf{F}} \tag{46}$$

2.13 FORCE FUNCTION IN TERMS OF LEGENDRE POLYNOMIALS

The quantities $v^{(i)}$ $(i = 2, \ldots, \infty)$ appearing in Eq. (2.12.1) can be expressed as

$$v^{(i)} = \frac{1}{m}\int \left(\frac{r}{R}\right)^i P_i(\cos\alpha)\rho\, d\tau \qquad (i = 2, \ldots, \infty) \tag{1}$$

where r is the distance from B^* to a generic point P of B (r must be smaller than R for all points of B), ρ is the mass density of B at P, α is the angle between the lines joining B^* to P and to $\bar{P}$ (see Fig. 2.13.1), $P_i(\cos\alpha)$ is a Legendre polynomial, to be defined presently, and $d\tau$ is the volume of a differential element of the figure occupied by B.

By definition, the *Legendre polynomials* $P_0(y)$, $P_1(y)$, ... are given by

$$P_0(y) \triangleq 1 \tag{2}$$

$$P_i(y) \triangleq \frac{1 \cdot 3 \cdot \cdots \cdot (2i-1)}{i!}\left[y^i - \frac{i(i-1)}{(2i-1)\cdot 2}y^{i-2} + \frac{i(i-1)(i-2)(i-3)}{(2i-1)(2i-3)\cdot 2\cdot 4}y^{i-4} + \cdots\right] \qquad (i = 1, \ldots, \infty) \tag{3}$$

where the last term in the bracketed series involves y^0 if i is even and y^1 if i is odd. For example,

$$P_1(y) = y \qquad P_2(y) = \frac{3y^2 - 1}{2} \qquad P_3(y) = \frac{5y^3 - 3y}{2} \tag{4}$$

The integrand in Eq. (1) depends on the position of $\bar{P}$ relative to B because it involves the angle α. To express $v^{(i)}$ in terms of integrals that are independent of the position of $\bar{P}$ and, therefore, can be evaluated as soon as the mass

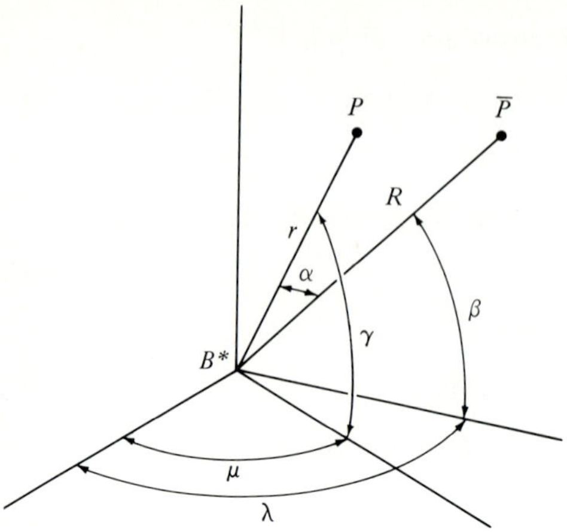

Figure 2.13.1

distribution of B has been specified, one works with *associated Legendre functions,* defined as

$$P_i^j(y) \triangleq \frac{(2i)!(1-y^2)^{j/2}}{2^i i!(i-j)!}\left[y^{i-j} - \frac{(i-j)(i-j-1)}{(2i-1)\cdot 2}y^{i-j-2} + \frac{(i-j)(i-j-1)(i-j-2)(i-j-3)}{(2i-1)(2i-3)\cdot 2\cdot 4}y^{i-j-4} + \cdots\right]$$

$$(i, j = 0, 1, \ldots, \infty) \quad (5)$$

where the last term in the bracketed series involves y^0 if $i-j$ is even and y^1 if $i-j$ is odd. For example,

$$P_1^{\,1}(y) = (1-y^2)^{1/2} \qquad P_2^{\,1}(y) = 3y(1-y^2)^{1/2} \qquad P_2^{\,2}(y) = 3(1-y^2) \quad (6)$$

and, as can be seen by reference to Eq. (3),

$$P_i^{\,0}(y) = P_i(y) \qquad (i = 0, \ldots, \infty) \quad (7)$$

The mass distribution of B is taken into consideration via constants C_{ij} and S_{ij} defined as

$$C_{i0} \triangleq \frac{1}{m}\int\left(\frac{r}{R_B}\right)^i P_i\,(\sin\gamma)\rho\, d\tau \qquad (i = 2, \ldots, \infty) \quad (8)$$

$$C_{ij} \triangleq \frac{2}{m}\frac{(i-j)!}{(i+j)!}\int\left(\frac{r}{R_B}\right)^i P_i^j\,(\sin\gamma)\cos j\mu\ \rho d\tau$$

$$(i = 2, \ldots, \infty;\ j = 1, \ldots, \infty) \quad (9)$$

$$S_{ij} \triangleq \frac{2}{m}\frac{(i-j)!}{(i+j)!}\int\left(\frac{r}{R_B}\right)^i P_i^j(\sin\gamma)\sin j\mu\ \rho d\tau$$
$$(i = 2, \ldots, \infty;\ j = 1, \ldots, \infty) \quad (10)$$

where μ, γ, and r are the spherical coordinates of P shown in Fig. 2.13.1, and R_B, an arbitrary quantity having the dimensions of length, is introduced to render C_{ij} and S_{ij} dimensionless. Expressed in terms of the quantities just defined, $v^{(i)}$ is given by

$$v^{(i)} = \left(\frac{R_B}{R}\right)^i \sum_{j=0}^{i} P_i^j(\sin\beta)(C_{ij}\cos j\lambda + S_{ij}\sin j\lambda) \qquad (i = 2, \ldots, \infty) \quad (11)$$

where λ, β, and R are the spherical coordinates of $\bar{P}$ shown in Fig. 2.13.1.

Substitution from Eq. (11) into Eq. (2.12.1) leads directly to

$$V = \frac{G\bar{m}m}{R}\left\{1 + \sum_{i=2}^{\infty}\left[\left(\frac{R_B}{R}\right)^i \sum_{j=0}^{i} P_i^j(\sin\beta)(C_{ij}\cos j\lambda + S_{ij}\sin j\lambda)\right]\right\} + C \quad (12)$$

Here V is written in the form adopted as a standard representation by the International Astronomical Union for the case when B is the Earth, in which context R_B is taken to be the Earth's mean equatorial radius. Since the series converges only for $R > R_B$, it fails to converge for points in a certain region of space above the surface of the Earth, this region varying in thickness from zero at the equator to approximately 20 km at the poles.

If B is axisymmetric and the symmetry axis of B is the line $\gamma = \pi/2$ in Fig. 2.13.1, then $S_{ij} = 0$ $(i, j = 0, \ldots, \infty)$, $C_{ij} = 0$ $(i, j = 2, \ldots, \infty)$, and the symbol J_i is traditionally introduced as

$$J_i \triangleq -C_{i0} \qquad (i = 2, \ldots, \infty) \quad (13)$$

Only the $j = 0$ term in Eq. (12) survives under these circumstances, so that, referring to Eq. (7), one can write

$$V = \frac{G\bar{m}m}{R}\left[1 - \sum_{i=2}^{\infty}\left(\frac{R_B}{R}\right)^i J_i P_i(\sin\beta)\right] + C \quad (14)$$

Derivations If p is the distance between P and $\bar{P}$ in Fig. 2.13.1, then $p^2 = R^2 - 2rR\cos\alpha + r^2$ and

$$V \underset{(2.11.5)}{=} \frac{G\bar{m}}{R}\int\left[1 - 2\frac{r}{R}\cos\alpha + \left(\frac{r}{R}\right)^2\right]^{-1/2} \rho\, d\tau + C \quad (15)$$

Use of the identity†

$$(1 - 2yx + y^2)^{-1/2} = \sum_{i=0}^{\infty} y^i P_i(x) \quad (16)$$

which applies if $|y| < 1$, permits the replacement of Eq. (15) with

†O. D. Kellogg, *Foundations of Potential Theory*, Ungar Publishing Company, New York, 1929, p. 128.

$$V = \frac{G\overline{m}}{R}\sum_{i=0}^{\infty}\int\left(\frac{r}{R}\right)^i P_i(\cos\alpha)\,\rho\,d\tau + C \tag{17}$$

if $r/R < 1$ for all points of B. The first two terms of Eq. (17) simplify since [see Eqs. (2) and (4)] $P_0(\cos\alpha) = 1$ and $P_1(\cos\alpha) = \cos\alpha$, so that

$$\int\left(\frac{r}{R}\right)^0 P_0(\cos\alpha)\,\rho\,d\tau = \int\rho\,d\tau = m \tag{18}$$

and

$$\int\left(\frac{r}{R}\right)^1 P_1(\cos\alpha)\,\rho\,d\tau = \frac{1}{R}\int r\cos\alpha\,\rho\,d\tau \tag{19}$$

The last integral vanishes by virtue of the definition of B^* as the mass center of B. Consequently,

$$V \underset{(17\text{-}19)}{=} \frac{G\overline{m}m}{R}\left[1 + \sum_{i=2}^{\infty}\frac{1}{m}\int\left(\frac{r}{R}\right)^i P_i(\cos\alpha)\,\rho\,d\tau\right] + C \tag{20}$$

Equation (1) follows directly from this relationship and Eq. (2.12.1).

To establish the validity of Eq. (11), we first form $\cos\alpha$ by dot-multiplying a unit vector directed from B^* to P with a unit vector directed from B^* to $\overline{P}$, which gives

$$\cos\alpha = \sin\beta\,\sin\gamma + \cos\beta\,\cos\gamma\,\cos(\lambda - \mu) \tag{21}$$

and then take advantage of the fact that, with the aid of Eq. (21), one may write†

$$P_i(\cos\alpha) = P_i(\sin\beta)\,P_i(\sin\gamma) + 2\sum_{j=1}^{i}\frac{(i-j)!}{(i+j)!}P_i^{\,j}(\sin\beta)\,P_i^{\,j}(\sin\gamma)\cos[j(\lambda-\mu)] \qquad (i = 2, \ldots, \infty) \tag{22}$$

so that substitution into Eq. (1) yields

$$\begin{aligned} v^{(i)} &= \frac{P_i(\sin\beta)}{m}\int\left(\frac{r}{R}\right)^i P_i(\sin\gamma)\rho\,d\tau \\ &\quad + \frac{2}{m}\sum_{j=1}^{i}\frac{(i-j)!}{(i+j)!}P_i^{\,j}(\sin\beta)\left[\cos j\lambda\int\left(\frac{r}{R}\right)^i P_i^{\,j}(\sin\gamma)\cos j\mu\,\rho d\tau\right. \\ &\quad \left. + \sin j\lambda\int\left(\frac{r}{R}\right)^i P_i^{\,j}(\sin\gamma)\sin j\mu\,\rho d\tau\right] \\ &\underset{(8\text{-}10)}{=} \left(\frac{R_B}{R}\right)^i P_i(\sin\beta)\,C_{i0} \\ &\quad + \left(\frac{R_B}{R}\right)^i \sum_{j=1}^{i} P_i^{\,j}(\sin\beta)(C_{ij}\cos j\lambda + S_{ij}\sin j\lambda) \qquad (i = 2, \ldots, \infty) \end{aligned} \tag{23}$$

which, in view of Eq. (7), is seen to be equivalent to Eq. (11).

†W.E. Byerly, *An Elementary Treatise on Fourier Series and Spherical, Cylindrical, and Ellipsoidal Harmonics,* Ginn and Company, Boston, 1893, p. 211.

Example An approximate force function V is to be constructed for the gravitational interaction of a uniform circular ring B of radius r and mass m and a particle $\bar{P}$ of mass $\bar{m}$, the particle being situated as shown in Fig. 2.13.2, where X_1, X_2, X_3 are mutually perpendicular lines.

Since the ring is axisymmetric, Eq. (14) may be used. This requires evaluation of J_i and $P_i(\sin\beta)$ for $i = 2, 3, \ldots, \infty$. We shall consider only $i = 2, 3, 4$.

From Eq. (13),

$$J_2 = -C_{20} \underset{(8)}{=} -\frac{1}{m}\int\left(\frac{r}{R_B}\right)^2 P_2(\sin\gamma)\rho\, d\tau \tag{24}$$

Setting $\gamma = 0$ (see Figs. 2.13.1 and 2.13.2), we have

$$P_2(\sin\gamma) = P_2(0) \underset{(4)}{=} -\frac{1}{2} \tag{25}$$

so that

$$J_2 \underset{(24,25)}{=} -\frac{1}{m}\int\left(\frac{r}{R_B}\right)^2\left(-\frac{1}{2}\right)\rho\, d\tau = \frac{1}{2}\left(\frac{r}{R_B}\right)^2 \tag{26}$$

while

$$P_2(\sin\beta) \underset{(4)}{=} \frac{1}{2}(3\sin^2\beta - 1) \tag{27}$$

Proceeding similarly, one finds that $J_3 = 0$ and that

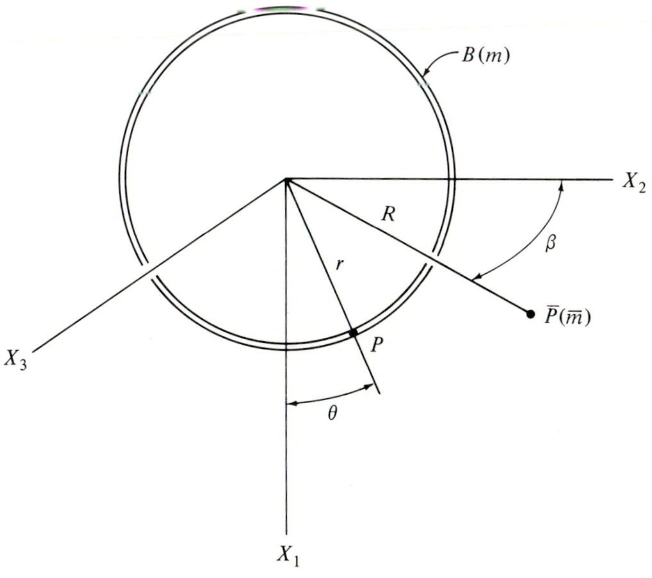

Figure 2.13.2

$$J_4 = -\frac{3}{4}\left(\frac{r}{R_B}\right)^4 \tag{28}$$

while

$$P_4(\sin\beta) \underset{(3)}{=} \frac{1}{8}(35\sin^4\beta - 30\sin^2\beta + 3) \tag{29}$$

Hence,

$$V \underset{(14)}{\approx} \frac{G\bar{m}m}{R}\left[1 - \frac{1}{4}\left(\frac{r}{R}\right)^2(3\sin^2\beta - 1) + \frac{3}{32}\left(\frac{r}{R}\right)^4(35\sin^4\beta - 30\sin^2\beta + 3)\right] + C \tag{30}$$

2.14 FORCE FUNCTION FOR TWO SMALL BODIES

When the distance R between the mass centers B^* and $\bar{B}^*$ of two (not necessarily rigid) bodies B and $\bar{B}$ exceeds in the case of each body the greatest distance from the mass center to any point of the body, so that the resultant of the system of gravitational forces exerted on B by $\bar{B}$ can be approximated by a force $\widetilde{\mathbf{F}}$ defined as [see Eq. (2.4.14)]

$$\widetilde{\mathbf{F}} \triangleq -\frac{G\bar{m}m}{R^2}[\mathbf{a}_1 + \mathbf{f}^{(2)} + \bar{\mathbf{f}}^{(2)}] \tag{1}$$

where G is the universal gravitational constant, m and $\bar{m}$ are the masses of B and $\bar{B}$, $\mathbf{a}_1$ is a unit vector directed from $\bar{B}^*$ to B^*, and $\mathbf{f}^{(2)}$ and $\bar{\mathbf{f}}^{(2)}$ are given by Eqs. (2.4.2) and (2.4.3), then a force function $\widetilde{V}$ such that

$$\widetilde{\mathbf{F}} = \nabla_{\mathbf{R}}\widetilde{V} \tag{2}$$

where $\mathbf{R} = R\mathbf{a}_1$, is given by

$$\widetilde{V} = \frac{G\bar{m}m}{R}[1 + v^{(2)} + \bar{v}^{(2)}] + C \tag{3}$$

with

$$v^{(2)} = \frac{1}{2mR^2}[\mathrm{tr}(\mathbf{I}) - 3I_{11}] \tag{4}$$

and

$$\bar{v}^{(2)} = \frac{1}{2\bar{m}R^2}[\mathrm{tr}(\bar{\mathbf{I}}) - 3\bar{I}_{11}] \tag{5}$$

where $\mathrm{tr}(\mathbf{I})$ and $\mathrm{tr}(\bar{\mathbf{I}})$ are the traces of the central inertia dyadics of B and $\bar{B}$, respectively, while I_{11} and $\bar{I}_{11}$ are, respectively, the moments of inertia of B and $\bar{B}$ about the line connecting B^* and $\bar{B}^*$.

Derivation Differentiating Eq. (3) with respect to R and then proceeding as in the derivation of Eq. (2.12.13), one arrives at

$$\nabla_{\mathbf{R}}\widetilde{V} = -\frac{G\overline{m}m}{R^2}\left(\mathbf{a}_1 + \frac{1}{mR^2}\left\{\frac{3}{2}\left[\mathrm{tr}(\mathbf{I}) - 5\mathbf{a}_1\cdot\mathbf{I}\cdot\mathbf{a}_1\right]\mathbf{a}_1 + 3\mathbf{I}\cdot\mathbf{a}_1\right\}\right.$$
$$\left. + \frac{1}{mR^2}\left\{\frac{3}{2}\left[\mathrm{tr}(\overline{\mathbf{I}}) - 5\mathbf{a}_1\cdot\overline{\mathbf{I}}\cdot\mathbf{a}_1\right]\mathbf{a}_1 + 3\overline{\mathbf{I}}\cdot\mathbf{a}_1\right\}\right) \tag{6}$$

and use of Eqs. (2.4.2) and (2.4.3) then gives

$$\nabla_{\mathbf{R}}\widetilde{V} = -\frac{G\overline{m}m}{R^2}\left[\mathbf{a}_1 + \mathbf{f}^{(2)} + \overline{\mathbf{f}}^{(2)}\right] = \underset{(1)}{\widetilde{\mathbf{F}}} \tag{7}$$

Example The example in Sec. 2.4 deals with an approximation $\widetilde{\mathbf{F}}$ to the gravitational force exerted on a rectangular parallelepiped by an oblate spheroid. To construct a force function $\widetilde{V}$ that satisfies Eq. (2), note that the moments of inertia I_{11} and $\overline{I}_{11}$ can be expressed as

$$I_{11} = \frac{m\alpha^2}{12}\left[2 - \epsilon^2 + (\mathbf{a}_1\cdot\mathbf{b}_1)^2\epsilon^2\right] \tag{8}$$

$$\overline{I}_{11} = \frac{\overline{m}\,\overline{\alpha}^2}{5}\left[2 - \overline{\epsilon}^2 + (\mathbf{a}_1\cdot\overline{\mathbf{b}}_1)^2\overline{\epsilon}^2\right] \tag{9}$$

while

$$\mathrm{tr}(\mathbf{I}) = \frac{m\alpha^2}{6}(3 - \epsilon^2) \tag{10}$$

and

$$\mathrm{tr}(\overline{\mathbf{I}}) = \frac{2\overline{m}\,\overline{\alpha}^2}{5}(3 - \overline{\epsilon}^2) \tag{11}$$

$\widetilde{V}$, formed by substituting into Eqs. (3)–(5), is given by

$$\widetilde{V} = \frac{G\overline{m}m}{R}\left\{1 + \frac{\alpha^2\epsilon^2}{24R^2}\left[1 - 3(\mathbf{a}_1\cdot\mathbf{b}_1)^2\right]\right.$$
$$\left. + \frac{\overline{\alpha}^2\overline{\epsilon}^2}{10R^2}\left[1 - 3(\mathbf{a}_1\cdot\mathbf{b}_1)^2\right]\right\} + C \tag{12}$$

and it may be verified that differentiation of $\widetilde{V}$ with respect to $\mathbf{R}$ [using Eqs. (2.9.8) and (2.9.9)] leads to $\widetilde{\mathbf{F}}$ as given by Eqs. (2.4.14) together with Eqs. (2.4.26) and (2.4.29).

2.15 FORCE FUNCTIONS FOR A CENTROBARIC BODY

Since for purposes of dealing with gravitational effects a centrobaric body (see Sec. 2.5) may be replaced with a particle situated at the mass center of the body, force functions suitable for dealing with the gravitational interaction of a centrobaric body and a particle, a centrobaric body and any body whatsoever, or a centrobaric body and a remote body can be obtained directly from Eqs. (2.10.2), (2.11.2), (2.11.5), (2.12.1), (2.13.12), or (2.13.14).

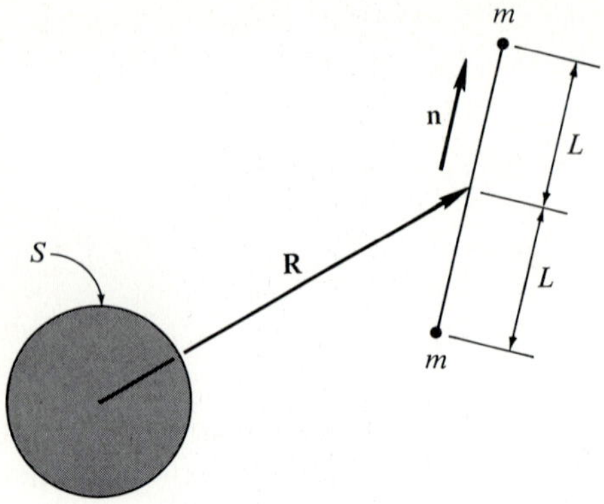

Figure 2.15.1

Example The gravitational force **F** exerted by a uniform sphere S of mass M on a dumbbell of mass $2m$ (see Fig. 2.15.1) can be expressed as

$$\mathbf{F} = \nabla_{\mathbf{R}} V \tag{1}$$

with

$$V \underset{(2.11.2)}{=} GMm\{[(\mathbf{R} + L\mathbf{n})^2]^{-1/2} + [(\mathbf{R} - L\mathbf{n})^2]^{-1/2}\} + C \tag{2}$$

where **n** is a unit vector directed as shown in Fig. 2.15.1.

2.16 FORCE FUNCTION FOR A BODY AND A SMALL BODY

When the distance between the mass centers B^* and $\bar{B}^*$ of two bodies B and $\bar{B}$ exceeds the greatest distance from B^* to any point of B, and a force function $V(\mathbf{p})$ for the force exerted by $\bar{B}$ on a particle of unit mass at a point P situated as shown in Fig. 2.16.1 is available, then there exists a function $\tilde{V}(\mathbf{R})$, where $\mathbf{R}$ is the position vector of B^* relative to $\bar{B}^*$, such that the resultant of all gravitational forces exerted by $\bar{B}$ on B can be approximated by a force $\tilde{\mathbf{F}}$ expressed as

$$\tilde{\mathbf{F}} = \nabla \tilde{V}(\mathbf{R}) \tag{1}$$

where the symbol ∇ denotes differentiation with respect to **R**. The function $\tilde{V}(\mathbf{R})$ is given by†

$$\tilde{V}(\mathbf{R}) = mV(\mathbf{R}) - \frac{1}{2}\mathbf{I} : \nabla\nabla V(\mathbf{R}) + C \tag{2}$$

where m is the mass of B, **I** is the central inertia dyadic of B, and C is an arbitrary constant.

†The double dot product in Eq. (2) is defined such that, for two dyads $\mathbf{u}_1\mathbf{u}_2$ and $\mathbf{v}_1\mathbf{v}_2$, $(\mathbf{u}_1\mathbf{u}_2) : (\mathbf{v}_1\mathbf{v}_2) = (\mathbf{u}_1 \cdot \mathbf{v}_1)(\mathbf{u}_2 \cdot \mathbf{v}_2)$; and it obeys the distributive law when applied to dyadics:

$$(\mathbf{a}_1\mathbf{a}_2 + \mathbf{b}_1\mathbf{b}_2 + \cdots) : (\mathbf{A}_1\mathbf{A}_2 + \mathbf{B}_1\mathbf{B}_2 + \cdots) = \mathbf{a}_1\mathbf{a}_2 : \mathbf{A}_1\mathbf{A}_2 + \mathbf{a}_1\mathbf{a}_2 : \mathbf{B}_1\mathbf{B}_2 + \cdots$$
$$+ \mathbf{b}_1\mathbf{b}_2 : \mathbf{A}_1\mathbf{A}_2 + \mathbf{b}_1\mathbf{b}_2 : \mathbf{B}_1\mathbf{B}_2 + \cdots$$
$$+ \cdots$$

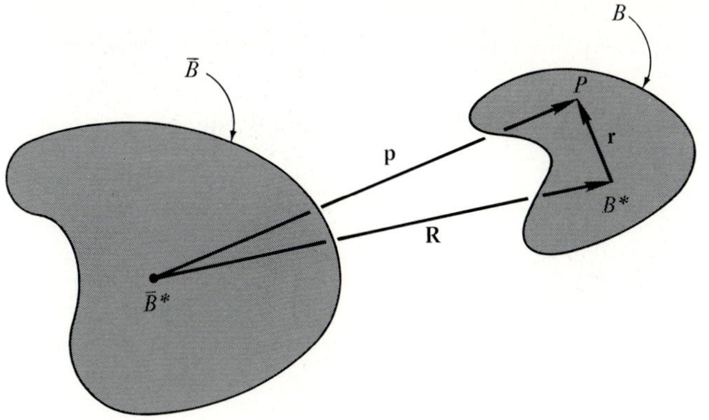

Figure 2.16.1

Derivations The gravitational force $d\mathbf{F}$ exerted by $\overline{B}$ on a differential element of B at P can be expressed as

$$d\mathbf{F} \underset{(2.11.1)}{=} \nabla_{\mathbf{p}} V \rho \, d\tau \tag{3}$$

where ρ is the mass density of B at P while $d\tau$ is the volume of the element. Consequently, the resultant $\mathbf{F}$ of all gravitational forces exerted by $\overline{B}$ on B is given by

$$\mathbf{F} = \int \nabla_{\mathbf{p}} V \rho \, d\tau \tag{4}$$

To obtain the desired approximation $\widetilde{\mathbf{F}}$ to $\mathbf{F}$, one can expand $V(\mathbf{p})$ in a Taylor series about $\mathbf{p} = \mathbf{R}$, retaining only terms up to and including those of degree 3 in $|\mathbf{r}|$, where $\mathbf{r} \triangleq \mathbf{p} - \mathbf{R}$ (see Fig. 2.16.1); differentiate with respect to $\mathbf{p}$; and then carry out the integration indicated in Eq. (4). These tasks are facilitated by introducing a function $W(\mathbf{r}) \triangleq V(\mathbf{r} + \mathbf{R})$, which permits one to write

$$\nabla_{\mathbf{p}} V \underset{(2.9.5)}{=} \nabla_{\mathbf{r}} W \tag{5}$$

and

$$\mathbf{F} \underset{(4,5)}{=} \int \nabla_{\mathbf{r}} W \rho \, d\tau \tag{6}$$

If $\mathbf{n}_1$, $\mathbf{n}_2$, $\mathbf{n}_3$ are any mutually perpendicular unit vectors and $r_i \triangleq \mathbf{r} \cdot \mathbf{n}_i$ $(i = 1, 2, 3)$, the Taylor series expansion of W about $\mathbf{r} = 0$ can be written (using the summation convention)

$$W = W(0) + r_i W_{,i} + \frac{1}{2!} r_i r_j W_{,ij} + \frac{1}{3!} r_i r_j r_k W_{,ijk} + \cdots \tag{7}$$

where

$$W_{,i} \triangleq \left.\frac{\partial W}{\partial r_i}\right|_{\mathbf{r}=0} \quad (i = 1, 2, 3) \qquad W_{,ij} \triangleq \left.\frac{\partial^2 W}{\partial r_i\, \partial r_j}\right|_{\mathbf{r}=0} \quad (i, j = 1, 2, 3) \tag{8}$$

and so forth. Differentiating Eq. (7) with respect to r_l, one obtains

$$\begin{aligned}\frac{\partial W}{\partial r_l} \underset{(7,8)}{=} 0 &+ \delta_{il} W_{,i} + \frac{1}{2!}(\delta_{il} r_j + r_i \delta_{jl}) W_{,ij} \\ &+ \frac{1}{3!}(\delta_{il} r_j r_k + r_i \delta_{jl} r_k + r_i r_j \delta_{kl}) W_{,ijk} + \cdots \\ &= W_{,l} + r_i W_{,il} + \frac{1}{2!} r_i r_j W_{,ijl} + \cdots \end{aligned} \tag{9}$$

Hence,

$$\nabla_{\mathbf{r}} W \underset{(2.9.1)}{=} (\nabla_{\mathbf{r}} W)_{\mathbf{r}=0} + \mathbf{r} \cdot (\nabla_{\mathbf{r}} \nabla_{\mathbf{r}} W)_{\mathbf{r}=0} + \frac{1}{2} \mathbf{rr} : (\nabla_{\mathbf{r}} \nabla_{\mathbf{r}} \nabla_{\mathbf{r}} W)_{\mathbf{r}=0} + \cdots \tag{10}$$

Substituting into Eq. (6) only the terms here displayed explicitly and noting that

$$\int \rho \, d\tau = m \qquad \int \mathbf{r}\, \rho \, d\tau = 0 \qquad \int \mathbf{rr}\, \rho \, d\tau = \frac{1}{2} \operatorname{tr}(\mathbf{I})\mathbf{U} - \mathbf{I} \tag{11}$$

where $\mathbf{U}$ is the unit dyadic, one can now write

$$\widetilde{\mathbf{F}} = m(\nabla_{\mathbf{r}} W)_{\mathbf{r}=0} + \frac{1}{2}\left[\frac{1}{2} \operatorname{tr}(\mathbf{I})\,\mathbf{U} - \mathbf{I}\right] : (\nabla_{\mathbf{r}} \nabla_{\mathbf{r}} \nabla_{\mathbf{r}} W)_{\mathbf{r}=0} \tag{12}$$

or, after using Eq. (5),

$$\widetilde{\mathbf{F}} = \left(\nabla_{\mathbf{p}} \left\{mV + \frac{1}{2}\left[\frac{1}{2} \operatorname{tr}(\mathbf{I})\mathbf{U} - \mathbf{I}\right] : \nabla_{\mathbf{p}} \nabla_{\mathbf{p}} V\right\} + C\right)_{\mathbf{p}=\mathbf{R}} \tag{13}$$

where C is an arbitrary constant. Now, differentiating $V(\mathbf{p})$ with respect to $\mathbf{p}$ and then setting $\mathbf{p}$ equal to $\mathbf{R}$ is precisely the same as differentiating $V(\mathbf{R})$ with respect to $\mathbf{R}$. Hence, if ∇ denotes differentiation with respect to $\mathbf{R}$, then

$$\widetilde{\mathbf{F}} = \nabla \left\{mV(\mathbf{R}) + \frac{1}{2}\left[\frac{1}{2} \operatorname{tr}(\mathbf{I})\mathbf{U} - \mathbf{I}\right] : \nabla\nabla V(\mathbf{R}) + C\right\} \tag{14}$$

and, since

$$\mathbf{U} : \nabla\nabla V(\mathbf{R}) = \nabla \cdot \nabla V(\mathbf{R}) = \nabla^2 V(\mathbf{R}) \underset{(2.11.7)}{=} 0 \tag{15}$$

Eq. (1) follows immediately if Eq. (2) is used to form $\widetilde{V}(\mathbf{R})$.

Example When $V(\mathbf{R})$ is available in the form of an explicit function $V^*(R, \lambda, \beta)$ of the spherical coordinates R, λ, β shown in Fig. 2.16.2, where $\mathbf{n}_1$, $\mathbf{n}_2$, $\mathbf{n}_3$ are any mutually perpendicular unit vectors fixed in a

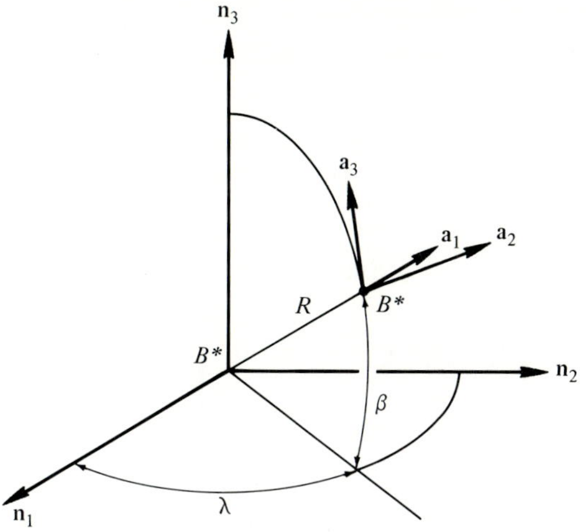

Figure 2.16.2

reference frame N, one can formulate $\widetilde{\mathbf{F}}$ as follows: let $\mathbf{a}_1$, $\mathbf{a}_2$, $\mathbf{a}_3$ be mutually perpendicular unit vectors pointing, respectively, in the directions in which B^* moves when R, λ, β increase one at a time. Define I_{jk} as

$$I_{jk} \triangleq \mathbf{a}_j \cdot \mathbf{I} \cdot \mathbf{a}_k \qquad (j, k = 1, 2, 3) \tag{16}$$

and Q_{ij} $(i, j = 1, 2, 3)$ and $\widetilde{V}^*(R, \lambda, \beta)$ as

$$Q_{11} \triangleq \frac{\partial^2 V^*}{\partial R^2} \tag{17}$$

$$Q_{22} \triangleq \frac{1}{R}\frac{\partial V^*}{\partial R} + \frac{\sec^2\beta}{R^2}\frac{\partial^2 V^*}{\partial \lambda^2} - \frac{\tan\beta}{R^2}\frac{\partial V^*}{\partial \beta} \tag{18}$$

$$Q_{33} \triangleq \frac{1}{R}\frac{\partial V^*}{\partial R} + \frac{1}{R^2}\frac{\partial^2 V^*}{\partial \beta^2} \tag{19}$$

$$Q_{12} = Q_{21} = \frac{\sec\beta}{R}\left(-\frac{1}{R}\frac{\partial V^*}{\partial \lambda} + \frac{\partial^2 V^*}{\partial R\,\partial \lambda}\right) \tag{20}$$

$$Q_{23} = Q_{32} = \frac{\sec\beta}{R^2}\left(\tan\beta\,\frac{\partial V^*}{\partial \lambda} + \frac{\partial^2 V^*}{\partial \beta\,\partial \lambda}\right) \tag{21}$$

$$Q_{31} = Q_{13} = \frac{1}{R}\left(-\frac{1}{R}\frac{\partial V^*}{\partial \beta} + \frac{\partial^2 V^*}{\partial R\,\partial \beta}\right) \tag{22}$$

and

$$\begin{aligned}\widetilde{V}^*(R, \lambda, \beta) \triangleq\ & mV^* - \tfrac{1}{2}(I_{11}Q_{11} + I_{22}Q_{22} + I_{33}Q_{33}) \\ & - (I_{12}Q_{12} + I_{23}Q_{23} + I_{31}Q_{31}) + C\end{aligned} \tag{23}$$

Then

$$\tilde{\mathbf{F}} = \mathbf{a}_1 \frac{\partial \tilde{V}^*}{\partial R} + \mathbf{a}_2 \frac{\sec\beta}{R} \frac{\partial \tilde{V}^*}{\partial \lambda} + \mathbf{a}_3 \frac{1}{R} \frac{\partial \tilde{V}^*}{\partial \beta} \tag{24}$$

To establish the validity of Eq. (24), it is helpful to note that the partial derivatives of $\mathbf{a}_1$, $\mathbf{a}_2$, $\mathbf{a}_3$ with respect to R, λ, β in N are given by

$$\begin{bmatrix} \dfrac{\partial \mathbf{a}_1}{\partial R} & \dfrac{\partial \mathbf{a}_1}{\partial \lambda} & \dfrac{\partial \mathbf{a}_1}{\partial \beta} \\ \dfrac{\partial \mathbf{a}_2}{\partial R} & \dfrac{\partial \mathbf{a}_2}{\partial \lambda} & \dfrac{\partial \mathbf{a}_2}{\partial \beta} \\ \dfrac{\partial \mathbf{a}_3}{\partial R} & \dfrac{\partial \mathbf{a}_3}{\partial \lambda} & \dfrac{\partial \mathbf{a}_3}{\partial \beta} \end{bmatrix} = \begin{bmatrix} 0 & \mathrm{c}\beta\, \mathbf{a}_2 & \mathbf{a}_3 \\ 0 & -\mathrm{c}\beta\, \mathbf{a}_1 + \mathrm{s}\beta\, \mathbf{a}_3 & 0 \\ 0 & -\mathrm{s}\beta\, \mathbf{a}_2 & -\mathbf{a}_1 \end{bmatrix} \tag{25}$$

Also, proceeding as in the example in Sec. 2.9, one can verify that, if $V(\mathbf{R}) = V^*(R, \lambda, \beta)$, then

$$\nabla V(\mathbf{R}) = \mathbf{a}_1 \frac{\partial V^*}{\partial R} + \mathbf{a}_2 \frac{1}{R\,\mathrm{c}\beta} \frac{\partial V^*}{\partial \lambda} + \mathbf{a}_3 \frac{1}{R} \frac{\partial V^*}{\partial \beta} \tag{26}$$

Differentiating once more with respect to $\mathbf{R}$ and using Eq. (25), one then finds that

$$\nabla\nabla V(\mathbf{R}) = \sum_{i=1}^{3} \sum_{j=1}^{3} Q_{ij}\, \mathbf{a}_i \mathbf{a}_j \tag{27}$$

Similarly, the inertia dyadic $\mathbf{I}$ can be expressed as

$$\mathbf{I} \underset{(16)}{=} \sum_{j=1}^{3} \sum_{k=1}^{3} I_{jk}\, \mathbf{a}_j \mathbf{a}_k \tag{28}$$

Hence,

$$\begin{aligned} \mathbf{I} : \nabla\nabla V(\mathbf{R}) \underset{(27,28)}{=} & \; I_{11} Q_{11} + I_{22} Q_{22} + I_{33} Q_{33} \\ & + 2(I_{12} Q_{12} + I_{23} Q_{23} + I_{31} Q_{31}) \end{aligned} \tag{29}$$

With V^* in place of $V(\mathbf{R})$, one thus obtains from Eq. (2),

$$\begin{aligned} \tilde{V}(\mathbf{R}) = mV^* & - \frac{1}{2}(I_{11} Q_{11} + I_{22} Q_{22} + I_{33} Q_{33}) \\ & - (I_{12} Q_{12} + I_{23} Q_{23} + I_{31} Q_{31}) + C \end{aligned} \tag{30}$$

Moreover, defining $\tilde{V}^*(R, \lambda, \beta)$ as the right-hand member of this equation, which yields Eq. (23), one can write [compare with Eq. (26)]

$$\nabla \tilde{V}(\mathbf{R}) = \mathbf{a}_1 \frac{\partial \tilde{V}^*}{\partial R} + \mathbf{a}_2 \frac{\sec\beta}{R} \frac{\partial \tilde{V}^*}{\partial \lambda} + \mathbf{a}_3 \frac{\partial \tilde{V}^*}{\partial \beta} \tag{31}$$

Substitution into Eq. (1) then produces Eq. (24).

2.17 FORCE FUNCTION EXPRESSIONS FOR THE MOMENT EXERTED ON A BODY BY A PARTICLE

Force functions (see Secs. 2.10–2.16) can be used in connection with gravitational moments. For the interaction of a particle and a (not necessarily rigid) body, the moment **M** given by Eq. (2.6.1) can be expressed as

$$\mathbf{M} = -\mathbf{R} \times \nabla V \tag{1}$$

where $V(\mathbf{R})$ is given by Eq. (2.11.2) or by Eq. (2.11.5) and $\nabla V \triangleq \nabla_{\mathbf{R}} V$; or, if W is defined as $W(\mathbf{a}_1, R) \triangleq V[\mathbf{R}(\mathbf{a}_1, R)]$, **M** is given also by

$$\mathbf{M} = -\mathbf{a}_1 \times \frac{\partial W}{\partial \mathbf{a}_1} \tag{2}$$

Similarly, for $\widetilde{\mathbf{M}}$ as defined in Eq. (2.6.3), one can write

$$\widetilde{\mathbf{M}} = -\mathbf{R} \times \nabla \widetilde{V} \tag{3}$$

where $\widetilde{V}$ is given by Eq. (2.12.13); or $\widetilde{\mathbf{M}}$ can be expressed as

$$\widetilde{\mathbf{M}} = -\mathbf{a}_1 \times \frac{\partial \widetilde{W}}{\partial \mathbf{a}_1} \tag{4}$$

where $\widetilde{W}$ is a function of $\mathbf{a}_1$ and R, defined in terms of the moment of inertia I_{11} of B about the line connecting $\bar{P}$ and B^* as

$$\widetilde{W} \triangleq -\frac{3G\bar{m}I_{11}}{2R^3} + C \tag{5}$$

Derivations Equation (1) is obtained by substituting from Eq. (2.11.1) into Eq. (2.6.1); and Eq. (2) then follows if one uses Eq. (2.9.13).

Referring to Eqs. (2.3.6) and (2.3.3), one can write

$$-R\mathbf{a}_1 \times \widetilde{\mathbf{F}} = \frac{3G\bar{m}}{R^3}\mathbf{a}_1 \times \mathbf{I} \cdot \mathbf{a}_1 \underset{(2.6.3)}{=} \widetilde{\mathbf{M}} \tag{6}$$

and, replacing $R\mathbf{a}_1$ with **R**, one then arrives at Eq. (3) after using Eq. (2.12.12) to eliminate $\widetilde{\mathbf{F}}$.

If $\widetilde{\widetilde{W}}$ is defined as

$$\widetilde{\widetilde{W}}(\mathbf{a}_1, R) \triangleq \widetilde{V}[\mathbf{R}(\mathbf{a}_1, R)] \tag{7}$$

then

$$\widetilde{\widetilde{W}} \underset{(2.12.13)}{=} \frac{G\bar{m}m}{R}\left\{1 + \frac{1}{2mR^2}[\mathrm{tr}(\mathbf{I}) - 3I_{11}]\right\} + C \tag{8}$$

and

$$\mathbf{R} \times \nabla\widetilde{V} \underset{(2.9.13)}{=} \mathbf{a}_1 \times \frac{\partial \widetilde{\widetilde{W}}}{\partial \mathbf{a}_1} \underset{(8)}{=} \mathbf{a}_1 \times \frac{\partial}{\partial \mathbf{a}_1}\left(\frac{G\bar{m}m}{R}\left\{1 + \frac{1}{2mR^2}[\mathrm{tr}(\mathbf{I}) - 3I_{11}]\right\} + C\right) \tag{9}$$

or, since tr(**I**) is independent of $\mathbf{a}_1$,

$$\mathbf{R} \times \nabla\widetilde{V} = \mathbf{a}_1 \times \frac{\partial}{\partial \mathbf{a}_1}\left(-\frac{3G\overline{m}I_{11}}{2R^3} + C\right) \underset{(5)}{=} \mathbf{a}_1 \times \frac{\partial \widetilde{W}}{\partial \mathbf{a}_1} \tag{10}$$

Consequently,

$$\widetilde{\mathbf{M}} \underset{(3)}{=} -\mathbf{a}_1 \times \frac{\partial \widetilde{W}}{\partial \mathbf{a}_1} \tag{11}$$

in agreement with Eq. (4).

Example In Fig. 2.17.1, $\mathbf{a}_1$, $\mathbf{a}_2$, $\mathbf{a}_3$ and $\mathbf{b}_1$, $\mathbf{b}_2$, $\mathbf{b}_3$ are dextral sets of orthogonal unit vectors, $\overline{P}$ is a particle of mass $\overline{m}$, and B is a rigid body of mass m. The unit vector $\mathbf{a}_1$ is chosen such that the position vector $\mathbf{R}$ of the mass center B^* of B relative to $\overline{P}$ is given by $\mathbf{R} = R\mathbf{a}_1$; and $\mathbf{b}_1$, $\mathbf{b}_2$, $\mathbf{b}_3$ are fixed in B.

If θ_1, θ_2, θ_3 are body-two orientation angles (see Sec. 1.7) for B in a reference frame in which $\mathbf{a}_1$, $\mathbf{a}_2$, $\mathbf{a}_3$ are fixed, one can express $\mathbf{R}$ as

$$\mathbf{R}(R, \theta_1, \theta_2, \theta_3) \underset{(1.7.31)}{=} R(c_2\,\mathbf{b}_1 + s_2 s_3\,\mathbf{b}_2 + s_2 c_3\,\mathbf{b}_3) \tag{12}$$

Given a force function $V(\mathbf{R})$ for the gravitational force exerted on B by $\overline{P}$, one can, therefore, define a function $W(R, \theta_1, \theta_2, \theta_3)$ as

$$W(R, \theta_1, \theta_2, \theta_3) \triangleq V[\mathbf{R}(R, \theta_2, \theta_3)] \tag{13}$$

The moment $\mathbf{M}$ exerted by $\overline{P}$ on B about B^* then can be expressed in terms of partial derivatives of W with respect to θ_2 and θ_3 by proceeding as follows.

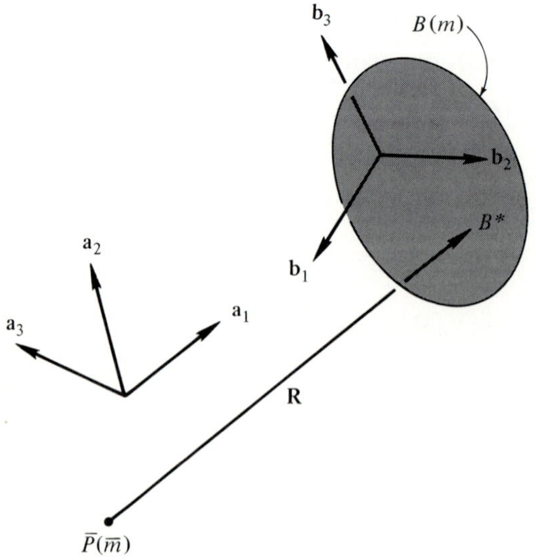

Figure 2.17.1

Partial differentiation of **R** with respect to R, θ_2, and θ_3 in B yields

$$\frac{\partial \mathbf{R}}{\partial R} \underset{(12)}{=} c_2 \mathbf{b}_1 + s_2 s_3 \mathbf{b}_2 + s_2 c_3 \mathbf{b}_3 \tag{14}$$

$$\frac{\partial \mathbf{R}}{\partial \theta_2} \underset{(12)}{=} R(-s_2 \mathbf{b}_1 + c_2 s_3 \mathbf{b}_2 + c_2 c_3 \mathbf{b}_3) \tag{15}$$

$$\frac{\partial \mathbf{R}}{\partial \theta_3} \underset{(12)}{=} R(s_2 c_3\, \mathbf{b}_1 - s_2 s_3\, \mathbf{b}_3) \tag{16}$$

From Eq. (2.9.6),

$$\frac{\partial W}{\partial R} = \nabla V \cdot \frac{\partial \mathbf{R}}{\partial R} \underset{(14)}{=} c_2 \nabla V \cdot \mathbf{b}_1 + s_2 s_3 \nabla V \cdot \mathbf{b}_2 + s_2 c_3 \nabla V \cdot \mathbf{b}_3 \tag{17}$$

$$\frac{\partial W}{\partial \theta_2} = \nabla V \cdot \frac{\partial \mathbf{R}}{\partial \theta_2} \underset{(15)}{=} R(-s_2 \nabla V \cdot \mathbf{b}_1 + c_2 s_3 \nabla V \cdot \mathbf{b}_2 + c_2 c_3 \nabla V \cdot \mathbf{b}_3) \tag{18}$$

$$\frac{\partial W}{\partial \theta_3} = \nabla V \cdot \frac{\partial \mathbf{R}}{\partial \theta_3} \underset{(16)}{=} R(s_2 c_3 \nabla V \cdot \mathbf{b}_2 - s_2 s_3 \nabla V \cdot \mathbf{b}_3) \tag{19}$$

Equations (17)–(19) may be solved for the dot products $\nabla V \cdot \mathbf{b}_1$, $\nabla V \cdot \mathbf{b}_2$, $\nabla V \cdot \mathbf{b}_3$; and ∇V then can be expressed as

$$\begin{aligned}\nabla V &= \nabla V \cdot \mathbf{b}_1 \mathbf{b}_1 + \nabla V \cdot \mathbf{b}_2 \mathbf{b}_2 + \nabla V \cdot \mathbf{b}_3 \mathbf{b}_3 \\ &= \left(\frac{\partial W}{\partial R} c_2 - \frac{1}{R}\frac{\partial W}{\partial \theta_2} s_2\right)\mathbf{b}_1 + \left(\frac{\partial W}{\partial R} s_2 s_3 + \frac{1}{R}\frac{\partial W}{\partial \theta_2} c_2 s_3 + \frac{1}{R}\frac{\partial W}{\partial \theta_3}\frac{c_3}{s_2}\right)\mathbf{b}_2 \\ &\quad + \left(\frac{\partial W}{\partial R} s_2 c_3 + \frac{1}{R}\frac{\partial W}{\partial \theta_2} c_2 c_3 - \frac{1}{R}\frac{\partial W}{\partial \theta_3}\frac{s_3}{s_2}\right)\mathbf{b}_3\end{aligned} \tag{20}$$

Consequently,

$$\begin{aligned}\mathbf{M} \underset{(1)}{=} -\mathbf{R} \times \nabla V \underset{(12,20)}{=} \frac{\partial W}{\partial \theta_3}\mathbf{b}_1 + \left(\frac{\partial W}{\partial \theta_2} c_3 - \frac{\partial W}{\partial \theta_3}\frac{c_2 s_3}{s_2}\right)\mathbf{b}_2 \\ - \left(\frac{\partial W}{\partial \theta_3}\frac{c_2 c_3}{s_2} + \frac{\partial W}{\partial \theta_2} s_3\right)\mathbf{b}_3\end{aligned} \tag{21}$$

This equation does not involve θ_1 explicitly; but, referred to $\mathbf{a}_1$, $\mathbf{a}_2$, $\mathbf{a}_3$, rather than to $\mathbf{b}_1$, $\mathbf{b}_2$, $\mathbf{b}_3$, **M** is given by [use Eq. (1.7.31)]

$$\mathbf{M} = \left(c_1 \frac{\partial W}{\partial \theta_2} + \frac{s_1}{s_2}\frac{\partial W}{\partial \theta_3}\right)\mathbf{a}_2 + \left(s_1 \frac{\partial W}{\partial \theta_2} - \frac{c_1}{s_2}\frac{\partial W}{\partial \theta_3}\right)\mathbf{a}_3 \tag{22}$$

which brings the dependence of **M** on θ_1 into evidence.

2.18 FORCE FUNCTION EXPRESSION FOR THE MOMENT EXERTED ON A SMALL BODY BY A BODY

When the distance between the mass centers B^* and $\bar{B}^*$ of two bodies B and $\bar{B}$ exceeds the greatest distance from B^* to any point of B, and a force function $V(\mathbf{p})$ for the forces exerted by $\bar{B}$ on a particle of unit mass at a point P situated as shown in Fig. 2.16.1 is available, then the system of gravitational forces exerted by $\bar{B}$ on B produces a moment about B^* that is given approximately by

$$\widetilde{\mathbf{M}} = -\mathbf{I} \underset{\cdot}{\times} \nabla\nabla V(\mathbf{R}) \tag{1}$$

where $\mathbf{I}$ is the central inertia dyadic of B, $\mathbf{R}$ is the position vector of B^* relative to $\bar{B}^*$, and ∇ denotes differentiation with respect to $\mathbf{R}$.†

Derivation The system of gravitational forces exerted by $\bar{B}$ on B produces a moment $\mathbf{M}$ about B^* such that

$$\mathbf{M} = \int \mathbf{r} \times \nabla_{\mathbf{p}} V \, \rho \, d\tau \tag{2}$$

Expanding $\nabla_{\mathbf{p}} V$ in a Taylor series by proceeding as in the derivation of Eq. (2.16.10), one obtains

$$\nabla_{\mathbf{p}} V = \nabla V(\mathbf{R}) + \mathbf{r} \cdot \nabla\nabla V(\mathbf{R}) + \cdots \tag{3}$$

so that, retaining only terms displayed explicitly, one arrives at

$$\widetilde{\mathbf{M}} \underset{(2)}{=} \int \mathbf{r} \times \nabla V(\mathbf{R}) \rho \, d\tau + \int \mathbf{r} \times [\mathbf{r} \cdot \nabla\nabla V(\mathbf{R})] \rho \, d\tau \tag{4}$$

The first integral vanishes because $\mathbf{r}$ originates at the mass center of B^*, and

$$\mathbf{r} \times [\mathbf{r} \cdot \nabla\nabla V(\mathbf{R})] = \mathbf{r}\mathbf{r} \underset{\cdot}{\times} \nabla\nabla V(\mathbf{R}) \tag{5}$$

Hence,

$$\widetilde{\mathbf{M}} = \left(\int \mathbf{r}\mathbf{r} \rho \, d\tau \right) \underset{\cdot}{\times} \nabla\nabla V(\mathbf{R}) = \left[\frac{1}{2} \operatorname{tr}(\mathbf{I}) \mathbf{U} - \mathbf{I} \right] \underset{\cdot}{\times} \nabla\nabla V(\mathbf{R}) \tag{6}$$

But $\nabla\nabla V(\mathbf{R})$ is a symmetric dyadic, and the cross-dot product of $\mathbf{U}$ and any symmetric dyadic is equal to zero. Hence Eq. (6) reduces to Eq. (1).

Example: When $V(\mathbf{R})$ is available in the form of an explicit function $V^*(R, \lambda, \beta)$ of the spherical coordinates R, λ, β shown in Fig. 2.16.2, one can find $\widetilde{\mathbf{M}}$ as follows.

†The cross-dot product in Eq. (1) is defined such that, for two dyads $\mathbf{u}_1\mathbf{u}_2$ and $\mathbf{v}_1\mathbf{v}_2$, $(\mathbf{u}_1\mathbf{u}_2) \underset{\cdot}{\times} (\mathbf{v}_1\mathbf{v}_2) \overset{\Delta}{=} (\mathbf{u}_1 \times \mathbf{v}_1)(\mathbf{u}_2 \cdot \mathbf{v}_2)$; it obeys the distributive law when applied to dyadics:

$$\begin{aligned}(\mathbf{a}_1\mathbf{a}_2 + \mathbf{b}_1\mathbf{b}_2 + \cdots) \underset{\cdot}{\times} (\mathbf{A}_1\mathbf{A}_2 + \mathbf{B}_1\mathbf{B}_2 + \cdots) &= \mathbf{a}_1\mathbf{a}_2 \underset{\cdot}{\times} \mathbf{A}_1\mathbf{A}_2 + \mathbf{a}_1\mathbf{a}_2 \underset{\cdot}{\times} \mathbf{B}_1\mathbf{B}_2 + \cdots \\ &\quad + \mathbf{b}_1\mathbf{b}_2 \underset{\cdot}{\times} \mathbf{A}_1\mathbf{A}_2 + \mathbf{b}_2\mathbf{b}_2 \underset{\cdot}{\times} \mathbf{B}_1\mathbf{B}_2 + \cdots \\ &\quad + \cdots\end{aligned}$$

Differentiate Eq. (2.16.2) to verify that

$$\nabla\nabla V(\mathbf{R}) = \sum_{i=1}^{3}\sum_{j=1}^{3} Q_{ij}\mathbf{a}_i\mathbf{a}_j \tag{7}$$

where Q_{ij} $(i, j = 1, 2, 3)$ are given in Eqs. (2.16.17)–(2.16.22). Next, let

$$I_{jk} \triangleq \mathbf{a}_j \cdot \mathbf{I} \cdot \mathbf{a}_k \qquad (j, k = 1, 2, 3) \tag{8}$$

Then

$$\begin{aligned}\underset{(1)}{\widetilde{\mathbf{M}}} = {} & [I_{12}Q_{31} - Q_{12}I_{31} + (I_{22} - I_{33})Q_{23} - (Q_{22} - Q_{33})I_{23}]\mathbf{a}_1 \\ & + [I_{23}Q_{12} - Q_{23}I_{12} + (I_{33} - I_{11})Q_{31} - (Q_{33} - Q_{11})I_{31}]\mathbf{a}_2 \\ & + [I_{31}Q_{23} - Q_{31}I_{23} + (I_{11} - I_{22})Q_{12} - (Q_{11} - Q_{22})I_{12}]\mathbf{a}_3\end{aligned} \tag{9}$$

CHAPTER

THREE

SIMPLE SPACECRAFT

This chapter deals with two kinds of spacecraft, namely, those that can be modeled as single rigid bodies and those forming simple gyrostats. Such spacecraft have played an important role in the exploration of space and, by studying their behavior, one can learn a great deal that is of interest also in connection with more complex spacecraft.

Throughout this chapter, the formulation of dynamical equations of motion is based on the angular momentum principle. Hence, knowledge of Lagrange's equations or other methods of analytical mechanics is not needed. In the first five sections, spacecraft consisting of a single rigid body are studied, primarily with a view to determining the effects of gravity and orbit eccentricity on the attitude motions of such vehicles. The remaining sections, devoted to gyrostats, contain material that is central in connection with dual-spin spacecraft.

3.1 ROTATIONAL MOTION OF A TORQUE-FREE, AXISYMMETRIC RIGID BODY

When an axisymmetric rigid body B (see Fig. 3.1.1) is subjected to the action of forces whose resultant moment about the mass center B^* of B is equal to zero, the rotational motion of B in a Newtonian reference frame N can be described in terms of two motions of simple rotation (see Sec. 1.1), each performed with an angular velocity of constant magnitude. One of these is the motion of B in a reference frame C in which the angular momentum $\mathbf{H}$ of B relative to B^* in N (a vector fixed in N) and the symmetry axis of B are fixed; the other is the motion of C in N. The associated angular velocities are called the angular velocity of *spin* and the angular

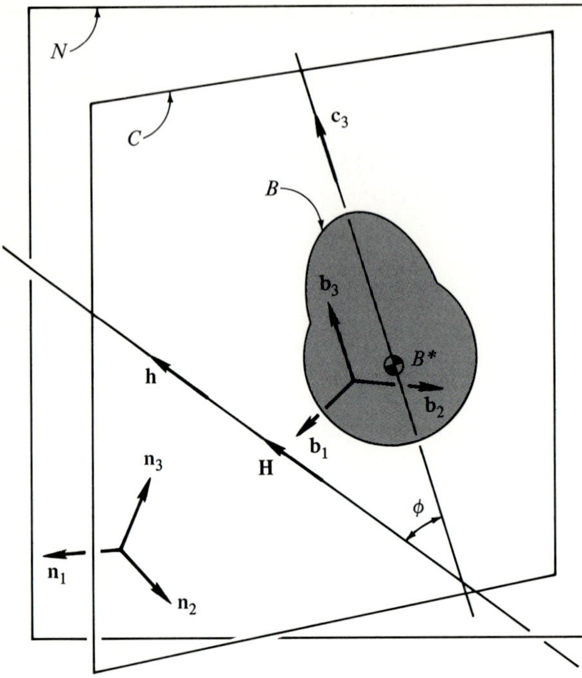

Figure 3.1.1

velocity of *precession*, respectively, and can be expressed as

$$^{C}\boldsymbol{\omega}^{B} = s\,\mathbf{c}_3 \tag{1}$$

and

$$^{N}\boldsymbol{\omega}^{C} = p\,\mathbf{h} \tag{2}$$

where $\mathbf{h}$ is a unit vector having the same direction as $\mathbf{H}$ (see Fig. 3.1.1); $\mathbf{c}_3$ is a unit vector parallel to the symmetry axis of B; s and p, called, respectively, the *spin speed* and the *precession speed*, are defined as

$$s \triangleq \left(1 - \frac{J}{I}\right)\hat{\omega}_3 \tag{3}$$

and

$$p \triangleq \frac{H}{I} \tag{4}$$

where H is the (constant) magnitude of $\mathbf{H}$; I and J are the moments of inertia of B about lines passing through B^* and respectively perpendicular and parallel to $\mathbf{c}_3$; and $\hat{\omega}_3$ is the value of $^{N}\boldsymbol{\omega}^{B} \cdot \mathbf{c}_3$ at time $t = 0$.

To describe the orientation of B in N for $t \geq 0$, one can let $\mathbf{n}_1$, $\mathbf{n}_2$, $\mathbf{n}_3$ be a dextral set of orthogonal unit vectors fixed in N, choosing these such that $\mathbf{n}_3 = \mathbf{c}_3$ at

$t = 0$ and $\mathbf{n}_2$ has the same direction as $\mathbf{H} \times \mathbf{n}_3$; and introduce $\mathbf{b}_1, \mathbf{b}_2, \mathbf{b}_3$ as a set of unit vectors fixed in B, with $\mathbf{b}_i = \mathbf{n}_i$ ($i = 1, 2, 3$) at $t = 0$ (see Fig. 3.1.1). The Euler parameters $\epsilon_1, \ldots, \epsilon_4$ (see Sec. 1.3) relating $\mathbf{n}_1, \mathbf{n}_2, \mathbf{n}_3$ to $\mathbf{b}_1, \mathbf{b}_2, \mathbf{b}_3$ then can be expressed as

$$\epsilon_1 = -\sin\phi \sin\frac{pt}{2}\cos\frac{st}{2} \tag{5}$$

$$\epsilon_2 = \sin\phi \sin\frac{pt}{2}\sin\frac{st}{2} \tag{6}$$

$$\epsilon_3 = \cos\phi \sin\frac{pt}{2}\cos\frac{st}{2} + \cos\frac{pt}{2}\sin\frac{st}{2} \tag{7}$$

$$\epsilon_4 = -\cos\phi \sin\frac{pt}{2}\sin\frac{st}{2} + \cos\frac{pt}{2}\cos\frac{st}{2} \tag{8}$$

where ϕ, the angle between $\mathbf{h}$ and $\mathbf{c}_3$ (see Fig. 3.1.1), is given by

$$\phi = \cos^{-1}\frac{\hat{\omega}_3 J}{H} = \cos^{-1}\frac{Js}{(I - J)p} \qquad 0 \le \phi \le \pi \tag{9}$$

The rotational motion of B can be described also by reference to the rolling motion of a spheroid σ on a plane π, the so-called *Poinsot construction*. Specifically, let σ be a spheroid whose center O is fixed in N, whose principal semidiameters are

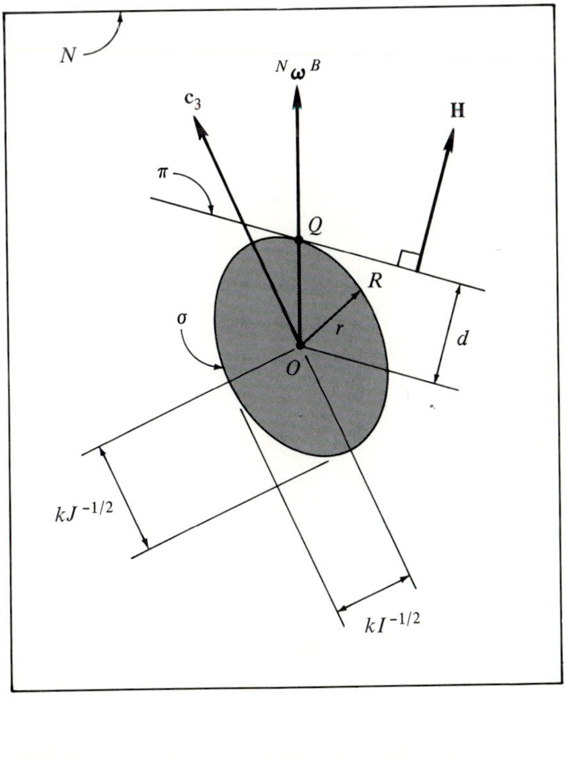

Figure 3.1.2

equal to $kI^{-1/2}$ and $kJ^{-1/2}$, where k is any constant, and whose symmetry axis (axis of revolution) is at all times parallel to $\mathbf{c}_3$ (see Fig. 3.1.2); and let π be the tangent plane to σ at Q, the point of intersection of the surface of σ with the ray that passes through O and has the same direction as ${}^N\boldsymbol{\omega}^B$ at some instant t^*. Then π is a unique plane, fixed in N at a distance $d = k({}^N\boldsymbol{\omega}^B \cdot \mathbf{H})^{1/2}/H$ from O and perpendicular to $\mathbf{H}$; and the rotational motion of B in N can be duplicated by a rolling motion of σ on π. The loci traced out by Q on σ and on π during such rolling are called, respectively, the *polhode* and the *herpolhode*.

Finally, one can also establish a correspondence between the rotational motion of B in N and a rolling motion of one cone upon another. To this end, let α be a right-circular cone whose vertex O is fixed in N, whose semivertex angle α^* is equal to the angle between ${}^N\boldsymbol{\omega}^B$ and $\mathbf{H}$, and whose axis is parallel to $\mathbf{H}$, as shown in Fig. 3.1.3; and let β be a right-circular conical surface whose vertex coincides with O, whose semivertex angle β^* is equal to the angle between ${}^N\boldsymbol{\omega}^B$ and $\mathbf{c}_3$, and whose axis is at all times parallel to $\mathbf{c}_3$, as shown in Fig. 3.1.3, which applies if σ is a prolate spheroid. When σ is oblate, refer to Fig. 3.1.4. The cones α and β are called respectively the *space cone* and the *body cone*, and the rotational motion of B in N can be duplicated by a rolling motion of the body cone on the space cone.

Derivations In Fig. 3.1.5, $\mathbf{b}_1$, $\mathbf{b}_2$, $\mathbf{b}_3$ form a dextral set of orthogonal unit vectors fixed in B, with $\mathbf{b}_3$ parallel to the symmetry axis of B; and $\mathbf{c}_1$, $\mathbf{c}_2$, $\mathbf{c}_3$ form a similar

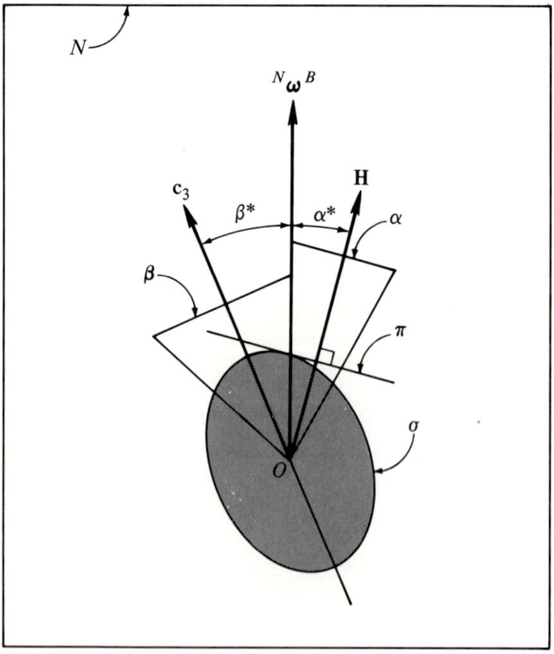

Figure 3.1.3

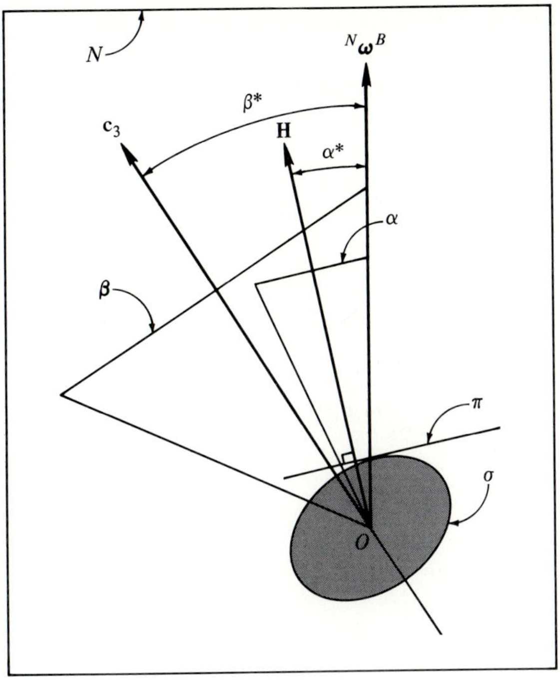

Figure 3.1.4

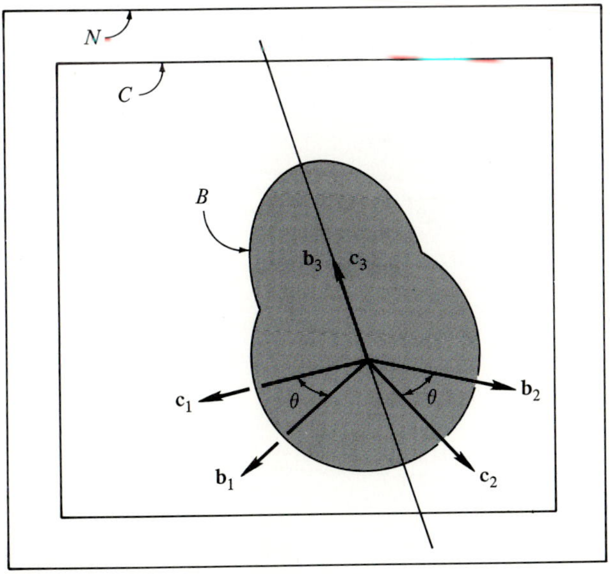

Figure 3.1.5

set fixed in a reference frame C, with $\mathbf{c}_3 = \mathbf{b}_3$ for all values of the time t and $\mathbf{c}_i = \mathbf{b}_i$ ($i = 1, 2$) at $t = 0$. The motion of B in C is thus necessarily a motion of simple rotation (see Sec. 1.1), and the angular velocity of B in C can be expressed as

$$^C\boldsymbol{\omega}^B \underset{(1.11.5)}{=} \dot{\theta}\,\mathbf{c}_3 \tag{10}$$

where θ is the angle between $\mathbf{b}_1$ and $\mathbf{c}_1$, as shown in Fig. 3.1.5.

If ω_i is defined as

$$\omega_i \triangleq {}^N\boldsymbol{\omega}^B \cdot \mathbf{c}_i \qquad (i = 1, 2, 3) \tag{11}$$

then the angular velocity of C in N and the angular momentum of B relative to B^* in N can be expressed as

$$^N\boldsymbol{\omega}^C \underset{(1.16.1)}{=} {}^N\boldsymbol{\omega}^B + {}^B\boldsymbol{\omega}^C \underset{(1.11.2)}{=} {}^N\boldsymbol{\omega}^B - {}^C\boldsymbol{\omega}^B \underset{(10,11)}{=} \omega_1\mathbf{c}_1 + \omega_2\mathbf{c}_2 + (\omega_3 - \dot{\theta})\mathbf{c}_3 \tag{12}$$

and

$$\mathbf{H} = H\mathbf{h} = I\omega_1\mathbf{c}_1 + I\omega_2\mathbf{c}_2 + J\omega_3\mathbf{c}_3 \tag{13}$$

Differentiating $\mathbf{H}$ with respect to t in N, one obtains

$$\begin{aligned}\frac{^N d\mathbf{H}}{dt} &\underset{(1.11.8)}{=} \frac{^C d\mathbf{H}}{dt} + {}^N\boldsymbol{\omega}^C \times \mathbf{H} \\ &\underset{(12,13)}{=} \{I\dot{\omega}_1 + [I\dot{\theta} + (J - I)\omega_3]\omega_2\}\mathbf{c}_1 \\ &\quad + \{I\dot{\omega}_2 - [I\dot{\theta} + (J - I)\omega_3]\omega_1\}\mathbf{c}_2 + J\dot{\omega}_3\mathbf{c}_3\end{aligned} \tag{14}$$

and, in accordance with the angular momentum principle, the coefficients of $\mathbf{c}_1$, $\mathbf{c}_2$, and $\mathbf{c}_3$ in this equation must vanish. Moreover, $\dot{\theta}$ may be chosen at will. If one takes

$$\dot{\theta} = \left(1 - \frac{J}{I}\right)\omega_3 \tag{15}$$

in order to simplify the coefficients of $\mathbf{c}_1$ and $\mathbf{c}_2$, one obtains from Eq. (14),

$$\dot{\omega}_i = 0 \qquad (i = 1, 2, 3) \tag{16}$$

and, after defining $\hat{\omega}_i$ as

$$\hat{\omega}_i \triangleq ({}^N\boldsymbol{\omega}^B \cdot \mathbf{c}_i)_{t=0} \qquad (i = 1, 2, 3) \tag{17}$$

one can write

$$\omega_i \underset{(16,17)}{=} \hat{\omega}_i \qquad (i = 1, 2, 3) \tag{18}$$

and

$$^N\boldsymbol{\omega}^C \underset{(12,15,18)}{=} \hat{\omega}_1\mathbf{c}_1 + \hat{\omega}_2\mathbf{c}_2 + \frac{J}{I}\hat{\omega}_3\mathbf{c}_3 \underset{(13)}{=} \frac{H\mathbf{h}}{I} \underset{(4)}{=} p\mathbf{h} \tag{19}$$

which is Eq. (2). Furthermore,

$$ {}^{C}\boldsymbol{\omega}^{B} \underset{(10,15,18)}{=} \left(1 - \frac{J}{I}\right)\hat{\omega}_3\mathbf{c}_3 \underset{(3)}{=} s\mathbf{c}_3 \tag{20} $$

in agreement with Eq. (1).

Since **H** can be expressed as

$$ \mathbf{H} \underset{(13,18)}{=} I\hat{\omega}_1\mathbf{c}_1 + I\hat{\omega}_2\mathbf{c}_2 + J\hat{\omega}_3\mathbf{c}_3 \tag{21} $$

this vector has a fixed orientation in C; and the angle ϕ between **H** and $\mathbf{c}_3$ is given by

$$ \phi = \cos^{-1}(\mathbf{h} \cdot \mathbf{c}_3) \underset{(21)}{=} \cos^{-1}\frac{J\hat{\omega}_3}{H} \underset{(3,4)}{=} \cos^{-1}\frac{Js}{(I - J)p} \tag{22} $$

which is Eq. (9).

The unit vectors $\mathbf{c}_1$ and $\mathbf{c}_2$ always can be chosen such that $\mathbf{c}_1 = \mathbf{n}_1$ and $\mathbf{c}_2 = \mathbf{n}_2$ at $t = 0$. As the motion of C in N is a motion of simple rotation, a matrix L such that at time t

$$ [\mathbf{c}_1 \quad \mathbf{c}_2 \quad \mathbf{c}_3] = [\mathbf{n}_1 \quad \mathbf{n}_2 \quad \mathbf{n}_3]\,L \tag{23} $$

therefore can be formed by using Eqs. (1.2.23)–(1.2.31) with

$$ \theta = pt \qquad \lambda_1 = -\sin\phi \qquad \lambda_2 = 0 \qquad \lambda_3 = \cos\phi \tag{24} $$

Similarly, the motion of B in C is such that one can write

$$ [\mathbf{b}_1 \quad \mathbf{b}_2 \quad \mathbf{b}_3] = [\mathbf{c}_1 \quad \mathbf{c}_2 \quad \mathbf{c}_3]\,M \tag{25} $$

with

$$ M \underset{(1.2.37)}{=} \begin{bmatrix} \cos st & -\sin st & 0 \\ \sin st & \cos st & 0 \\ 0 & 0 & 1 \end{bmatrix} \tag{26} $$

From Eqs. (23) and (25) it then follows that

$$ [\mathbf{b}_1 \quad \mathbf{b}_2 \quad \mathbf{b}_3] = [\mathbf{n}_1 \quad \mathbf{n}_2 \quad \mathbf{n}_3]\,W \tag{27} $$

where $W = LM$; and, once the elements W_{ij} $(i, j = 1, 2, 3)$ of W have been found, $\epsilon_1, \ldots, \epsilon_4$ can be formed by letting W_{ij} play the part of C_{ij} in Eqs. (1.3.15)–(1.3.18), which leads to Eqs. (5)–(8).

In connection with the Poinsot construction, two observations regarding the spheroid σ are in order. First, if **r** is the position vector from point O (see Fig. 3.1.2) to a point R on the surface of σ, then the equation of this surface can be expressed as

$$ f(\mathbf{r}) \triangleq \mathbf{r} \cdot \mathbf{K} \cdot \mathbf{r} - k^2 = 0 \tag{28} $$

where **K** is the inertia dyadic of B for B^*. One way to see this is to express **r** and **K** as

$$ \mathbf{r} = x_1\mathbf{c}_1 + x_2\mathbf{c}_2 + x_3\mathbf{c}_3 \tag{29} $$

and

$$ \mathbf{K} = I\mathbf{c}_1\mathbf{c}_1 + I\mathbf{c}_2\mathbf{c}_2 + J\mathbf{c}_3\mathbf{c}_3 \tag{30} $$

and then to carry out the dot-multiplications indicated in Eq. (28), thus obtaining

$$Ix_1^2 + Ix_2^2 + Jx_3^2 = k^2 \tag{31}$$

which is the Cartesian form of the equation of a spheroidal surface whose axis of symmetry is parallel to $\mathbf{c}_3$ and whose principal semidiameters are equal to $kI^{-1/2}$ and $kJ^{-1/2}$. Second, the plane that is tangent to σ at R is normal to the vector $\nabla_{\mathbf{r}}f$; and, if $\mathbf{U}$ is a unit dyadic,

$$\nabla_{\mathbf{r}}f \underset{(28)}{=} \mathbf{U} \cdot \mathbf{K} \cdot \mathbf{r} + \mathbf{r} \cdot \mathbf{K} \cdot \mathbf{U} = 2\mathbf{K} \cdot \mathbf{r} \tag{32}$$

Now let $\mathbf{q}$ be the position vector of Q (see Fig. 3.1.2) relative to O, so that $\mathbf{q}$ can be expressed as

$$\mathbf{q} = \lambda\, {}^N\boldsymbol{\omega}^B \tag{33}$$

where λ is a positive scalar. Then the plane that is tangent to σ at Q is normal to the vector $\mathbf{v}$ given by

$$\mathbf{v} \underset{(32)}{=} 2\mathbf{K} \cdot \mathbf{q} \underset{(33)}{=} 2\lambda\mathbf{K} \cdot {}^N\boldsymbol{\omega}^B \tag{34}$$

But $\mathbf{K} \cdot {}^N\boldsymbol{\omega}^B = \mathbf{H}$, and $\mathbf{H}$ has a fixed orientation in N. Thus, the plane that is tangent to σ at Q is seen to have the same orientation in N regardless of the choice of t^*. Furthermore, since Q is a point of the surface of σ,

$$\mathbf{q} \cdot \mathbf{K} \cdot \mathbf{q} - k^2 \underset{(28)}{=} 0 \tag{35}$$

$$\lambda^2\, {}^N\boldsymbol{\omega}^B \cdot \mathbf{K} \cdot {}^N\boldsymbol{\omega}^B \underset{(33,35)}{=} k^2 \tag{36}$$

and

$$\lambda \underset{(36)}{=} k({}^N\boldsymbol{\omega}^B \cdot \mathbf{K} \cdot {}^N\boldsymbol{\omega}^B)^{-1/2} = k({}^N\boldsymbol{\omega}^B \cdot \mathbf{H})^{-1/2} \tag{37}$$

The distance d from O to π (see Fig. 3.1.2) thus can be expressed as

$$d = \frac{|\mathbf{q} \cdot \mathbf{H}|}{H} \underset{(33)}{=} \frac{\lambda|{}^N\boldsymbol{\omega}^B \cdot \mathbf{H}|}{H} \underset{(37)}{=} \frac{k({}^N\boldsymbol{\omega}^B \cdot \mathbf{H})^{1/2}}{H} \tag{38}$$

which shows that this distance is independent of the choice of t^*, for ${}^N\boldsymbol{\omega}^B \cdot \mathbf{H}$ and H are constants [see Eqs. (11), (16), and (21)]. One must conclude, therefore, that π is a unique plane fixed in N. Moreover, if σ is made to move in such a way that ${}^N\boldsymbol{\omega}^\sigma = {}^N\boldsymbol{\omega}^B$, then σ duplicates the rotational motion of B in N; and σ is seen to be rolling on π because the velocity of the point of σ that is in contact with π is given by

$${}^N\boldsymbol{\omega}^\sigma \times \mathbf{q} \underset{(33)}{=} \lambda\, {}^N\boldsymbol{\omega}^B \times {}^N\boldsymbol{\omega}^B = 0 \tag{39}$$

Finally, to prove that the rotational motion of B in N can be duplicated by a rolling motion of the body cone β on the space cone α, it will be shown that β rolls on α if β moves in such a way that ${}^N\boldsymbol{\omega}^\beta = {}^N\boldsymbol{\omega}^B$.

Three vectors play important parts in connection with α and β, namely, $\mathbf{h}$, $\mathbf{c}_3$, and ${}^N\boldsymbol{\omega}^B$. These are related to each other because

$$ {}^N\boldsymbol{\omega}^B \underset{(1.16.1)}{=} {}^N\boldsymbol{\omega}^C + {}^C\boldsymbol{\omega}^B \underset{(1\text{-}3)}{=} p\mathbf{h} + \left(1 - \frac{J}{I}\right)\hat{\omega}_3\mathbf{c}_3 \tag{40} $$

Now, p is intrinsically positive [see Eq. (4)], and $\mathbf{c}_3$ always can be chosen such that $\hat{\omega}_3$ is positive. When this is done, the angle between $\mathbf{h}$ and $\mathbf{c}_3$ cannot exceed 90°; and, if σ is a prolate spheroid, that is, if $J < I$, then $\mathbf{h}$, $\mathbf{c}_3$, and ${}^N\boldsymbol{\omega}^B$ are oriented relative to each other as shown in Fig. 3.1.6, whereas, if σ is an oblate spheroid, that is, if $J > I$, then Fig. 3.1.7 applies. By construction, the cones α and β thus have a contact line parallel to ${}^N\boldsymbol{\omega}^B$, as shown in Figs. 3.1.3 and 3.1.4. If R is a point of β lying on this line, and if $\mathbf{r}$ is the position vector of R relative to O, then

$$ \mathbf{r} = \mu\,{}^N\boldsymbol{\omega}^B \tag{41} $$

where μ is some scalar; and, when β moves in such a way that ${}^N\boldsymbol{\omega}^\beta = {}^N\boldsymbol{\omega}^B$, then the velocity of R in N is given by

$$ \mathbf{r} \times {}^N\boldsymbol{\omega}^\beta \underset{(41)}{=} \mu\,{}^N\boldsymbol{\omega}^B \times {}^N\boldsymbol{\omega}^B = 0 \tag{42} $$

which means that β is rolling on α.

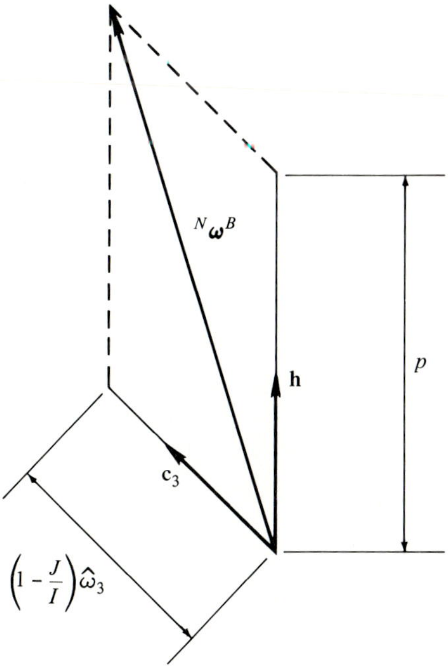

Figure 3.1.6

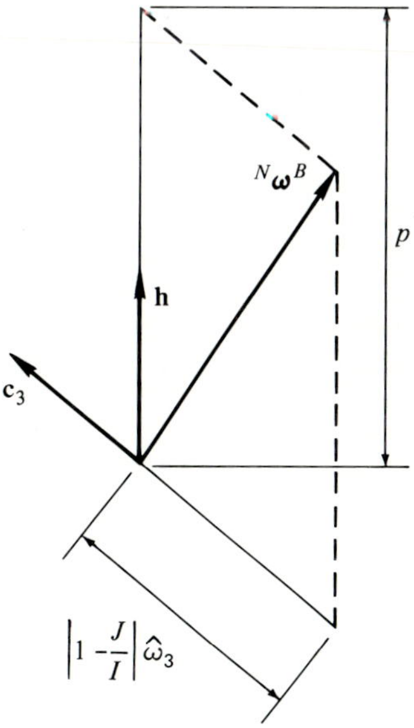

Figure 3.1.7

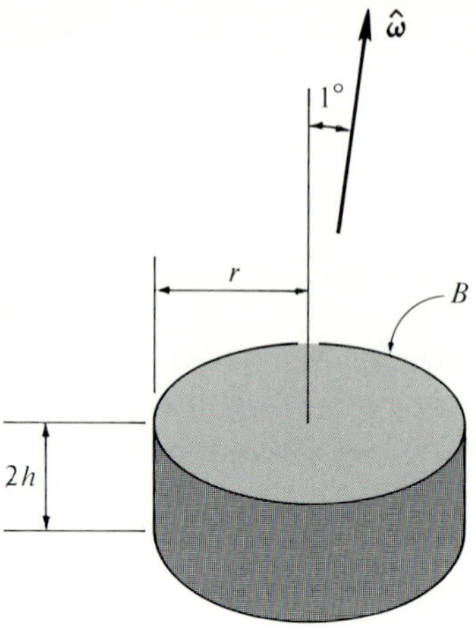

Figure 3.1.8

Example A solid right-circular cylinder B is to be set into motion in such a way that $\hat{\boldsymbol{\omega}}$, the initial angular velocity of B, is parallel to the cylinder's symmetry axis. However, an "injection" error of as much as 1° (see Fig. 3.1.8) is anticipated, and the angle between the symmetry axis at time t and the (fixed) line with which the symmetry axis coincides at $t = 0$ may, therefore, differ from zero, acquiring a maximum value δ that depends on the "shape factor" h/r. To study this dependence, δ is to be plotted versus h/r for $0 \leq h/r \leq 2$.

The symmetry axis of B has at each instant the same orientation as the symmetry axis of the body cone β (see Figs. 3.1.3 and 3.1.4). Hence,

$$\delta = 2\phi \underset{(9)}{=} 2 \cos^{-1} \frac{\hat{\omega}_3 J}{H} \tag{43}$$

where H can be expressed as

$$H \underset{(21)}{=} [(I\hat{\omega}_2)^2 + (J\hat{\omega}_3)^2]^{1/2} \tag{44}$$

since the unit vectors $\mathbf{b}_1$ and $\mathbf{b}_2$, and thus also $\mathbf{c}_1$ and $\mathbf{c}_2$, can be introduced in such a way that $\hat{\omega}_1 = 0$. Consequently,

$$\delta \underset{(43,44)}{=} 2 \cos^{-1}\left[1 + \left(\frac{I}{J}\right)^2 \left(\frac{\hat{\omega}_2}{\hat{\omega}_3}\right)^2\right]^{-1/2} = 2 \tan^{-1}\left(\frac{I}{J}\frac{\hat{\omega}_2}{\hat{\omega}_3}\right) \tag{45}$$

Furthermore,

$$\frac{\hat{\omega}_2}{\hat{\omega}_3} = \tan 1° = 0.017455 \tag{46}$$

and, if M is the mass of B, then

$$I = \frac{M}{12}(3r^2 + 4h^2) \tag{47}$$

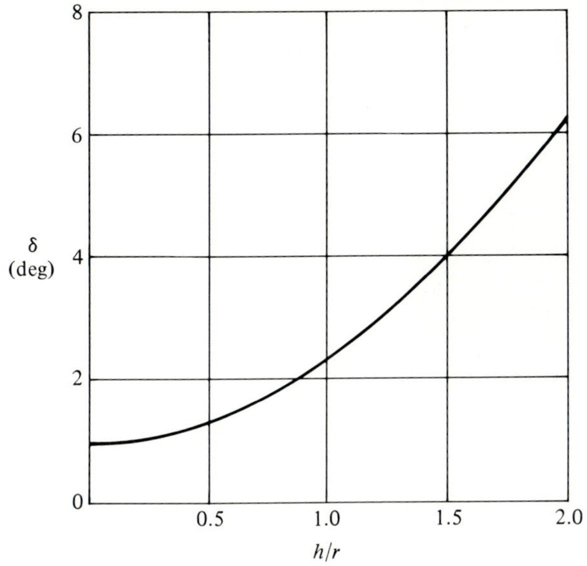

Figure 3.1.9

while

$$J = \frac{Mr^2}{2} \tag{48}$$

Thus,

$$\delta \underset{(45)}{=} 2 \tan^{-1} \left\{ 0.017455 \left[\frac{1}{2} + \frac{2}{3} \left(\frac{h}{r} \right)^2 \right] \right\} \tag{49}$$

Figure 3.1.9 shows the desired plot of this relationship.

3.2 EFFECT OF A GRAVITATIONAL MOMENT ON AN AXISYMMETRIC RIGID BODY IN A CIRCULAR ORBIT

In Fig. 3.2.1, O designates a point fixed in a Newtonian reference frame N, B^* is the mass center of an axisymmetric rigid body B, R is the distance between O and B^*, and $\mathbf{a}_1$ is a unit vector. B is presumed to be subjected to the action of a force system S such that **F**, the resultant of S, and **M**, the total moment of S about B^*, are given by

$$\mathbf{F} = -\mu m R^{-2} \mathbf{a}_1 \tag{1}$$

and

$$\mathbf{M} = \frac{3\mu}{R^3} \mathbf{a}_1 \times \mathbf{I} \cdot \mathbf{a}_1 \tag{2}$$

where μ is a constant, m is the mass of B, and **I** is the central inertia dyadic of B.

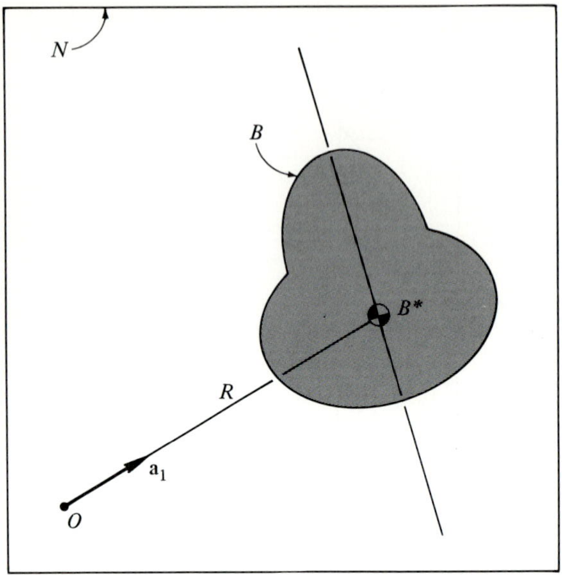

Figure 3.2.1

[If μ is set equal to $G\overline{m}$, $\mathbf{F}$ becomes equal to $\hat{\mathbf{F}}$ as defined in Eq. (2.3.5), while $\mathbf{M}$ becomes equal to $\widetilde{\mathbf{M}}$ as defined in Eq. (2.6.3). The force system S is then approximately equivalent to the system of gravitational forces exerted on B by a particle of mass $\overline{m}$ at point O.]

Due to the action of S, the rotational motion of B in N depends on the translational motion of B^* in N, but not vice versa, because R appears in Eq. (2), but no variable involving the orientation of B in N appears in Eq. (1); and one possible motion of B^* is motion in a circular orbit that is fixed in N, has a radius R, and is described with an angular speed Ω given by

$$\Omega = (\mu R^{-3})^{1/2} \tag{3}$$

When B^* moves in this fashion, the rotational motion of B in N proceeds in such a way that

$$2\dot{\epsilon}_1 = \epsilon_2(\hat{\omega}_3 - s + \Omega) - \epsilon_3\omega_2 + \epsilon_4\omega_1 \tag{4}$$

$$2\dot{\epsilon}_2 = \epsilon_3\omega_1 + \epsilon_4\omega_2 - \epsilon_1(\hat{\omega}_3 - s + \Omega) \tag{5}$$

$$2\dot{\epsilon}_3 = \epsilon_4(\hat{\omega}_3 - s - \Omega) + \epsilon_1\omega_2 - \epsilon_2\omega_1 \tag{6}$$

$$2\dot{\epsilon}_4 = -\epsilon_1\omega_1 - \epsilon_2\omega_2 - \epsilon_3(\hat{\omega}_3 - s - \Omega) \tag{7}$$

$$\dot{\omega}_1 = -s\omega_2 + (1 - JI^{-1})[\omega_2\hat{\omega}_3 - 12\Omega^2(\epsilon_1\epsilon_2 - \epsilon_3\epsilon_4)(\epsilon_3\epsilon_1 + \epsilon_2\epsilon_4)] \tag{8}$$

$$\dot{\omega}_2 = s\omega_1 - (1 - JI^{-1})[\omega_1\hat{\omega}_3 - 6\Omega^2(\epsilon_3\epsilon_1 + \epsilon_2\epsilon_4)(1 - 2\epsilon_2{}^2 - 2\epsilon_3{}^2)] \tag{9}$$

$$\omega_3 = \hat{\omega}_3, \text{ a constant} \tag{10}$$

where ϵ_i $(i = 1, \ldots, 4)$, ω_j $(j = 1, 2, 3)$, s, I, and J are defined as follows: Let $\mathbf{c}_1$, $\mathbf{c}_2$, $\mathbf{c}_3$ be a dextral set of orthogonal unit vectors such that $\mathbf{c}_3$ is parallel to the

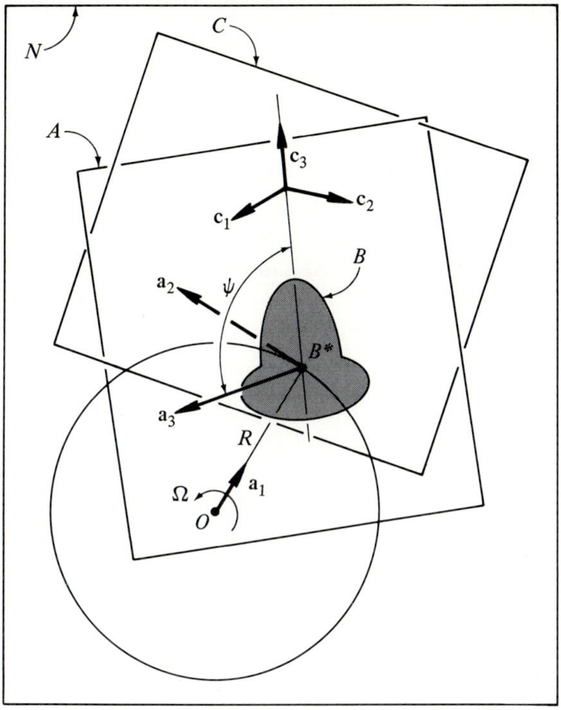

Figure 3.2.2

symmetry axis of B ($\mathbf{c}_3$ is thus fixed in B, but $\mathbf{c}_1$ and $\mathbf{c}_2$ need not be fixed in B), and introduce unit vectors $\mathbf{a}_2$ and $\mathbf{a}_3$ (see Fig. 3.2.2) such that $\mathbf{a}_1$, $\mathbf{a}_2$, $\mathbf{a}_3$ form a dextral, orthogonal set with $\mathbf{a}_2 = ({}^N d\mathbf{a}_1/dt)\Omega^{-1}$, so that $\mathbf{a}_3$ is normal to the orbit plane; let $C_{ij} = \mathbf{a}_i \cdot \mathbf{c}_j$ $(i, j = 1, 2, 3)$, and form ϵ_i $(i = 1, \ldots, 4)$ in accordance with Eqs. (1.3.15)–(1.3.18), thus making ϵ_i $(i = 1, \ldots, 4)$ Euler parameters which determine the orientation of a reference frame C, in which $\mathbf{c}_1$, $\mathbf{c}_2$, $\mathbf{c}_3$ are fixed, relative to a reference frame A in which $\mathbf{a}_1$, $\mathbf{a}_2$, $\mathbf{a}_3$ are fixed. Next, denoting the angular velocities of B in N and in C by ${}^N\boldsymbol{\omega}^B$ and ${}^C\boldsymbol{\omega}^B$, respectively, let

$$\omega_j \triangleq {}^N\boldsymbol{\omega}^B \cdot \mathbf{c}_j \qquad (j = 1, 2, 3) \tag{11}$$

and

$$s \triangleq {}^C\boldsymbol{\omega}^B \cdot \mathbf{c}_3 \tag{12}$$

which means that $|s|$ is equal to the magnitude of ${}^C\boldsymbol{\omega}^B$, since ${}^C\boldsymbol{\omega}^B$ is necessarily parallel to $\mathbf{c}_3$. Finally, let

$$I \triangleq \mathbf{c}_1 \cdot \mathbf{I} \cdot \mathbf{c}_1 = \mathbf{c}_2 \cdot \mathbf{I} \cdot \mathbf{c}_2 \tag{13}$$

and

$$J \triangleq \mathbf{c}_3 \cdot \mathbf{I} \cdot \mathbf{c}_3 \tag{14}$$

In Eqs. (4)–(9), s and $\hat{\omega}_3$ are "free" quantities; that is, s can be taken to be any desired function of time and $\hat{\omega}_3$ can have any constant value. As for $\epsilon_1, \ldots, \epsilon_4$,

ω_1, and ω_2, the nonlinearity and strong coupling of the equations prevent one from finding general expressions for these in terms of finite numbers of elementary functions. However, the equations are written in a form that is convenient for purposes of numerical integration, and they possess particular solutions of practical interest. Hence, they can furnish useful information about effects of **M** on the rotational motion of B in N. Suppose, for example, that ${}^N\boldsymbol{\omega}^B$ is parallel to $\mathbf{c}_3$ at $t = 0$. Then, if B were torque-free, $\mathbf{c}_3$ would remain fixed in N for $t > 0$ (see Sec. 3.1), and the angle ψ between $\mathbf{c}_3$ and $\mathbf{a}_3$ (see Fig. 3.2.2) would thus remain equal to its initial value, say $\hat{\psi}$. The behavior of ψ in the presence of **M** can be studied by means of numerical solutions of Eqs. (4)–(9), since

$$\psi = \cos^{-1}(\mathbf{a}_3 \cdot \mathbf{c}_3) \underset{(1.3.14)}{=} \cos^{-1}(1 - 2\epsilon_1^2 - 2\epsilon_2^2) \tag{15}$$

However, to carry out such a solution, one must first (in addition to specifying s and $\hat{\omega}_3$) choose an initial orientation of $\mathbf{c}_1$, $\mathbf{c}_2$, $\mathbf{c}_3$ relative to $\mathbf{a}_1$, $\mathbf{a}_2$, $\mathbf{a}_3$ in order to be able to give meaningful initial values to ϵ_i $(i = 1, \ldots, 4)$. (ω_1 and ω_2 are initially equal to zero, by hypothesis.) For instance, if at $t = 0$ these two sets of unit vectors are oriented relative to each other as shown in Fig. 3.2.3, then the initial values of ϵ_i $(i = 1, \ldots, 4)$ are, respectively [see Eqs. (1.3.1)–(1.3.3)], 0, $\sin(\hat{\psi}/2)$, 0, and $\cos(\hat{\psi}/2)$.

Figure 3.2.4, obtained by using these initial conditions with $\hat{\psi} = 0.1$ rad, shows plots of ψ versus the number of orbits traversed by B^*, for various values of the dimensionless parameters x and y defined as

$$x \triangleq JI^{-1} - 1 \qquad y \triangleq \hat{\omega}_3\Omega^{-1} - 1 \tag{16}$$

The fact that in some of these plots ψ departs markedly from $\hat{\psi}$ means that **M** can have a profound effect on the rotational motion of B under certain circumstances. This conclusion is supported by the following considerations: B can move in such a way that $\mathbf{c}_3 = \mathbf{a}_3$ for all t and ${}^N\boldsymbol{\omega}^B = \hat{\omega}_3\,\mathbf{a}_3$, a motion that corresponds to

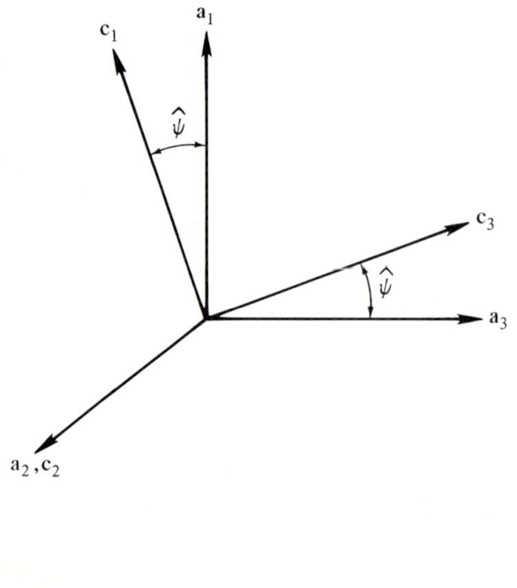

Figure 3.2.3

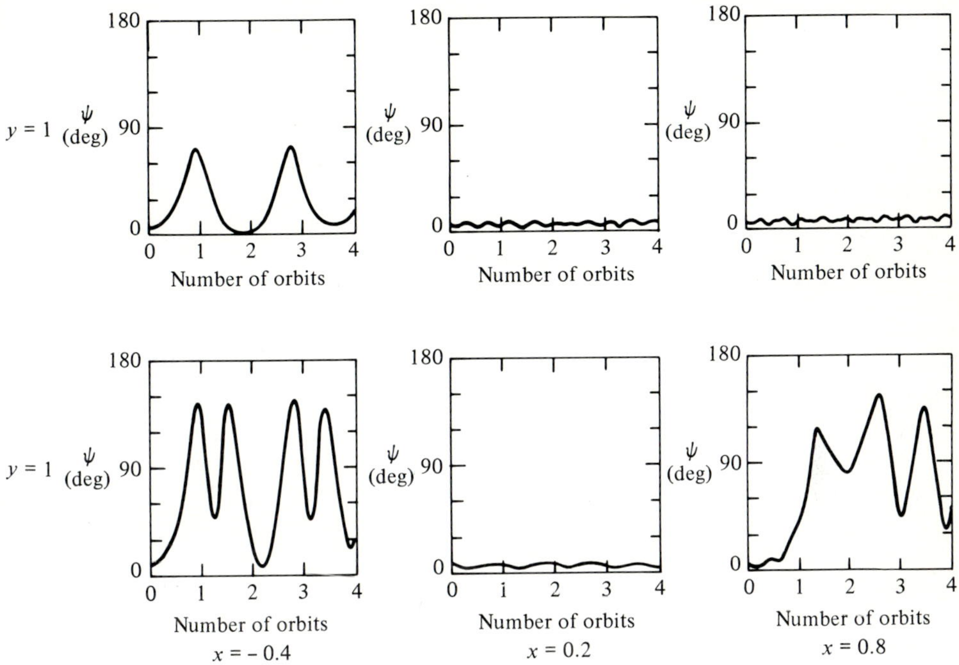

Figure 3.2.4

$$s = \hat{\omega}_3 - \Omega \tag{17}$$

and

$$\epsilon_1 = \epsilon_2 = \epsilon_3 = \omega_1 = \omega_2 = 0 \qquad \epsilon_4 = 1 \tag{18}$$

which is a particular solution of Eqs. (4)–(9); and this solution is unstable whenever J/I and $\hat{\omega}_3/\Omega$ have values such that the point with coordinates x and y as defined in Eqs. (16) lies in a shaded region of Fig. 3.2.5, that is, in a region such that at least one of the following inequalities is satisfied:

$$f_1(x, y) \triangleq 1 + 3x + [x + y(1 + x)]^2 < 0 \tag{19}$$

$$f_2(x, y) \triangleq [x + y(1 + x)][4x + y(1 + x)] < 0 \tag{20}$$

$$f_3(x, y) \triangleq f_1^2 - 4f_2 < 0 \tag{21}$$

The points in Fig. 3.2.5 corresponding to the plots in Fig. 3.2.4 are circled, and it can be seen that large values of ψ are encountered precisely when x and y have values such that the motion under consideration is unstable.

Figures 3.2.4 and 3.2.5 both suggest that the effect of **M** on the motion of B can be reduced by increasing $|\hat{\omega}_3/\Omega|$. This is the idea underlying *spin stabilization* of satellites. More specifically, if in addition to $|\hat{\omega}_3/\Omega|$ the initial values of $|\hat{\omega}_3/\omega_1|$ and $|\hat{\omega}_3/\omega_2|$ are sufficiently large, and if the symmetry axis of B is not initially

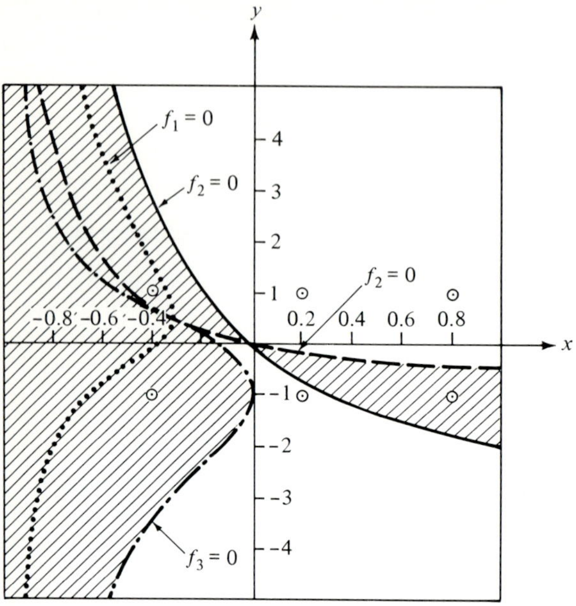

Figure 3.2.5

normal to the orbital plane, then the orientation of the symmetry axis in N changes slowly, in the following way: The plane determined by the orbit normal and the symmetry axis rotates nearly uniformly about the orbit normal, turning in N through an angle $\Delta\sigma$ given by

$$\Delta\sigma \approx 3\pi \frac{\Omega}{\hat{\omega}_3} (IJ^{-1} - 1) \cos \hat{\psi} \qquad \text{rad} \tag{22}$$

during each orbital revolution of B^* in N; and the angle between the orbit normal and the symmetry axis remains nearly constant.

Derivations The kinematical Eqs. (4)–(7) will be obtained by using Eq. (1.13.8) with

$$\omega \triangleq [{}^A\boldsymbol{\omega}^C \cdot \mathbf{c}_1 \quad {}^A\boldsymbol{\omega}^C \cdot \mathbf{c}_2 \quad {}^A\boldsymbol{\omega}^C \cdot \mathbf{c}_3] \tag{23}$$

and ${}^A\boldsymbol{\omega}^C \cdot \mathbf{c}_i$ $(i = 1, 2, 3)$ expressed in terms of ϵ_k $(k = 1, \ldots, 4)$, ω_j $(j = 1, 2, 3)$ as defined in Eq. (11), and s as defined in Eq. (12).

Note, first, that

$$ {}^A\boldsymbol{\omega}^C \underset{(1.16.1)}{=} {}^A\boldsymbol{\omega}^N + {}^N\boldsymbol{\omega}^B + {}^B\boldsymbol{\omega}^C \underset{(1.11.2)}{=} -{}^N\boldsymbol{\omega}^A + {}^N\boldsymbol{\omega}^B - {}^C\boldsymbol{\omega}^B \tag{24}$$

Next, resolve ${}^N\boldsymbol{\omega}^A$, ${}^N\boldsymbol{\omega}^B$, and ${}^C\boldsymbol{\omega}^B$ into components parallel to $\mathbf{c}_i$ $(i = 1, 2, 3)$ by taking advantage of the following facts: The motion of A in N is one of simple rotation such that

$$
{}^N\boldsymbol{\omega}^A \underset{(1.11.5)}{=} \Omega \mathbf{a}_3 \underset{(1.3.12-1.3.14)}{=} \Omega[2(\epsilon_3\epsilon_1 - \epsilon_2\epsilon_4)\mathbf{c}_1 + 2(\epsilon_2\epsilon_3 + \epsilon_1\epsilon_4)\mathbf{c}_2 + (1 - 2\epsilon_1{}^2 - 2\epsilon_2{}^2)\mathbf{c}_3] \tag{25}
$$

In accordance with Eq. (11),

$$
{}^N\boldsymbol{\omega}^B = \omega_1\mathbf{c}_1 + \omega_2\mathbf{c}_2 + \omega_3\mathbf{c}_3 \tag{26}
$$

and, since ${}^C\boldsymbol{\omega}^B$ is necessarily parallel to $\mathbf{c}_3$,

$$
{}^C\boldsymbol{\omega}^B \underset{(12)}{=} s\mathbf{c}_3 \tag{27}
$$

Substitution from Eqs. (25)–(27) into Eq. (24) and subsequent dot-multiplications with $\mathbf{c}_i$ $(i = 1, 2, 3)$ thus yield

$$
{}^A\boldsymbol{\omega}^C \cdot \mathbf{c}_1 = -2\Omega(\epsilon_3\epsilon_1 - \epsilon_2\epsilon_4) + \omega_1 \tag{28}
$$

$$
{}^A\boldsymbol{\omega}^C \cdot \mathbf{c}_2 = -2\Omega(\epsilon_2\epsilon_3 + \epsilon_1\epsilon_4) + \omega_2 \tag{29}
$$

$$
{}^A\boldsymbol{\omega}^C \cdot \mathbf{c}_3 = -\Omega(1 - 2\epsilon_1{}^2 - 2\epsilon_2{}^2) + \omega_3 - s \tag{30}
$$

Using Eq. (1.13.8) together with Eq. (23), one can, therefore, write, for example,

$$
\begin{aligned}
\dot{\epsilon}_1 &= \frac{1}{2}\left({}^A\boldsymbol{\omega}^C \cdot \mathbf{c}_1\epsilon_4 - {}^A\boldsymbol{\omega}^C \cdot \mathbf{c}_2\epsilon_3 + {}^A\boldsymbol{\omega}^C \cdot \mathbf{c}_3\epsilon_2\right) \\
&\underset{(28-30)}{=} \frac{1}{2}\Big\{\epsilon_2[-\Omega(1 - 2\epsilon_1{}^2 - 2\epsilon_2{}^2) + \omega_3 - s] + \epsilon_3[2\Omega(\epsilon_2\epsilon_3 + \epsilon_1\epsilon_4) - \omega_2] \\
&\qquad + \epsilon_4[-2\Omega(\epsilon_3\epsilon_1 - \epsilon_2\epsilon_4) + \omega_1]\Big\} \\
&= \frac{1}{2}\Big[\epsilon_2(\omega_3 - s) - \epsilon_3\omega_2 + \epsilon_4\omega_1 \\
&\qquad + \Omega(-\epsilon_2 + 2\epsilon_2\epsilon_1{}^2 + 2\epsilon_2{}^3 + 2\epsilon_2\epsilon_3{}^2 + 2\epsilon_2\epsilon_4{}^2)\Big] \\
&= \frac{1}{2}\Big[\epsilon_2(\omega_3 - s) - \epsilon_3\omega_2 + \epsilon_4\omega_1 \\
&\qquad + \Omega\epsilon_2(-1 + 2\epsilon_1{}^2 + 2\epsilon_2{}^2 + 2\epsilon_3{}^2 + 2\epsilon_4{}^2)\Big] \\
&\underset{(1.3.4)}{=} \frac{1}{2}\left[\epsilon_2(\omega_3 - s + \Omega) - \epsilon_3\omega_2 + \epsilon_4\omega_1\right]
\end{aligned} \tag{31}
$$

This equation differs from Eq. (4) only because it contains ω_3 rather than $\hat{\omega}_3$. Equations differing from Eqs. (5)–(7) in the same way are obtained similarly. Hence, the derivation of Eqs. (4)–(7) will be complete once it has been established that ω_3 must remain constant. This, as will now be shown, can be done by using the angular momentum principle.

The angular momentum $\mathbf{H}$ of B relative to B^* in N can be expressed as

$$
\mathbf{H} = I\omega_1\mathbf{c}_1 + I\omega_2\mathbf{c}_2 + J\omega_3\mathbf{c}_3 \tag{32}
$$

and, differentiating $\mathbf{H}$ with respect to t in N, one obtains

$$\frac{{}^N d\mathbf{H}}{dt} \underset{(1.11.8)}{=} \frac{{}^C d\mathbf{H}}{dt} + {}^N\boldsymbol{\omega}^C \times \mathbf{H} \underset{(1.16.1)}{=} \frac{{}^N d\mathbf{H}}{dt} + ({}^N\boldsymbol{\omega}^B + {}^B\boldsymbol{\omega}^C) \times \mathbf{H}$$

$$\underset{(32,26,27)}{=} [I(\dot{\omega}_1 + s\omega_2) + (J - I)\omega_2\omega_3]\mathbf{c}_1$$

$$+ [I(\dot{\omega}_2 - s\omega_1) - (J - I)\omega_3\omega_1]\mathbf{c}_2 + J\dot{\omega}_3\mathbf{c}_3 \tag{33}$$

To resolve **M** [see Eq. (2)] into components parallel to $\mathbf{c}_i$ $(i = 1, 2, 3)$, note that, from Eqs. (1.3.6)–(1.3.8),

$$\mathbf{a}_1 = (1 - 2\epsilon_2^2 - 2\epsilon_3^2)\mathbf{c}_1 + 2(\epsilon_1\epsilon_2 - \epsilon_3\epsilon_4)\mathbf{c}_2 + 2(\epsilon_3\epsilon_1 + \epsilon_2\epsilon_4)\mathbf{c}_3 \tag{34}$$

so that

$$\mathbf{a}_1 \times \mathbf{I} \cdot \mathbf{a}_1 = [4(\epsilon_1\epsilon_2 - \epsilon_3\epsilon_4)(\epsilon_3\epsilon_1 + \epsilon_2\epsilon_4)(J - I)]\mathbf{c}_1$$

$$+ [2(\epsilon_3\epsilon_1 + \epsilon_2\epsilon_4)(1 - 2\epsilon_2^2 - 2\epsilon_3^2)(I - J)]\mathbf{c}_2 \tag{35}$$

M can then be formed by substituting from this equation into Eq. (2); and, after using Eq. (3) to eliminate μ, one obtains the following equations by substituting from Eq. (33) into ${}^N d\mathbf{H}/dt = \mathbf{M}$:

$$I(\dot{\omega}_1 + s\omega_2) + (J - I)\omega_2\omega_3 = 12\Omega^2(\epsilon_1\epsilon_2 - \epsilon_3\epsilon_4)(\epsilon_3\epsilon_1 + \epsilon_2\epsilon_4)(J - I) \tag{36}$$

$$I(\dot{\omega}_2 - s\omega_1) - (J - I)\omega_3\omega_1 = 6\Omega^2(\epsilon_3\epsilon_1 + \epsilon_2\epsilon_4)(1 - 2\epsilon_2^2 - 2\epsilon_3^2)(I - J) \tag{37}$$

$$\dot{\omega}_3 = 0 \tag{38}$$

The last of these requires that ω_3 be a constant, say $\hat{\omega}_3$, and the first two are thus equivalent to Eqs. (8) and (9).

To show that Eqs. (18) comprise an unstable solution of Eqs. (4)–(9) under certain conditions, we introduce functions $\tilde{\epsilon}_i(t)$ $(i = 1, \ldots, 4)$ and $\tilde{\omega}_j(t)$ $(j = 1, 2)$ by expressing ϵ_i $(i = 1, \ldots, 4)$ and ω_j $(j = 1, 2)$ as

$$\epsilon_i = \tilde{\epsilon}_i \qquad (i = 1, 2, 3) \tag{39}$$

$$\epsilon_4 = 1 + \tilde{\epsilon}_4 \tag{40}$$

$$\omega_j = \tilde{\omega}_j \qquad (j = 1, 2, 3) \tag{41}$$

and then substitute into Eqs. (4)–(9), dropping all terms which are nonlinear in these functions. With x and y as in Eqs. (16), and, with s as in Eq. (17), this leads to the linearized variational system

$$\dot{\tilde{\epsilon}}_1 = \Omega\tilde{\epsilon}_2 + \tfrac{1}{2}\tilde{\omega}_1 \tag{42}$$

$$\dot{\tilde{\epsilon}}_2 = \tfrac{1}{2}\tilde{\omega}_2 - \Omega\tilde{\epsilon}_1 \tag{43}$$

$$\dot{\tilde{\epsilon}}_3 = \dot{\tilde{\epsilon}}_4 = 0 \tag{44}$$

$$\dot{\tilde{\omega}}_1 = -[y + x(1 + y)]\Omega\tilde{\omega}_2 \tag{45}$$

$$\dot{\tilde{\omega}}_2 = [y + x(1 + y)]\Omega\tilde{\omega}_1 - 6x\Omega^2\tilde{\epsilon}_2 \tag{46}$$

or, in matrix form,

$$\dot{z} = zA \tag{47}$$

where

$$z \triangleq [\tilde{\epsilon}_1 \quad \tilde{\epsilon}_2 \quad \tilde{\epsilon}_3 \quad \tilde{\epsilon}_4 \quad \tilde{\omega}_1 \quad \tilde{\omega}_2] \tag{48}$$

and

$$A \triangleq \begin{bmatrix} 0 & -\Omega & 0 & 0 & 0 & 0 \\ \Omega & 0 & 0 & 0 & 0 & -6x\Omega^2 \\ 0 & 0 & 0 & 0 & 0 & 0 \\ 0 & 0 & 0 & 0 & 0 & 0 \\ 1/2 & 0 & 0 & 0 & 0 & Q\Omega \\ 0 & 1/2 & 0 & 0 & -Q\Omega & 0 \end{bmatrix} \tag{49}$$

with

$$Q \triangleq x + y(1 + x) \tag{50}$$

The eigenvalues of A are the values of λ satisfying the characteristic equation

$$|A - \lambda U| = 0 \tag{51}$$

where U is the 6×6 unit matrix. In detail, this equation is

$$\lambda^2(\lambda^4 + 2b\lambda^2 + c) = 0 \tag{52}$$

with

$$b = \frac{1}{2}\left(1 + Q^2 + 3x\right)\Omega^2 \qquad c = Q(Q + 3x)\Omega^4 \tag{53}$$

Consequently, the six eigenvalues of A are

$$0, \quad 0, \quad \pm[-b \pm (b^2 - c)^{1/2}]^{1/2} \tag{54}$$

Now, Eqs. (18) comprise an unstable solution of Eqs. (4)–(9) whenever any eigenvalue of A has a positive real part; and this occurs whenever any one of the following conditions is satisfied:

$$b < 0 \qquad c < 0 \qquad b^2 - c < 0 \tag{55}$$

Expressing these inequalities in terms of x and y by reference to Eqs. (50) and (53), one arrives at the inequalities (19)–(21).

To obtain an approximate description of the motion of the symmetry axis of B in N that is valid when $|\hat{\omega}_3|$ is large in comparison with Ω and with the initial values of $|\omega_1|$ and $|\omega_2|$, we take advantage of the fact that s [see Eq. (12)] can be chosen freely. That is, we require that $\mathbf{c}_1$ remain normal to $\mathbf{a}_3$, and hence in the $\mathbf{a}_1$–$\mathbf{a}_2$ plane, as indicated in Fig. 3.2.6. The orientation of the symmetry axis of B in A then can be discussed in terms of the angle ϕ between $\mathbf{a}_1$ and $\mathbf{c}_1$ and the angle ψ between $\mathbf{a}_3$ and $\mathbf{c}_3$; and the motion in N of the plane determined by $\mathbf{a}_3$ and $\mathbf{c}_3$ depends only on the angle σ between $\mathbf{c}_1$ and a line L fixed in the orbit plane, as shown in Fig. 3.2.6.

When the Ω-dependent terms in Eqs. (8) and (9) are omitted (on the grounds that $\Omega << |\omega_3|$), these two equations possess the integral

$$\omega_1^2 + \omega_2^2 = \hat{\omega}_1^2 + \hat{\omega}_2^2 \tag{56}$$

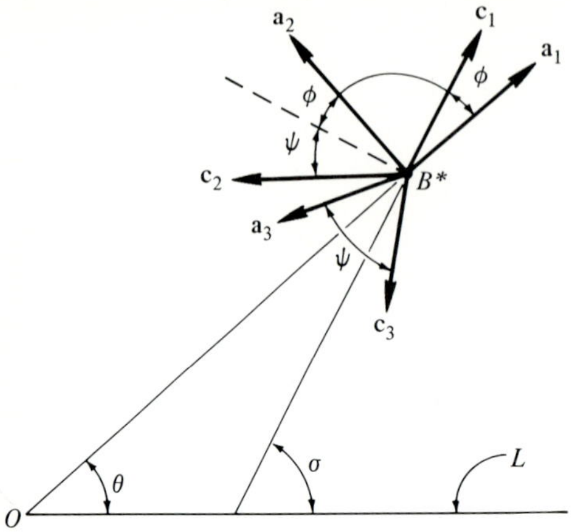

Figure 3.2.6

where $\hat{\omega}_1$ and $\hat{\omega}_2$ are the initial values of ω_1 and ω_2. This shows that $|\omega_1|$ and $|\omega_2|$ remain small in comparison with $|\hat{\omega}_3|$ if $|\hat{\omega}_1|$ and $|\hat{\omega}_2|$ meet this requirement. Accordingly, the angular momentum can be written, approximately, as

$$\mathbf{H} \underset{(32)}{\approx} J\hat{\omega}_3 \mathbf{c}_3 \tag{57}$$

so that

$$\frac{{}^N d\mathbf{H}}{dt} \approx J\hat{\omega}_3 \frac{{}^N d\mathbf{c}_3}{dt} \underset{(1.11.9)}{=} J\hat{\omega}_3 \, {}^N\boldsymbol{\omega}^C \times \mathbf{c}_3 = J\hat{\omega}_3(\dot{\sigma}\mathbf{a}_3 + \dot{\psi}\mathbf{c}_1) \times \mathbf{c}_3 \tag{58}$$

or

$$\frac{{}^N d\mathbf{H}}{dt} \approx J\hat{\omega}_3(\dot{\sigma} \sin \psi \, \mathbf{c}_1 - \dot{\psi}\mathbf{c}_2) \tag{59}$$

For the unit vector $\mathbf{a}_1$ one can write

$$\mathbf{a}_1 = \cos \phi \, \mathbf{c}_1 - \sin \phi \, (\cos \psi \, \mathbf{c}_2 - \sin \psi \, \mathbf{c}_3) \tag{60}$$

and this, together with Eqs. (3), (13), and (14), permits one to express $\mathbf{M}$ as

$$\mathbf{M} \underset{(2)}{=} 3\Omega^2(I - J) \sin \psi \sin \phi \, (\sin \phi \cos \psi \, \mathbf{c}_1 + \cos \phi \, \mathbf{c}_2) \tag{61}$$

Substitution into ${}^N d\mathbf{H}/dt = \mathbf{M}$ thus leads to

$$\hat{\omega}_3 J\dot{\sigma} \approx \frac{3}{2} \Omega^2(I - J) \cos \psi \{1 - \cos [2(\sigma - \theta)]\} \tag{62}$$

and

$$-\hat{\omega}_3 J\dot{\psi} \approx \frac{3}{2} \Omega^2(I - J) \sin \psi \sin [2(\sigma - \theta)] \tag{63}$$

where θ is the angle between line L and line OB^* (Fig. 3.2.6), so that $\dot{\theta} = \Omega$ and $\dot{\sigma}$ and $\dot{\psi}$ can be expressed as

$$\dot{\sigma} = \frac{d\sigma}{d\theta}\dot{\theta} = \frac{d\sigma}{d\theta}\Omega \qquad \dot{\psi} = \frac{d\psi}{d\theta}\Omega \tag{64}$$

Consequently,

$$\frac{d\sigma}{d\theta} \underset{(62,64)}{\approx} \frac{3}{2}\frac{\Omega}{\hat{\omega}_3}\left(\frac{I}{J} - 1\right)\cos\psi\{1 - \cos[2(\sigma - \theta)]\} \tag{65}$$

and

$$\frac{d\psi}{d\theta} \underset{(63,64)}{\approx} \frac{3}{2}\frac{\Omega}{\hat{\omega}_3}\left(1 - \frac{I}{J}\right)\sin\psi\sin[2(\sigma - \theta)] \tag{66}$$

These equations show that σ and ψ remain nearly constant when $\Omega/\hat{\omega}_3$ is sufficiently small. Hence one may expect to obtain meaningful results by "averaging" over one orbital revolution of B^* in N, that is, by integrating both sides of each equation with respect to θ, from 0 to 2π, keeping everything but θ constant. This leads to

$$\frac{d\sigma}{d\theta} \underset{(65)}{\approx} \frac{3}{2}\frac{\Omega}{\hat{\omega}_3}\left(\frac{I}{J} - 1\right)\cos\psi \tag{67}$$

and

$$\frac{d\psi}{d\theta} \underset{(66)}{\approx} 0 \tag{68}$$

from which it follows that

$$\psi \underset{(68)}{\approx} \hat{\psi}, \text{ a constant} \tag{69}$$

and that

$$\frac{d\sigma}{d\theta} \underset{(67,69)}{\approx} \frac{3}{2}\frac{\Omega}{\hat{\omega}_3}\left(\frac{I}{J} - 1\right)\cos\hat{\psi} \tag{70}$$

This, in turn, means that

$$\sigma \approx \frac{3}{2}\frac{\Omega}{\hat{\omega}_3}\left(\frac{I}{J} - 1\right)(\cos\hat{\psi})\,\theta \tag{71}$$

which shows that σ changes by an amount $\Delta\sigma$ given by Eq. (22) when θ increases by 2π rad.

Example B can move in such a way that $\mathbf{c}_3$ remains perpendicular to $\mathbf{a}_1$, the angle ψ between $\mathbf{c}_3$ and $\mathbf{a}_3$ (see Fig. 3.2.2) has a constant value, say $\hat{\psi}$, and the angular velocity of B in A is given by

$$^{A}\boldsymbol{\omega}^{B} = Z\Omega\mathbf{c}_3 \tag{72}$$

where Z is a constant. However, this motion is possible only if $\hat{\psi}$ and Z are related properly to I and J. To verify this claim, and to discover the relationship between $\hat{\psi}$, Z, I, and J that must be satisfied, orient $\mathbf{c}_1$ and $\mathbf{c}_2$ as shown in Fig. 3.2.3, form $C_{ij} = \mathbf{a}_i \cdot \mathbf{c}_j (i, j = 1, 2, 3)$, and use Eqs. (1.3.15)–(1.3.18) to determine ϵ_i $(i = 1, \ldots, 4)$:

$$\epsilon_1 = \sin\frac{\hat{\psi}}{2} \qquad \epsilon_2 = \epsilon_3 = 0 \qquad \epsilon_4 = \cos\frac{\hat{\psi}}{2} \tag{73}$$

Next, find ω_j ($j = 1, 2, 3$) for the given motion by writing

$$^N\boldsymbol{\omega}^B = {}^N\boldsymbol{\omega}^A + {}^A\boldsymbol{\omega}^B \underset{(72)}{=} \Omega\mathbf{a}_3 + Z\Omega\mathbf{c}_3 \tag{74}$$

and, from Eq. (11),

$$\omega_j = (\Omega\mathbf{a}_3 + Z\Omega\mathbf{c}_3)\cdot\mathbf{c}_j \qquad (j = 1, 2, 3) \tag{75}$$

or

$$\omega_1 = 0 \qquad \omega_2 = \Omega\sin\hat{\psi} \qquad \omega_3 = \Omega\cos\hat{\psi} + Z\Omega \tag{76}$$

Now substitute from Eqs. (73) and (76) into Eqs. (4)–(10); Eqs. (4), (7), and (9) are satisfied identically; Eqs. (5) and (6) each require

$$\Omega\cos\hat{\psi} - \hat{\omega}_3 + s = 0 \tag{77}$$

and, from Eqs. (8) and (10),

$$s = (1 - JI^{-1})\hat{\omega}_3 \tag{78}$$

Finally, eliminate $\hat{\omega}_3$ and s from the first of these equations by using the remaining two, which gives the desired relationship between $\hat{\psi}$, Z, I, and J, namely,

$$Z = (JI^{-1} - 1)\cos\hat{\psi} \tag{79}$$

3.3 INFLUENCE OF ORBIT ECCENTRICITY ON THE ROTATIONAL MOTION OF AN AXISYMMETRIC RIGID BODY

Under the circumstances described at the beginning of Sec. 3.2, B^* can move also in an elliptic orbit that is fixed in N, has an eccentricity e and major semidiameter a, and is traversed in a time T related to μ and a by

$$n \triangleq 2\pi T^{-1} = (\mu a^{-3})^{1/2} \tag{1}$$

The polar coordinates, R and θ, of B^* (see Fig. 3.3.1) satisfy the equation

$$R = a(1 - e^2)(1 + e\cos\theta)^{-1} \tag{2}$$

and θ is governed by the differential equation

$$\Omega \triangleq \dot{\theta} = na^2(1 - e^2)^{1/2}R^{-2} \tag{3}$$

while the rotational motion of B in N proceeds in such a way that

$$2\dot{\epsilon}_1 = \epsilon_2(\hat{\omega}_3 - s + \Omega) - \epsilon_3\omega_2 + \epsilon_4\omega_1 \tag{4}$$

$$2\dot{\epsilon}_2 = \epsilon_3\omega_1 + \epsilon_4\omega_2 - \epsilon_1(\hat{\omega}_3 - s + \Omega) \tag{5}$$

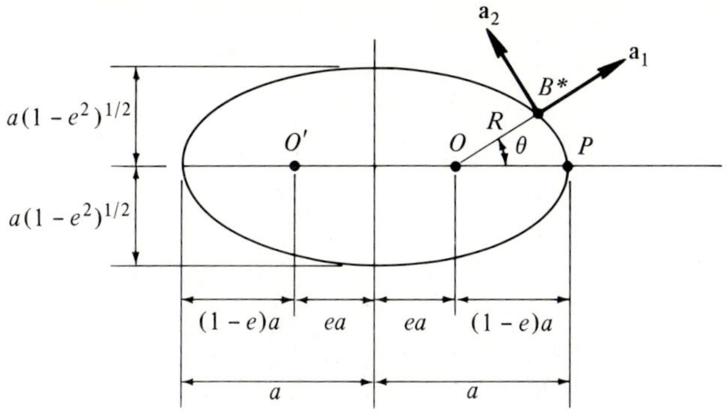

Figure 3.3.1

$$2\dot{\epsilon}_3 = \epsilon_4(\hat{\omega}_3 - s - \Omega) + \epsilon_1\omega_2 - \epsilon_2\omega_1 \tag{6}$$

$$2\dot{\epsilon}_4 = -\epsilon_1\omega_1 - \epsilon_2\omega_2 - \epsilon_3(\hat{\omega}_3 - s - \Omega) \tag{7}$$

$$\dot{\omega}_1 = -s\omega_2 + (1 - JI^{-1})[\omega_2\hat{\omega}_3 - 12n^2a^3R^{-3}(\epsilon_1\epsilon_2 - \epsilon_3\epsilon_4)(\epsilon_3\epsilon_1 + \epsilon_2\epsilon_4)] \tag{8}$$

$$\dot{\omega}_2 = s\omega_1 - (1 - JI^{-1})[\omega_1\hat{\omega}_3 - 6n^2a^3R^{-3}(\epsilon_3\epsilon_1 + \epsilon_2\epsilon_4)(1 - 2\epsilon_2^2 - 2\epsilon_3^2)] \tag{9}$$

$$\omega_3 = \hat{\omega}_3, \text{ a constant} \tag{10}$$

where ϵ_i $(i = 1, \ldots, 4)$, ω_j $(j = 1, 2, 3)$, s, I, and J are defined as in Sec. 3.2, with Ω given by Eq. (3) rather than by Eq. (3.2.3). Thus, Eqs. (3.2.11)–(3.2.14) apply, s may be any desired function of time, and $\hat{\omega}_3$ can have any constant value.

One way to study the effect of e on the rotational motion of B is to compare the results of numerical solutions of Eqs. (2)–(9) corresponding to various values of e, but satisfying identical initial conditions. Suppose, for example, that, at $t = 0$, B^* is at the periapsis P (see Fig. 3.3.1), ${}^N\boldsymbol{\omega}^B$ is parallel to $\mathbf{c}_3$, and $\mathbf{c}_1$, $\mathbf{c}_2$, $\mathbf{c}_3$ are oriented relative to $\mathbf{a}_1$, $\mathbf{a}_2$, $\mathbf{a}_3$ as shown in Fig. 3.2.3, so that the initial values of θ, ϵ_i $(i = 1, \ldots, 4)$ and ω_j $(j = 1, 2)$ are, respectively, 0, 0, $\sin(\hat{\psi}/2)$, 0, $\cos(\hat{\psi}/2)$, 0, 0, the last six of which are the same as those used in connection with Fig. 3.2.4. Then, with $J/I = 1.2$, $\hat{\omega}_3/n = 2$, and $\hat{\psi} = 0.1$ rad, one obtains for $e = 0, 0.1$, and 0.5 the plots shown in Fig. 3.3.2, where, as in Sec. 3.2, ψ is the angle between $\mathbf{a}_3$ and $\mathbf{c}_3$. [The plot for $e = 0$ is simply a magnified version of the plot corresponding to $x = 0.2$ and $y = 1$ in Fig. 3.2.4, since the present problem becomes the earlier one when $e = 0$ and $R = a$, in which case, from Eq. (3), $\Omega = n$, so that $\hat{\omega}_3/n = \hat{\omega}_3/\Omega$, and, in accordance with Eqs. (3.2.16), $x = 1.2 - 1.0 = 0.2$ and $y = 2 - 1 = 1$.] Evidently, the behavior of ψ depends intimately on the value of e. A stability analysis analogous to the one performed in connection with Eqs. (3.2.18) sheds further light on this dependence. Specifically, B can once again move in such a way that $\mathbf{c}_3$ remains parallel to $\mathbf{a}_3$ and ${}^N\boldsymbol{\omega}^B = \hat{\omega}_3\mathbf{a}_3$, for this motion corresponds to Eqs. (3.2.17) and (3.2.18), which comprise a particular so-

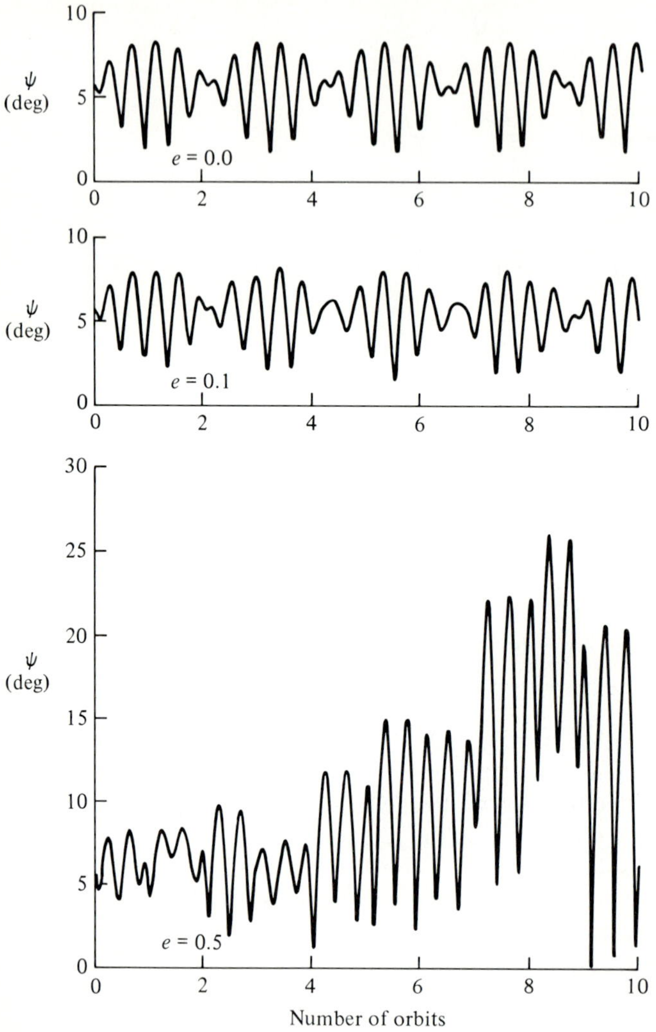

Figure 3.3.2

lution of Eqs. (4)–(9), regardless of the value of e. But the stability of this motion does depend on e, as well as on $\bar{x}$ and $\bar{y}$, defined by analogy to Eqs. (3.2.16) as

$$\bar{x} \triangleq JI^{-1} - 1 \qquad \bar{y} \triangleq \hat{\omega}_3 n^{-1} - 1 \tag{11}$$

This dependence is illustrated in Fig. 3.3.3, where a cross means that Eqs. (3.2.17) and (3.2.18) constitute an unstable solution of Eqs. (4)–(9) for the corresponding values of $\bar{x}$, $\bar{y}$, and e. The chart for $e = 0$ deals with precisely the same situation as does Fig. 3.2.5 and thus serves as a check on the procedure used to obtain the

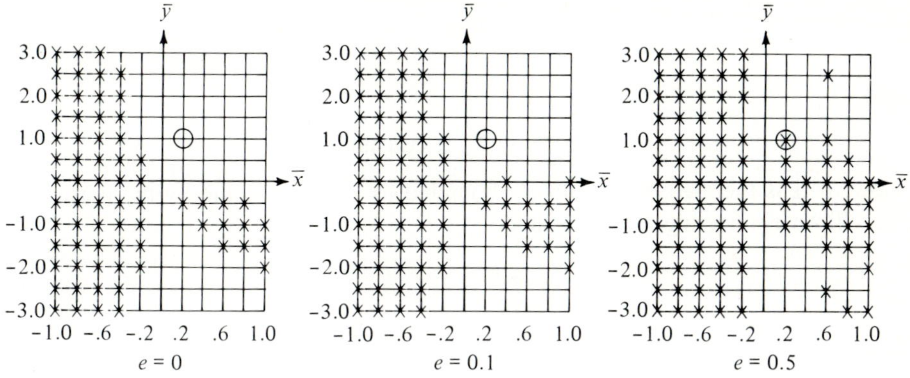

Figure 3.3.3

present results. The point of each chart corresponding to a plot in Fig. 3.3.2 is circled, and it appears once again that large values of ψ are associated with instability of the motion under consideration. Finally, the fact that successive charts contain progressively more crosses means that regions of instability in the $\bar{x}$–$\bar{y}$ plane grow with increasing e. Moreover, when $e \neq 0$, instability can be associated with points in the first quadrant of the $\bar{x}$–$\bar{y}$ plane, which is not the case when $e = 0$.

The eccentricity also makes itself felt during motions of the kind considered at the end of Sec. 3.2, that is, motions such that $|\hat{\omega}_3/n|$ and the initial values of $|\hat{\omega}_3/\omega_1|$ and $|\hat{\omega}_3/\omega_2|$ are large. If the symmetry axis of B is not initially normal to the orbital plane, then the plane determined by the orbit normal and the symmetry axis again rotates nearly uniformly about the orbit normal, but it now turns in N during each orbital revolution of B^* through an angle $\Delta\sigma$ given by

$$\Delta\sigma \approx 3\pi \frac{n}{\hat{\omega}_3}(IJ^{-1} - 1)(1 - e^2)^{-3/2} \cos \hat{\psi} \qquad \text{rad} \tag{12}$$

The angle ψ between the orbit normal and the symmetry axis remains nearly constant, as before.

Derivations Equation (2) is the equation of an ellipse such that the sum of the distances from B^* to O and to O' is equal to $2a$. The area A enclosed by the ellipse and the radius of curvature ρ at point P are given by

$$A = \pi a^2(1 - e^2)^{1/2} \tag{13}$$

and

$$\rho = a(1 - e^2) \tag{14}$$

respectively.

The $\mathbf{a}_2$ component of the acceleration of B^* in N is equal to zero if the resultant of the forces acting on B is given by Eq. (3.2.1). Thus,

$$\frac{d}{dt}(R^2\dot{\theta}) = 0 \tag{15}$$

and

$$R^2\dot{\theta} = C, \text{ a constant} \tag{16}$$

Now, if α is the area of the surface bounded by OP, OB^*, and by the portion of the ellipse connecting P to B^*, then

$$\frac{d\alpha}{dt} = \frac{1}{2}R^2\dot{\theta} \tag{17}$$

Hence,

$$\frac{d\alpha}{dt} \underset{(16,17)}{=} \frac{C}{2} \tag{18}$$

and the area A swept out by OB^* during the time T required for one orbital revolution of B^* is given by

$$A \underset{(18)}{=} \frac{CT}{2} \tag{19}$$

Thus,

$$C \underset{(19,13)}{=} 2\pi T^{-1}a^2(1 - e^2)^{1/2} \tag{20}$$

and

$$\dot{\theta} \underset{(16,20)}{=} 2\pi T^{-1}a^2(1 - e^2)^{1/2}R^{-2} \tag{21}$$

or

$$\dot{\theta} \underset{(1)}{=} na^2(1 - e^2)^{1/2}R^{-2} \tag{22}$$

in agreement with Eq. (3).

When B^* is at P, the $\mathbf{a}_1$ component of the acceleration of B^* in N is equal to $-R^2\dot{\theta}^2\rho^{-1}\mathbf{a}_1$, so that

$$-mR^2\dot{\theta}^2\rho^{-1} \underset{(3.2.1)}{=} -\mu mR^{-2} \tag{23}$$

or

$$\dot{\theta}^2 = \mu\rho R^{-4} \underset{(14)}{=} \mu a(1 - e^2)R^{-4} \tag{24}$$

But, when B^* is at P,

$$R \underset{(2)}{=} a(1 - e) \tag{25}$$

and

$$\dot{\theta} \underset{(22,25)}{=} n(1 - e^2)^{1/2}(1 - e)^{-2} \tag{26}$$

Consequently,

$$n^2 \underset{(24\text{–}26)}{=} \mu a^{-3} \tag{27}$$

from which Eq. (1) follows immediately.

Equations (4)–(10) may be obtained in precisely the same way as Eqs. (3.2.4)–(3.2.10). To show that Eqs. (3.2.18) comprise an unstable solution of Eqs. (4)–(9) under certain conditions, we again introduce functions $\widetilde{\epsilon}_i(t)$ $(i = 1, \ldots, 4)$ and $\widetilde{\omega}_j(t)$ $(j = 1, 2)$ by expressing ϵ_i $(i = 1, \ldots, 4)$ and ω_j $(j = 1, 2)$ as

$$\epsilon_i = \widetilde{\epsilon}_i \qquad (i = 1, 2, 3) \tag{28}$$

$$\epsilon_4 = 1 + \widetilde{\epsilon}_4 \tag{29}$$

$$\omega_j = \widetilde{\omega}_j \qquad (j = 1, 2) \tag{30}$$

and take

$$s = \hat{\omega}_3 - \Omega \tag{31}$$

Substitution into Eqs. (4)–(9) then leads to the linearized variational system

$$\dot{\widetilde{\epsilon}}_1 = \Omega \widetilde{\epsilon}_2 + 2^{-1} \widetilde{\omega}_1 \tag{32}$$

$$\dot{\widetilde{\epsilon}}_2 = 2^{-1} \widetilde{\omega}_2 - \Omega \widetilde{\epsilon}_1 \tag{33}$$

$$\dot{\widetilde{\epsilon}}_3 = \dot{\widetilde{\epsilon}}_4 = 0 \tag{34}$$

$$\dot{\widetilde{\omega}}_1 = [\Omega - (1 + \bar{x})(1 + \bar{y})n] \widetilde{\omega}_2 \tag{35}$$

$$\dot{\widetilde{\omega}}_2 = -[\Omega - (1 + \bar{x})(1 + \bar{y})n] \widetilde{\omega}_1 - 6\bar{x}n^2a^3R^{-3} \widetilde{\epsilon}_2 \tag{36}$$

where $\bar{x}$ and $\bar{y}$ are given by Eqs. (11). This system of equations differs from its counterpart in Sec. 3.2 in one important respect: it is a system with time-dependent, rather than constant, coefficients, since Ω is a function of R [see Eq. (3)] and R is time-dependent. In fact, the coefficients are *periodic* functions of t, of period T [see Eq. (1)].

The linearized variational system can be expressed in matrix form as

$$\dot{z} = zA \tag{37}$$

where

$$z \triangleq [\widetilde{\epsilon}_1 \quad \widetilde{\epsilon}_2 \quad \widetilde{\epsilon}_3 \quad \widetilde{\epsilon}_4 \quad \widetilde{\omega}_1 \quad \widetilde{\omega}_2] \tag{38}$$

and

$$A(t) \triangleq \begin{bmatrix} 0 & -\Omega & 0 & 0 & 0 & 0 \\ \Omega & 0 & 0 & 0 & 0 & W \\ 0 & 0 & 0 & 0 & 0 & 0 \\ 0 & 0 & 0 & 0 & 0 & 0 \\ 2^{-1} & 0 & 0 & 0 & 0 & -V \\ 0 & 2^{-1} & 0 & 0 & V & 0 \end{bmatrix} \tag{39}$$

with

$$V \triangleq \Omega - (1 + \bar{x})(1 + \bar{y})n \qquad W \triangleq -6\bar{x}n^2a^3R^{-3} \tag{40}$$

And, if $H(t)$ is now defined as the 6×6 matrix which satisfies the differential equation

$$\dot{H} = HA \tag{41}$$

and the initial condition

$$H(0) = U \tag{42}$$

where U is the 6×6 unit matrix, then Eqs. (3.2.18) comprise an unstable solution of Eqs. (4)–(9) whenever the modulus of any eigenvalue of $H(T)$ exceeds unity. The plots in Fig. 3.3.3 can thus be constructed by performing simultaneous numerical integrations of Eqs. (3) and (41), using as initial conditions $\theta(0) = 0$ and Eq. (42), and terminating the integrations at $t = T$, with T given by Eq. (1); finding the six eigenvalues of $H(T)$; and examining the modulus of each eigenvalue.

Equation (12) can be obtained by proceeding as in the derivation of Eq. (3.2.22) up to Eq. (3.2.61), which must be replaced with

$$\mathbf{M} = 3n^2a^3R^{-3}(I - J)\sin\psi\sin\phi\,(\sin\phi\cos\psi\,\mathbf{c}_1 + \cos\phi\,\mathbf{c}_2) \tag{43}$$

because Eq. (1), rather than Eq. (3.2.3), must now be used to eliminate μ. Applying the angular momentum principle, one then obtains

$$\hat{\omega}_3 J\dot{\sigma} \approx \frac{3}{2}n^2a^3R^{-3}(I - J)\cos\psi\,\{1 - \cos[2(\sigma - \theta)]\} \tag{44}$$

and

$$-\hat{\omega}_3 J\dot{\psi} \approx \frac{3}{2}n^2a^3R^{-3}(I - J)\sin\psi\sin[2(\sigma - \theta)] \tag{45}$$

Moreover,

$$\dot{\sigma} = \frac{d\sigma}{d\theta}\dot{\theta} \underset{(3)}{=} \frac{d\sigma}{d\theta}na^2(1 - e^2)^{1/2}R^{-2} \tag{46}$$

and

$$\dot{\psi} = \frac{d\psi}{d\theta}na^2(1 - e^2)^{1/2}R^{-2} \tag{47}$$

Consequently,

$$\frac{d\sigma}{d\theta} \underset{(44,46)}{\approx} \frac{3naR^{-1}}{2\hat{\omega}_3}\left(\frac{I}{J} - 1\right)(1 - e^2)^{-1/2}\cos\psi\,\{1 - \cos[2(\sigma - \theta)]\} \tag{48}$$

and

$$\frac{d\psi}{d\sigma} \underset{(45,46)}{\approx} \frac{3naR^{-1}}{2\hat{\omega}_3}\left(\frac{I}{J} - 1\right)(1 - e^2)^{-1/2}\sin\psi\sin[2(\sigma - \theta)] \tag{49}$$

or, after elimination of R by reference to Eq. (2),

$$\frac{d\sigma}{d\theta} \approx \frac{3}{2}\frac{n}{\hat{\omega}_3}\left(\frac{I}{J} - 1\right)(1 - e^2)^{-3/2} \cos\psi(1 + e\cos\theta)\{1 - \cos[2(\sigma - \theta)]\} \quad (50)$$

and

$$\frac{d\psi}{d\theta} \approx \frac{3}{2}\frac{n}{\hat{\omega}_3}\left(\frac{I}{J} - 1\right)(1 - e^2)^{-3/2} \sin\psi(1 + e\cos\theta)\sin[2(\sigma - \theta)] \quad (51)$$

"Averaging" by integrating with respect to θ, from 0 to 2π, now yields

$$\frac{d\sigma}{d\theta} \underset{(50)}{\approx} \frac{3}{2}\frac{n}{\hat{\omega}_3}\left(\frac{I}{J} - 1\right)(1 - e^2)^{-3/2} \cos\psi \quad (52)$$

and

$$\frac{d\psi}{d\theta} \underset{(51)}{\approx} 0 \quad (53)$$

so that

$$\psi \approx \hat{\psi}, \text{ a constant} \quad (54)$$

and

$$\sigma \underset{(52)}{\approx} \frac{3}{2}\frac{n}{\hat{\omega}_3}\left(\frac{I}{J} - 1\right)(1 - e^2)^{-3/2} (\cos\hat{\psi})\theta \quad (55)$$

It follows that σ changes by an amount $\Delta\sigma$ given by Eq. (12) when θ increases by 2π rad.

Example Applying Eq. (12) to the motion of the Earth by taking

$$\frac{n}{\hat{\omega}_3} \approx \frac{1}{365} \qquad IJ^{-1} \approx \frac{305}{304} \qquad e \approx 0.017 \qquad \hat{\psi} \approx 23.5° \quad (56)$$

one obtains

$$\Delta\sigma \approx 2.5\pi \times 10^{-5} \text{ rad} \quad (57)$$

Correspondingly, the time required for the line of intersection of the Earth's equatorial plane with the ecliptic plane to turn through 2π rad in N is approximately 80,000 years. Astronomical observations indicate that 26,000 years is a more nearly correct value. This discrepancy is attributable to leaving the effect of the Moon on the Earth's rotation out of account.

3.4 TORQUE-FREE MOTION OF AN UNSYMMETRIC RIGID BODY

When an unsymmetric rigid body B is subjected to the action of forces whose resultant moment about the mass center B^* of B is equal to zero, the angular velocity of B in a Newtonian reference frame N can be described as follows: Let $\mathbf{b}_1$, $\mathbf{b}_2$, $\mathbf{b}_3$ be a dextral set of unit vectors parallel to the central principal axes of inertia of B, numbering the unit vectors in such a way that

$$I_1 > I_2 > I_3 \tag{1}$$

where I_i $(i = 1, 2, 3)$ are defined in terms of the inertia dyadic $\mathbf{I}$ of B for B^* as

$$I_i \triangleq \mathbf{b}_i \cdot \mathbf{I} \cdot \mathbf{b}_i \qquad (i = 1, 2, 3) \tag{2}$$

Next, let

$$\omega_i \triangleq \boldsymbol{\omega} \cdot \mathbf{b}_i \qquad (i = 1, 2, 3) \tag{3}$$

where $\boldsymbol{\omega}$ is the angular velocity of B in N, and define h and e as

$$h \triangleq I_1^2 \hat{\omega}_1^2 + I_2^2 \hat{\omega}_2^2 + I_3^2 \hat{\omega}_3^2 \tag{4}$$

$$e \triangleq I_1 \hat{\omega}_1^2 + I_2 \hat{\omega}_2^2 + I_3 \hat{\omega}_3^2 \tag{5}$$

where $\hat{\omega}_i$ is the value of ω_i at time $t = 0$. Then, if two of $\hat{\omega}_i$ $(i = 1, 2, 3)$ are equal to zero, one has for all $t \geq 0$,

$$\omega_1 = \hat{\omega}_1 \qquad \omega_2 = \omega_3 = 0 \qquad \text{if } \hat{\omega}_2 = \hat{\omega}_3 = 0 \tag{6}$$

$$\omega_2 = \hat{\omega}_2 \qquad \omega_3 = \omega_1 = 0 \qquad \text{if } \hat{\omega}_3 = \hat{\omega}_1 = 0 \tag{7}$$

$$\omega_3 = \hat{\omega}_3 \qquad \omega_1 = \omega_2 = 0 \qquad \text{if } \hat{\omega}_1 = \hat{\omega}_2 = 0 \tag{8}$$

whereas, when at least two of $\hat{\omega}_1$, $\hat{\omega}_2$, $\hat{\omega}_3$ differ from zero, three possibilities present themselves. Specifically, let

$$z_1 \triangleq \frac{h - e(I_2 + I_3)}{I_2 I_3} \qquad z_2 \triangleq \frac{h - e(I_3 + I_1)}{I_3 I_1} \qquad z_3 \triangleq \frac{h - e(I_1 + I_2)}{I_1 I_2} \tag{9}$$

If

$$\frac{h}{e} > I_2 \tag{10}$$

then

$$\omega_1^2 = \frac{h - eI_3}{I_1(I_1 - I_3)} dn^2 \, [p(t - t_0), k] \tag{11}$$

$$\omega_2^2 = \frac{h - eI_1}{I_2(I_2 - I_1)} sn^2 \, [p(t - t_0), k] \tag{12}$$

$$\omega_3^2 = \frac{h - eI_1}{I_3(I_3 - I_1)} cn^2 \, [p(t - t_0), k] \tag{13}$$

where dn, sn, and cn, are Jacobian elliptic functions,† p and k are defined as

†See, e.g., W. Flügge, *Handbook of Engineering Mechanics*, McGraw-Hill, New York, 1962, p. **15**–24.

$$p \triangleq (z_1 - z_2)^{1/2} \qquad k \triangleq (z_3 - z_2)^{1/2} p^{-1} \tag{14}$$

and t_0 is a constant that can be expressed in terms of $\hat{\omega}_i$ and I_i ($i = 1, 2, 3$) by setting t equal to zero in any one of Eqs. (11)–(13). For example, from Eq. (12),

$$\hat{\omega}_2{}^2 = \frac{h - eI_1}{I_2(I_2 - I_1)} sn^2(-pt_0, k) \tag{15}$$

If [by contrast with (10)]

$$\frac{h}{e} < I_2 \tag{16}$$

then Eqs. (11)–(13) give way to

$$\omega_1{}^2 = \frac{h - eI_3}{I_1(I_1 - I_3)} cn^2\,[p(t - t_0), k] \tag{17}$$

$$\omega_2{}^2 = \frac{h - eI_3}{I_2(I_2 - I_3)} sn^2\,[p(t - t_0), k] \tag{18}$$

$$\omega_3{}^2 = \frac{h - eI_1}{I_3(I_3 - I_1)} dn^2\,[p(t - t_0), k] \tag{19}$$

with

$$p \triangleq (z_3 - z_2)^{1/2} \qquad k \triangleq (z_1 - z_2)^{1/2} p^{-1} \tag{20}$$

and Eq. (15) may be replaced with

$$\hat{\omega}_2{}^2 = \frac{h - eI_3}{I_2(I_2 - I_3)} sn^2(-pt_0, k) \tag{21}$$

Finally, if

$$\frac{h}{e} = I_2 \tag{22}$$

then

$$\omega_1{}^2 = \frac{I_2 I_3 p^2}{(I_1 - I_2)(I_3 - I_1)} \operatorname{csch}^2[p(t - t_0)] \tag{23}$$

$$\omega_2{}^2 = \frac{I_3 I_1 p^2}{(I_2 - I_3)(I_1 - I_2)} \operatorname{ctnh}^2[p(t - t_0)] \tag{24}$$

$$\omega_3{}^2 = \frac{I_1 I_2 p^2}{(I_3 - I_1)(I_2 - I_3)} \operatorname{csch}^2[p(t - t_0)] \tag{25}$$

where

$$p \triangleq (z_1 - z_2)^{1/2} \tag{26}$$

and t_0 may be found from

$$\hat{\omega}_2^{\,2} = \frac{I_3 I_1 p^2}{(I_2 - I_3)(I_1 - I_2)} \operatorname{ctnh}^2(-pt_0) \tag{27}$$

To choose appropriate signs for ω_i $(i = 1, 2, 3)$, one can use knowledge of the signs of any two of $\hat{\omega}_i$ $(i = 1, 2, 3)$ that differ from zero, in conjunction with Table 3.4.1, which shows all permissible combinations of signs of (nonvanishing) ω_i $(i = 1, 2, 3)$. For example, suppose $h/e > I_2$, $\hat{\omega}_1 > 0$, $\hat{\omega}_2 < 0$, and $\hat{\omega}_3 = 0$. Then, from Eqs. (11)–(13) and the first row of Table 3.4.1,

$$\omega_1 = \left[\frac{h - eI_3}{I_1(I_1 - I_3)}\right]^{1/2} dn\,[p(t - t_0), k] \tag{28}$$

$$\omega_2 = -\left[\frac{h - eI_1}{I_2(I_2 - I_1)}\right]^{1/2} sn\,[p(t - t_0), k] \tag{29}$$

$$\omega_3 = \left[\frac{h - eI_1}{I_3(I_3 - I_1)}\right]^{1/2} cn\,[p(t - t_0), k] \tag{30}$$

At times, it may be necessary to reorient $\mathbf{b}_i$ $(i = 1, 2, 3)$ before one can find suitable entries in Table 3.4.1. For instance, if $\hat{\omega}_1$, $\hat{\omega}_2$, $\hat{\omega}_3$ all are positive for a given choice of $\mathbf{b}_i$ $(i = 1, 2, 3)$, then Table 3.4.1 contains no corresponding row; but $\mathbf{b}_i$ can be replaced with $-\mathbf{b}_i$ $(i = 1, 2, 3)$, in which case $\hat{\omega}_i$ is replaced with $-\hat{\omega}_i$ $(i = 1, 2, 3)$, and the last row of Table 3.4.1 accommodates this possibility.

Once ω_i $(i = 1, 2, 3)$ are known as functions of t, space-three angles θ_i $(i = 1, 2, 3)$ (see Sec. 1.7) relating $\mathbf{b}_i$ $(i = 1, 2, 3)$ to a dextral set of orthogonal unit vectors $\mathbf{n}_i$ $(i = 1, 2, 3)$ fixed in N can be found as follows: When two of $\hat{\omega}_1$, $\hat{\omega}_2$, $\hat{\omega}_3$ are equal to zero, choose $\mathbf{n}_i$ such that $\mathbf{n}_i = \mathbf{b}_i$ $(i = 1, 2, 3)$ at $t = 0$. Then

$$\theta_1 = \hat{\omega}_1 t \qquad \theta_2 = \theta_3 = 0 \qquad \text{if } \hat{\omega}_2 = \hat{\omega}_3 = 0 \tag{31}$$

$$\theta_2 = \hat{\omega}_2 t \qquad \theta_3 = \theta_1 = 0 \qquad \text{if } \hat{\omega}_3 = \hat{\omega}_1 = 0 \tag{32}$$

$$\theta_3 = \hat{\omega}_3 t \qquad \theta_1 = \theta_2 = 0 \qquad \text{if } \hat{\omega}_1 = \hat{\omega}_2 = 0 \tag{33}$$

When at least two of $\hat{\omega}_1$, $\hat{\omega}_2$, $\hat{\omega}_3$ differ from zero, choose $\mathbf{n}_3$ such that

$$\mathbf{n}_3 = \mathbf{H} h^{-1/2} \tag{34}$$

where $\mathbf{H}$ is the angular momentum of B with respect to B^* in N. Then take

$$\theta_2 = \sin^{-1}(-I_1 \omega_1 h^{-1/2}) \qquad -\frac{\pi}{2} < \theta_2 < \frac{\pi}{2} \tag{35}$$

Table 3.4.1

ω_1	ω_2	ω_3
+	−	+
+	+	−
−	+	+
−	−	−

Next, let

$$\alpha \triangleq \sin^{-1} \frac{I_2\omega_2 h^{-1/2}}{\cos\theta_2} \qquad -\frac{\pi}{2} \le \alpha \le \frac{\pi}{2} \tag{36}$$

and take

$$\theta_1 = \begin{cases} \alpha & \text{if } \omega_3 \ge 0 \\ \pi - \alpha & \text{if } \omega_3 < 0 \end{cases} \tag{37}$$

Finally, determine θ_3 by solving the differential equation

$$\dot{\theta}_3 = h^{1/2} \frac{e - I_1\omega_1^{\,2}}{h - I_1^{\,2}\omega_1^{\,2}} \tag{38}$$

The rotational motion of B can be described also by reference to the rolling motion of an ellipsoid σ on a plane π, the so-called "Poinsot construction." Let σ be an ellipsoid whose center O is fixed in N, whose principal semidiameters are equal to $kI_j^{-1/2}$ $(j = 1, 2, 3)$, where k is a constant, and whose principal axes are at all times parallel to the principal axes of inertia of B for B^*; and let π be the plane that is tangent to σ at Q, the point of intersection of the surface of σ with the ray that passes through O and has the same direction as $\boldsymbol{\omega}$ at some instant t^*. Then π is a unique plane, fixed in N at a distance $d = k(\boldsymbol{\omega} \cdot \mathbf{H})^{1/2}/h^2$ from O and perpendicular to $\mathbf{H}$; and the rotational motion of B in N can be duplicated by a rolling motion of σ on π. The loci traced out by Q on σ and on π during such rolling are called, respectively, the *polhode* and the *herpolhode*.

Equations (31)–(33) describe motions during which a principal axis of inertia of B for B^* plays the part of an axis of rotation (see Sec. 1.1). Such a motion is stable if and only if the axis of rotation is the axis of maximum or minimum moment of inertia of B for B^*. That is, if $\mathbf{b}_i = \mathbf{n}_i$ $(i = 1, 2, 3)$ at $t = 0$, where $\mathbf{n}_i$ $(i = 1, 2, 3)$ are fixed in N, then the angle between $\mathbf{b}_1$ and $\mathbf{n}_1$ can be kept arbitrarily small for $t > 0$ by making $\hat{\omega}_2/\hat{\omega}_1$ and $\hat{\omega}_3/\hat{\omega}_1$ sufficiently small; the angle between $\mathbf{b}_3$ and $\mathbf{n}_3$ can be kept arbitrarily small for $t > 0$ by making $\hat{\omega}_1/\hat{\omega}_3$ and $\hat{\omega}_2/\hat{\omega}_3$ sufficiently small; but the angle between $\mathbf{b}_2$ and $\mathbf{n}_2$ cannot be kept arbitrarily small for $t > 0$ by making $\hat{\omega}_3/\hat{\omega}_2$ and $\hat{\omega}_1/\hat{\omega}_2$ sufficiently small.

Derivations. In accordance with the angular momentum principle, ω_i $(i = 1, 2, 3)$ are governed by

$$I_1\dot{\omega}_1 - (I_2 - I_3)\omega_2\omega_3 = 0 \tag{39}$$

$$I_2\dot{\omega}_2 - (I_3 - I_1)\omega_3\omega_1 = 0 \tag{40}$$

$$I_3\dot{\omega}_3 - (I_1 - I_2)\omega_1\omega_2 = 0 \tag{41}$$

It can be verified by inspection that Eqs. (6)–(8) furnish solutions of Eqs. (39)–(41).

Multiplication of Eqs. (39), (40), and (41) by $I_1\omega_1$, $I_2\omega_2$, and $I_3\omega_3$, respectively, and subsequent addition yield the *angular momentum integral,* that is, the equation

$$I_1^2\omega_1^2 + I_2^2\omega_2^2 + I_3^2\omega_3^2 = h \tag{42}$$

where h is the constant defined in Eq. (4). Similarly, multiplication of Eqs. (39), (40), and (41) with ω_1, ω_2, and ω_3, respectively, and subsequent addition lead to the *kinetic energy integral,*

$$I_1\omega_1^2 + I_2\omega_2^2 + I_3\omega_3^2 = e \tag{43}$$

where e is the constant defined in Eq. (5).

We now define z as

$$z \triangleq -(\omega_1^2 + \omega_2^2 + \omega_3^2) \tag{44}$$

and note that

$$\dot{z} = -2(\omega_1\dot{\omega}_1 + \omega_2\dot{\omega}_2 + \omega_3\dot{\omega}_3) \tag{45}$$

or, after using Eqs. (39), (40), and (41) to eliminate $\dot{\omega}_1$, $\dot{\omega}_2$, $\dot{\omega}_3$,

$$\dot{z} = 2\triangle \frac{\omega_1\omega_2\omega_3}{I_1 I_2 I_3} \tag{46}$$

where $\triangle$ is defined as

$$\triangle \triangleq \begin{vmatrix} 1 & 1 & 1 \\ I_1 & I_2 & I_3 \\ I_1^2 & I_2^2 & I_3^2 \end{vmatrix} = (I_1 - I_2)(I_2 - I_3)(I_3 - I_1) \tag{47}$$

Moreover, solving Eqs. (42), (43), and (44) for ω_1^2, ω_2^2, and ω_3^2, we have

$$\omega_1^2 = \frac{I_2 I_3}{(I_1 - I_2)(I_3 - I_1)}(z - z_1) \tag{48}$$

$$\omega_2^2 = \frac{I_3 I_1}{(I_2 - I_3)(I_1 - I_2)}(z - z_2) \tag{49}$$

$$\omega_3^2 = \frac{I_1 I_2}{(I_3 - I_1)(I_2 - I_3)}(z - z_3) \tag{50}$$

where z_i $(i = 1, 2, 3)$ are given by Eqs. (9). Substituting from Eqs. (48)–(50) into Eq. (46), we thus arrive at

$$\dot{z} = 2[(z - z_1)(z - z_2)(z - z_3)]^{1/2} \tag{51}$$

The nature of the solution of Eq. (51) depends on the relative magnitudes of z_1, z_2, z_3. These, as will now be shown, depend, in turn, on the relative magnitudes of h/e and I_2.

Eliminating ω_1^2 from Eqs. (42) and (43), one can write

$$I_2(I_1 - I_2)\omega_2^2 + I_3(I_1 - I_3)\omega_3^2 = eI_1 - h \tag{52}$$

which shows, in view of (1), that

$$eI_1 - h > 0 \tag{53}$$

Similarly, eliminating ω_3^2, one finds

$$eI_3 - h < 0 \tag{54}$$

Now, from Eqs. (9),

$$z_1 - z_2 = \frac{(I_1 - I_2)(h - eI_3)}{I_1 I_2 I_3} \tag{55}$$

$$z_2 - z_3 = \frac{(I_2 - I_3)(h - eI_1)}{I_1 I_2 I_3} \tag{56}$$

$$z_3 - z_1 = \frac{(I_3 - I_1)(h - eI_2)}{I_1 I_2 I_3} \tag{57}$$

Hence,

$$z_1 - z_2 > 0 \qquad z_2 - z_3 < 0 \tag{58}$$

and

$$z_3 - z_1 \begin{cases} < 0 & \text{if } \dfrac{h}{e} > I_2 \\ > 0 & \text{if } \dfrac{h}{e} < I_2 \\ = 0 & \text{if } \dfrac{h}{e} = I_2 \end{cases} \tag{59}$$

Table 3.4.2 shows z_1, z_2, and z_3 ordered in accordance with (58) and (59) in relation to h/e.

Returning to Eq. (51), we note that the general solution of the equation

$$\dot{z} = 2[(z - y_1)(z - y_2)(z - y_3)]^{1/2} \tag{60}$$

can be expressed as

$$z = y_1 + (y_2 - y_1)\, sn^2[p(t - t_0), k] \tag{61}$$

provided y_i ($i = 1, 2, 3$) be constants such that

$$y_1 < y_2 < y_3 \tag{62}$$

and p and k be defined as

Table 3.4.2

$\frac{h}{e} > I_2$	$z_1 > z_3 > z_2$
$\frac{h}{e} < I_2$	$z_3 > z_1 > z_2$
$\frac{h}{e} = I_2$	$z_3 = z_1 > z_2$

$$p \stackrel{\Delta}{=} (y_3 - y_1)^{1/2} \qquad k \stackrel{\Delta}{=} (y_2 - y_1)^{1/2} p^{-1} \tag{63}$$

Hence, if $h/e > I_2$, we take

$$y_1 = z_2 \qquad y_2 = z_3 \qquad y_3 = z_1 \tag{64}$$

to bring the first row of Table 3.4.2 into accord with (62). From Eqs. (61) and (64) we then have

$$z = z_2 + (z_3 - z_2)\, sn^2[p(t - t_0), k] \tag{65}$$

with

$$p \underset{(63,64)}{=} (z_1 - z_2)^{1/2} \qquad k \underset{(63,64)}{=} (z_3 - z_2)^{1/2} p^{-1} \tag{66}$$

in agreement with Eqs. (14). Furthermore,

$$\omega_1{}^2 \underset{(48,65,66)}{=} \frac{I_2 I_3 (z_2 - z_1)}{(I_1 - I_2)(I_3 - I_1)} \{1 - k^2 sn^2[p(t - t_0), k]\} \tag{67}$$

$$\omega_2{}^2 \underset{(49,65)}{=} \frac{I_3 I_1 (z_3 - z_2)}{(I_2 - I_3)(I_1 - I_2)}\, sn^2[p(t - t_0), k] \tag{68}$$

$$\omega_3{}^2 \underset{(50,65)}{=} \frac{I_1 I_2 (z_2 - z_3)}{(I_3 - I_1)(I_2 - I_3)} \{1 - sn^2[p(t - t_0), k]\} \tag{69}$$

Eliminating z_i ($i = 1, 2, 3$) by using Eqs. (9), and noting that

$$1 - k^2 sn^2[p(t - t_0), k] = dn^2[p(t - t_0), k] \tag{70}$$

while

$$1 - sn^2[p(t - t_0), k] = cn^2[p(t - t_0), k] \tag{71}$$

we thus arrive at Eqs. (11)–(13).

For $h/e < I_2$, one obtains Eqs. (17)–(20) by proceeding similarly after replacing Eqs. (64) with

$$y_1 = z_2 \qquad y_2 = z_1 \qquad y_3 = z_3 \tag{72}$$

in order to bring the second row of Table 3.4.2 into accord with (62).

Finally, if $h/e = I_2$, Eq. (51) becomes (see Table 3.4.2)

$$\dot{z} = 2(z - z_1)(z - z_2)^{1/2} \tag{73}$$

which possesses the integral

$$\ln\left[\frac{(z - z_2)^{1/2} - p}{(z - z_2)^{1/2} + p}\right] = 2p(t - t_0) \tag{74}$$

where p is given by Eq. (26). It follows that

$$z - z_2 = p^2 \left\{\frac{1 + \exp[2p(t - t_0)]}{1 - \exp[2p(t - t_0)]}\right\}^2 = p^2 \operatorname{ctnh}^2[p(t - t_0)] \tag{75}$$

and from this that

$$z - z_1 = z - z_3 = p^2\{\text{ctnh}^2[p(t - t_0) - 1]\} = p^2\text{csch}^2[p(t - t_0)] \quad (76)$$

Equations (23)–(25) can now be obtained by substituting into Eqs. (48)–(50).

To deal with the question of appropriate signs for ω_i $(i = 1, 2, 3)$, consider first Eqs. (11)–(13). Taking square roots of both sides of each of these, one can write

$$\omega_1 = \pm \left[\frac{h - eI_3}{I_1(I_1 - I_3)}\right]^{1/2} dn[p(t - t_0), k] \quad (77)$$

$$\omega_2 = \pm \left[\frac{h - eI_1}{I_2(I_2 - I_1)}\right]^{1/2} sn[p(t - t_0), k] \quad (78)$$

$$\omega_3 = \pm \left[\frac{h - eI_1}{I_3(I_3 - I_1)}\right]^{1/2} cn[p(t - t_0), k] \quad (79)$$

and, differentiating with respect to t,

$$\dot{\omega}_1 = \mp pk^2\left[\frac{h - eI_3}{I_1(I_1 - I_3)}\right]^{1/2} sn[p(t - t_0), k]\, cn[p(t - t_0), k] \quad (80)$$

$$\dot{\omega}_2 = \pm p \left[\frac{h - eI_1}{I_2(I_2 - I_1)}\right]^{1/2} cn[p(t - t_0), k]\, dn[p(t - t_0), k] \quad (81)$$

$$\dot{\omega}_3 = \mp p \left[\frac{h - eI_1}{I_3(I_3 - I_1)}\right]^{1/2} sn[p(t - t_0), k]\, cn[p(t - t_0), k] \quad (82)$$

If one now substitutes into Eqs. (39)–(41), using, for example, the upper signs for ω_i and $\dot{\omega}_i$ $(i = 1, 2, 3)$, and eliminating p and k by reference to Eqs. (9) and (14), one finds that Eq. (40) is satisfied identically while Eqs. (39) and (41) require that

$$(-1 - 1)\, sn[p(t - t_0), k]\, cn[p(t - t_0), k] = 0 \quad (83)$$

which cannot be satisfied for all t. Hence, one is dealing with an inadmissible combination of signs. If, however, one uses any of the combinations of signs shown in Table 3.4.1, each of Eqs. (39)–(41) is satisfied identically. Moreover, the same conclusion emerges when one works with Eqs. (17)–(19) or (23)–(25) in place of Eqs. (11)–(13).

Turning to the determination of the space-three orientation angles θ_i $(i = 1, 2, 3)$ (see Sec. 1.7) relating $\mathbf{b}_i$ $(i = 1, 2, 3)$ to a dextral set of orthogonal unit vectors $\mathbf{n}_i$ $(i = 1, 2, 3)$ fixed in N, we refer to Eqs. (1.17.2) and (1.17.4) to write the following differential equations governing θ_i $(i = 1, 2, 3)$:

$$\dot{\theta}_1 = \omega_1 + \frac{\omega_2 s_1 s_2 + \omega_3 c_1 s_2}{c_2} \quad (84)$$

$$\dot{\theta}_2 = \omega_2 c_1 - \omega_3 s_1 \quad (85)$$

$$\dot{\theta}_3 = \frac{\omega_2 s_1 + \omega_3 c_1}{c_2} \quad (86)$$

If $\mathbf{n}_i = \mathbf{b}_i$ $(i = 1, 2, 3)$ at $t = 0$, then $\theta_i(t)$ $(i = 1, 2, 3)$ must satisfy, in addition to Eqs. (84)–(86), the initial conditions

$$\theta_i(0) = 0 \qquad (i = 1, 2, 3) \tag{87}$$

Suppose now that two of $\hat{\omega}_i$ $(i = 1, 2, 3)$ are equal to zero, so that Eqs. (6)–(8) describe ω_i $(i = 1, 2, 3)$. Then it can be verified by inspection that Eqs. (31)–(33) furnish corresponding expressions for θ_i $(i = 1, 2, 3)$ that satisfy Eqs. (84)–(87).

When at least two of $\hat{\omega}_1$, $\hat{\omega}_2$, $\hat{\omega}_3$ differ from zero, then $\mathbf{H}$, the angular momentum of B with respect to B^*, can be expressed as

$$\mathbf{H} = I_1\omega_1\mathbf{b}_1 + I_2\omega_2\mathbf{b}_2 + I_3\omega_3\mathbf{b}_3 \tag{88}$$

which shows that h as defined in Eq. (4) is equal to $\mathbf{H}^2$. Since during torque-free motion $\mathbf{H}$ has a fixed orientation in N, one can, therefore, fix a unit vector $\mathbf{n}_3$ in N by making

$$\mathbf{n}_3 \triangleq \mathbf{H}h^{-1/2} \underset{(88)}{=} (I_1\omega_1\mathbf{b}_1 + I_2\omega_2\mathbf{b}_2 + I_3\omega_3\mathbf{b}_3)h^{-1/2} \tag{89}$$

If C_{ij} is now defined as

$$C_{ij} \triangleq \mathbf{n}_i \cdot \mathbf{b}_j \qquad (i, j = 1, 2, 3) \tag{90}$$

one has

$$C_{3j} \underset{(89)}{=} I_j\omega_j h^{-1/2} \qquad (j = 1, 2, 3) \tag{91}$$

and since $|C_{31}| \neq 1$ [see Eqs. (11), (17), and (23)], one can find θ_1 and θ_2 by using Eqs. (1.7.2)–(1.7.4), thus arriving at Eqs. (35)–(37). As for θ_3, note that

$$s_1c_2 \underset{(1.7.1)}{=} C_{32} \underset{(91)}{=} I_2\omega_2 h^{-1/2} \tag{92}$$

while

$$c_1c_2 \underset{(1.17.1)}{=} C_{33} \underset{(91)}{=} I_3\omega_3 h^{-1/2} \tag{93}$$

Hence,

$$s_1 \underset{(92)}{=} \frac{I_2\omega_2}{c_2h^{1/2}} \qquad c_1 \underset{(43)}{=} \frac{I_3\omega_3}{c_2h^{1/2}} \tag{94}$$

and

$$\dot{\theta}_3 \underset{(86)}{=} \frac{I_2\omega_2^2 + I_3\omega_3^2}{c_2^2h^{1/2}} \underset{(35)}{=} \frac{I_2\omega_2^2 + I_3\omega_3^2}{(1 - I_1^2\omega_1^2h^{-1})h^{1/2}} \underset{(43)}{=} h^{1/2}\frac{e - I_1\omega_1^2}{h - I_1^2\omega_1^2} \tag{95}$$

in agreement with Eq. (38).

The validity of the Poinsot construction may be established by proceeding as in Sec. 3.1, with the exception that Eqs. (3.1.29)–(3.1.31) must be replaced with

$$\mathbf{r} = x_1\mathbf{b}_1 + x_2\mathbf{b}_2 + x_3\mathbf{b}_3 \tag{96}$$

$$\mathbf{K} = I_1\mathbf{b}_1\mathbf{b}_1 + I_2\mathbf{b}_2\mathbf{b}_2 + I_3\mathbf{b}_3\mathbf{b}_3 \tag{97}$$

and
$$I_1 x_1^2 + I_2 x_2^2 + I_3 x_3^2 = k^2 \tag{98}$$

respectively.

In preparation for proving that the motions described by Eqs. (31) and (33) are stable, whereas those corresponding to Eqs. (32) are unstable, we refer to Eq. (88) to note that the angle between **H** and $\mathbf{b}_1$ can be expressed as

$$\measuredangle(\mathbf{H}, \mathbf{b}_1) = \cos^{-1}\left[1 + \left(\frac{I_2\omega_2}{I_1\omega_1}\right)^2 + \left(\frac{I_3\omega_3}{I_1\omega_1}\right)^2\right]^{-1/2} \tag{99}$$

and if $\mathbf{b}_1 = \mathbf{n}_1$ at $t = 0$, then the angle between **H** and $\mathbf{n}_1$ is given by

$$\measuredangle(\mathbf{H}, \mathbf{n}_1) \underset{(99)}{=} \cos^{-1}\left[1 + \left(\frac{I_2\hat{\omega}_2}{I_1\hat{\omega}_1}\right)^2 + \left(\frac{I_3\hat{\omega}_3}{I_1\hat{\omega}_1}\right)^2\right]^{-1/2} \tag{100}$$

Now, in order for the angle between $\mathbf{b}_1$ and $\mathbf{n}_1$ to be small, it is necessary and sufficient that $\measuredangle(\mathbf{H}, \mathbf{b}_1)$ and $\measuredangle(\mathbf{H}, \mathbf{n}_1)$ each be small. Hence the stability of the motion described by Eqs. (31) hinges upon the relationship between the magnitudes of $\hat{\omega}_2/\hat{\omega}_1$ and $\hat{\omega}_3/\hat{\omega}_1$, on the one hand, and the magnitudes of $\measuredangle(\mathbf{H}, \mathbf{b}_1)$ and $\measuredangle(\mathbf{H}, \mathbf{n}_1)$, on the other hand.

As regards $\measuredangle(\mathbf{H}, \mathbf{n}_1)$, it follows immediately from Eq. (100) that this angle becomes small when $\hat{\omega}_2/\hat{\omega}_1$ and $\hat{\omega}_3/\hat{\omega}_1$ become small. To prove that this is the case also for $\measuredangle(\mathbf{H}, \mathbf{b}_1)$, we introduce ϵ_i $(i = 1, 2)$ as

$$\epsilon_i \triangleq \frac{\hat{\omega}_i}{\hat{\omega}_1} \qquad (i = 2, 3) \tag{101}$$

and use Eqs. (4) and (5) to write

$$\frac{h}{e} = \frac{I_1^2 + I_2^2\epsilon_2^2 + I_3^2\epsilon_3^2}{I_1 + I_2\epsilon_2^2 + I_3\epsilon_3^2} \tag{102}$$

which shows that h/e becomes nearly equal to I_1 when ϵ_2 and ϵ_3 become small. Since $I_1 > I_2$ by (1), this means that $h/e > I_2$ for small ϵ_2 and ϵ_3, and we may thus use Eqs. (11)–(13) to study ω_i $(i = 1, 2, 3)$. Now, using Eqs. (4), (5), and (101), we find that

$$\frac{h - eI_3}{I_1(I_1 - I_3)} = \left[1 + \frac{I_2(I_2 - I_3)}{I_1(I_1 - I_3)}\,\epsilon_2^2\right]\hat{\omega}_1^2 \tag{103}$$

$$\frac{h - eI_1}{I_2(I_2 - I_1)} = \left[\epsilon_2^2 + \frac{I_3(I_3 - I_1)}{I_2(I_2 - I_1)}\,\epsilon_3^2\right]\hat{\omega}_1^2 \tag{104}$$

$$\frac{h - eI_2}{I_3(I_3 - I_1)} = \left[\frac{I_2(I_2 - I_1)}{I_3(I_3 - I_1)}\,\epsilon_2^2 + \epsilon_3^2\right]\hat{\omega}_1^2 \tag{105}$$

From Eqs. (11)–(13) it can thus be seen that ω_1^2 does not become small, but ω_2^2 and ω_3^2 do, when ϵ_2 and ϵ_3 become small; and Eq. (99) therefore implies that $\measuredangle(\mathbf{H}, \mathbf{b}_1)$, like $\measuredangle(\mathbf{H}, \mathbf{n}_1)$, becomes small when $\hat{\omega}_2/\hat{\omega}_1$ and $\hat{\omega}_3/\hat{\omega}_1$ become small. Consequently, Eqs. (31) describe a stable motion.

The stability of the motion corresponding to Eqs. (33) can be established by proceeding similarly. To deal with Eqs. (32), we introduce η_i as

$$\eta_i \stackrel{\Delta}{=} \frac{\hat{\omega}_i}{\hat{\omega}_2} \qquad (i = 1, 3) \tag{106}$$

and use Eqs. (4) and (5) to show that h/e becomes nearly equal to I_2 when η_1 and η_3 become small, so that one cannot tell at a glance whether Eqs. (11)–(13) or Eqs. (17)–(19) should be used to determine the character of ω_i $(i = 1, 2, 3)$. However, the coefficients of the elliptic functions in the expressions for ω_1^2 in Eqs. (11) and (17) are the same, as are those in the expressions for ω_3^2 in Eqs. (13) and (19). Hence, it is unnecessary to make a choice, and we note that ω_1^2 and ω_3^2 depend on

$$\frac{h - eI_3}{I_1(I_1 - I_3)} \underset{(4,5)}{=} \left[\frac{I_2(I_2 - I_3)}{I_1(I_1 - I_2)} + \eta_1^2\right]\hat{\omega}_2^2 \tag{107}$$

and on

$$\frac{h - eI_1}{I_3(I_3 - I_1)} \underset{(4,5)}{=} \left[\frac{I_2(I_2 - I_1)}{I_3(I_3 - I_1)} + \eta_3^2\right]\hat{\omega}_2^2 \tag{108}$$

respectively. Clearly, these quantities cannot be made small by making η_1 and η_3 small. Consequently, the angle between $\mathbf{H}$ and $\mathbf{b}_2$ cannot be kept arbitrarily small for $t > 0$ by making $\hat{\omega}_3/\hat{\omega}_2$ and $\hat{\omega}_1/\hat{\omega}_2$ sufficiently small, and it follows that the motion described by Eqs. (32) is unstable.

Example As an alternative to the procedure just employed to establish the instability of "rotation about the intermediate axis," as the motion described by Eqs. (32) is sometimes called, one can proceed directly from the differential equations (39)–(41) to the linearized variational system associated with Eqs. (32) by introducing $\tilde{\omega}_i$ $(i = 1, 2, 3)$ such that

$$\omega_1 = \tilde{\omega}_1 \qquad \omega_2 = \hat{\omega}_2 + \tilde{\omega}_2 \qquad \omega_3 = \tilde{\omega}_3 \tag{109}$$

and linearizing in $\tilde{\omega}_i$ $(i = 1, 2, 3)$, which gives

$$\begin{bmatrix} \dot{\tilde{\omega}}_1 \\ \dot{\tilde{\omega}}_2 \\ \dot{\tilde{\omega}}_3 \end{bmatrix} = \begin{bmatrix} 0 & 0 & \dfrac{\hat{\omega}_2(I_2 - I_3)}{I_1} \\ 0 & 0 & 0 \\ \dfrac{\hat{\omega}_2(I_1 - I_2)}{I_3} & 0 & 0 \end{bmatrix} \begin{bmatrix} \tilde{\omega}_1 \\ \tilde{\omega}_2 \\ \tilde{\omega}_3 \end{bmatrix} \tag{110}$$

The characteristic equation for this system is

$$\lambda\left[\lambda^2 - \frac{(I_1 - I_2)(I_2 - I_3)}{I_3 I_1}\hat{\omega}_2^2\right] = 0 \tag{111}$$

and it follows from (1) that this equation possesses one positive root.

The stability of rotations about the axes of maximum and minimum moment of inertia of B for B^* can also be established without reference to explicit expressions for $\omega_i(t)$ $(i = 1, 2, 3)$. For example, when attempting to prove that Eqs. (31) describe a stable motion, one can introduce Z as

$$Z \triangleq eI_1 - h \underset{(42,43)}{=} I_2(I_1 - I_2)\omega_2^2 + I_3(I_1 - I_3)\omega_3^2 \tag{112}$$

and note that Z is a constant (since e, I_1, and h are constants), so that throughout any motion,

$$I_2(I_1 - I_2)\omega_2^2 + I_3(I_1 - I_3)\omega_3^2 = I_2(I_1 - I_2)\hat{\omega}_2^2 + I_3(I_1 - I_3)\hat{\omega}_3^2 \tag{113}$$

The right-hand member of this equation can be made arbitrarily small by making $\hat{\omega}_2$ and $\hat{\omega}_3$ sufficiently small. The left-hand member vanishes only when ω_2 and ω_3 each vanish, since $I_1 - I_2$ and $I_1 - I_3$ are positive by (1). Hence ω_2 and ω_3 can be kept arbitrarily small by making $\hat{\omega}_2$ and $\hat{\omega}_3$ sufficiently small, a fact previously established by using Eqs. (11)–(13).

3.5 EFFECT OF A GRAVITATIONAL MOMENT ON AN UNSYMMETRIC RIGID BODY IN A CIRCULAR ORBIT

In Fig. 3.5.1, O designates a point fixed in a Newtonian reference frame N, B^* is the mass center of an unsymmetric rigid body B, R is the distance between O and B^*, and $\mathbf{a}_1$ is a unit vector. B is presumed to be subjected to the action of a force system S such that $\mathbf{F}$, the resultant of S, and $\mathbf{M}$, the total moment of S about B^*, are given by

$$\mathbf{F} = -\frac{\mu m}{R^2}\mathbf{a}_1 \tag{1}$$

and

$$\mathbf{M} = \frac{3\mu}{R^3}\mathbf{a}_1 \times \mathbf{I} \cdot \mathbf{a}_1 \tag{2}$$

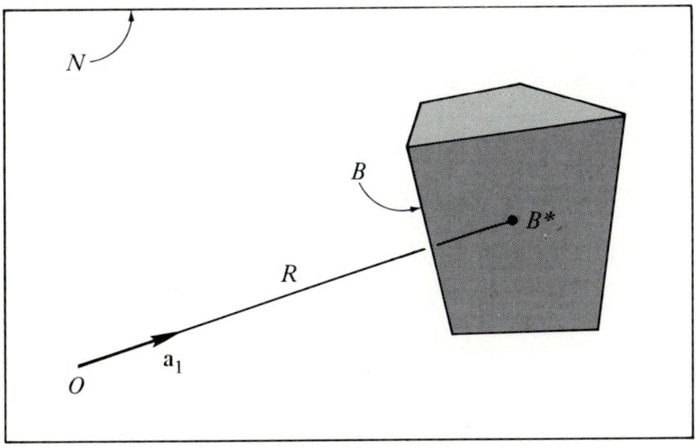

Figure 3.5.1

where μ is a constant, m is the mass of B, and $\mathbf{I}$ is the central inertia dyadic of B. [If μ is set equal to $G\overline{m}$, $\mathbf{F}$ becomes equal to $\hat{\mathbf{F}}$ as defined in Eq. (2.3.5) while $\mathbf{M}$ becomes equal to $\widetilde{\mathbf{M}}$ as defined in Eq. (2.6.3). The force system S is thus seen to be approximately equivalent to the system of gravitational forces exerted on B by a particle of mass $\overline{m}$ at point O.]

Due to the action of S, the rotational motion of B in N depends on the translational motion of B^* in N, but not vice versa, because R appears in Eq. (2), but no variable involving the orientation of B in N appears in Eq. (1). One possible motion for B^* is motion in a circular orbit that is fixed in N, has a radius R, and is described with an angular speed Ω given by

$$\Omega = (\mu R^{-3})^{1/2} \tag{3}$$

When B^* moves in this fashion, the rotational motion of B in N proceeds in such a way that

$$\dot{C}_{31} = C_{32}\omega_3 - C_{33}\omega_2 \tag{4}$$

$$\dot{C}_{32} = C_{33}\omega_1 - C_{31}\omega_3 \tag{5}$$

$$\dot{C}_{33} = C_{31}\omega_2 - C_{32}\omega_1 \tag{6}$$

$$\dot{C}_{11} = C_{12}\omega_3 - C_{13}\omega_2 + \Omega(C_{13}C_{32} - C_{12}C_{33}) \tag{7}$$

$$\dot{C}_{12} = C_{13}\omega_1 - C_{11}\omega_3 + \Omega(C_{11}C_{33} - C_{13}C_{31}) \tag{8}$$

$$\dot{C}_{13} = C_{11}\omega_2 - C_{12}\omega_1 + \Omega(C_{12}C_{31} - C_{11}C_{32}) \tag{9}$$

$$\dot{\omega}_1 = K_1(\omega_2\omega_3 - 3\Omega^2 C_{12}C_{13}) \tag{10}$$

$$\dot{\omega}_2 = K_2(\omega_3\omega_1 - 3\Omega^2 C_{13}C_{11}) \tag{11}$$

$$\dot{\omega}_3 = K_3(\omega_1\omega_2 - 3\Omega^2 C_{11}C_{12}) \tag{12}$$

where C_{ij}, ω_i, and K_i $(i, j = 1, 2, 3)$ are defined as follows: Let $\mathbf{b}_i$ $(i = 1, 2, 3)$ be a dextral set of unit vectors parallel to the central principal axes of inertia of B, and introduce unit vectors $\mathbf{a}_2$ and $\mathbf{a}_3$ (see Fig. 3.5.2) such that $\mathbf{a}_1$, $\mathbf{a}_2$, $\mathbf{a}_3$ form a dextral, orthogonal set with $\mathbf{a}_2 = ({}^N d\mathbf{a}_1/dt)\Omega^{-1}$, so that $\mathbf{a}_3$ is normal to the orbit plane; let

$$C_{ij} \triangleq \mathbf{a}_i \cdot \mathbf{b}_j \qquad (i, j = 1, 2, 3) \tag{13}$$

and take

$$\omega_i \triangleq {}^N\boldsymbol{\omega}^B \cdot \mathbf{b}_i \qquad (i = 1, 2, 3) \tag{14}$$

where ${}^N\boldsymbol{\omega}^B$ denotes the angular velocity of B in N. Finally, with

$$I_j \triangleq \mathbf{b}_j \cdot \mathbf{I} \cdot \mathbf{b}_j \qquad (j = 1, 2, 3) \tag{15}$$

let

$$K_1 \triangleq \frac{I_2 - I_3}{I_1} \qquad K_2 \triangleq \frac{I_3 - I_1}{I_2} \qquad K_3 \triangleq \frac{I_1 - I_2}{I_3} \tag{16}$$

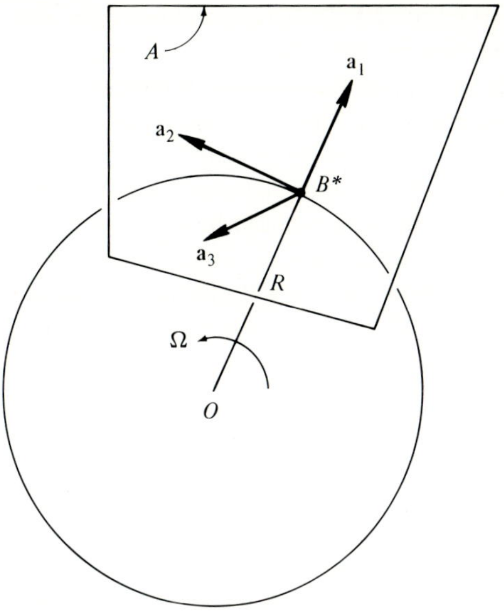

Figure 3.5.2

noting that K_i ($i = 1, 2, 3$) are not independent of each other, any one of the three being expressible in terms of the remaining two. For example,

$$K_3 = -\frac{K_1 + K_2}{1 + K_1 K_2} \tag{17}$$

Moreover,

$$-1 < K_i < 1 \qquad (i = 1, 2, 3) \tag{18}$$

The nonlinearity and strong coupling of Eqs. (4)–(12) prevent one from obtaining the general solution of these equations in closed form. However, the equations are written in a form that is convenient for purposes of numerical integration, and there exist four integrals of the equations, namely,

$$C_{11}C_{31} + C_{12}C_{32} + C_{13}C_{33} = 0 \tag{19}$$

$$C_{i1}^2 + C_{i2}^2 + C_{i3}^2 = 1 \qquad (i = 1, 3) \tag{20}$$

and

$$Z \triangleq \frac{1}{2}[(\omega_1 - \Omega C_{31})^2 I_1 + (\omega_2 - \Omega C_{32})^2 I_2 + (\omega_3 - \Omega C_{33})^2 I_3]$$

$$+ \frac{\Omega^2}{2}[-K_1 I_1 C_{32}^2 + K_2 I_2 (C_{31}^2 + 3C_{13}^2) - 3K_3 I_3 C_{12}^2] = \text{constant} \tag{21}$$

which can be used to check on the accuracy of such integrations. Furthermore, Eqs. (4)–(12) possess particular solutions of practical interest. Hence, they can furnish useful information about effects of **M** on the rotational motion of B in N. Consider, for example, motions during which B remains at rest in a reference frame A

in which B^* and $\mathbf{a}_i$ $(i = 1, 2, 3)$ are fixed (see Fig. 3.5.2). Such a motion is possible if and only if each principal axis of inertia of B for B^* is parallel to one of $\mathbf{a}_i$ $(i = 1, 2, 3)$, and Eqs. (4)–(12) can be used to prove that it is a stable motion when K_1 and K_2 have values corresponding to the unshaded region of the stability chart shown in Fig. 3.5.3, whereas it is an unstable motion when K_1 and K_2 have values corresponding to the cross-hatched portions of this diagram. As used here, the term *stable* means that $C_{11} - 1$, C_{12}, C_{13}, C_{31}, C_{32}, and $C_{33} - 1$ can be kept arbitrarily small for $t > 0$ by making these quantities, as well as ω_1, ω_2, and $\omega_3 - \Omega$, sufficiently small at $t = 0$. Now, when stability is defined in this way, the rotational motion under consideration is unstable for all K_1 and K_2 in the absence of $\mathbf{M}$. Hence, one may say that $\mathbf{M}$ can have a stabilizing effect. Suppose, however, that one calls the motion stable when the angle between $\mathbf{a}_3$ and $\mathbf{b}_3$ can be kept arbitrarily small for $t > 0$ by making this angle, as well as ω_1 and ω_2, sufficiently small at $t = 0$, this being the definition of stability that corresponds to the one used in connection with torque-free motion in Sec. 3.4. Then, if $I_3 > I_1$ and $I_3 > I_2$ or if $I_3 < I_1$ and $I_3 < I_2$, the motion is stable when $\mathbf{M} = 0$, but it can be unstable when $\mathbf{M} \neq 0$. In this sense, $\mathbf{M}$ can have a destabilizing effect. No matter which of these two definitions of stability is used, the stability of the motion in question can be rather tenuous. That is, even if all stability requirements are met, if the principal axes of inertia of B for B^* are nearly parallel to $\mathbf{a}_i$ $(i = 1, 2, 3)$ at $t = 0$, and if $|{}^A\boldsymbol{\omega}^B|$ is small at $t = 0$, it may occur that the principal axes become markedly misaligned from $\mathbf{a}_i$ $(i = 1, 2, 3)$ at some time subsequent to $t = 0$. This is illustrated in Fig. 3.5.4,

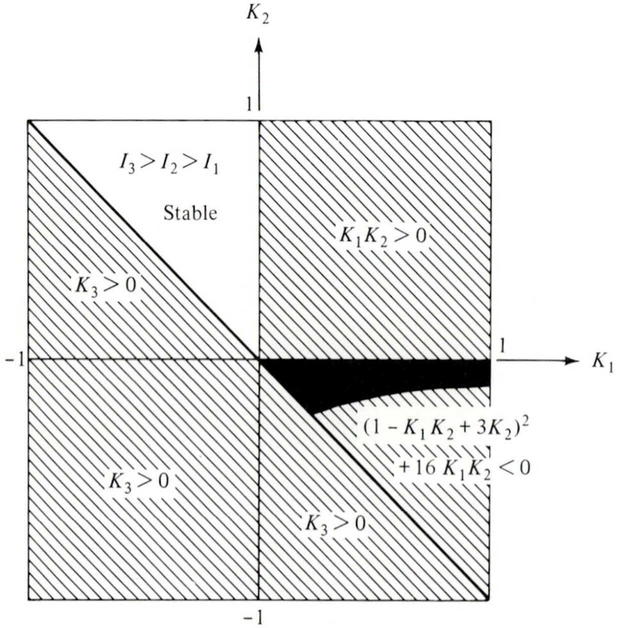

Figure 3.5.3

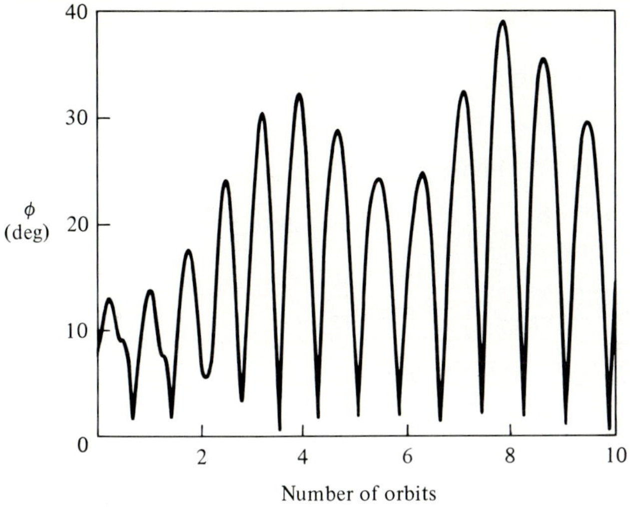

Figure 3.5.4

which shows a plot of the angle ϕ between $\mathbf{a}_3$ and $\mathbf{b}_3$ as a function of the number of orbits traversed by B^* subsequent to $t = 0$. This plot was obtained by integrating Eqs. (4)–(12) numerically with $K_1 = -0.5$, $K_2 = 0.9$ (and, hence, $I_3 > I_1$, $I_3 > I_2$), and

$$C_{11}(0) = 0.9924 \qquad C_{12}(0) = -0.0868 \qquad C_{13}(0) = 0.0872 \tag{22}$$

$$C_{31}(0) = -0.0789 \qquad C_{32}(0) = 0.0944 \qquad C_{33}(0) = 0.9924 \tag{23}$$

$$\omega_1(0) = \omega_2(0) = 0.1\Omega \qquad \omega_3(0) = 1.1\Omega \tag{24}$$

so that the initial values of C_{ij} $(i = 1, 3; j = 1, 2, 3)$ and ω_i $(i = 1, 2, 3)$ differ relatively little from the values of the corresponding quantities during a motion that is stable by either definition of stability, namely,

$$C_{11} = 1.000 \qquad C_{12} = 0 \qquad C_{13} = 0 \tag{25}$$

$$C_{31} = 0 \qquad C_{32} = 0 \qquad C_{33} = 1.000 \tag{26}$$

$$\omega_1 = \omega_2 = 0 \qquad \omega_3 = \Omega \tag{27}$$

Nevertheless, ϕ is seen to acquire values larger than five times its initial value, which shows that a relatively small disturbance can have a quite pronounced effect on a so-called stable motion.

Derivations The angular velocity of B in A can be expressed as

$$^A\boldsymbol{\omega}^B \underset{(1.16.1)}{=} {^A\boldsymbol{\omega}^N} + {^N\boldsymbol{\omega}^B} \underset{(1.11.2)}{=} -{^N\boldsymbol{\omega}^A} + {^N\boldsymbol{\omega}^B} \underset{(1.11.5)}{=} -\Omega\mathbf{a}_3 + {^N\boldsymbol{\omega}^B} \tag{28}$$

Consequently,

$$ {}^A\boldsymbol{\omega}^B \cdot \mathbf{b}_1 \underset{(13,14)}{=} -\Omega C_{31} + \omega_1 \tag{29} $$

$$ {}^A\boldsymbol{\omega}^B \cdot \mathbf{b}_2 = -\Omega C_{32} + \omega_2 \tag{30} $$

$$ {}^A\boldsymbol{\omega}^B \cdot \mathbf{b}_3 = -\Omega C_{33} + \omega_3 \tag{31} $$

With $i = 3$ and $j = 1$, Eq. (1.10.1) now gives

$$ \dot{C}_{31} = C_{32}\,{}^A\boldsymbol{\omega}^B \cdot \mathbf{b}_3 - C_{33}\,{}^A\boldsymbol{\omega}^B \cdot \mathbf{b}_2 \underset{(30,31)}{=} C_{32}(-\Omega C_{33} + \omega_3) - C_{33}(-\Omega C_{32} + \omega_2) $$

$$ = C_{32}\omega_3 - C_{33}\omega_2 \tag{32} $$

which is Eq. (4), while for $i = 1$ and $j = 3$ one has

$$ \begin{aligned} \dot{C}_{13} &= C_{11}\,{}^A\boldsymbol{\omega}^B \cdot \mathbf{b}_2 - C_{12}\,{}^A\boldsymbol{\omega}^B \cdot \mathbf{b}_1 \\ &\underset{(29,30)}{=} C_{11}(-\Omega C_{32} + \omega_2) - C_{12}(-\Omega C_{31} + \omega_1) \\ &= C_{11}\omega_2 - C_{12}\omega_1 + \Omega(C_{12}C_{31} - C_{11}C_{32}) \end{aligned} \tag{33} $$

in agreement with Eq. (9). Equations (5)–(8) are obtained similarly.

The angular momentum $\mathbf{H}$ of B relative to B^* in N can be expressed as

$$ \mathbf{H} = I_1\omega_1\mathbf{b}_1 + I_2\omega_2\mathbf{b}_2 + I_3\omega_3\mathbf{b}_3 \tag{34} $$

and differentiation with respect to t in N gives

$$ \frac{{}^N d\mathbf{H}}{dt} \underset{(1.11.8)}{=} \frac{{}^B d\mathbf{H}}{dt} + {}^N\boldsymbol{\omega}^B \times \mathbf{H} $$

$$ \underset{(16)}{=} I_1(\dot{\omega}_1 - K_1\omega_2\omega_3)\mathbf{b}_1 + I_2(\dot{\omega}_2 - K_2\omega_3\omega_1)\mathbf{b}_2 + I_3(\dot{\omega}_3 - K_3\omega_1\omega_2)\mathbf{b}_3 \tag{35} $$

while $\mathbf{M}$, resolved into components parallel to $\mathbf{b}_i$ $(i = 1, 2, 3)$ by reference to Eqs. (2), (3), (16), and (2.6.8), is given by

$$ \mathbf{M} = -3\Omega^2(I_1K_1C_{12}C_{13}\mathbf{b}_1 + I_2K_2C_{13}C_{11}\mathbf{b}_2 + I_3K_3C_{11}C_{12}\mathbf{b}_3) \tag{36} $$

Substitution from Eqs. (35) and (36) into ${}^N d\mathbf{H}/dt = \mathbf{M}$ then yields Eqs. (10)–(12).

Equations (16) can be written

$$ K_1 = \frac{I_2/I_3 - 1}{I_1/I_3} \qquad K_2 = \frac{1 - I_1/I_3}{I_2/I_3} \tag{37} $$

$$ K_3 = \frac{I_1}{I_3} - \frac{I_2}{I_3} \tag{38} $$

Solution of Eqs. (37) for I_1/I_3 and I_2/I_3 gives

$$ \frac{I_1}{I_3} = \frac{1 - K_2}{1 + K_1K_2} \qquad \frac{I_2}{I_3} = \frac{1 + K_1}{1 + K_1K_2} \tag{39} $$

and substitution from Eqs. (39) into Eq. (38) yields Eq. (17).

The inequalities (18) follow from the fact that the sum of two principal moments of inertia must exceed the third, while the difference between them must be smaller than the third principal moment of inertia. Thus, for example,

$$I_1 + I_2 > I_3 \tag{40}$$

and

$$I_2 - I_3 < I_1 \tag{41}$$

so that

$$\frac{I_2 - I_3}{I_1} \underset{(40)}{>} -1 \tag{42}$$

and

$$\frac{I_2 - I_3}{I_1} \underset{(41)}{<} 1 \tag{43}$$

in agreement with (18) for $i = 1$.

Equation (19) is obtained by setting $i = 1$ and $j = 3$ in Eq. (1.2.14); and taking $i = j = 1$ and $i = j = 3$ in Eq. (1.2.14) leads to Eqs. (20). The validity of Eq. (21) may be verified by differentiating the left-hand member with respect to t, eliminating $\dot{C}_{ij}$ and $\dot{\omega}_j$ ($i = 1, 3; j = 1, 2, 3$) by using Eqs. (4)–(12), and observing that the result vanishes identically. However, this does not reveal the origin of the equation, which will now be discussed.

When B remains at rest in A, the acceleration $\mathbf{a}$ of a generic particle P of B in N is given by

$$\mathbf{a} = \Omega^2 \mathbf{a}_3 \times [\mathbf{a}_3 \times (R\mathbf{a}_1 + \mathbf{r})] \tag{44}$$

where $\mathbf{r}$ is the position vector of P relative to B^* (see Fig. 3.5.5). This acceleration is "centripetal" in the sense that it has the same direction as the vector $\overrightarrow{PQ}$, where Q is the foot of the perpendicular dropped from P onto the line that passes through O and is parallel to $\mathbf{a}_3$. A "centrifugal" force $d\mathbf{F}$ can be associated with P by letting

$$d\mathbf{F} \triangleq -\mathbf{a}\rho \, d\tau \tag{45}$$

where ρ is the mass density of B at P and $d\tau$ is the volume of a differential element

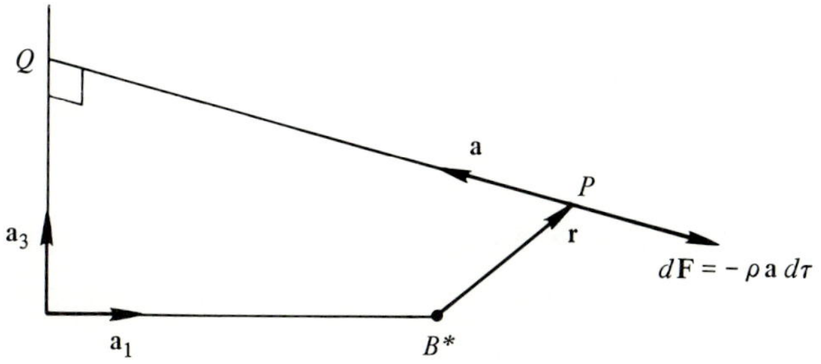

Figure 3.5.5

of B, and requiring that this force be applied at P (see Fig. 3.5.5). The sum of the moments about B^* of all such forces is then called the *centrifugal moment* and is given by

$$\begin{aligned}\mathbf{M}^* &\triangleq \int \mathbf{r} \times d\mathbf{F} \underset{(45)}{=} -\int \mathbf{r} \times \mathbf{a}\rho\, d\tau \\ &\underset{(44)}{=} -\Omega^2 \int \mathbf{r} \times \{\mathbf{a}_3 \times [\mathbf{a}_3 \times (R\mathbf{a}_1 + \mathbf{r})]\}\rho\, d\tau \\ &= -R\Omega^2\left(\int \mathbf{r}\rho\, d\tau\right) \times [\mathbf{a}_3 \times (\mathbf{a}_3 \times \mathbf{a}_1)] \\ &\quad -\Omega^2 \int \mathbf{r} \times [\mathbf{a}_3 \times (\mathbf{a}_3 \times \mathbf{r})]\rho\, d\tau \end{aligned} \tag{46}$$

The first of the last two integrals is equal to zero because $\mathbf{r}$ originates at the mass center B^* of B. As for the last integral,

$$\int \mathbf{r} \times [\mathbf{a}_3 \times (\mathbf{a}_3 \times \mathbf{r})]\rho\, d\tau = \int \mathbf{a}_3 \times \mathbf{r}\,\mathbf{r} \cdot \mathbf{a}_3\rho\, d\tau = \mathbf{a}_3 \times \mathbf{I} \cdot \mathbf{a}_3 \tag{47}$$

where, as before, $\mathbf{I}$ is the central inertia dyadic of B. Hence,

$$\mathbf{M}^* = -\Omega^2\mathbf{a}_3 \times \mathbf{I} \cdot \mathbf{a}_3 \tag{48}$$

and it may be noted that the centrifugal moment bears a striking resemblance to the gravitational moment $\mathbf{M}$ as given in Eq. (2.6.3). This resemblance suggests the possibility of expressing $\mathbf{M}^*$ in a form analogous to that employed in connection with $\mathbf{M}$ in Eq. (2.17.4). That is, one is motivated to seek a function W^* of $\mathbf{a}_3$ such that

$$\mathbf{M}^* = -\mathbf{a}_3 \times \frac{\partial W^*}{\partial \mathbf{a}_3} \tag{49}$$

Not surprisingly, this function exists and can be expressed as

$$\begin{aligned}W^* &= \frac{\Omega^2}{2}\mathbf{a}_3 \cdot \mathbf{I} \cdot \mathbf{a}_3 \underset{(13,15)}{=} \frac{\Omega^2}{2}(I_1 C_{31}{}^2 + I_2 C_{32}{}^2 + I_3 C_{33}{}^2) \\ &\underset{(20)}{=} \frac{\Omega^2}{2}[(I_1 - I_3)C_{31}{}^2 + (I_2 - I_3)C_{32}{}^2 + I_3] \\ &\underset{(16)}{=} -\frac{\Omega^2}{2}(K_2 I_2 C_{31}{}^2 - K_1 I_1 C_{32}{}^2 - I_3)\end{aligned} \tag{50}$$

Similarly, for $\mathbf{M}$ expressed as

$$\mathbf{M} = 3\Omega^2\mathbf{a}_1 \times \mathbf{I} \cdot \mathbf{a}_1 \tag{51}$$

one can introduce a function W of $\mathbf{a}_1$ such that

$$\mathbf{M} = \mathbf{a}_1 \times \frac{\partial W}{\partial \mathbf{a}_1} \tag{52}$$

by letting

$$W = -\frac{3\Omega^2}{2}\mathbf{a}_1 \cdot \mathbf{I} \cdot \mathbf{a}_1 \underset{(13,15)}{=} -\frac{3\Omega^2}{2}(I_1 C_{11}{}^2 + I_2 C_{12}{}^2 + I_3 C_{13}{}^2)$$

$$\underset{(20)}{=} -\frac{3\Omega^2}{2}[I_1 + (I_2 - I_1)C_{12}{}^2 + (I_3 - I_1)C_{13}{}^2]$$

$$\underset{(16)}{=} -\frac{3\Omega^2}{2}(I_1 - K_3 I_3 C_{12}{}^2 + K_2 I_2 C_{13}{}^2) \tag{53}$$

Finally, consider the rotational kinetic energy of B in A, that is, K defined as

$$K \triangleq \tfrac{1}{2}{}^A\boldsymbol{\omega}^B \cdot \mathbf{I} \cdot {}^A\boldsymbol{\omega}^B$$

$$\underset{(29,31)}{=} \tfrac{1}{2}[(\omega_1 - \Omega C_{31})^2 I_1 + (\omega_2 - \Omega C_{32})^2 I_2 + (\omega_3 - \Omega C_{33})^2 I_3] \tag{54}$$

The quantity Z defined in Eq. (21) now can be seen to be given by

$$Z = K - W - W^* + \frac{\Omega^2}{2}(I_3 - 3I_1) \tag{55}$$

The motivation for *subtracting* W and W^* from K is that energies are generally the negatives of force functions [see, for example, Eqs. (2.17.3) and (2.17.7)]. As for the last term in Eq. (55), this is the sum of the time-independent terms in the expressions for W and W^*, and it is included in Z in order that Z vanish whenever B remains at rest in A with each central principal axis of inertia of B parallel to one of $\mathbf{a}_i$ ($i = 1, 2, 3$). In conclusion, then, Z is simply a sum of three energies associated, respectively, with the motion of B in A, with the gravitational moment $\mathbf{M}$, and with the centrifugal moment $\mathbf{M}^*$. (Z can also be regarded as the "rotational" Hamiltonian for B in N.)

The centrifugal moment $\mathbf{M}^*$ is of interest also in connection with the question of the possibility of motions during which B remains at rest in A, for such a motion is possible if and only if

$$\mathbf{M} + \mathbf{M}^* = 0 \tag{56}$$

or, in view of Eqs. (48) and (51),

$$3\mathbf{a}_1 \times \mathbf{I} \cdot \mathbf{a}_1 - \mathbf{a}_3 \times \mathbf{I} \cdot \mathbf{a}_3 = 0 \tag{57}$$

If J_{kl} ($k, l = 1, 2, 3$) now are defined as

$$J_{kl} \triangleq \mathbf{a}_k \cdot \mathbf{I} \cdot \mathbf{a}_l \qquad (k, l = 1, 2, 3) \tag{58}$$

then

$$\mathbf{I} \cdot \mathbf{a}_1 = J_{11}\mathbf{a}_1 + J_{21}\mathbf{a}_2 + J_{31}\mathbf{a}_3 \tag{59}$$

and

$$\mathbf{a}_1 \times \mathbf{I} \cdot \mathbf{a}_1 = -J_{31}\mathbf{a}_2 + J_{21}\mathbf{a}_3 \tag{60}$$

Similarly,

$$\mathbf{a}_3 \times \mathbf{I} \cdot \mathbf{a}_3 = -J_{23}\mathbf{a}_1 + J_{13}\mathbf{a}_2 \tag{61}$$

and the three scalar equations corresponding to Eq. (57) are satisfied if and only if

$$J_{12} = J_{23} = J_{31} = 0 \tag{62}$$

which is the case if and only if each of $\mathbf{a}_i$ ($i = 1, 2, 3$) is parallel to a principal axis of inertia of B for B^*.

To show that a motion of the kind just considered is stable when K_1 and K_2 have values corresponding to the unshaded region of Fig. 3.5.3, we note that such a motion can be defined with precision by requiring that

$$\mathbf{b}_i = \mathbf{a}_i \qquad (i = 1, 2, 3) \tag{63}$$

for all t. Throughout the motion, one then has

$$C_{11} - 1 = C_{12} = C_{13} = C_{31} = C_{32} = C_{33} - 1 = 0 \tag{64}$$

$$\omega_1 = \omega_2 = \omega_3 - \Omega = 0 \tag{65}$$

Now consider the most general motion, that is, the motion that ensues if, at $t = 0$,

$$C_{ij} = \hat{C}_{ij} \qquad \omega_i = \hat{\omega}_i \qquad (i, j = 1, 2, 3) \tag{66}$$

Throughout this motion,

$$\begin{aligned}(\omega_1 - \Omega C_{31})^2 I_1 + (\omega_2 - \Omega C_{32})^2 I_2 + (\omega_3 - \Omega C_{33})^2 I_3 \\ + \Omega^2[-K_1 I_1 C_{32}^2 + K_2 I_2 (C_{31}^2 + 3C_{13}^2) - 3K_3 I_3 C_{12}^2] \\ \underset{(21)}{=} (\hat{\omega}_1 - \Omega\hat{C}_{31})^2 I_1 + (\hat{\omega}_2 - \Omega\hat{C}_{32})^2 I_2 + (\hat{\omega}_3 - \Omega\hat{C}_{33})^2 I_3 \\ + \Omega^2[-K_1 I_1 \hat{C}_{32}^2 + K_2 I_2 (\hat{C}_{31}^2 + 3\hat{C}_{13}^2) - 3K_3 I_3 \hat{C}_{12}^2]\end{aligned} \tag{67}$$

The right-hand member of this equation vanishes if

$$\hat{C}_{12} = \hat{C}_{13} = \hat{C}_{31} = \hat{C}_{32} = \hat{C}_{33} - 1 = 0 \tag{68}$$

and

$$\hat{\omega}_1 = \hat{\omega}_2 = \hat{\omega}_3 - \Omega = 0 \tag{69}$$

Hence, the left-hand member can be kept arbitrarily small by making $\hat{C}_{12}$, $\hat{C}_{13}$, $\hat{C}_{31}$, $\hat{C}_{32}$, $\hat{C}_{33} - 1$, $\hat{\omega}_1$, $\hat{\omega}_2$, and $\hat{\omega}_3 - \Omega$ sufficiently small. Now, if

$$K_1 < 0 \qquad K_2 > 0 \qquad K_3 < 0 \tag{70}$$

then the left-hand member is either positive or equal to zero. Hence, under these circumstances, C_{12}, C_{13}, C_{31}, and C_{32} can be kept arbitrarily small by the same means which keep the left-hand member small. As for $C_{11} - 1$ and $C_{33} - 1$, it follows from what has just been established and from Eq. (20) that these quantities can be kept arbitrarily small by making $\hat{C}_{11} - 1$, as well as $\hat{C}_{12}$, $\hat{C}_{13}$, $\hat{C}_{31}$, $\hat{C}_{32}$, and $\hat{C}_{33} - 1$, sufficiently small. What remains to be verified is that the inequalities (70) are satisfied in the unshaded portion of Fig. 3.5.3. For the first two of these, this is self-evident, and to see that the third is satisfied, one needs only to refer to Eq. (17) after noting that $K_1 + K_2 > 0$ and $1 + K_1 K_2 > 0$ in the region of interest.

For values of K_1 and K_2 belonging to the cross-hatched portion of Fig. 3.5.3, the instability of the motion described by Eqs. (64) may be established by working with linearized forms of Eqs. (4)–(12) and (19). That is, C_{ij} and ω_j ($i = 1, 3$; $j = 1, 2, 3$) are expressed as

$$C_{11} = 1 + \widetilde{C}_{11} \qquad C_{33} = 1 + \widetilde{C}_{33} \tag{71}$$

$$C_{12} = \widetilde{C}_{12} \qquad C_{13} = \widetilde{C}_{13} \qquad C_{31} = \widetilde{C}_{31} \qquad C_{32} = \widetilde{C}_{32} \tag{72}$$

$$\omega_1 = \widetilde{\omega}_1 \qquad \omega_2 = \widetilde{\omega}_2 \qquad \omega_3 = \Omega + \widetilde{\omega}_3 \tag{73}$$

and the equations obtained by substituting these expressions into Eqs. (4)–(12) and (19) are linearized, which gives

$$\dot{\widetilde{C}}_{31} = \widetilde{C}_{32}\Omega - \widetilde{\omega}_2 \qquad \dot{\widetilde{C}}_{32} = \widetilde{\omega}_1 - \widetilde{C}_{31}\Omega \qquad \dot{\widetilde{C}}_{33} = 0 \tag{74}$$

$$\dot{\widetilde{C}}_{11} = 0 \qquad \dot{\widetilde{C}}_{12} = -\widetilde{\omega}_3 + \Omega\widetilde{C}_{33} \qquad \dot{\widetilde{C}}_{13} = \widetilde{\omega}_2 - \Omega\widetilde{C}_{32} \tag{75}$$

$$\dot{\widetilde{\omega}}_1 = K_1\Omega\widetilde{\omega}_2 \qquad \dot{\widetilde{\omega}}_2 = K_2\Omega\widetilde{\omega}_1 - 3\Omega^2K_2\widetilde{C}_{13} \qquad \dot{\widetilde{\omega}}_3 = -3K_3\Omega^2\widetilde{C}_{12} \tag{76}$$

and

$$\widetilde{C}_{31} + \widetilde{C}_{13} \underset{(19)}{=} 0 \tag{77}$$

Solving the second of Eqs. (75) for $\widetilde{\omega}_3$, substituting into the third of Eqs. (76), and using the third of Eqs. (74), one now has

$$\ddot{\widetilde{C}}_{12} - 3\Omega^2K_3\widetilde{C}_{12} = 0 \tag{78}$$

while the first two of Eqs. (74) and (76), together with Eq. (77), permit one to write

$$\begin{bmatrix} \dot{\widetilde{C}}_{31} \\ \dot{\widetilde{C}}_{32} \\ \dot{\widetilde{\omega}}_1 \\ \dot{\widetilde{\omega}}_2 \end{bmatrix} = \begin{bmatrix} 0 & \Omega & 0 & -1 \\ -\Omega & 0 & 1 & 0 \\ 0 & 0 & 0 & \Omega K_1 \\ 3\Omega^2K_2 & 0 & \Omega K_2 & 0 \end{bmatrix} \begin{bmatrix} \widetilde{C}_{31} \\ \widetilde{C}_{32} \\ \widetilde{\omega}_1 \\ \widetilde{\omega}_2 \end{bmatrix} \tag{79}$$

The characteristic equations associated with Eqs. (78) and (79) are, respectively,

$$\lambda^2 - 3K_3\Omega^2 = 0 \tag{80}$$

and

$$\begin{vmatrix} -\mu & \Omega & 0 & -1 \\ -\Omega & -\mu & 1 & 0 \\ 0 & 0 & -\mu & \Omega K_1 \\ 3\Omega^2K_2 & 0 & \Omega K_2 & -\mu \end{vmatrix} = \mu^4 + (1 - K_1K_2 + 3K_2)\Omega^2\mu^2 - 4K_1K_2\Omega^4 = 0 \tag{81}$$

and these have roots with positive real parts if

$$K_3 > 0 \tag{82}$$

or if any one of the following is satisfied:

$$K_1 K_2 > 0 \tag{83}$$

$$1 - K_1 K_2 + 3K_2 < 0 \tag{84}$$

$$(1 - K_1 K_2 + 3K_2)^2 + 16K_1 K_2 < 0 \tag{85}$$

As may be seen by reference to Fig. 3.5.3, at least one of these conditions is satisfied at every point of the cross-hatched portions of this diagram. Consequently, Eqs. (64) describe an unstable motion whenever K_1 and K_2 have values belonging to such a region. The solid black region in Fig. 3.5.3 represents values of K_1 and K_2 for which we have not established either the stability or the instability of the motion under consideration.

Example B can move in such a way that $\mathbf{b}_3$ remains equal to $\mathbf{a}_3$ at all times while the angle θ between $\mathbf{a}_1$ and $\mathbf{b}_1$ (see Fig. 3.5.6) varies with time. It is to be shown that θ is either a periodic or a monotonic function of t.

During the motion in question,

$$\begin{bmatrix} C_{11} & C_{12} & C_{13} \\ C_{21} & C_{22} & C_{23} \\ C_{31} & C_{32} & C_{33} \end{bmatrix} \underset{(1.2.37)}{=} \begin{bmatrix} \cos\theta & -\sin\theta & 0 \\ \sin\theta & \cos\theta & 0 \\ 0 & 0 & 1 \end{bmatrix} \tag{86}$$

and

$$\omega_1 = \omega_2 = 0 \qquad \omega_3 = \Omega + \dot{\theta}$$

Equations (4)–(11) are satisfied identically, but Eq. (12) requires that

$$\ddot{\theta} + p^2 \sin\theta \cos\theta = 0 \tag{87}$$

where p^2, defined as

$$p^2 \triangleq -3\Omega^2 K_3 \underset{(16)}{=} 3\Omega^2 \frac{I_2 - I_1}{I_3} \tag{88}$$

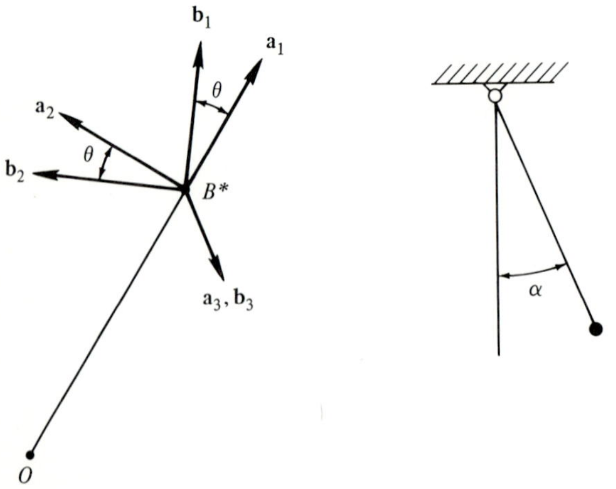

Figure 3.5.6

may be regarded as intrinsically nonnegative since the unit vectors $\mathbf{b}_1$ and $\mathbf{b}_2$ can always be chosen such that $I_2 \geq I_1$. Introducing α as

$$\alpha \triangleq 2\theta \tag{89}$$

one can replace Eq. (87) with

$$\ddot{\alpha} + p^2 \sin \alpha = 0 \tag{90}$$

which is recognizable as the equation of motion of a simple pendulum in a uniform gravitational field (see Fig. 3.5.6). Consequently, α is either a periodic or a monotonic function of t; and, since $\theta = \alpha/2$, the same is true for θ.

3.6 ANGULAR MOMENTUM, INERTIA TORQUE, AND KINETIC ENERGY OF A GYROSTAT

Figure 3.6.1 represents a simple gyrostat G formed by a rigid body A that carries an axisymmetric rotor B whose axis and mass center B^* are fixed in A. F designates an arbitrary frame of reference.

The angular momentum $\mathbf{H}_G$ of G in F relative to the mass center G^* of G, the inertia torque $\mathbf{T}_G$ of G in F, and the kinetic energy K_G of G in F each can be expressed as the sum of two quantities, one associated with a (fictitious) rigid body R that has the same mass distribution as G, but moves like A, and the other accounting for "internal" motions, that is, rotation of B relative to A. In other words, one can write

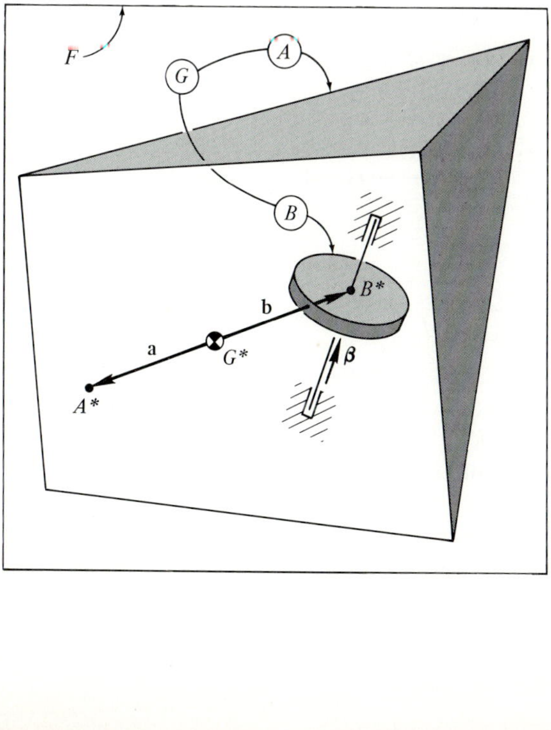

Figure 3.6.1

$$\mathbf{H}_G = \mathbf{H}_R + \mathbf{H}_I \tag{1}$$

$$\mathbf{T}_G = \mathbf{T}_R + \mathbf{T}_I \tag{2}$$

and

$$K_G = K_R + K_I \tag{3}$$

where $\mathbf{H}_R$ is the angular momentum of R in F relative to G^*, $\mathbf{T}_R$ is the inertia torque of R in F, K_R is the kinetic energy of R in F, and $\mathbf{H}_I$, $\mathbf{T}_I$, and K_I are given by

$$\mathbf{H}_I = J\,{}^A\omega^B\,\boldsymbol{\beta} \tag{4}$$

$$\mathbf{T}_I = -J({}^A\dot{\omega}^B\,\boldsymbol{\beta} + {}^A\omega^B\,{}^F\boldsymbol{\omega}^A \times \boldsymbol{\beta}) \tag{5}$$

and

$$K_I = \frac{J\,{}^A\omega^B}{2}({}^A\omega^B + 2\,{}^F\boldsymbol{\omega}^A \cdot \boldsymbol{\beta}) \tag{6}$$

with $\boldsymbol{\beta}$, ${}^A\omega^B$, ${}^F\boldsymbol{\omega}^A$, and J defined as follows: $\boldsymbol{\beta}$ is a unit vector parallel to the axis of symmetry of B (see Fig. 3.6.1); ${}^A\omega^B = {}^A\boldsymbol{\omega}^B \cdot \boldsymbol{\beta}$, where ${}^A\boldsymbol{\omega}^B$ is the angular velocity of B in A (see Sec. 1.11); ${}^F\boldsymbol{\omega}^A$ is the angular velocity of A in F; and J is the moment of inertia of B about the axis of symmetry of B.

K, the rotational kinetic energy of G in F, can be expressed as

$$K = \frac{1}{2}(\mathbf{H}_G \cdot \mathbf{I}_G^{-1} \cdot \mathbf{H}_G + J^{-1}\mathbf{H}_I^2 - \mathbf{H}_I \cdot \mathbf{I}_G^{-1} \cdot \mathbf{H}_I) \tag{7}$$

where $\mathbf{I}_G$ is the central inertia dyadic of G; and this result can be used to show that K must lie between two values that depend on the magnitudes of $\mathbf{H}_G$ and $\mathbf{H}_I$, as well as on J, $I_{\min}$ and $I_{\max}$, the latter two being the minimum moment of inertia of G and the largest central moment of inertia of G, respectively. Specifically,

$$K_{\min} \le K \le K_{\max} \tag{8}$$

where

$$K_{\min} = \frac{1}{2}\left(\frac{\mathbf{H}_G^2}{I_{\max}} + \frac{\mathbf{H}_I^2}{J} - \mathbf{H}_I \cdot \mathbf{I}_G^{-1} \cdot \mathbf{H}_I\right) \tag{9}$$

and

$$K_{\max} = \frac{1}{2}\left(\frac{\mathbf{H}_G^2}{I_{\min}} + \frac{\mathbf{H}_I^2}{J} - \mathbf{H}_I \cdot \mathbf{I}_G^{-1} \cdot \mathbf{H}_I\right) \tag{10}$$

Whenever B is at rest in A, G is said to be moving quasi-rigidly in F. Under these circumstances, $K = K_{\min}$ if and only if ${}^F\boldsymbol{\omega}^A$ is parallel to the axis of maximum central moment of inertia of G, and $K = K_{\max}$ if and only if ${}^F\boldsymbol{\omega}^A$ is parallel to the axis of minimum moment of inertia of G.

Derivations Let m_A and m_B be the masses of A and B, respectively, and let $\mathbf{a}$ and $\mathbf{b}$ be the position vectors (see Fig. 3.6.1) from G^*, the mass center of G, to A^* and

B^*, the mass centers of A and B, respectively. Then

$$m_A \mathbf{a} + m_B \mathbf{b} = 0 \tag{11}$$

and $\mathbf{H}_A$, the angular momentum of A in F relative to G^*, is given by

$$\mathbf{H}_A = \mathbf{I}_A \cdot {}^F\boldsymbol{\omega}^A + m_A \mathbf{a} \times {}^F\mathbf{v}^{A^*} \tag{12}$$

where $\mathbf{I}_A$ is the central inertia dyadic of A and ${}^F\mathbf{v}^{A^*}$ is the velocity of A^* in F. Similarly,

$$\mathbf{H}_B = \mathbf{I}_B \cdot {}^F\boldsymbol{\omega}^B + m_B \mathbf{b} \times {}^F\mathbf{v}^{B^*} \tag{13}$$

so that $\mathbf{H}_G$, which is the sum of $\mathbf{H}_A$ and $\mathbf{H}_B$, can be expressed as

$$\mathbf{H}_G = \mathbf{I}_A \cdot {}^F\boldsymbol{\omega}^A + \mathbf{I}_B \cdot {}^F\boldsymbol{\omega}^B + m_A \mathbf{a} \times {}^F\mathbf{v}^{A^*} + m_B \mathbf{b} \times {}^F\mathbf{v}^{B^*} \tag{14}$$

Now,

$$ {}^F\boldsymbol{\omega}^B \underset{(1.16.1)}{=} {}^F\boldsymbol{\omega}^A + {}^A\boldsymbol{\omega}^B \tag{15}$$

$$ {}^F\mathbf{v}^{A^*} = {}^F\mathbf{v}^{G^*} + {}^F\boldsymbol{\omega}^A \times \mathbf{a} \tag{16}$$

$$ {}^F\mathbf{v}^{B^*} = {}^F\mathbf{v}^{G^*} + {}^F\boldsymbol{\omega}^A \times \mathbf{b} \tag{17}$$

Consequently,

$$\begin{aligned}
\mathbf{H}_G &\underset{(14)}{=} \mathbf{I}_A \cdot {}^F\boldsymbol{\omega}^A + \mathbf{I}_B \underset{(15)}{\cdot} {}^F\boldsymbol{\omega}^A + \mathbf{I}_B \cdot {}^A\boldsymbol{\omega}^B + m_A \underset{(16)}{\mathbf{a}} \times {}^F\mathbf{v}^{G^*} \\
&\quad + m_A \mathbf{a} \times ({}^F\boldsymbol{\omega}^A \times \mathbf{a}) + m_B \underset{(17)}{\mathbf{b}} \times {}^F\mathbf{v}^{G^*} + m_B \mathbf{b} \times ({}^F\boldsymbol{\omega}^A \times \mathbf{b}) \\
&= [\mathbf{I}_A + m_A(\mathbf{a} \cdot \mathbf{a}\mathbf{U} - \mathbf{a}\mathbf{a}) + \mathbf{I}_B + m_B(\mathbf{b} \cdot \mathbf{b}\mathbf{U} - \mathbf{b}\mathbf{b})] \cdot {}^F\boldsymbol{\omega}^A \\
&\quad + (m_A \mathbf{a} + m_B \mathbf{b}) \cdot {}^F\mathbf{v}^{G^*} + \mathbf{I}_B \cdot {}^A\boldsymbol{\omega}^B
\end{aligned} \tag{18}$$

The quantity within the square brackets is the central inertia dyadic of R. Hence, the first line of the right-hand member of Eq. (18) is $\mathbf{H}_R$, the angular momentum of R in F relative to G^*; the first term in the second line vanishes in accordance with Eq. (11), and, since B is axisymmetric and ${}^A\boldsymbol{\omega}^B = {}^A\omega^B \boldsymbol{\beta}$, the last term is equal to $J\,{}^A\omega^B \boldsymbol{\beta}$. Substitution from Eq. (18) into Eq. (1) thus leads immediately to Eq. (4).

The inertia torques $\mathbf{T}_G$ and $\mathbf{T}_R$ can be expressed as

$$\mathbf{T}_G = -\frac{{}^F d\mathbf{H}_G}{dt} \qquad \mathbf{T}_R = -\frac{{}^F d\mathbf{H}_R}{dt} \tag{19}$$

Differentiation of Eq. (1) with respect to t therefore produces

$$\begin{aligned}
\mathbf{T}_G &= \mathbf{T}_R - \frac{{}^F d\mathbf{H}_I}{dt} \underset{(4)}{=} \mathbf{T}_R - J\left({}^A\dot{\omega}^B \boldsymbol{\beta} + {}^A\omega^B \frac{{}^F d\boldsymbol{\beta}}{dt}\right) \\
&= \mathbf{T}_R - J({}^A\dot{\omega}^B \boldsymbol{\beta} + {}^A\omega^B\, {}^F\boldsymbol{\omega}^A \underset{(1.11.9)}{\times} \boldsymbol{\beta})
\end{aligned} \tag{20}$$

and substitution into Eq. (2) establishes the validity of Eq. (5).

Forming $2K_G$ as the sum of contributions from A and B, one has

$$2K_G = m_A({}^F\mathbf{v}^{A*})^2 + {}^F\boldsymbol{\omega}^A \cdot \mathbf{I}_A \cdot {}^F\boldsymbol{\omega}^A + m_B({}^F\mathbf{v}^{B*})^2 + {}^F\boldsymbol{\omega}^B \cdot \mathbf{I}_B \cdot {}^F\boldsymbol{\omega}^B \tag{21}$$

or, after using Eqs. (15)–(17) to eliminate ${}^F\boldsymbol{\omega}^B$, ${}^F\mathbf{v}^{A*}$, and ${}^F\mathbf{v}^{B*}$,

$$\begin{aligned}2K_G = {} & m_A({}^F\mathbf{v}^{G*})^2 + m_A({}^F\boldsymbol{\omega}^A \times \mathbf{a})^2 + 2m_A{}^F\mathbf{v}^{G*} \cdot {}^F\boldsymbol{\omega}^A \times \mathbf{a} \\ & + m_B({}^F\mathbf{v}^{G*})^2 + m_B({}^F\boldsymbol{\omega}^A \times \mathbf{b})^2 + 2m_B{}^F\mathbf{v}^{G*} \cdot {}^F\boldsymbol{\omega}^A \times \mathbf{b} \\ & + {}^F\boldsymbol{\omega}^A \cdot \mathbf{I}_A \cdot {}^F\boldsymbol{\omega}^A + {}^F\boldsymbol{\omega}^A \cdot \mathbf{I}_B \cdot {}^F\boldsymbol{\omega}^A + {}^A\boldsymbol{\omega}^B \cdot \mathbf{I}_B \cdot {}^A\boldsymbol{\omega}^B \\ & + 2{}^F\boldsymbol{\omega}^A \cdot \mathbf{I}_B \cdot {}^A\boldsymbol{\omega}^B\end{aligned} \tag{22}$$

Now,

$$\begin{aligned}({}^F\boldsymbol{\omega}^A \times \mathbf{a})^2 &= ({}^F\boldsymbol{\omega}^A \times \mathbf{a}) \cdot {}^F\boldsymbol{\omega}^A \times \mathbf{a} = ({}^F\boldsymbol{\omega}^A \times \mathbf{a}) \times {}^F\boldsymbol{\omega}^A \cdot \mathbf{a} \\ &= ({}^F\boldsymbol{\omega}^A)^2\, \mathbf{a} \cdot \mathbf{a} - {}^F\boldsymbol{\omega}^A \cdot \mathbf{a}\mathbf{a} \cdot {}^F\boldsymbol{\omega}^A \\ &= {}^F\boldsymbol{\omega}^A \cdot (\mathbf{a} \cdot \mathbf{a}\mathbf{U} - \mathbf{a}\mathbf{a}) \cdot {}^F\boldsymbol{\omega}^A\end{aligned} \tag{23}$$

and, similarly,

$$({}^F\boldsymbol{\omega}^A \times \mathbf{b})^2 = {}^F\boldsymbol{\omega}^A \cdot (\mathbf{b} \cdot \mathbf{b}\mathbf{U} - \mathbf{b}\mathbf{b}) \cdot {}^F\boldsymbol{\omega}^A \tag{24}$$

Hence,

$$\begin{aligned}2K_G \underset{(22)}{=} {} & (m_A + m_B)({}^F\mathbf{v}^{G*})^2 \\ & + {}^F\boldsymbol{\omega}^A \cdot [\mathbf{I}_A + m_A(\mathbf{a} \cdot \mathbf{a}\mathbf{U} - \mathbf{a}\mathbf{a}) + \mathbf{I}_B + m_B(\mathbf{b} \cdot \mathbf{b}\mathbf{U} - \mathbf{b}\mathbf{b})] \cdot {}^F\boldsymbol{\omega}^A \\ & + 2{}^F\mathbf{v}^{G*} \cdot (m_A\mathbf{a} + m_B\mathbf{b}) \cdot {}^F\boldsymbol{\omega}^A + 2{}^F\boldsymbol{\omega}^A \cdot \mathbf{I}_B \cdot {}^A\boldsymbol{\omega}^B + {}^A\boldsymbol{\omega}^B \cdot \mathbf{I}_B \cdot {}^A\boldsymbol{\omega}^B\end{aligned} \tag{25}$$

The first two lines of the right-hand member of this equation form $2K_R$; the first term in the third line vanishes in accordance with Eq. (11), while the second term is equal to $2J\,{}^A\boldsymbol{\omega}^B \cdot \boldsymbol{\beta}$; and the last term in the equation is equal to $J({}^A\omega^B)^2$. Equation (6) thus follows from Eqs. (3) and Eq. (25). To establish the validity of Eq. (7), we begin by writing

$$\begin{aligned}2K \triangleq 2K_G - (m_A + m_B)({}^F\mathbf{v}^{G*})^2 &\underset{(3)}{=} 2K_R + 2K_I - (m_A + m_B)({}^F\mathbf{v}^{G*})^2 \\ &= {}^F\boldsymbol{\omega}^A \cdot \mathbf{I}_G \cdot {}^F\boldsymbol{\omega}^A \underset{(6)}{+} J({}^A\omega^B)^2 + 2{}^F\boldsymbol{\omega}^A \cdot (J\,{}^A\omega^B\,\boldsymbol{\beta})\end{aligned} \tag{26}$$

Next, to eliminate angular velocities, we note that

$$\mathbf{H}_R = {}^F\boldsymbol{\omega}^A \cdot \mathbf{I}_G \tag{27}$$

so that

$$\mathbf{H}_G \underset{(1)}{=} {}^F\boldsymbol{\omega}^A \underset{(27)}{\cdot} \mathbf{I}_G + \mathbf{H}_I \tag{28}$$

or

$${}^F\boldsymbol{\omega}^A \underset{(28)}{=} (\mathbf{H}_G - \mathbf{H}_I) \cdot \mathbf{I}_G^{-1} \tag{29}$$

Also,

$$J({}^A\omega^B)^2 \underset{(4)}{=} \frac{\mathbf{H}_I^{\,2}}{J} \qquad J\,{}^A\omega^B\,\boldsymbol{\beta} \underset{(4)}{=} \mathbf{H}_I \tag{30}$$

Consequently,

$$2K \underset{(26)}{=} (\mathbf{H}_G - \mathbf{H}_I) \cdot \mathbf{I}_G^{-1} \cdot \underset{(29)}{\mathbf{I}_G} \cdot (\mathbf{H}_G - \mathbf{H}_I) \cdot \mathbf{I}_G^{-1} + \underset{(30)}{\frac{\mathbf{H}_I^2}{J}}$$

$$+ 2(\mathbf{H}_G - \underset{(29)}{\mathbf{H}_I)} \cdot \mathbf{I}_G^{-1} \cdot \underset{(30)}{\mathbf{H}_I}$$

$$= (\mathbf{H}_G - \mathbf{H}_I) \cdot (\mathbf{H}_G \cdot \mathbf{I}_G^{-1} - \mathbf{H}_I \cdot \mathbf{I}_G^{-1}) + \frac{\mathbf{H}_I^2}{J}$$

$$+ 2\mathbf{H}_G \cdot \mathbf{I}_G^{-1} \cdot \mathbf{H}_I - 2\mathbf{H}_I \cdot \mathbf{I}_G^{-1} \cdot \mathbf{H}_I \tag{31}$$

and this reduces to Eq. (7) when one takes advantage of the fact that $\mathbf{I}_G^{-1}$ is a symmetric dyadic.

To establish the validity of (8), it suffices to show that

$$\frac{\mathbf{H}_G^2}{I_{\max}} \le \mathbf{H}_G \cdot \mathbf{I}_G^{-1} \cdot \mathbf{H}_G \le \frac{\mathbf{H}_G^2}{I_{\min}} \tag{32}$$

To this end, mutually perpendicular unit vectors $\mathbf{a}_1$, $\mathbf{a}_2$, $\mathbf{a}_3$ are introduced, with $\mathbf{a}_1$ and $\mathbf{a}_3$ parallel, respectively, to the axis of maximum and minimum central moment of inertia of G; and a unit vector $\boldsymbol{\alpha}$ is defined as

$$\boldsymbol{\alpha} \triangleq \mathbf{H}_G\,|\mathbf{H}_G|^{-1} \tag{33}$$

where $|\mathbf{H}_G|$ denotes the magnitude of $\mathbf{H}_G$. If, in addition, α_j and I_j are defined as

$$\alpha_j \triangleq \boldsymbol{\alpha} \cdot \mathbf{a}_j \qquad I_j \triangleq \mathbf{a}_j \cdot \mathbf{I}_G \cdot \mathbf{a}_j \qquad (j = 1, 2, 3) \tag{34}$$

one then has

$$\alpha_1^2 + \alpha_2^2 + \alpha_3^2 = 1 \tag{35}$$

and

$$I_{\max} = I_1 > I_2 > I_3 = I_{\min} \tag{36}$$

Moreover, it follows from Eqs. (33) and (34) that

$$\mathbf{H}_G = |\mathbf{H}_G|(\alpha_1\mathbf{a}_1 + \alpha_2\mathbf{a}_2 + \alpha_3\mathbf{a}_3) \tag{37}$$

and that

$$\mathbf{I}_G^{-1} = \mathbf{a}_1\mathbf{a}_1 I_1^{-1} + \mathbf{a}_2\mathbf{a}_2 I_2^{-1} + \mathbf{a}_3\mathbf{a}_3 I_3^{-1} \tag{38}$$

Hence,

$$\mathbf{H}_G \cdot \mathbf{I}_G^{-1} \cdot \mathbf{H}_G \underset{(37,38)}{=} \mathbf{H}_G^2(\alpha_1^2 I_1^{-1} + \alpha_2^2 I_2^{-1} + \alpha_3^2 I_3^{-1}) \tag{39}$$

or, after elimination of α_1 by reference to Eq. (35),

$$\mathbf{H}_G \cdot \mathbf{I}_G^{-1} \cdot \mathbf{H}_G = \frac{\mathbf{H}_G^2}{I_1}\left[1 + \left(\frac{I_1}{I_2} - 1\right)\alpha_2^2 + \left(\frac{I_1}{I_3} - 1\right)\alpha_3^2\right] \tag{40}$$

In view of (36), Eq. (40) shows that $\mathbf{H}_G \cdot \mathbf{I}_G^{-1} \cdot \mathbf{H}_G \ge \mathbf{H}_G^2 I_1^{-1}$, which is equivalent to the left-hand portion of (32). Similarly, elimination of α_3 from Eq. (39) establishes the validity of the right-hand portion of (32).

When G is moving quasi-rigidly in F, that is, when B is at rest in A (so that $\mathbf{H}_I = 0$), then

$$\mathbf{H}_G \underset{(28)}{=} {}^F\boldsymbol{\omega}^A \cdot \mathbf{I}_G \tag{41}$$

$$2K \underset{(7)}{=} \mathbf{H}_G \cdot \mathbf{I}_G{}^{-1} \cdot \mathbf{H}_G \underset{(41)}{=} {}^F\boldsymbol{\omega}^A \cdot \mathbf{I}_G \cdot {}^F\boldsymbol{\omega}^A \tag{42}$$

and, if ω_i is defined as

$$\omega_i \triangleq \mathbf{a}_i \cdot {}^F\boldsymbol{\omega}^A \qquad (i = 1, 2, 3) \tag{43}$$

one can write

$$\mathbf{H}_G{}^2 \underset{(43)}{=} I_1{}^2\omega_1{}^2 + I_2{}^2\omega_2{}^2 + I_3{}^2\omega_3{}^2 \tag{44}$$

and

$$2K = I_1\omega_1{}^2 + I_2\omega_2{}^2 + I_3\omega_3{}^2 \tag{45}$$

It follows that

$$\omega_1{}^2 \underset{(44)}{=} \frac{\mathbf{H}_G{}^2 - I_2{}^2\omega_2{}^2 - I_3{}^2\omega_3{}^2}{I_1{}^2} \tag{46}$$

and

$$\omega_3{}^2 \underset{(44)}{=} \frac{\mathbf{H}_G{}^2 - I_1{}^2\omega_1{}^2 - I_2{}^2\omega_2{}^2}{I_3{}^2} \tag{47}$$

so that, if ω_1 is eliminated from Eq. (45) with the aid of Eq. (46), one has

$$2K = \frac{\mathbf{H}_G{}^2}{I_1}\left[1 + \frac{I_2(I_1 - I_2)\omega_2{}^2 + I_3(I_1 - I_3)\omega_3{}^2}{\mathbf{H}_G{}^2}\right] \tag{48}$$

whereas elimination of ω_3 by reference to Eq. (47) gives

$$2K = \frac{\mathbf{H}_G{}^2}{I_3}\left[1 + \frac{I_1(I_1 - I_3)\omega_1{}^2 + I_2(I_2 - I_3)\omega_2{}^2}{\mathbf{H}_G{}^2}\right] \tag{49}$$

Moreover, when $\mathbf{H}_I = 0$, then

$$2K_{\text{min}} \underset{(9)}{=} \frac{\mathbf{H}_G{}^2}{I_{\text{max}}} \underset{(36)}{=} \frac{\mathbf{H}_G{}^2}{I_1} \tag{50}$$

and

$$2K_{\text{max}} \underset{(10)}{=} \frac{\mathbf{H}_G{}^2}{I_{\text{min}}} \underset{(36)}{=} \frac{\mathbf{H}_G{}^2}{I_3} \tag{51}$$

Suppose now that $K = K_{\text{min}}$. Then, from Eqs. (48) and (50),

$$I_2(I_1 - I_2)\omega_2{}^2 + I_3(I_1 - I_3)\omega_3{}^2 = 0 \tag{52}$$

and, since the coefficients of ω_2^2 and ω_3^2 are intrinsically positive by (36), ω_2 and ω_3 must be equal to zero, which means that ${}^F\boldsymbol{\omega}^A$ is parallel to $\mathbf{a}_1$, that is, to the axis of maximum central moment of inertia of G; and conversely, if ${}^F\boldsymbol{\omega}^A$ is parallel to $\mathbf{a}_1$, then $\omega_2 = \omega_3 = 0$ and, from Eqs. (48) and (50), $K = K_{\text{min}}$. Similarly, if $K = K_{\text{max}}$, it follows from Eqs. (49) and (51) that ${}^F\boldsymbol{\omega}^A$ is parallel to $\mathbf{a}_3$, and hence to the axis of minimum moment of inertia of G; and conversely, if ${}^F\boldsymbol{\omega}^A$ is parallel to $\mathbf{a}_3$, then $\omega_1 = \omega_2 = 0$ and, from Eqs. (49) and (51), $K = K_{\text{max}}$.

Example A simple gyrostat G, formed by a rigid body A that carries an axisymmetric rotor B whose axis and mass center are fixed in A, is at rest in a Newtonian reference frame N, and no forces act on A or B except those exerted by A on B and vice versa. It is to be shown that any rotation of B relative to A, brought about, say, by means of a motor each of whose parts belongs either to A or B, is accompanied by a rotation of A in N about an axis that is fixed both in A and N, passes through the mass center G^* of G, and is parallel to the unit vector $\mathbf{n}$ given by

$$\mathbf{n} = \pm \frac{\boldsymbol{\beta} \cdot \mathbf{I}_G^{-1}}{|\boldsymbol{\beta} \cdot \mathbf{I}_G^{-1}|} \tag{53}$$

where $\mathbf{I}_G$ is the central inertia dyadic of G and $\boldsymbol{\beta}$ is a unit vector parallel to the axis of B.

So long as no external forces act on G, G^* remains at rest in N; and $\mathbf{H}_G$, the angular momentum of G in N relative to G^*, remains equal to zero. Consequently,

$$\mathbf{H}_R + \mathbf{H}_I \underset{(1)}{=} 0 \tag{54}$$

Now,

$$\mathbf{H}_R = {}^N\boldsymbol{\omega}^A \cdot \mathbf{I}_G \tag{55}$$

and

$$\mathbf{H}_I \underset{(4)}{=} J\,{}^A\omega^B\,\boldsymbol{\beta} \tag{56}$$

Hence,

$${}^N\boldsymbol{\omega}^A \cdot \mathbf{I}_G + J\,{}^A\omega^B\,\boldsymbol{\beta} = 0 \tag{57}$$

from which it follows that

$${}^N\boldsymbol{\omega}^A = -J\,{}^A\omega^B\,\boldsymbol{\beta} \cdot \mathbf{I}_G^{-1} \tag{58}$$

Let $\mathbf{n}$ be the unit vector defined as

$$\mathbf{n} \triangleq \frac{{}^N\boldsymbol{\omega}^A}{|{}^N\boldsymbol{\omega}^A|} \underset{(58)}{=} \pm \frac{\boldsymbol{\beta} \cdot \mathbf{I}_G^{-1}}{|\boldsymbol{\beta} \cdot \mathbf{I}_G^{-1}|} \tag{59}$$

(The sign ambiguity arises from the fact that ${}^A\omega^B$ can be positive or negative.) Since $\boldsymbol{\beta}$ and $\mathbf{I}_G$ are independent of time t in A, the derivative of $\mathbf{n}$ with respect to t in A vanishes. It follows that the derivative of $\mathbf{n}$ with respect to t in N also vanishes, for

$$\frac{{}^N d\mathbf{n}}{dt} \underset{(1.11.8)}{=} \frac{{}^A d\mathbf{n}}{dt} + {}^N\boldsymbol{\omega}^A \times \mathbf{n} = 0 + {}^N\boldsymbol{\omega}^A \times \frac{{}^N\boldsymbol{\omega}^A}{|{}^N\boldsymbol{\omega}^A|} = 0 \tag{60}$$

In other words, $\mathbf{n}$ is fixed both in A and in N; and, since ${}^N\boldsymbol{\omega}^A$ is parallel to $\mathbf{n}$, ${}^N\boldsymbol{\omega}^A$ has a fixed orientation in both A and N. The only motions for which this is true and G^* remains fixed in N are rotations of A in N about an axis that passes through G^* and is parallel to $\mathbf{n}$.

3.7 DYNAMICAL EQUATIONS FOR A SIMPLE GYROSTAT

In Fig. 3.7.1, G designates a simple gyrostat consisting of a rigid body A and an axisymmetric rotor B whose axis and mass center, B^*, are fixed in A; $\mathbf{a}_1$, $\mathbf{a}_2$, $\mathbf{a}_3$ form a dextral set of mutually perpendicular unit vectors fixed in A, each parallel to a central principal axis of inertia of G. The axis of B is parallel to a unit vector $\boldsymbol{\beta}$, and G has central principal moments of inertia I_1, I_2, I_3, while B has an axial moment of inertia J.

After defining ${}^A\omega^B$, ω_i, β_i, and M_i as

$${}^A\omega^B \triangleq {}^A\boldsymbol{\omega}^B \cdot \boldsymbol{\beta} \tag{1}$$

$$\omega_i \triangleq {}^N\boldsymbol{\omega}^A \cdot \mathbf{a}_i \qquad (i = 1, 2, 3) \tag{2}$$

$$\beta_i \triangleq \boldsymbol{\beta} \cdot \mathbf{a}_i \qquad (i = 1, 2, 3) \tag{3}$$

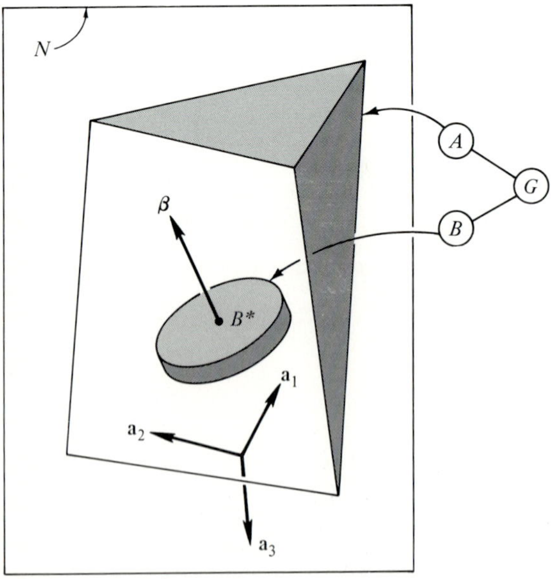

Figure 3.7.1

and

$$M_i \triangleq \mathbf{M} \cdot \mathbf{a}_i \qquad (i = 1, 2, 3) \tag{4}$$

where ${}^A\boldsymbol{\omega}^B$ is the angular velocity of B in A, ${}^N\boldsymbol{\omega}^A$ is the angular velocity of A in a Newtonian reference frame N, and $\mathbf{M}$ is the moment about the mass center G^* of G of all forces acting on G, one can write the following set of coupled differential equations governing ${}^A\omega^B$ and ω_i $(i = 1, 2, 3)$:

$$\begin{aligned}{}^A\dot{\omega}^B\left[1 - J\left(\frac{\beta_1^2}{I_1} + \frac{\beta_2^2}{I_2} + \frac{\beta_3^2}{I_3}\right)\right] &= \frac{\boldsymbol{\beta} \cdot \mathbf{M}^{A/B}}{J} \\ &+ \frac{\beta_1}{I_1}[(I_3 - I_2)\omega_2\omega_3 - J\,{}^A\omega^B(\beta_2\omega_3 - \beta_3\omega_2) - M_1] \\ &+ \frac{\beta_2}{I_2}[(I_1 - I_3)\omega_3\omega_1 - J\,{}^A\omega^B(\beta_3\omega_1 - \beta_1\omega_3) - M_2] \\ &+ \frac{\beta_3}{I_3}[(I_2 - I_1)\omega_1\omega_2 - J\,{}^A\omega^B(\beta_1\omega_2 - \beta_2\omega_1) - M_3]\end{aligned} \tag{5}$$

where $\mathbf{M}^{A/B}$ is the moment about B^* of all forces exerted by A on B; and

$$I_1\dot{\omega}_1 = (I_2 - I_3)\omega_2\omega_3 - J[{}^A\dot{\omega}^B\beta_1 - {}^A\omega^B(\beta_2\omega_3 - \beta_3\omega_2)] + M_1 \tag{6}$$

$$I_2\dot{\omega}_2 = (I_3 - I_1)\omega_3\omega_1 - J[{}^A\dot{\omega}^B\beta_2 - {}^A\omega^B(\beta_3\omega_1 - \beta_1\omega_3)] + M_2 \tag{7}$$

$$I_3\dot{\omega}_3 = (I_1 - I_2)\omega_1\omega_2 - J[{}^A\dot{\omega}^B\beta_3 - {}^A\omega^B(\beta_1\omega_2 - \beta_2\omega_1)] + M_3 \tag{8}$$

If the axis of B is parallel to a central principal axis of inertia of G, the complexity of these equations is reduced considerably; and further simplifications can be made if B is completely free to rotate relative to A or if B is made to rotate relative to A with a constant angular speed, say by means of a motor each part of which belongs either to A or to B. Under these circumstances, and with $\boldsymbol{\beta} = \mathbf{a}_3$, the differential equations governing ω_1, ω_2, ω_3 can be written

$$I_1\dot{\omega}_1 = (I_2 - \bar{I}_3)\omega_2\omega_3 - J\sigma\omega_2 + M_1 \tag{9}$$

$$I_2\dot{\omega}_2 = (\bar{I}_3 - I_1)\omega_3\omega_1 + J\sigma\omega_1 + M_2 \tag{10}$$

$$\bar{I}_3\dot{\omega}_3 = (I_1 - I_2)\omega_1\omega_2 + M_3 \tag{11}$$

where $\bar{I}_3$ and σ are given in Table 3.7.1, with ${}^A\hat{\omega}^B$ and $\hat{\omega}_3$ denoting the initial values of ${}^A\omega^B$ and ω_3, respectively.

Table 3.7.1

Rotor motion	$\bar{I}_3$	σ
Free	$I_3 - J$	$\hat{\omega}_3 + {}^A\hat{\omega}^B$
${}^A\omega^B = {}^A\hat{\omega}^B$	I_3	${}^A\hat{\omega}^B$

Derivations The inertia torque $\mathbf{T}_B$ of B in N is given by

$$\mathbf{T}_B = -{}^N\boldsymbol{\alpha}^B \cdot \mathbf{I}_B - {}^N\boldsymbol{\omega}^B \times \mathbf{I}_B \cdot {}^N\boldsymbol{\omega}^B \tag{12}$$

where ${}^N\boldsymbol{\alpha}^B$ and ${}^N\boldsymbol{\omega}^B$ are, respectively, the angular acceleration and the angular velocity of B in N, and $\mathbf{I}_B$, the central inertia dyadic of B, can be expressed as

$$\begin{aligned}\mathbf{I}_B &= K(\boldsymbol{\gamma}\boldsymbol{\gamma} + \boldsymbol{\delta}\boldsymbol{\delta}) + J\boldsymbol{\beta}\boldsymbol{\beta} \\ &= K\mathbf{U} + (J - K)\boldsymbol{\beta}\boldsymbol{\beta}\end{aligned} \tag{13}$$

Here K is the transverse central moment of inertia of B, $\boldsymbol{\gamma}$ and $\boldsymbol{\delta}$ are unit vectors such that $\boldsymbol{\beta}$, $\boldsymbol{\gamma}$, and $\boldsymbol{\delta}$ are mutually perpendicular, and $\mathbf{U}$ is the unit dyadic. Hence,

$${}^N\boldsymbol{\alpha}^B \cdot \mathbf{I}_B \underset{(13)}{=} K\,{}^N\boldsymbol{\alpha}^B + (J - K)\,{}^N\boldsymbol{\alpha}^B \cdot \boldsymbol{\beta}\boldsymbol{\beta} \tag{14}$$

and

$${}^N\boldsymbol{\omega}^B \times \mathbf{I}_B \cdot {}^N\boldsymbol{\omega}^B \underset{(13)}{=} (J - K)\,{}^N\boldsymbol{\omega}^B \times \boldsymbol{\beta}\boldsymbol{\beta} \cdot {}^N\boldsymbol{\omega}^B \tag{15}$$

from which it follows that

$$\mathbf{T}_B \cdot \boldsymbol{\beta} \underset{(12)}{=} -J\,{}^N\boldsymbol{\alpha}^B \cdot \boldsymbol{\beta} \tag{16}$$

Since, by D'Alembert's principle,

$$\mathbf{T}_B \cdot \boldsymbol{\beta} + \mathbf{M}^{A/B} \cdot \boldsymbol{\beta} = 0 \tag{17}$$

one thus has

$$J\,{}^N\boldsymbol{\alpha}^B \cdot \boldsymbol{\beta} \underset{(16,17)}{=} \mathbf{M}^{A/B} \cdot \boldsymbol{\beta} \tag{18}$$

Now,

$$\begin{aligned}{}^N\boldsymbol{\alpha}^B &\underset{(1.20.1)}{=} \frac{{}^Nd\,{}^N\boldsymbol{\omega}^B}{dt} \underset{(1.16.1)}{=} \frac{{}^Nd}{dt}({}^N\boldsymbol{\omega}^A + {}^A\omega^B\boldsymbol{\beta}) \\ &= \underset{(1.20.1)}{{}^N\boldsymbol{\alpha}^A} + {}^A\dot{\omega}^B\boldsymbol{\beta} + {}^A\omega^B\,{}^N\boldsymbol{\omega}^A \underset{(1.11.9)}{\times} \boldsymbol{\beta}\end{aligned} \tag{19}$$

Consequently,

$$J(\underset{(19)}{{}^N\boldsymbol{\alpha}^A \cdot \boldsymbol{\beta}} + {}^A\dot{\omega}^B) \underset{(18)}{=} \boldsymbol{\beta} \cdot \mathbf{M}^{A/B} \tag{20}$$

and, since

$${}^N\boldsymbol{\alpha}^A \cdot \boldsymbol{\beta} \underset{(2)}{=} \dot{\omega}_1\beta_1 + \dot{\omega}_2\beta_2 + \dot{\omega}_3\beta_3 \tag{21}$$

one can write

$$J(\dot{\omega}_1\beta_1 + \dot{\omega}_2\beta_2 + \dot{\omega}_3\beta_3 + {}^A\dot{\omega}^B) \underset{(20)}{=} \boldsymbol{\beta} \cdot \mathbf{M}^{A/B} \tag{22}$$

$\mathbf{T}_R$, the inertia torque of a rigid body R that has the same mass distribution as G but moves like A, is given by

$$\mathbf{T}_R = -\,{}^N\boldsymbol{\alpha}^A \cdot \mathbf{I}_G - {}^N\boldsymbol{\omega}^A \times \mathbf{I}_G \cdot {}^N\boldsymbol{\omega}^A \tag{23}$$

Hence, $\mathbf{T}_G$, the inertia torque of G in N, can be expressed with the aid of Eqs. (3.6.2) and (3.6.5) as

$$\mathbf{T}_G = -\,{}^N\boldsymbol{\alpha}^A \cdot \mathbf{I}_G - {}^N\boldsymbol{\omega}^A \times \mathbf{I}_G \cdot {}^N\boldsymbol{\omega}^A - J({}^A\dot{\omega}^B \boldsymbol{\beta} + {}^A\omega^B\, {}^N\boldsymbol{\omega}^A \times \boldsymbol{\beta}) \tag{24}$$

Carrying out the operations indicated in Eq. (24) after noting that

$$\mathbf{I}_G = I_1\mathbf{a}_1\mathbf{a}_1 + I_2\mathbf{a}_2\mathbf{a}_2 + I_3\mathbf{a}_3\mathbf{a}_3 \tag{25}$$

and then appealing to D'Alembert's principle to write

$$\mathbf{T}_G + \mathbf{M} = 0 \tag{26}$$

one arrives at Eqs. (6)–(8); and solution of these equations for $\dot{\omega}_1$, $\dot{\omega}_2$, $\dot{\omega}_3$, followed by substitution into Eq. (22), produces Eq. (5).

Suppose now that $\boldsymbol{\beta} = \mathbf{a}_3$, so that [see Eq. (3)]

$$\beta_1 = \beta_2 = 0 \qquad \beta_3 = 1 \tag{27}$$

Then

$${}^A\dot{\omega}^B\left(1 - \frac{J}{I_3}\right) \underset{(5)}{=} \frac{\boldsymbol{\beta}\cdot\mathbf{M}^{A/B}}{J} + \frac{I_2 - I_1}{I_3}\,\omega_1\omega_2 - \frac{M_3}{I_3} \tag{28}$$

and

$$I_1\dot{\omega}_1 \underset{(6)}{=} (I_2 - I_3)\omega_2\omega_3 - J\,{}^A\omega^B\,\omega_2 + M_1 \tag{29}$$

$$I_2\dot{\omega}_2 \underset{(7)}{=} (I_3 - I_1)\omega_3\omega_1 + J\,{}^A\omega^B\,\omega_1 + M_2 \tag{30}$$

$$I_3\dot{\omega}_3 \underset{(8)}{=} (I_1 - I_2)\omega_1\omega_2 - J\,{}^A\dot{\omega}^B + M_3 \tag{31}$$

If B is completely free to rotate relative to A, that is, if $\boldsymbol{\beta}\cdot\mathbf{M}^{A/B} = 0$, then [see Eqs. (22) and (27)]

$$\dot{\omega}_3 + {}^A\dot{\omega}^B = 0 \tag{32}$$

which implies that

$$\omega_3 + {}^A\omega^B = \hat{\omega}_3 + {}^A\hat{\omega}^B \tag{33}$$

where $\hat{\omega}_3$ and ${}^A\hat{\omega}^B$ are the initial values of ω_3 and ${}^A\omega^B$, respectively. Hence,

$$\begin{aligned} I_1\dot{\omega}_1 &\underset{(29)}{=} (I_2 - I_3)\omega_2\omega_3 - J(\hat{\omega}_3 + {}^A\hat{\omega}^B - \omega_3)\omega_2 + M_1 \\ &= (I_2 - I_3 + J)\omega_2\omega_3 - J(\hat{\omega}_3 + {}^A\hat{\omega}^B) + M_1 \end{aligned} \tag{34}$$

Similarly, from Eqs. (30) and (33),

$$I_2\dot{\omega}_2 = (I_3 - J - I_1)\omega_3\omega_1 + J(\hat{\omega}_3 + {}^A\hat{\omega}^B) + M_2 \tag{35}$$

Furthermore,

$$^A\dot{\omega}^B \underset{(28)}{=} \frac{I_2 - I_1}{I_3 - J}\,\omega_1\omega_2 - \frac{M_3}{I_3 - J} \tag{36}$$

and substitution into Eq. (31) yields

$$(I_3 - J)\dot{\omega}_3 = (I_1 - I_2)\omega_1\omega_2 + M_3 \tag{37}$$

Replacing $I_3 - J$ with $\bar{I}_3$ and $\hat{\omega}_3 + {}^A\hat{\omega}^B$ with σ in Eqs. (34), (35), and (37), one arrives at Eqs. (9), (10), and (11), respectively. On the other hand, if B is made to rotate relative to A with the constant angular speed ${}^A\hat{\omega}^B$, then Eqs. (29)–(31) become Eqs. (9)–(11) when I_3 is replaced with $\bar{I}_3$, ${}^A\hat{\omega}^B$ is replaced with σ, and ${}^A\dot{\omega}^B$ is set equal to zero. Thus Eqs. (9)–(11) apply in both cases, provided $\bar{I}_3$ and σ be interpreted in accordance with Table 3.7.1.

Example In Fig. 3.7.2, O designates a point fixed in N, R is the distance between O and G^*, and $\mathbf{e}_1$ is a unit vector. G is presumed to be subjected to the action of a force system S such that $\mathbf{F}$, the resultant of S, and $\mathbf{M}$, the total moment of S about G^*, are given by

$$\mathbf{F} = -\mu m R^{-2}\mathbf{e}_1 \tag{38}$$

and

$$\mathbf{M} = \frac{3\mu}{R^3}\,\mathbf{e}_1 \times \mathbf{I}_G \cdot \mathbf{e}_1 \tag{39}$$

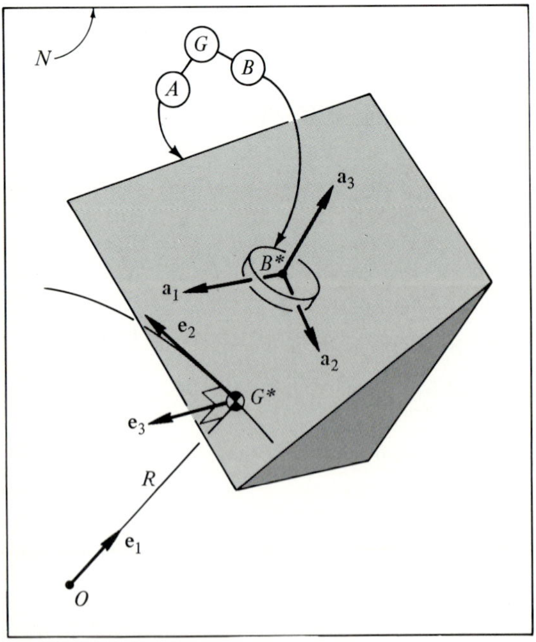

Figure 3.7.2

where μ is a constant, m is the mass of G, and $\mathbf{I}_G$ is the central inertia dyadic of G. [If μ is set equal to $G\overline{m}$, where G is the universal gravitational constant, then $\mathbf{F}$ becomes equal to $\hat{\mathbf{F}}$ as defined in Eq. (2.3.5) while $\mathbf{M}$ becomes equal to $\widetilde{\mathbf{M}}$ as defined in Eq. (2.6.3). The force system S is thus approximately equivalent to the system of gravitational forces exerted on G by a particle of mass $\overline{m}$ at point O.]

Due to the action of S, the rotational motion of A in N depends on the translational motion of G^* in N, but not vice versa, because R appears in Eq. (39), but no variable involving the orientation of A in N appears in Eq. (38); and one possible motion of G^* is motion in a circular orbit that is fixed in N, has a radius R, and is described with an angular speed Ω given by

$$\Omega = (\mu R^{-3})^{1/2} \tag{40}$$

If G^* moves in this fashion, and if G is an axisymmetric gyrostat whose axis of symmetry is parallel to that of B, that is, if

$$I_1 = I_2 \tag{41}$$

then dynamical equations governing the rotational motion of A in N can be expressed in the form

$$\dot{\omega}_1 = f_1(\theta_1, \theta_2, \theta_3, \omega_2) \tag{42}$$

$$\dot{\omega}_2 = f_2(\theta_1, \theta_2, \theta_3, \omega_1) \tag{43}$$

$$\omega_3 = \hat{\omega}_3 \tag{44}$$

where $\hat{\omega}_3$ is the permanent value of ω_3, and $\theta_1, \theta_2, \theta_3$ are body-three orientation angles (see Sec. 1.7) relating $\mathbf{a}_i$ to $\mathbf{e}_i$ ($i = 1, 2, 3$), with $\mathbf{e}_2 \triangleq ({}^N d\mathbf{e}_1/dt)\Omega^{-1}$ and $\mathbf{e}_3 = \mathbf{e}_1 \times \mathbf{e}_2$, so that $\mathbf{e}_3$ is normal to the orbit plane (see Fig. 3.7.2). The functions f_1 and f_2 are to be determined.

To express $\mathbf{M}$ [see Eq. (39)] in terms of $\mathbf{a}_1, \mathbf{a}_2, \mathbf{a}_3$, note that

$$\mathbf{e}_1 \underset{(1.7.11)}{=} c_2c_3\mathbf{a}_1 - c_2s_3\mathbf{a}_2 + s_2\mathbf{a}_3 \tag{45}$$

Next, with

$$\mathbf{I}_G = I_1(\mathbf{a}_1\mathbf{a}_1 + \mathbf{a}_2\mathbf{a}_2) + I_3\mathbf{a}_3\mathbf{a}_3 \tag{46}$$

carry out the multiplication indicated in Eq. (39) to obtain, with the aid of Eq. (40),

$$\mathbf{M} = 3\Omega^2(I_1 - I_3)s_2c_2(s_3\mathbf{a}_1 + c_3\mathbf{a}_2) \tag{47}$$

Next, use Eqs. (11) and (41) to verify Eq. (44). Finally, write

$$I_1\dot{\omega}_1 - (I_2 - \bar{I}_3)\omega_2\hat{\omega}_3 + J_3\sigma\omega_2 \underset{(9)}{=} 3\Omega^2(I_1 - I_3)s_2c_2s_3 \tag{48}$$

and

$$I_2\dot{\omega}_2 - (\bar{I}_3 - I_1)\hat{\omega}_3\omega_1 - J_3\sigma\omega_1 \underset{(10)}{=} 3\Omega^2(I_1 - I_3)s_2c_2c_3 \tag{49}$$

and compare these equations with Eqs. (42) and (43) to conclude that

$$f_1 = \left[\left(1 - \frac{\bar{I}_3}{I_1}\right)\hat{\omega}_3 - \frac{J_3}{I_1}\sigma\right]\omega_2 + 3\Omega^2\left(1 - \frac{I_3}{I_1}\right)s_2c_2s_3 \tag{50}$$

and

$$f_2 = -\left[\left(1 - \frac{\bar{I}_3}{I_1}\right)\hat{\omega}_3 - \frac{J_3}{I_1}\sigma\right]\omega_1 + 3\Omega^2\left(1 - \frac{I_3}{I_1}\right)s_2c_2c_3 \tag{51}$$

3.8 ROTATIONAL MOTION OF A TORQUE-FREE, AXISYMMETRIC GYROSTAT

In Fig. 3.8.1, A designates a rigid body whose central inertia ellipsoid E_A (not shown) may have three unequal principal diameters, whereas B is a rigid body whose central inertia ellipsoid E_B is an ellipsoid of revolution. B is connected to A in such a way that both the mass center B^* of B and the axis of revolution of E_B are fixed in A, but B can rotate relative to A about this axis. The body G formed by A and B is thus a simple gyrostat; and, if the inertia ellipsoid E_G of G for the mass center G^* of G is an ellipsoid of revolution, G is called a simple axisymmetric gyrostat.

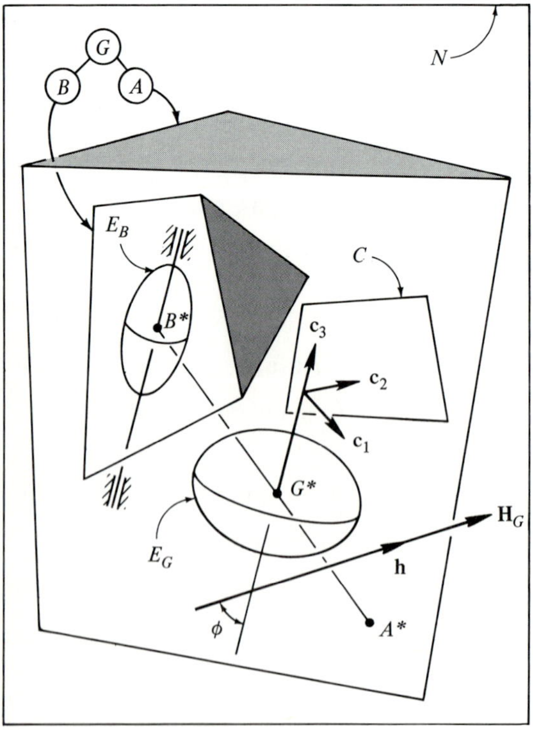

Figure 3.8.1

If the axes of revolution of E_B and E_G are parallel to each other and G is subjected to the action of forces whose resultant moment about G^* is equal to zero, the rotational motion of G in a Newtonian reference frame N proceeds much like that of a single, axisymmetric rigid body (see Sec. 3.1); that is, while B rotates relative to A with, in general, variable angular speed, the rotational motion of A in N can be described in terms of two motions of simple rotation (see Sec. 1.1). One of these is the motion of A in a reference frame C in which the angular momentum $\mathbf{H}_G$ of G relative to G^* in N (a vector fixed in N) and the symmetry axis of E_G are fixed; the other is the motion of C in N. The associated angular velocities are called, respectively, the angular velocity of *spin* and the angular velocity of *precession,* and they can be expressed as

$$^C\boldsymbol{\omega}^A = s\,\mathbf{c}_3 \tag{1}$$

and

$$^N\boldsymbol{\omega}^C = p\,\mathbf{h} \tag{2}$$

Here $\mathbf{h}$ is a unit vector having the same direction as $\mathbf{H}_G$ (see Fig. 3.8.1); $\mathbf{c}_3$ is a unit vector parallel to the symmetry axis of E_G; s and p, called the *spin speed* and the *precession speed,* respectively, are given by

$$s = \frac{I - J}{J}\frac{J\hat{\omega}_3 + K\hat{r}}{I} - \frac{K}{J}r \tag{3}$$

and

$$p = \frac{H}{I} \tag{4}$$

where H is the (constant) magnitude of $\mathbf{H}_G$; I and J are the moments of inertia of G about lines passing through G^* and, respectively, perpendicular and parallel to $\mathbf{c}_3$; K is the moment of inertia of B about the symmetry axis of B; $r \triangleq {}^A\boldsymbol{\omega}^B \cdot \mathbf{c}_3$; $\hat{r}$ is the value of r at $t = 0$; and $\hat{\omega}_3$ is the value of $^N\boldsymbol{\omega}^A \cdot \mathbf{c}_3$ at $t = 0$. The quantity r satisfies the differential equation

$$\dot{r} = \frac{J}{K(J - K)}\mathbf{c}_3 \cdot \mathbf{M}^{A/B} \tag{5}$$

where $\mathbf{M}^{A/B}$ is the moment about B^* of all forces exerted by A on B.

Throughout the motion of G, the angle ϕ between $\mathbf{c}_3$ and $\mathbf{H}_G$ remains constant and is given by

$$\phi = \cos^{-1}\left(\frac{J\hat{\omega}_3 + K\hat{r}}{H}\right) \tag{6}$$

Derivations In Fig. 3.8.1, $\mathbf{c}_1$ and $\mathbf{c}_2$ are unit vectors not fixed either in A or in B, but such that $\mathbf{c}_1 \times \mathbf{c}_2 = \mathbf{c}_3$. If C is a reference frame in which $\mathbf{c}_1$, $\mathbf{c}_2$, and $\mathbf{c}_3$ are fixed, then, since $\mathbf{c}_3$ is permanently parallel to the axis of revolution of E_B, and since this axis is fixed in A, the angular velocity of A in C is necessarily parallel to $\mathbf{c}_3$ and, therefore, can be expressed as

$$^{C}\boldsymbol{\omega}^{A} = s\,\mathbf{c}_3 \tag{7}$$

Similarly, as B is constrained to rotate relative to A about a line parallel to $\mathbf{c}_3$, the angular velocity of B in A can be written

$$^{A}\boldsymbol{\omega}^{B} = r\,\mathbf{c}_3 \tag{8}$$

Finally, the angular velocity of C in N always can be expressed in terms of functions p_1, p_2, and p_3 of t as

$$^{N}\boldsymbol{\omega}^{C} = p_1\mathbf{c}_1 + p_2\mathbf{c}_2 + p_3\mathbf{c}_3 \tag{9}$$

Consequently,

$$^{N}\boldsymbol{\omega}^{A} \underset{(1.16.1)}{=} {}^{N}\boldsymbol{\omega}^{C} + {}^{C}\boldsymbol{\omega}^{A} \underset{(7,9)}{=} p_1\mathbf{c}_1 + p_2\mathbf{c}_2 + (p_3 + s)\mathbf{c}_3 \tag{10}$$

and

$$^{N}\boldsymbol{\alpha}^{A} \underset{(1.20.1)}{=} \frac{^{N}d}{dt}\,{}^{N}\boldsymbol{\omega}^{A} \underset{(10)}{=} (\dot{p}_1 + sp_2)\mathbf{c}_1 + (\dot{p}_2 - sp_1)\mathbf{c}_2 + (\dot{p}_3 + \dot{s})\mathbf{c}_3 \tag{11}$$

$\mathbf{T}_R$, the inertia torque of R in N, where R is a rigid body having the same mass distribution as G but moving like A, is thus given by

$$\begin{aligned} \mathbf{T}_R &= -\,{}^{N}\boldsymbol{\alpha}^{A} \cdot \mathbf{I}_G - {}^{N}\boldsymbol{\omega}^{A} \times \mathbf{I}_G \cdot {}^{N}\boldsymbol{\omega}^{A} \\ &= -\,\{I\dot{p}_1 + [(J - I)p_3 + Js]p_2\}\mathbf{c}_1 \\ &\quad - \{I\dot{p}_2 - [(J - I)p_3 + Js]p_1\}\mathbf{c}_2 - J(\dot{p}_3 + \dot{s})\mathbf{c}_3 \end{aligned} \tag{12}$$

while $\mathbf{T}_I$, the internal inertia torque of G, can be written

$$\begin{aligned} \mathbf{T}_I &\underset{(3.6.5)}{=} -\,K(\dot{r}\,\mathbf{c}_3 + r\,{}^{N}\boldsymbol{\omega}^{A} \times \mathbf{c}_3) \\ &= -K[\dot{r}\,\mathbf{c}_3 + r(p_2\mathbf{c}_1 - p_1\mathbf{c}_2)] \end{aligned} \tag{13}$$

If G is subjected to the action of forces whose resultant moment about G^* is equal to zero, the inertia torque of G in N, that is, the sum of $\mathbf{T}_R$ and $\mathbf{T}_I$ [see Eq. (3.6.2)], vanishes identically. Hence, Eqs. (12) and (13) lead directly to

$$I\dot{p}_1 + [(J - I)p_3 + Js + Kr]p_2 = 0 \tag{14}$$

$$I\dot{p}_2 - [(J - I)p_3 + Js + Kr]p_1 = 0 \tag{15}$$

$$J(\dot{p}_3 + \dot{s}) + K\dot{r} = 0 \tag{16}$$

The quantity s, introduced in Eq. (7), may be chosen at will. A choice that simplifies subsequent analysis is

$$s = \frac{(I - J)p_3 - Kr}{J} \tag{17}$$

because this permits one to replace Eqs. (14)–(16) with

$$\dot{p}_1 = \dot{p}_2 = \dot{p}_3 = 0 \tag{18}$$

from which it follows that

$$p_1 = \hat{p}_1 \qquad p_2 = \hat{p}_2 \qquad p_3 = \hat{p}_3 \tag{19}$$

where $\hat{p}_i$ denotes the initial value of p_i $(i = 1, 2, 3)$.

$\mathbf{H}_R$, the angular momentum of R in N relative to G^*, is given by

$$\begin{aligned}\mathbf{H}_R &\underset{(10)}{=} I(p_1\mathbf{c}_1 + p_2\mathbf{c}_2) + J(p_3 + s)\mathbf{c}_3 \\ &\underset{(17,19)}{=} I(\hat{p}_1\mathbf{c}_1 + \hat{p}_2\mathbf{c}_2) + (I\hat{p}_3 - Kr)\mathbf{c}_3\end{aligned} \tag{20}$$

while $\mathbf{H}_I$, the internal angular momentum of G, can be expressed as

$$\mathbf{H}_I \underset{(3.6.4)}{=} \underset{(8)}{Kr\mathbf{c}_3} \tag{21}$$

Hence,

$$\mathbf{H}_G \underset{(3.6.1,20,21)}{=} I(\hat{p}_1\mathbf{c}_1 + \hat{p}_2\mathbf{c}_2 + \hat{p}_3\mathbf{c}_3) \tag{22}$$

Also,

$${}^N\boldsymbol{\omega}^C \underset{(9,19)}{=} \hat{p}_1\mathbf{c}_1 + \hat{p}_2\mathbf{c}_2 + \hat{p}_3\mathbf{c}_3 \tag{23}$$

Thus, ${}^N\boldsymbol{\omega}^C$ is seen to be parallel to $\mathbf{H}_G$; and, if $\mathbf{h}$ is a unit vector having the same direction as $\mathbf{H}_G$, and H denotes the magnitude of $\mathbf{H}_G$, then

$${}^N\boldsymbol{\omega}^C = \frac{H}{I}\mathbf{h} \tag{24}$$

which establishes the validity of Eqs. (2) and (4).

$\hat{\omega}_3$, the value of ${}^N\boldsymbol{\omega}^A \cdot \mathbf{c}_3$ at $t = 0$, is given by

$$\hat{\omega}_3 \underset{(10)}{=} \hat{p}_3 + \hat{s} \underset{(17)}{=} \frac{I\hat{p}_3 - K\hat{r}}{J} \tag{25}$$

Also,

$$s \underset{(17)}{=} \frac{(I - J)\hat{p}_3 - Kr}{\underset{(19)}{J}} \tag{26}$$

Solving Eq. (25) for $\hat{p}_3$ and substituting into Eq. (26) produces Eq. (3).

Let $\mathbf{T}_B$ denote the inertia torque of B in N. Then, since E_B is an ellipsoid of revolution whose axis is parallel to $\mathbf{c}_3$,

$$\mathbf{c}_3 \cdot \mathbf{T}_B = -\mathbf{c}_3 \cdot {}^N\boldsymbol{\alpha}^B K \tag{27}$$

Moreover,

$$\mathbf{c}_3 \cdot {}^N\boldsymbol{\alpha}^B \underset{(1.20.1)}{=} \mathbf{c}_3 \cdot \frac{{}^N d\,{}^N\boldsymbol{\omega}^B}{dt} = \dot{p}_3 + \dot{s} + \dot{r} \underset{(17,18)}{=} \dot{r}\,\frac{J-K}{J} \tag{28}$$

Hence, if $\mathbf{M}^{A/B}$ is the moment about B^* of all forces exerted by A on B, so that

$$\mathbf{c}_3 \cdot (\mathbf{M}^{A/B} + \mathbf{T}_B) = 0 \tag{29}$$

then

$$\mathbf{c}_3 \cdot \mathbf{M}^{A/B} \underset{(27,28)}{-} \frac{\dot{r}(J-K)K}{J} \underset{(29)}{=} 0 \tag{30}$$

which is equivalent to Eq. (5).

By definition,

$$\phi \triangleq \cos^{-1}(\mathbf{h} \cdot \mathbf{c}_3) \tag{31}$$

where

$$\mathbf{h} = \frac{\mathbf{H}_G}{H} \underset{(22)}{=} \frac{I}{H}(\hat{p}_1\mathbf{c}_1 + \hat{p}_2\mathbf{c}_2 + \hat{p}_3\mathbf{c}_3) \tag{32}$$

Hence,

$$\phi = \cos^{-1}\left(\frac{I\hat{p}_3}{H}\right) \underset{(25)}{=} \cos^{-1}\left(\frac{J\hat{\omega}_3 + K\hat{r}}{H}\right) \tag{33}$$

in agreement with Eq. (6).

Example Figure 3.8.2 represents an axisymmetric gyrostat G consisting of two rigid bodies A and B. A is formed by removing from a rigid, uniform right-circular cylinder of radius R and height $3R$ the material within a right-circular cylinder of radius $R/4$ and height R; and B is a rigid, uniform, right-circular cylinder, made of the same material as A and completely filling the cavity in A.

At a certain instant, say $t = 0$, the angular velocity of A in a Newtonian reference frame N has components of magnitude 3Ω and 8Ω, directed as shown in Fig. 3.8.2, and B is at rest in A. If B is now made to rotate relative to A in a suitable manner for $0 \leq t \leq T$, then the motion of A in N for $t \geq T$ is a motion of simple rotation (see Sec. 1.1), performed with a constant angular speed $n\Omega$. The number n and the way in which B must move relative to A are to be determined.

The moments of inertia I, J, and K are

$$I = mR^2 \qquad J = \frac{mR^2}{2} \qquad K = \frac{mR^2}{1536} \tag{34}$$

where m is the mass of G. The central angular momentum of G thus has the magnitude

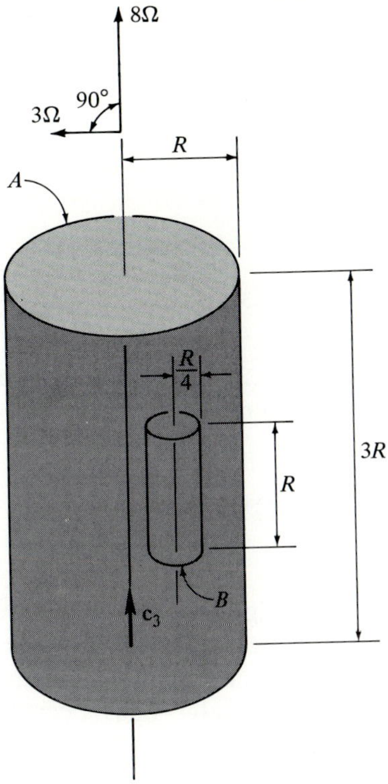

Figure 3.8.2

$$H = [(3\Omega I)^2 + (8\Omega J)^2]^{1/2} = 5mR^2\Omega \tag{35}$$

so that

$$p \underset{(4)}{=} 5\Omega \tag{36}$$

and

$$s \underset{(3)}{=} 4\Omega - \frac{r}{768} \tag{37}$$

The motion of A in N is a simple rotational motion for $t \geq T$ if $s = 0$ for $t \geq T$; and the associated angular speed is then equal to p. Hence, $n = 5$ and, for $t \geq T$, B must move relative to A in such a way that

$$r \underset{(37)}{=} 4(768)\Omega = 3072\Omega \tag{38}$$

In other words, for $t \geq T$, the angular velocity of B in A must satisfy the equation

$$^A\boldsymbol{\omega}^B \cdot \mathbf{c}_3 \underset{(8)}{=} 3072\Omega \tag{39}$$

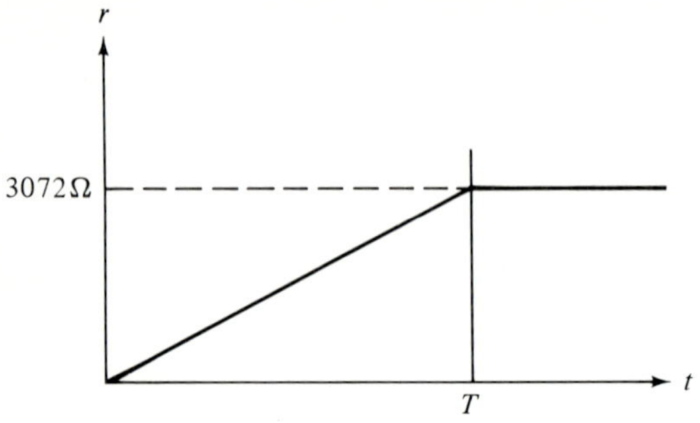

Figure 3.8.3

where $\mathbf{c}_3$ is a unit vector directed as shown in Fig. 3.8.2. The time history of r depicted in Fig. 3.8.3 thus serves to accomplish the desired objective.

3.9 REORIENTATION OF A TORQUE-FREE GYROSTAT INITIALLY AT REST

Figure 3.9.1 represents a simple gyrostat G formed by a rigid body A that carries an axisymmetric rotor B whose axis and mass center B^* are fixed in A. No forces act on A or B except those exerted by A on B and vice versa, and G has the following inertia properties: A and B have masses m_A and m_B, respectively. A has a central inertia dyadic $\mathbf{I}_A$, and B has an axial moment of inertia J and a central transverse moment of inertia K.

Subsequent to any instant at which A and B are at rest in a Newtonian reference frame N, A can be made to perform a simple rotational motion (see Sec. 1.1) in N. This is accomplished by making B rotate relative to A, say by means of a motor each part of which belongs either to A or B; and, by orienting the axis of B suitably in A, one can make the axis of rotation of A in N parallel to any desired unit vector $\boldsymbol{\nu}$. To this end, designate as $\boldsymbol{\beta}$ a unit vector parallel to the axis of B; let $\mathbf{c}$ be the position vector of B^* relative to A^*; and introduce $\mathbf{L}$, $\widetilde{\boldsymbol{\beta}}$, and $\mathbf{M}$ as

$$\mathbf{L} \triangleq \mathbf{I}_A + K\mathbf{U} + \frac{m_A m_B}{m_A + m_B}(\mathbf{c}^2\mathbf{U} - \mathbf{cc}) \tag{1}$$

$$\widetilde{\boldsymbol{\beta}} \triangleq \frac{\boldsymbol{\nu} \cdot \mathbf{L}}{|\boldsymbol{\nu} \cdot \mathbf{L}|} \tag{2}$$

and

$$\mathbf{M} \triangleq \mathbf{L} + (J - K)\widetilde{\boldsymbol{\beta}}\widetilde{\boldsymbol{\beta}} \tag{3}$$

Then take

$$\boldsymbol{\beta} = -\widetilde{\boldsymbol{\beta}}\ \mathrm{sgn}(\widetilde{\boldsymbol{\beta}} \cdot \mathbf{M}^{-1} \cdot \boldsymbol{\nu}) \tag{4}$$

By choosing a suitable value for ${}^A\theta^B$, the angle of rotation of B in A, one can cause ${}^N\theta^A$, the angle of rotation of A in N, to have any desired value. Specifically, if ${}^A\theta^B$

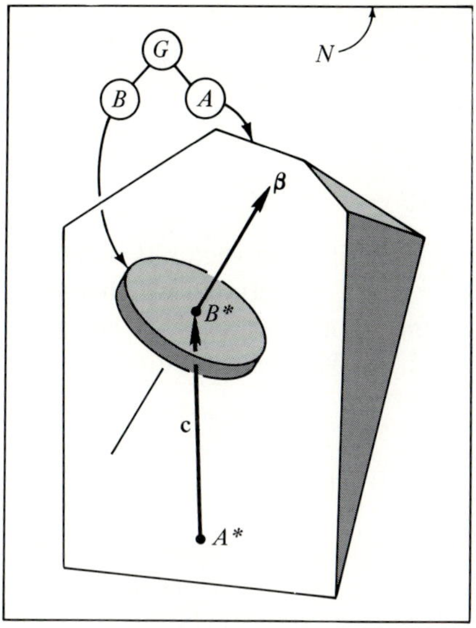

Figure 3.9.1

and ${}^N\theta^A$ are regarded as positive for rotations that are right-handed with respect to $\boldsymbol{\beta}$ and $\boldsymbol{\nu}$, respectively, and if both angles are taken equal to zero at an instant at which A and B are at rest in N, then

$$ {}^A\theta^B = \frac{{}^N\theta^A}{J|\widetilde{\boldsymbol{\beta}} \cdot \mathbf{M}^{-1}|} \tag{5} $$

Derivations As was shown in the example in Sec. 3.6, every rotation of B relative to A is accompanied by a simple rotation of A in N such that the associated axis of rotation is parallel to a unit vector $\boldsymbol{\nu}$ that satisfies the equation

$$ \boldsymbol{\nu} + \frac{\boldsymbol{\beta} \cdot \mathbf{I}_G^{-1}}{|\boldsymbol{\beta} \cdot \mathbf{I}_G^{-1}|} \underset{(3.6.53)}{=} 0 \tag{6} $$

where $\mathbf{I}_G$ is the central inertia dyadic of G. Scalar multiplication of this equation with $\mathbf{I}_G$ produces

$$ \boldsymbol{\nu} \cdot \mathbf{I}_G + \frac{\boldsymbol{\beta}}{|\boldsymbol{\beta} \cdot \mathbf{I}_G^{-1}|} = 0 \tag{7} $$

Now, $\mathbf{I}_G$ can be expressed as

$$ \mathbf{I}_G = \mathbf{I}_A + K\mathbf{U} + \frac{m_A m_B}{m_A + m_B}(\mathbf{c}^2\mathbf{U} - \mathbf{cc}) + (J - K)\boldsymbol{\beta\beta} \tag{8} $$

so that

$$ \mathbf{I}_G \underset{(1)}{=} \mathbf{L} + (J - K)\boldsymbol{\beta\beta} \tag{9} $$

Hence,

$$\boldsymbol{\nu} \cdot \mathbf{L} + (J - K)\boldsymbol{\nu} \cdot \boldsymbol{\beta}\boldsymbol{\beta} + \frac{\boldsymbol{\beta}}{|\boldsymbol{\beta} \cdot \mathbf{I}_G^{-1}|} \underset{(7,9)}{=} 0 \tag{10}$$

or, equivalently,

$$\boldsymbol{\beta} = -\boldsymbol{\nu} \cdot \mathbf{L} \frac{|\boldsymbol{\beta} \cdot \mathbf{I}_G^{-1}|}{1 + |\boldsymbol{\beta} \cdot \mathbf{I}_G^{-1}|(J - K)\boldsymbol{\nu} \cdot \boldsymbol{\beta}} \tag{11}$$

which shows that $\boldsymbol{\beta}$ is parallel to $\boldsymbol{\nu} \cdot \mathbf{L}$. Consequently, if a unit vector $\widetilde{\boldsymbol{\beta}}$ is defined as in Eq. (2), then $\boldsymbol{\beta}$ is equal either to $\widetilde{\boldsymbol{\beta}}$ or to $-\widetilde{\boldsymbol{\beta}}$. To determine which sign is appropriate, define $\mathbf{M}$ as in Eq. (3) and note that, both with $\boldsymbol{\beta} = \widetilde{\boldsymbol{\beta}}$ and with $\boldsymbol{\beta} = -\widetilde{\boldsymbol{\beta}}$, it follows from Eqs. (3) and (9) that $\mathbf{I}_G = \mathbf{M}$. Next, multiply Eq. (6) scalarly with $\boldsymbol{\nu}$, obtaining

$$1 + \frac{\boldsymbol{\beta} \cdot \mathbf{I}_G^{-1} \cdot \boldsymbol{\nu}}{|\boldsymbol{\beta} \cdot \mathbf{I}_G^{-1}|} = 0 \tag{12}$$

which shows that $\boldsymbol{\beta} \cdot \mathbf{I}_G^{-1} \cdot \boldsymbol{\nu}$ is necessarily negative. Consequently, if $\widetilde{\boldsymbol{\beta}} \cdot \mathbf{I}_G^{-1} \cdot \boldsymbol{\nu}$ is negative, then $\boldsymbol{\beta} = \widetilde{\boldsymbol{\beta}}$, whereas, if $\widetilde{\boldsymbol{\beta}} \cdot \mathbf{I}_G^{-1} \cdot \boldsymbol{\nu}$ is positive, then $\boldsymbol{\beta} = -\widetilde{\boldsymbol{\beta}}$. This is precisely what Eq. (4) asserts. In the example in Sec. 3.6, it was shown also that

$$^N\boldsymbol{\omega}^A \underset{(3.6.58)}{=} -J\,^A\omega^B\, \boldsymbol{\beta} \cdot \mathbf{I}_G^{-1} \tag{13}$$

where $^A\omega^B \triangleq {^A\boldsymbol{\omega}^B} \cdot \boldsymbol{\beta}$. Hence, if $^N\omega^A$ is defined as $^N\omega^A \triangleq {^N\boldsymbol{\omega}^A} \cdot \boldsymbol{\nu}$, then scalar multiplication of Eq. (13) with $\boldsymbol{\nu}$ yields

$$\begin{aligned} ^N\omega^A &= -J\,^A\omega^B\, \boldsymbol{\beta} \cdot \mathbf{I}_G^{-1} \cdot \boldsymbol{\nu} \\ &\underset{(6)}{=} J\,^A\omega^B \frac{(\boldsymbol{\beta} \cdot \mathbf{I}_G^{-1})^2}{|\boldsymbol{\beta} \cdot \mathbf{I}_G^{-1}|} = J\,^A\omega^B |\boldsymbol{\beta} \cdot \mathbf{I}_G^{-1}| \end{aligned} \tag{14}$$

so that (recall that $\mathbf{M} = \mathbf{I}_G$)

$$^A\omega^B \underset{(14)}{=} \frac{^N\omega^A}{J|\widetilde{\boldsymbol{\beta}} \cdot \mathbf{M}^{-1}|} \tag{15}$$

Equation (5) follows directly from this relationship and from the definitions of $^A\theta^B$ and $^N\theta^A$.

Example Figure 3.9.2 shows a simple gyrostat G formed by a uniform rectangular block A of mass 1200 kg and a thin circular disk B of mass 20 kg. A has sides of length 3 m, 4 m, 7 m, and B has a radius of 1 m. The center of B is situated at a corner of A.

A unit vector $\boldsymbol{\beta}$ parallel to the axis of B and an angle $^A\theta^B$ are to be found such that a rotation of B in A characterized by the vector $^A\theta^B\, \boldsymbol{\beta}$ will cause A to acquire the same change in orientation in a Newtonian reference frame N that results from subjecting A successively to the two rotations characterized by the vectors $(\pi/2)\mathbf{n}_1$ rad and $(\pi/2)\mathbf{n}_3$ rad, where $\mathbf{n}_1$ and $\mathbf{n}_3$ are unit vectors fixed in N and directed as shown in Fig. 3.9.2.

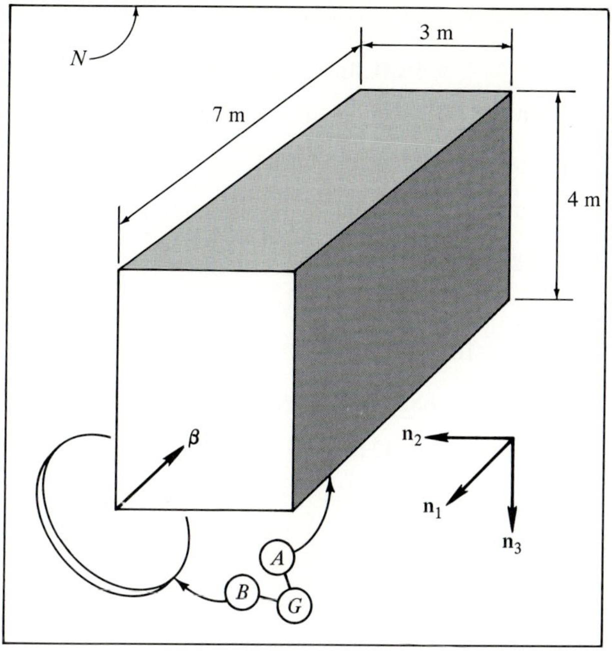

Figure 3.9.2

To determine a single rotation that is equivalent to the two successive rotations, introduce the Rodrigues vectors (see Sec. 1.4) associated with the two rotations, expressing these as (see Sec. 1.6)

$$^{N}\boldsymbol{\rho}^{\bar{A}} = \mathbf{n}_1 \qquad ^{\bar{A}}\boldsymbol{\rho}^{A} = \mathbf{n}_3 \tag{16}$$

where $\bar{A}$ is a fictitious body which moves like A during the first rotation, but remains fixed in N while A performs the second rotation. Then $^{N}\boldsymbol{\rho}^{A}$, the Rodrigues vector for a single rotation of A in N that produces the same orientation change resulting from the two successive rotations, is given by

$$^{N}\boldsymbol{\rho}^{A} \underset{(1.6.5)}{=} \frac{^{N}\boldsymbol{\rho}^{\bar{A}} + {}^{\bar{A}}\boldsymbol{\rho}^{A} + {}^{\bar{A}}\boldsymbol{\rho}^{A} \times {}^{N}\boldsymbol{\rho}^{\bar{A}}}{1 - {}^{N}\boldsymbol{\rho}^{\bar{A}} \cdot {}^{\bar{A}}\boldsymbol{\rho}^{A}} = \mathbf{n}_1 + \mathbf{n}_2 + \mathbf{n}_3 \tag{17}$$

Hence, if $\boldsymbol{\nu}$ is a unit vector and $^{N}\theta^{A}$ is an angle such that the vector $^{N}\theta^{A}\ \boldsymbol{\nu}$ characterizes the single rotation, then

$$\mathbf{n}_1 + \mathbf{n}_2 + \mathbf{n}_3 \underset{(1.4.1)}{=} \boldsymbol{\nu} \tan({}^{N}\theta^{A}/2) \tag{18}$$

and this equation is satisfied if

$$^{N}\theta^{A} = \frac{2\pi}{3} \text{ rad} \qquad \boldsymbol{\nu} = \frac{\mathbf{n}_1 + \mathbf{n}_2 + \mathbf{n}_3}{\sqrt{3}} \tag{19}$$

The central inertia dyadic of A can be written

$$\mathbf{I}_A = 2500\,\mathbf{n}_1\mathbf{n}_1 + 6500\,\mathbf{n}_2\mathbf{n}_2 + 5800\,\mathbf{n}_3\mathbf{n}_3 \tag{20}$$

and the position vector $\mathbf{c}$ from the mass center of A to that of B is

$$\mathbf{c} = 3.5\,\mathbf{n}_1 + 1.5\,\mathbf{n}_2 + 2.0\,\mathbf{n}_3 \tag{21}$$

while the axial and central transverse moments of inertia of B are

$$J = 10 \text{ kg} \cdot \text{m}^2 \qquad K = 5 \text{ kg} \cdot \text{m}^2 \tag{22}$$

Hence,

$$\begin{aligned}\mathbf{L} \underset{(1)}{=} & \ 2628\,\mathbf{n}_1\mathbf{n}_1 - 103.3\,\mathbf{n}_1\mathbf{n}_2 - 137.7\,\mathbf{n}_1\mathbf{n}_3 \\ & - 103.3\,\mathbf{n}_2\mathbf{n}_1 + 6825\,\mathbf{n}_2\mathbf{n}_2 - 59.02\,\mathbf{n}_2\mathbf{n}_3 \\ & - 137.7\,\mathbf{n}_3\mathbf{n}_1 - 59.02\,\mathbf{n}_3\mathbf{n}_2 + 6090\,\mathbf{n}_3\mathbf{n}_3\end{aligned} \tag{23}$$

and

$$\boldsymbol{\nu} \cdot \mathbf{L} \underset{(19)}{=} 1378\,\mathbf{n}_1 + 3847\,\mathbf{n}_2 + 3402\,\mathbf{n}_3 \tag{24}$$

so that

$$\widetilde{\boldsymbol{\beta}} \underset{(2)}{=} 0.2592\,\mathbf{n}_1 + 0.7235\,\mathbf{n}_2 + 0.6398\,\mathbf{n}_3 \tag{25}$$

and

$$\begin{aligned}\mathbf{M} \underset{(3)}{=} & \ 2628\,\mathbf{n}_1\mathbf{n}_1 - 102.4\,\mathbf{n}_1\mathbf{n}_2 - 136.9\,\mathbf{n}_1\mathbf{n}_3 \\ & - 102.4\,\mathbf{n}_2\mathbf{n}_1 + 6828\,\mathbf{n}_2\mathbf{n}_2 - 56.71\,\mathbf{n}_2\mathbf{n}_3 \\ & - 136.9\,\mathbf{n}_3\mathbf{n}_1 - 56.71\,\mathbf{n}_3\mathbf{n}_2 + 6092\,\mathbf{n}_3\mathbf{n}_3\end{aligned} \tag{26}$$

from which it follows that

$$\widetilde{\boldsymbol{\beta}} \cdot \mathbf{M}^{-1} = 1.085 \times 10^{-4}(\mathbf{n}_1 + \mathbf{n}_2 + \mathbf{n}_3) \tag{27}$$

Consequently, $\widetilde{\boldsymbol{\beta}} \cdot \mathbf{M}^{-1} \cdot \boldsymbol{\nu}$ is positive, and the desired unit vector is given by

$$\boldsymbol{\beta} \underset{(4)}{=} -(0.2592\,\mathbf{n}_1 + 0.7235\,\mathbf{n}_2 + 0.6398\,\mathbf{n}_3) \tag{28}$$

while

$${}^A\theta^B \underset{(5)}{=} \frac{2\pi}{3 \times 10 \times 1.085 \times 10^{-4} \times \sqrt{3}} = 1114 \text{ rad} \tag{29}$$

3.10 EFFECT OF A GRAVITATIONAL MOMENT ON AN AXISYMMETRIC GYROSTAT IN A CIRCULAR ORBIT

In Fig. 3.10.1, O designates a point fixed in a Newtonian reference frame N, G^* is the mass center of a gyrostat G formed by a rigid body A and an axisymmetric rotor B whose axis and mass center B^* are fixed in A, R is the distance between O

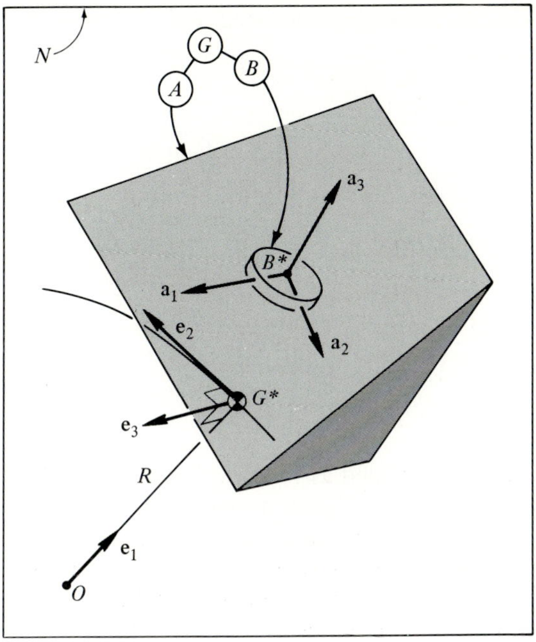

Figure 3.10.1

and G^*, and $\mathbf{e}_1$ is a unit vector. G is a simple axisymmetric gyrostat; that is, the central inertia ellipsoid of G is an ellipsoid of revolution; and the symmetry axis of G is parallel to the rotor axis. G is presumed to be subjected to the action of a force system S such that $\mathbf{F}$, the resultant of S, and $\mathbf{M}$, the total moment of S about G^*, are given by

$$\mathbf{F} = -\mu m R^{-2}\mathbf{e}_1 \tag{1}$$

and

$$\mathbf{M} = \frac{3\mu}{R^3}\mathbf{e}_1 \times \mathbf{I}_G \cdot \mathbf{e}_1 \tag{2}$$

where μ is a constant, m is the mass of G, and $\mathbf{I}_G$ is the central inertia dyadic of G. [If μ is set equal to $G\overline{m}$, $\mathbf{F}$ becomes equal to $\hat{\mathbf{F}}$ as defined in Eq. (2.3.5) while $\mathbf{M}$ becomes equal to $\widetilde{\mathbf{M}}$ as defined in Eq. (2.6.3). The force system S is thus approximately equivalent to the system of gravitational forces exerted on G by a particle of mass $\overline{m}$ at O.]

Due to the action of S, the rotational motion of A in N depends on the translational motion of G^* in N, but not vice versa, because R appears in Eq. (2), but no variable involving the orientation of A in N appears in Eq. (1). One possible motion of G^* is motion in a circular orbit that is fixed in N, has a radius R, and is described with an angular speed Ω given by

$$\Omega = (\mu R^{-3})^{1/2} \tag{3}$$

When G^* moves in this fashion and B either is completely free to rotate relative to

A or is made to rotate relative to A with a constant angular speed, then the rotational motion of A in N proceeds in such a way that

$$\dot{\theta}_1 = \frac{\omega_1 c_3 - \omega_2 s_3 + \Omega c_1 s_2}{c_2} \tag{4}$$

$$\dot{\theta}_2 = \omega_1 s_3 + \omega_2 c_3 - \Omega s_1 \tag{5}$$

$$\dot{\theta}_3 = \frac{-\omega_1 s_2 c_3 + \omega_2 s_2 s_3 + \omega_3 c_2 - \Omega c_1}{c_2} \tag{6}$$

$$\dot{\omega}_1 = \alpha\Omega\omega_2 + \beta\Omega^2 s_2 c_2 s_3 \tag{7}$$

$$\dot{\omega}_2 = -\alpha\Omega\omega_1 + \beta\Omega^2 s_2 c_2 c_3 \tag{8}$$

$$\omega_3 = \hat{\omega}_3, \text{ a constant} \tag{9}$$

where α, β, θ_i, ω_i, s_i, c_i $(i = 1, 2, 3)$ are defined as follows: Let $\mathbf{a}_1$, $\mathbf{a}_2$, $\mathbf{a}_3$ be a dextral set of orthogonal unit vectors fixed in A such that $\mathbf{a}_3$ is parallel to the symmetry axes of G and B ($\mathbf{a}_1$, $\mathbf{a}_2$, $\mathbf{a}_3$ are thus parallel to central principal axes of inertia of G); and introduce unit vectors $\mathbf{e}_2$ and $\mathbf{e}_3$ (see Fig. 3.10.1) such that $\mathbf{e}_1$, $\mathbf{e}_2$, $\mathbf{e}_3$ form a dextral, orthogonal set with $\mathbf{e}_2 = ({}^N d\mathbf{e}_1/dt)\Omega^{-1}$, so that $\mathbf{e}_3$ is normal to the orbit plane. Next, let θ_1, θ_2, θ_3 be body-three orientation angles (see Sec. 1.7) relating $\mathbf{a}_i$ to $\mathbf{e}_i$ $(i = 1, 2, 3)$; define s_i and c_i as $s_i \triangleq \sin\theta_i$, $c_i \triangleq \cos\theta_i$ $(i = 1, 2, 3)$; and take

$$\omega_i \triangleq {}^N\boldsymbol{\omega}^A \cdot \mathbf{a}_i \qquad (i = 1, 2, 3) \tag{10}$$

Finally, let

$$\alpha \triangleq \left[\left(1 - \frac{\bar{I}_3}{I_1}\right)\hat{\omega}_3 - \frac{J}{I_1}\sigma\right]\Omega^{-1} \tag{11}$$

and

$$\beta \triangleq 3\left(1 - \frac{I_3}{I_1}\right) \tag{12}$$

where I_1, I_3, $\bar{I}_3$, J, and σ have the same meaning as in Sec. 3.7.

G can move in such a way that the symmetry axis of G remains normal to the orbit plane while A rotates with a constant angular speed $\hat{\omega}_3$ in N and B rotates relative to A with a constant angular speed ${}^A\hat{\omega}^B$. This motion is unstable if x and y, given by

$$x = \frac{I_3}{I_1} - 1 \qquad y = \frac{(\hat{\omega}_3 - \Omega)I_3 + {}^A\hat{\omega}^B J}{\Omega I_3} \tag{13}$$

have values such that the point with coordinates x and y lies in one of the shaded regions of Fig. 3.2.5, these being regions in which at least one of the inequalities (3.2.19)–(3.2.21) is satisfied.

Derivations If E is a reference frame in which $\mathbf{e}_1$, $\mathbf{e}_2$, $\mathbf{e}_3$ are fixed, then

$$^N\boldsymbol{\omega}^A \underset{(1.16.1)}{=} {^N\boldsymbol{\omega}^E} + {^E\boldsymbol{\omega}^A} \tag{14}$$

with

$$^N\boldsymbol{\omega}^E = \Omega\mathbf{e}_3$$

$$\underset{(1.7.11)}{=} \Omega[(-c_1s_2c_3 + s_3s_1)\mathbf{a}_1 + (c_1s_2s_3 + c_3s_1)\mathbf{a}_2 + c_1c_2\mathbf{a}_3] \tag{15}$$

and

$$^E\boldsymbol{\omega}^A \underset{(1.17.1)}{=} (\dot{\theta}_1c_2c_3 + \dot{\theta}_2s_3)\mathbf{a}_1 - (\dot{\theta}_1c_2s_3 - \dot{\theta}_2c_3)\mathbf{a}_2 + (\dot{\theta}_1s_2 + \dot{\theta}_3)\mathbf{a}_3 \tag{16}$$

so that, from Eq. (10),

$$\omega_1 = (\dot{\theta}_1c_2 - \Omega c_1s_2)c_3 + (\dot{\theta}_2 + \Omega s_1)s_3 \tag{17}$$

$$\omega_2 = -(\dot{\theta}_1c_2 - \Omega c_1s_2)s_3 + (\dot{\theta}_2 + \Omega s_1)c_3 \tag{18}$$

$$\omega_3 = \dot{\theta}_1s_2 + \dot{\theta}_3 + \Omega c_1c_2 \tag{19}$$

Solution of these equations for $\dot{\theta}_1$, $\dot{\theta}_2$, $\dot{\theta}_3$ produces Eqs. (4)–(6); and Eqs. (7)–(9) follow directly from Eqs. (3.7.42)–(3.7.44) together with Eqs. (3.7.50), (3.7.51), (11), and (12).

To deal with the stability of motions such that the symmetry axis of G remains normal to the orbit plane and A rotates with a constant angular speed $\hat{\omega}_3$ in N, begin by observing that Eqs. (4)–(9) are satisfied if

$$\theta_1 = \theta_2 = \omega_1 = \omega_2 = 0 \tag{20}$$

while

$$\theta_3 = rt \qquad \omega_3 = \hat{\omega}_3 \tag{21}$$

with

$$r \triangleq \hat{\omega}_3 - \Omega \tag{22}$$

Next, introduce perturbations $\widetilde{\theta}_1$, $\widetilde{\theta}_2$, and $\widetilde{\theta}_3$ by letting

$$\theta_1 = \widetilde{\theta}_1 \qquad \theta_2 = \widetilde{\theta}_2 \qquad \theta_3 = rt + \widetilde{\theta}_3 \tag{23}$$

and substitute into Eqs. (17) and (18), dropping all terms that are nonlinear in the perturbations, which gives

$$\omega_1 \underset{(17)}{=} (\dot{\widetilde{\theta}}_1 - \Omega\widetilde{\theta}_2)\cos rt + (\dot{\widetilde{\theta}}_2 + \Omega\widetilde{\theta}_1)\sin rt \tag{24}$$

$$\omega_2 \underset{(18)}{=} -(\dot{\widetilde{\theta}}_1 - \Omega\widetilde{\theta}_2)\sin rt + (\dot{\widetilde{\theta}}_2 + \Omega\widetilde{\theta}_1)\cos rt \tag{25}$$

Now use these equations to eliminate ω_1 and ω_2 from Eqs. (7) and (8), obtaining

$$\{\ddot{\widetilde{\theta}}_1 + [r - (1 + \alpha)\Omega]\dot{\widetilde{\theta}}_2 - (r - \alpha\Omega)\Omega\widetilde{\theta}_1\}\cos rt$$
$$+ \{\ddot{\widetilde{\theta}}_2 - [r - (1 + \alpha)\Omega]\dot{\widetilde{\theta}}_1 + [r - (\alpha + \beta)\Omega]\Omega\widetilde{\theta}_2\}\sin rt \underset{(7)}{=} 0 \tag{26}$$

and

$$\{\ddot{\tilde{\theta}}_1 + [r - (1 + \alpha)\Omega]\dot{\tilde{\theta}}_2 - (r - \alpha\Omega)\Omega\tilde{\theta}_1\}\sin rt$$
$$+ \{\ddot{\tilde{\theta}}_2 - [r - (1 + \alpha)\Omega]\dot{\tilde{\theta}}_1 + [r - (\alpha + \beta)\Omega]\Omega\tilde{\theta}_2\}\cos rt \underset{(8)}{=} 0 \quad (27)$$

The coefficients of sin rt and cos rt in these equations must vanish separately. Hence, one may write the matrix differential equation

$$A\ddot{v} + B\dot{v} + Cv = 0 \quad (28)$$

where A is the 2×2 unit matrix, while

$$B \triangleq \begin{bmatrix} 0 & r - (1 + \alpha)\Omega \\ -r - (1 + \alpha)\Omega & 0 \end{bmatrix}$$
$$C \triangleq \begin{bmatrix} (r - \alpha\Omega)\Omega & 0 \\ 0 & [r - (\alpha + \beta)\Omega]\Omega \end{bmatrix} \quad (29)$$

and

$$v \triangleq \begin{bmatrix} \tilde{\theta}_1 \\ \tilde{\theta}_2 \end{bmatrix} \quad (30)$$

The characteristic equation associated with Eq. (28) may be written by setting the determinant of the matrix $A\lambda^2 + B\lambda + C$ equal to zero:

$$\lambda^4 + \left[1 + \left(\frac{r}{\Omega} - \alpha\right)^2 - \beta\right]\Omega^2\lambda^2 + \left(\frac{r}{\Omega} - \alpha\right)\left(\frac{r}{\Omega} - \alpha - \beta\right)\Omega^4 = 0 \quad (31)$$

Now introduce Q and x as

$$Q \triangleq \frac{r}{\Omega} - \alpha \qquad x \triangleq -\frac{\beta}{3} \quad (32)$$

Then

$$\lambda^4 + (1 + Q^2 + 3x)\Omega^2\lambda^2 + Q(Q + 3x)\Omega^4 = 0 \quad (33)$$

or

$$\lambda^4 + 2b\lambda^2 + c = 0 \quad (34)$$

where

$$b \triangleq \frac{1}{2}(1 + Q^2 + 3x)\Omega^2 \qquad c \triangleq Q(Q + 3x)\Omega^4 \quad (35)$$

Moreover, if y is defined as

$$y \triangleq \frac{Q - x}{1 + x} \quad (36)$$

then

$$Q = x + y(1 + x) \quad (37)$$

Equations (35) and (37) are identical with Eqs. (3.2.53) and (3.2.50), respectively, a fact that will prove useful presently.

Equations (20) and (21) comprise an unstable solution of Eqs. (4)–(9) whenever any root of Eq. (34) has a positive real part, that is, whenever any one of the inequalities (3.2.55) is satisfied; and, as in Sec. 3.2, these inequalities imply the inequalities (3.2.19)–(3.2.21).

It remains to verify Eqs. (13). The first of these is an immediate consequence of the definitions of β and x, given in Eqs. (12) and (32), respectively. As for the second, note that

$$y \underset{(36,32)}{=} \frac{r/\Omega - \alpha + \beta/3}{1 - \beta/3} \underset{(11,12,22)}{=} \frac{\hat{\omega}_3}{\Omega}\frac{\bar{I}_3}{I_3} + \frac{J}{I_3}\frac{\sigma}{\Omega} - 1 \tag{38}$$

Now suppose that B is completely free to rotate relative to A. Then, in accordance with Table 3.7.1, p. 219,

$$y \underset{(38)}{=} \frac{\hat{\omega}_3}{\Omega}\left(1 - \frac{J}{I_3}\right) + \frac{J}{I_3}\left(\frac{\hat{\omega}_3}{\Omega} + \frac{{}^A\hat{\omega}^B}{\Omega}\right) - 1 \tag{39}$$

in agreement with the second of Eqs. (13). Similarly, if B is made to rotate relative to A with a constant angular speed, then Eq. (38) together with values for $\bar{I}_3$ and σ taken from Table 3.7.1 leads to the second of Eqs. (13).

Example The gyrostat G considered in the example in Sec. 3.8 and shown in Fig. 3.8.2 is to be placed in a circular orbit about the Earth in such a way that the inertial orientation of A remains fixed, $\mathbf{c}_3$ is normal to the orbit plane, and B rotates relative to A with a constant angular speed ${}^A\hat{\omega}^B$. Values of ${}^A\hat{\omega}^B$ for which such motions are unstable are to be determined.

The system of gravitational forces exerted on G by the Earth can be approximated with S; and, if all other forces, such as those arising from solar radiation, the Earth's magnetic field, etc., are regarded as negligible, the instability conditions (3.2.19)–(3.2.21) apply with x and y given by Eqs. (13), where

$$I_1 = mR^2 \qquad I_3 = \frac{mR^2}{2} \qquad J = \frac{mR^2}{1536} \tag{40}$$

and $\hat{\omega}_3 = 0$ [see Eqs. (9) and (10)]. Thus

$$x \underset{(13)}{=} -0.5 \qquad y \underset{(13)}{=} -1 + \frac{1}{768}\frac{{}^A\hat{\omega}^B}{\Omega} \tag{41}$$

The points of the line $x = -0.5$ that lie in a shaded portion of Fig. 3.2.5 have coordinates such that

$$-4.367 < y < 4.000 \tag{42}$$

Hence the motion under consideration is unstable if

$$-2819\Omega < {}^A\hat{\omega}^B < 3840\Omega \tag{43}$$

3.11 EFFECT OF A GRAVITATIONAL MOMENT ON AN UNSYMMETRIC GYROSTAT IN A CIRCULAR ORBIT

In Fig. 3.11.1, O designates a point fixed in a Newtonian reference frame N, G^* is the mass center of a gyrostat G formed by a rigid body A and an axisymmetric rotor B whose axis and mass center B^* are fixed in A, R is the distance between O and G^*, and $\mathbf{e}_1$ is a unit vector. The axis of B is parallel to a central principal axis of inertia of G, and G is presumed to be subjected to the action of a force system S such that $\mathbf{F}$, the resultant of S, and $\mathbf{M}$, the total moment of S about G^*, are given by

$$\mathbf{F} = -\mu m R^{-2}\mathbf{e}_1 \tag{1}$$

and

$$\mathbf{M} = -\frac{3\mu}{R^3}\mathbf{e}_1 \times \mathbf{I}_G \cdot \mathbf{e}_1 \tag{2}$$

where μ is a constant, m is the mass of G, and $\mathbf{I}_G$ is the central inertia dyadic of G. [If μ is set equal to $G\overline{m}$, $\mathbf{F}$ becomes equal to $\hat{\mathbf{F}}$ as defined in Eq. (2.3.5) while $\mathbf{M}$ becomes equal to $\widetilde{\mathbf{M}}$ as defined in Eq. (2.6.3). The force system S is thus approximately equivalent to the system of gravitational forces exerted on G by a particle of mass $\overline{m}$ at O.]

Due to the action of S, the rotational motion of A in N depends on the translational motion of G^* in N but not vice versa, because R appears in Eq. (2), but no variable involving the orientation of A in N appears in Eq. (1). One possible motion

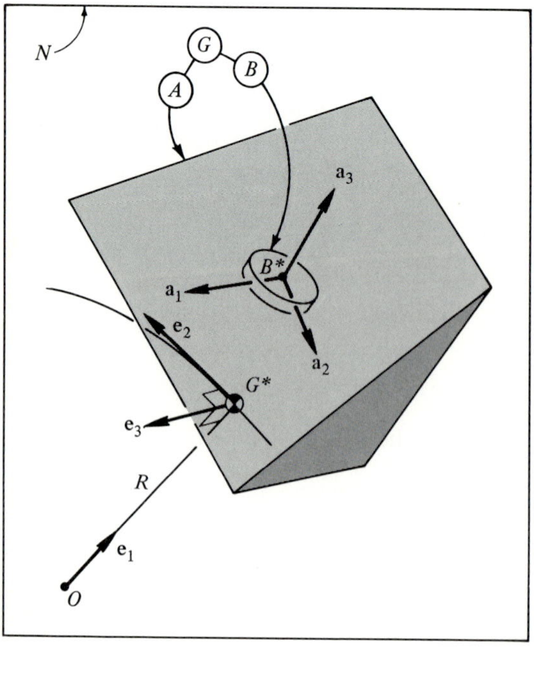

Figure 3.11.1

of G^* is motion in a circular orbit that is fixed in N, has a radius R, and is described with an angular speed Ω given by

$$\Omega = (\mu R^{-3})^{1/2} \tag{3}$$

When G^* moves in this fashion and B either is completely free to rotate relative to A or is made to rotate relative to A with a constant angular speed, then the rotational motion of A in N proceeds in such a way that

$$\dot{C}_{31} = C_{32}\omega_3 - C_{33}\omega_2 \tag{4}$$

$$\dot{C}_{32} = C_{33}\omega_1 - C_{31}\omega_3 \tag{5}$$

$$\dot{C}_{33} = C_{31}\omega_2 - C_{32}\omega_1 \tag{6}$$

$$\dot{C}_{11} = C_{12}\omega_3 - C_{13}\omega_2 + \Omega(C_{13}C_{32} - C_{12}C_{33}) \tag{7}$$

$$\dot{C}_{12} = C_{13}\omega_1 - C_{11}\omega_3 + \Omega(C_{11}C_{33} - C_{13}C_{31}) \tag{8}$$

$$\dot{C}_{13} = C_{11}\omega_2 - C_{12}\omega_1 + \Omega(C_{12}C_{31} - C_{11}C_{32}) \tag{9}$$

$$\dot{\omega}_1 = \overline{K}_1\omega_2\omega_3 - 3\Omega^2 K_1 C_{12}C_{13} - \sigma\omega_2\left(\frac{J}{I_1}\right) \tag{10}$$

$$\dot{\omega}_2 = \overline{K}_2\omega_3\omega_1 - 3\Omega^2 K_2 C_{13}C_{11} + \sigma\omega_1\left(\frac{J}{I_2}\right) \tag{11}$$

$$\dot{\omega}_3 = \overline{K}_3\omega_1\omega_2 - 3\Omega^2 \overline{K}_3 C_{11}C_{12} \tag{12}$$

where C_{ij}, ω_i, I_j, K_i, $\overline{K}_i$ $(i, j = 1, 2, 3)$, $\bar{I}_3$, J, and σ are defined as follows: Let $\mathbf{a}_i$ $(i = 1, 2, 3)$ be a dextral set of unit vectors parallel to the central principal axes of inertia of G, with $\mathbf{a}_3$ parallel to the axis of B ($\mathbf{a}_1$, $\mathbf{a}_2$, $\mathbf{a}_3$ are thus fixed in A); and introduce unit vectors $\mathbf{e}_2$ and $\mathbf{e}_3$ (see Fig. 3.11.1) such that $\mathbf{e}_1$, $\mathbf{e}_2$, $\mathbf{e}_3$ form a dextral, orthogonal set with $\mathbf{e}_2 \triangleq ({}^N d\mathbf{e}_1/dt)\Omega^{-1}$, so that $\mathbf{e}_3$ is normal to the orbit plane; let

$$C_{ij} \triangleq \mathbf{e}_i \cdot \mathbf{a}_j \qquad (i, j = 1, 2, 3) \tag{13}$$

and take

$$\omega_i = {}^N\boldsymbol{\omega}^A \cdot \mathbf{a}_i \qquad (i = 1, 2, 3) \tag{14}$$

Next, with

$$I_j \triangleq \mathbf{a}_j \cdot \mathbf{I}_G \cdot \mathbf{a}_j \qquad (j = 1, 2, 3) \tag{15}$$

let

$$K_1 \triangleq \frac{I_2 - I_3}{I_1} \qquad K_2 \triangleq \frac{I_3 - I_1}{I_2} \qquad K_3 \triangleq \frac{I_1 - I_2}{I_3} \tag{16}$$

and

$$\overline{K}_1 \triangleq \frac{I_2 - \bar{I}_3}{I_1} \qquad \overline{K}_2 \triangleq \frac{\bar{I}_3 - I_1}{I_2} \qquad \overline{K}_3 \triangleq \frac{I_1 - I_2}{\bar{I}_3} \tag{17}$$

Finally, let $\bar{I}_3$, J, and σ have the same meaning as in Sec. 3.7.

G can move in such a way that $\mathbf{a}_i = \mathbf{e}_i$ $(i = 1, 2, 3)$ while ${}^A\boldsymbol{\omega}^B \cdot \mathbf{a}_3$ has the constant value ${}^A\hat{\omega}^B$. This motion is unstable if

$$\bar{K}_3 > 0 \tag{18}$$

or if any one of the following is satisfied:

$$b < 0 \qquad c < 0 \qquad b^2 - c < 0 \tag{19}$$

where b and c are given by

$$b = \frac{1}{2}\left[1 - K_1K_2 + 3K_2 + \left(\frac{{}^A\hat{\omega}^B}{\Omega}\right)\frac{J(I_2K_2 - I_1K_1)}{I_1I_2}\left(\frac{{}^A\hat{\omega}^B}{\Omega}\right)^2\frac{J^2}{I_1I_2}\right] \tag{20}$$

and

$$c = -\left(K_1 - \frac{J\,{}^A\hat{\omega}^B}{I_1\Omega}\right)\left(4K_2 + \frac{J\,{}^A\hat{\omega}^B}{I_2\Omega}\right) \tag{21}$$

Derivations Equations (4)–(9) are identical with Eqs. (3.5.4)–(3.5.9) respectively. For the derivation of the latter, see Eqs. (3.5.28)–(3.5.33).

If $M_i \triangleq \mathbf{M} \cdot \mathbf{a}_i$ $(i = 1, 2, 3)$, where $\mathbf{M}$ is given by Eq. (2), then, making use of Eqs. (3), (16), and (2.6.8), one can write

$$M_1 = -3\Omega^2 I_1 K_1 C_{12} C_{13} \tag{22}$$

$$M_2 = -3\Omega^2 I_2 K_2 C_{13} C_{11} \tag{23}$$

$$M_3 = -3\Omega^2 I_3 K_3 C_{11} C_{12} \tag{24}$$

and substitution into Eqs. (3.7.9)–(3.7.11) leads directly to equations that can be seen to be equivalent to Eqs. (10)–(12), once Eqs. (17) have been brought into play.

When G moves in such a way that $\mathbf{a}_i = \mathbf{e}_i$ $(i = 1, 2, 3)$, then

$$C_{11} - 1 = C_{12} = C_{13} = C_{31} = C_{32} = C_{33} - 1 = 0 \tag{25}$$

and

$$\omega_1 = \omega_2 = \omega_3 - \Omega = 0 \tag{26}$$

To verify that this is a possible motion, note that Eqs. (4)–(12) are satisfied when C_{ij} and ω_j $(i = 1, 3; j = 1, 2, 3)$ have values compatible with Eqs. (25) and (26); and, to establish conditions under which the motion is unstable, introduce perturbations $\tilde{C}_{ij}$ and $\tilde{\omega}_j$ $(i = 1, 3; j = 1, 2, 3)$ as in Eqs. (3.5.71)–(3.5.73), substitute into Eqs. (4)–(12), and linearize the resulting equations, which produces

$$\dot{\tilde{C}}_{31} = \tilde{C}_{33}\Omega - \tilde{\omega}_2 \qquad \dot{\tilde{C}}_{32} = \tilde{\omega}_1 - \tilde{C}_{31}\Omega \qquad \dot{\tilde{C}}_{33} = 0 \tag{27}$$

$$\dot{\tilde{C}}_{11} = 0 \qquad \dot{\tilde{C}}_{12} = -\tilde{\omega}_3 + \Omega\tilde{C}_{33} \qquad \dot{\tilde{C}}_{13} = \tilde{\omega}_2 - \Omega\tilde{C}_{32} \tag{28}$$

$$\dot{\widetilde{\omega}}_1 = \overline{K}_1 \widetilde{\omega}_2 \Omega - \sigma \widetilde{\omega}_2 \frac{J_3}{I_1} \tag{29}$$

$$\dot{\widetilde{\omega}}_2 = \overline{K}_2 \Omega \widetilde{\omega}_1 - 3\Omega^2 K_2 \widetilde{C}_{13} + \sigma \widetilde{\omega}_1 \frac{J_3}{I_2} \tag{30}$$

$$\dot{\widetilde{\omega}}_3 = -3\Omega^2 \overline{K}_3 \widetilde{C}_{12} \tag{31}$$

Also, note that [see Eqs. (1.2.14) with $i = 1, j = 3$]

$$C_{11} C_{31} + C_{12} C_{32} + C_{13} C_{33} = 0 \tag{32}$$

so that, in view of Eqs. (3.5.71) and (3.5.72), one has after linearization,

$$\widetilde{C}_{31} + \widetilde{C}_{13} = 0 \tag{33}$$

Now solve the second of Eqs. (28) for $\widetilde{\omega}_3$, substitute into Eq. (31), and use the third of Eqs. (27) to obtain

$$\ddot{\widetilde{C}}_{12} - 3\Omega^2 \overline{K}_3 \widetilde{C}_{12} = 0 \tag{34}$$

and refer to the first two of Eqs. (27) and (28) together with Eq. (33) to write

$$\begin{bmatrix} \dot{\widetilde{C}}_{31} \\ \dot{\widetilde{C}}_{32} \\ \dot{\widetilde{\omega}}_1 \\ \dot{\widetilde{\omega}}_2 \end{bmatrix} = \begin{bmatrix} 0 & \Omega & 0 & -1 \\ -\Omega & 0 & 1 & 0 \\ 0 & 0 & 0 & \Omega K_1{}^* \\ 3\Omega^2 K_2 & 0 & \Omega K_2{}^* & 0 \end{bmatrix} \begin{bmatrix} \widetilde{C}_{31} \\ \widetilde{C}_{32} \\ \widetilde{\omega}_1 \\ \widetilde{\omega}_2 \end{bmatrix} \tag{35}$$

where $K_1{}^*$ and $K_2{}^*$ are defined as

$$K_1{}^* \triangleq \overline{K}_1 - \frac{\sigma J}{\Omega I_1} \qquad K_2{}^* \triangleq \overline{K}_2 + \frac{\sigma J}{\Omega I_2} \tag{36}$$

The characteristic equations associated with Eqs. (34) and (35) are, respectively,

$$\lambda^2 - 3\Omega^2 \overline{K}_3 = 0 \tag{37}$$

and

$$\begin{vmatrix} -\mu & \Omega & 0 & -1 \\ -\Omega & -\mu & 1 & 0 \\ 0 & 0 & -\mu & \Omega K_1{}^* \\ 3\Omega^2 K_2 & 0 & \Omega K_2{}^* & -\mu \end{vmatrix} = \mu^4 + (1 - K_1{}^* K_2{}^* + 3K_2)\Omega^2 \mu^2 - K_1{}^*(K_2{}^* + 3K_2)\Omega^4 = 0 \tag{38}$$

or

$$\mu^4 + 2b\mu^2 + c = 0 \tag{39}$$

where b and c are defined as

$$b \triangleq \frac{1}{2}(1 - K_1{}^* K_2{}^* + 3K_2)\Omega^2 \qquad c \triangleq -K_1{}^*(K_2{}^* + 3K_2)\Omega^4 \tag{40}$$

respectively. Equations (37) and (39) have roots with positive real parts if $\overline{K}_3 > 0$ of if any one of the following conditions is satisfied: $b > 0$, $c > 0$, $b^2 - c > 0$. Now, if B is free to rotate relative to A, then (see Table 3.7.1 and take $\hat{\omega}_3 = \Omega$)

$$K_1{}^* \underset{(36,17)}{=} \frac{I_2 - I_3 + J}{I_1} - \frac{(\Omega + {}^A\hat{\omega}^B)J}{\Omega I_1} \underset{(16)}{=} K_1 - \frac{{}^A\hat{\omega}^B J}{\Omega I_1} \tag{41}$$

and

$$K_2{}^* \underset{(36,17)}{=} \frac{I_3 - J - I_1}{I_2} + \frac{(\Omega + {}^A\hat{\omega}^B)J}{\Omega I_2} \underset{(16)}{=} K_2 + \frac{{}^A\hat{\omega}^B J}{\Omega I_2} \tag{42}$$

whereas, if B is made to rotate relative to A with a constant angular speed ${}^A\hat{\omega}^B$, then (see Table 3.7.1)

$$K_1{}^* \underset{(36,17)}{=} \frac{I_2 - I_3}{I_1} - \frac{{}^A\hat{\omega}^B J}{\Omega I_1} \underset{(16)}{=} K_1 - \frac{{}^A\hat{\omega}^B J}{\Omega I_1} \tag{43}$$

and

$$K_2{}^* \underset{(36,17)}{=} \frac{I_3 - I_1}{I_2} + \frac{{}^A\hat{\omega}^B J}{\Omega I_2} \underset{(16)}{=} K_2 + \frac{{}^A\hat{\omega}^B J}{\Omega I_2} \tag{44}$$

In other words, $K_1{}^*$ and $K_2{}^*$ are the same when B is free to rotate in A and when B is made to rotate relative to A with a constant angular speed. Finally, substitution from Eqs. (41) and (42) into Eqs. (40) produces Eqs. (20) and (21).

Example For a gyrostat such that $I_1 = 200 \text{ kg}\cdot\text{m}^2$, $I_2 = 1000 \text{ kg}\cdot\text{m}^2$, $I_3 = 1100 \text{ kg}\cdot\text{m}^2$, and $J = 50 \text{ kg}\cdot\text{m}^2$, the motion corresponding to Eqs. (25) and (26) is stable if B is kept at rest in A, for the gyrostat then behaves like a rigid body for which

$$K_1 \underset{(16)}{=} -0.5 \qquad K_2 \underset{(16)}{=} 0.9 \tag{45}$$

and Fig. 3.5.3 applies. However, as was pointed out in Sec. 3.5, the stability of the motion under consideration is rather tenuous; that is, relatively small disturbances can cause relatively large fluctuations in orientation. Hence it is intended to operate the gyrostat with ${}^A\hat{\omega}^B \neq 0$. In doing so, one must avoid values of ${}^A\hat{\omega}^B$ corresponding to which the motion is unstable. Such values are to be determined.

In accordance with Eqs. (20) and (21),

$$b = 2.075 + 0.1250\,\frac{{}^A\hat{\omega}^B}{\Omega} + 0.006250\left(\frac{{}^A\hat{\omega}^B}{\Omega}\right)^2 \tag{46}$$

and

$$c = 1.800 + 0.9250\,\frac{{}^A\hat{\omega}^B}{\Omega} + 0.01250\left(\frac{{}^A\hat{\omega}^B}{\Omega}\right)^2 \tag{47}$$

so that

$$b^2 - c = 2.506 - 0.4063 \frac{{}^A\hat{\omega}^B}{\Omega} + 0.02906 \left(\frac{{}^A\hat{\omega}^B}{\Omega}\right)^2 + 1.5625 \times 10^{-3} \left(\frac{{}^A\hat{\omega}^B}{\Omega}\right)^3 + 3.9063 \times 10^{-5} \left(\frac{{}^A\hat{\omega}^B}{\Omega}\right)^4 \quad (48)$$

Now,

$$(0.1250)^2 - 4(2.075)(0.006250) < 0$$

Hence, b cannot vanish; and, since b is positive for ${}^A\hat{\omega}^B/\Omega = 0$, it is positive for all ${}^A\hat{\omega}^B/\Omega$. The first of conditions (19) thus imposes no restrictions on ${}^A\hat{\omega}^B/\Omega$. On the other hand, c is negative for $-72 < {}^A\hat{\omega}^B/\Omega < -2$. Therefore, this range of values must be avoided. Finally, $b^2 - c$ is intrinsically positive so that the third of conditions (19) imposes no restrictions on ${}^A\hat{\omega}^B/\Omega$.

The instability associated with $-72 < {}^A\hat{\omega}^B/\Omega < -2$ comes to light clearly when one plots the angle ϕ between $\mathbf{e}_3$ and $\mathbf{a}_3$ as a function of the number of orbits traversed by G^* subsequent to $t = 0$, doing so for various values of ${}^A\hat{\omega}^B/\Omega$. Figure 3.11.2 shows such plots for a gyrostat whose rotor is made to

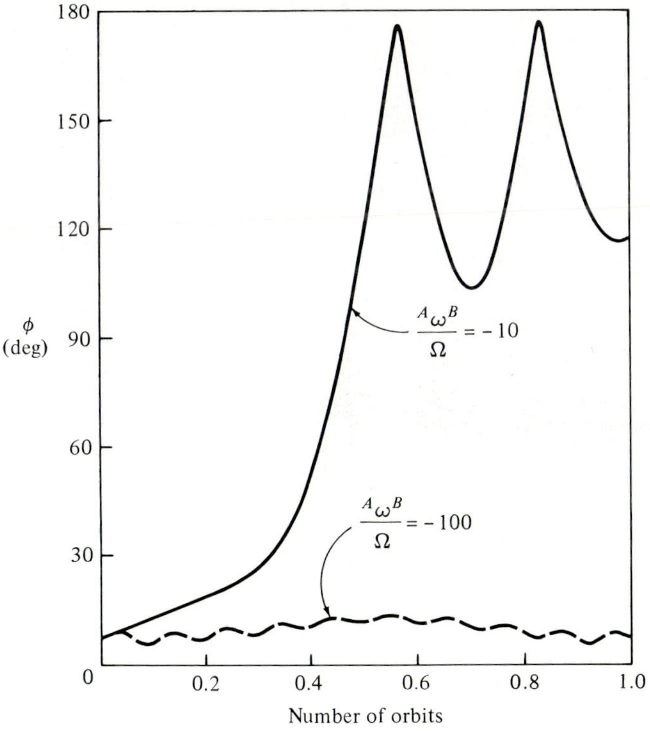

Figure 3.11.2

rotate relative to the carrier with a constant angular speed and which has inertia properties such that $I_1 = 200 \text{ kg} \cdot \text{m}^2$, $I_2 = 1000 \text{ kg} \cdot \text{m}^2$, $I_3 = 1100 \text{ kg} \cdot \text{m}^2$, and $J = 50 \text{ kg} \cdot \text{m}^2$, so that [see Eqs. (16)] $K_1 = -0.5$ and $K_2 = 0.9$. The initial conditions used to generate the plots in Fig. 3.11.2, given in Eqs. (3.5.22)–(3.5.24), represent a disturbance, at $t = 0$, of the motion corresponding to Eqs. (25) and (26); and the plots show that the disturbance has a far greater effect on ϕ when ${}^A\hat{\omega}^B/\Omega = -10$ than when ${}^A\hat{\omega}^B/\Omega = -100$.

CHAPTER

FOUR

COMPLEX SPACECRAFT

Whereas the formulation of dynamical equations of motion for *simple* spacecraft can be accomplished with essentially equal ease by means of any one of a number of methods, the task of formulating such equations for *complex* spacecraft can become prohibitively laborious unless a particularly efficacious method is employed. The first seven sections of this chapter address this important matter, and the methodology there set forth is then employed in the remaining four sections to show how to deal with spacecraft consisting either entirely or in part of elastic components. The last section, which contains an introduction to finite element analysis, indicates how this discipline can be brought to bear on problems involving large motions of deformable spacecraft.

4.1 GENERALIZED ACTIVE FORCES

Given a system S consisting of N particles $P_1, \ldots, P_N$, suppose that n generalized speeds are defined as in Eq. (1.21.1). Let $\mathbf{v}_r^{P_i}$ denote the rth partial velocity of P_i in A (see Sec. 1.21), and let $\mathbf{R}_i$ be the resultant of all contact and body forces acting on P_i. Then $F_1, \ldots, F_n$, called *generalized active forces for S* in A, are defined as

$$F_r \triangleq \sum_{i=1}^{N} \mathbf{v}_r^{P_i} \cdot \mathbf{R}_i \qquad (r = 1, \ldots, n) \tag{1}$$

Some forces that contribute to $\mathbf{R}_i$ $(i = 1, \ldots, N)$ make no contributions to F_r $(r = 1, \ldots, n)$. (Indeed, this is the principal motivation for introducing generalized forces.) For example, the total contribution to F_r of all contact forces exerted on particles of S across smooth surfaces of rigid bodies vanishes; and, if B is a rigid body belonging to S, the total contribution to F_r of all contact and gravitational forces exerted by all particles of B on each other is equal to zero.

If a set of contact and/or body forces acting on a rigid body B belonging to S is equivalent to a couple of torque $\mathbf{T}$ together with a force $\mathbf{R}$ applied at a point Q of B, then $(F_r)_B$, the contribution of this set of forces to F_r, is given by

$$(F_r)_B = \boldsymbol{\omega}_r \cdot \mathbf{T} + \mathbf{v}_r \cdot \mathbf{R} \qquad (r = 1, \ldots, n) \tag{2}$$

where $\boldsymbol{\omega}_r$ and $\mathbf{v}_r$ are, respectively, the rth partial angular velocity of B in A and the rth partial velocity of Q in A.

Derivations Let $\mathbf{C}$ be a contact force exerted on a particle P of S by a smooth rigid body B. Then, if $\mathbf{n}$ is a unit vector normal to the surface of B at P,

$$\mathbf{C} = C\mathbf{n} \tag{3}$$

where C is some scalar. Next, consider ${}^A\mathbf{v}^P$, the velocity of P in A. This can be expressed as

$${}^A\mathbf{v}^P = {}^A\mathbf{v}^{\bar{B}} + {}^B\mathbf{v}^P \tag{4}$$

where ${}^A\mathbf{v}^{\bar{B}}$ is the velocity in A of that point $\bar{B}$ of B that is in contact with P, ${}^B\mathbf{v}^P$ is the velocity of P in B, and ${}^B\mathbf{v}^P$ must be perpendicular to $\mathbf{n}$ if P is neither to lose contact with, nor to penetrate, B. Moreover,

$${}^B\mathbf{v}_r{}^P \cdot \mathbf{n} = 0 \tag{5}$$

because otherwise there can exist values of $u_1, \ldots, u_n$ such that ${}^B\mathbf{v}^P$ is not perpendicular to $\mathbf{n}$. Now suppose that B is a part of S. Then

$${}^A\mathbf{v}_r{}^P \underset{(4)}{=} {}^A\mathbf{v}_r{}^{\bar{B}} + {}^B\mathbf{v}_r{}^P \tag{6}$$

Consequently,

$$({}^A\mathbf{v}_r{}^P - {}^A\mathbf{v}_r{}^{\bar{B}}) \cdot \mathbf{n} \underset{(6)}{=} {}^B\mathbf{v}_r{}^P \cdot \mathbf{n} \underset{(5)}{=} 0 \tag{7}$$

and the contribution to F_r of the forces exerted on each other by P and B is [see Eq. (1)]

$${}^A\mathbf{v}_r{}^P \cdot (C\mathbf{n}) + {}^A\mathbf{v}_r{}^{\bar{B}} \cdot (-C\mathbf{n}) = C({}^A\mathbf{v}_r{}^P - {}^A\mathbf{v}_r{}^{\bar{B}}) \cdot \mathbf{n} \underset{(7)}{=} 0 \tag{8}$$

Alternatively, suppose that B is not a part of S. Then $u_1, \ldots, u_n$ always can be chosen in such a way that ${}^A\mathbf{v}^{\bar{B}}$ is independent of $u_1, \ldots, u_n$, which means that

$${}^A\mathbf{v}_r{}^P \underset{(4)}{=} {}^B\mathbf{v}_r{}^P \qquad (r = 1, \ldots, n) \tag{9}$$

and that the contribution to F_r of the contact force exerted by B on P is [see Eq. (1)]

$${}^A\mathbf{v}_r{}^P \cdot (C\mathbf{n}) \underset{(9)}{=} C\,{}^B\mathbf{v}_r{}^P \cdot \mathbf{n} \underset{(7)}{=} 0 \tag{10}$$

In both cases, therefore, the contact forces exerted across a smooth surface of a rigid body contribute nothing to F_r ($r = 1, \ldots, n$).

In Fig. 4.1.1, P_i and P_j designate particles of a rigid body B belonging to S, $\mathbf{R}_{ij}$ is the resultant of all contact and gravitational forces exerted on P_i by P_j, and $\mathbf{R}_{ji}$ is the resultant of all contact and gravitational forces exerted on P_j by P_i. To show that the total contribution of $\mathbf{R}_{ij}$ and $\mathbf{R}_{ji}$ to F_r ($r = 1, \ldots, n$) is equal to zero, it is helpful to note (see Prob. 1.31) that the partial velocities $\mathbf{v}_r^{P_i}$ and $\mathbf{v}_r^{P_j}$ of P_i and P_j in A are related to the partial angular velocity $\boldsymbol{\omega}_r$ of B in A by

$$\mathbf{v}_r^{P_j} = \mathbf{v}_r^{P_i} + \boldsymbol{\omega}_r \times \mathbf{p}_{ij} \tag{11}$$

where $\mathbf{p}_{ij}$ is the position vector from P_i to P_j.

The Law of Action and Reaction asserts that $\mathbf{R}_{ij}$ and $\mathbf{R}_{ji}$ have equal magnitudes and opposite directions, and that the lines of action of $\mathbf{R}_{ij}$ and $\mathbf{R}_{ji}$ coincide. Furthermore, the line of action of $\mathbf{R}_{ij}$ must pass through P_i, and that of $\mathbf{R}_{ji}$ through P_j. Consequently,

$$\mathbf{R}_{ji} = -\mathbf{R}_{ij} \tag{12}$$

and $\mathbf{R}_{ij}$ is parallel to $\mathbf{p}_{ij}$, so that

$$\mathbf{p}_{ij} \times \mathbf{R}_{ij} = 0 \tag{13}$$

The total contribution of $\mathbf{R}_{ij}$ and $\mathbf{R}_{ji}$ to F_r ($r = 1, \ldots, n$) is thus [see Eq. (1)]

$$\mathbf{v}_r^{P_i} \cdot \mathbf{R}_{ij} + \mathbf{v}_r^{P_j} \cdot \mathbf{R}_{ji} \underset{(12)}{=} (\mathbf{v}_r^{P_i} - \mathbf{v}_r^{P_j}) \cdot \mathbf{R}_{ij}$$

$$\underset{(11)}{=} -\boldsymbol{\omega}_r \cdot \mathbf{p}_{ij} \times \mathbf{R}_{ij} \underset{(13)}{=} 0 \tag{14}$$

To establish the validity of Eq. (2), we introduce forces $\mathbf{K}_1, \ldots, \mathbf{K}_{N'}$ acting on particles $P_1, \ldots, P_{N'}$ of a rigid body B (see Fig. 4.1.2), and let $\mathbf{p}_1, \ldots, \mathbf{p}_{N'}$ be the position vectors from a point Q fixed in B to $P_1, \ldots, P_{N'}$, respectively.

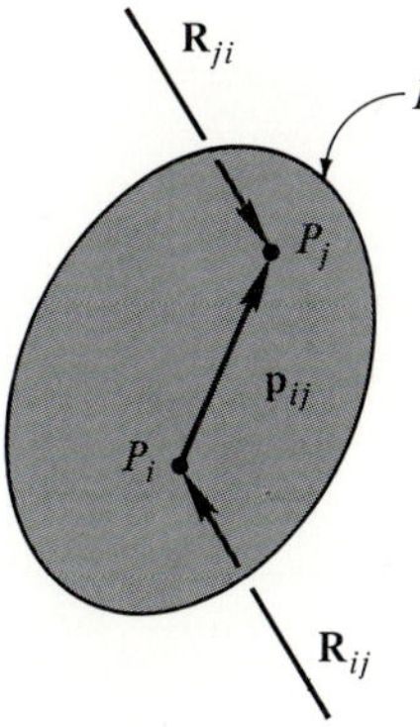

Figure 4.1.1

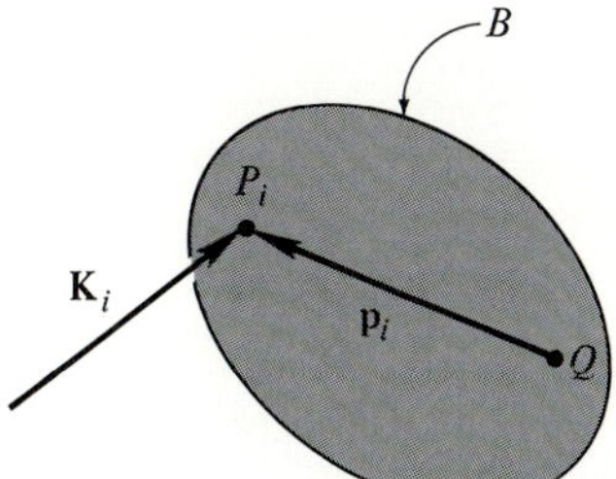

Figure 4.1.2

Then, by definition of equivalence, the set of forces $\mathbf{K}_1, \ldots, \mathbf{K}_{N'}$ is equivalent to a couple of torque $\mathbf{T}$ and a force $\mathbf{R}$ applied at Q if and only if

$$\mathbf{T} = \sum_{i=1}^{N'} \mathbf{p}_i \times \mathbf{K}_i \tag{15}$$

and

$$\mathbf{R} = \sum_{i=1}^{N'} \mathbf{K}_i \tag{16}$$

where $\mathbf{p}_i$ is the position vector from Q to P_i. Also by definition, the contribution of $\mathbf{K}_1, \ldots, \mathbf{K}_{N'}$ to the generalized active force F_r is

$$(F_r)_B = \sum_{i=1}^{N'} \mathbf{v}_r{}^{P_i} \cdot \mathbf{K}_i \qquad (r = 1, \ldots, n) \tag{17}$$

Now, referring once again to Prob. 1.31, one can write

$$\mathbf{v}_r{}^{P_i} = \mathbf{v}_r{}^{Q} + \boldsymbol{\omega}_r \times \mathbf{p}_i \tag{18}$$

Hence,

$$(F_r)_B \underset{(17,18)}{=} \sum_{i=1}^{N'} (\mathbf{v}_r{}^{Q} + \boldsymbol{\omega}_r \times \mathbf{p}_i) \cdot \mathbf{K}_i = \boldsymbol{\omega}_r \cdot \sum_{i=1}^{N'} \mathbf{p}_i \times \mathbf{K}_i + \mathbf{v}_r{}^{Q} \cdot \sum_{i=1}^{N'} \mathbf{K}_i \tag{19}$$

and, using Eqs. (15) and (16), one arrives at Eq. (2).

Example Figure 4.1.3 shows a uniform rod B of mass m and length L. B is free to move in a plane fixed in a reference frame N. $\bar{P}$ is a particle of mass $\bar{m}$, fixed in N, and q_1, q_2, and q_3 are generalized coordinates characterizing the configuration of B in N.

Suppose that the resultant force exerted by $\bar{P}$ on B is approximated with $\tilde{\mathbf{F}}$ as defined in Eq. (2.3.6) and that the moment about B^* of all forces exerted

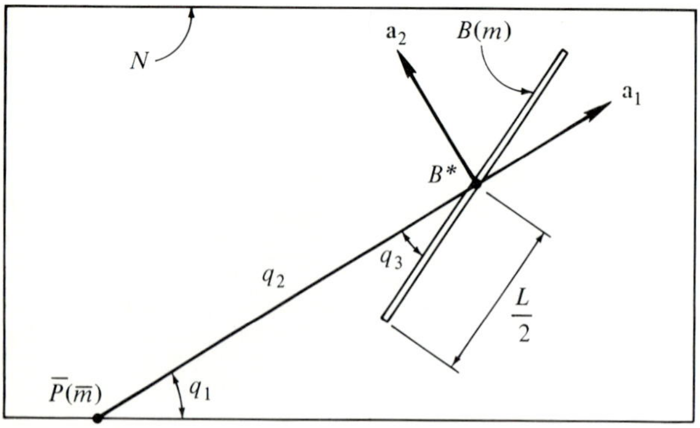

Figure 4.1.3

by $\overline{P}$ on B is approximated with $\widetilde{\mathbf{M}}$ as defined in Eq. (2.6.3). Then the set of forces exerted on B by $\overline{P}$ is equivalent to a couple of torque $\mathbf{T}$ together with a force $\mathbf{R}$ applied at B^*, with

$$\mathbf{R} \underset{(2.3.32)}{=} -\frac{G\overline{m}m}{{q_2}^2}\left\{\mathbf{a}_1\left[1+\frac{L^2}{8{q_2}^2}(2-3\sin^2 2q_3)\right]-\mathbf{a}_2\frac{L^2}{8{q_2}^2}\sin 2q_3\right\} \tag{20}$$

and

$$\mathbf{T} \underset{(2.6.3)}{=} \frac{3G\overline{m}}{{q_2}^3}\mathbf{a}_1 \times \mathbf{I}\cdot\mathbf{a}_1 \tag{21}$$

where $\mathbf{a}_1$ and $\mathbf{a}_2$ are unit vectors directed as shown in Fig. 4.1.3, and $\mathbf{I}$ is the central inertia dyadic of B. Hence, $\mathbf{T}$ can be expressed also as

$$\mathbf{T} = -\frac{G\overline{m}mL^2}{8{q_2}^3}\sin 2q_3\,\mathbf{a}_1 \times \mathbf{a}_2 \tag{22}$$

If u_1, u_2, u_3 are defined as

$$u_r \triangleq \dot{q}_r \qquad (r = 1, 2, 3) \tag{23}$$

then $\boldsymbol{\omega}$, the angular velocity of B in N, and $\mathbf{v}$, the velocity of B^* in N, are given by

$$\boldsymbol{\omega} = (u_1 + u_3)\mathbf{a}_1 \times \mathbf{a}_2 \qquad \mathbf{v} = u_2\mathbf{a}_1 + u_1 q_2\mathbf{a}_2 \tag{24}$$

so that the partial angular velocities of B in N and the partial velocities of B^* in N (see Sec. 1.21) are

$$\boldsymbol{\omega}_1 = \mathbf{a}_1 \times \mathbf{a}_2 \qquad \boldsymbol{\omega}_2 = 0 \qquad \boldsymbol{\omega}_3 = \mathbf{a}_1 \times \mathbf{a}_2 \tag{25}$$

and

$$\mathbf{v}_1 = q_2\mathbf{a}_2 \qquad \mathbf{v}_2 = \mathbf{a}_1 \qquad \mathbf{v}_3 = 0 \tag{26}$$

The contributions of forces exerted by $\overline{P}$ on B to the generalized active forces corresponding to u_1, u_2, u_3 are thus, respectively,

$$\begin{aligned}(F_1)_B &\underset{(2)}{=} \underset{(25)}{\mathbf{a}_1 \times \mathbf{a}_2} \cdot \underset{(22)}{\left(-\frac{G\overline{m}mL^2}{8{q_2}^3}\sin 2q_3\,\mathbf{a}_1 \times \mathbf{a}_2\right)} \\ &\quad + \underset{(26)}{q_2\mathbf{a}_2} \cdot \underset{(20)}{\left(-\frac{G\overline{m}m}{{q_2}^2}\left\{\mathbf{a}_1\left[1+\frac{L^2}{8{q_2}^2}(2-3\sin^2 2q_3)\right]\right.\right.} \\ &\quad \left.\left. - \mathbf{a}_2\frac{L^2}{8{q_2}^2}\sin 2q_3\right\}\right) = 0\end{aligned} \tag{27}$$

$$(F_2)_B \underset{(2)}{=} -\frac{G\overline{m}m}{{q_2}^2}\underset{(20,26)}{\left[1+\frac{L^2}{8{q_2}^2}(2-3\sin^2 2q_3)\right]} \tag{28}$$

$$(F_3)_B \underset{(2)}{=} -\frac{G\bar{m}mL^2}{\underset{(22,25)}{8q_2^{\,3}}} \sin 2q_3 \tag{29}$$

4.2 POTENTIAL ENERGY

If the configuration of a system S in a reference frame A is characterized by n generalized coordinates $q_1, \ldots, q_n$, there may exist a function W of $q_1, \ldots, q_n$, and t such that *all* n generalized active forces $F_1, \ldots, F_n$ for S in A (see Sec. 4.1) can be expressed as

$$F_r = -\frac{\partial W}{\partial q_r} \qquad (r = 1, \ldots, n) \tag{1}$$

In that event, W is called a *potential energy* of S in A. [One speaks of "a" rather than "the" potential energy of S in A because, if W satisfies Eq. (1), then so does $W + f$, where f is any function of t.]

In some situations, force functions discussed in Secs. 2.10–2.19 furnish potential energies. For example, when S consists of a single particle P of mass m moving in A under the action of the gravitational force exerted on P by a particle $\bar{P}$ of mass $\bar{m}$ fixed in A, then V as given in Eq. (2.10.2) can be regarded as the negative of a potential energy of P in A; and V also provides the negative of a potential energy if S consists of two particles moving in A under the action of their mutual gravitational attraction. In other situations, potential energies cannot be found so easily, but some of the force functions of Secs. 2.10–2.14 are, nevertheless, useful, for they produce contributions to potential energies. For example, if S consists of a rigid body B of mass m and two particles P_1 and P_2 of masses m_1 and m_2, respectively, and if these objects move in a reference frame A under the action of mutual gravitational attractions, then Eqs. (2.10.2) and (2.11.5) lead to a potential energy of S in the form

$$W = -G\left(m_1 m_2 p_0^{-1} + m_1 \int p_1^{-1} \rho\, d\tau + m_2 \int p_2^{-1} \rho\, d\tau\right) \tag{2}$$

where p_0 is the distance between P_1 and P_2, p_i is the distance from P_i $(i = 1, 2)$ to a generic point P of B, and ρ is the mass density of B at P.

Although force functions associated with approximations to gravitational forces and moments can at times be used in the formulation of a potential energy, it can occur that apparently relevant force functions exist when potential energies do not exist. For example, let S consist of a rigid body B of mass m moving under the action of gravitational forces exerted by a particle $\bar{P}$ of mass $\bar{m}$ fixed in a reference frame A, and suppose that the resultant gravitational force exerted on B by $\bar{P}$ is approximated with $\tilde{\mathbf{F}}$ as defined in Eq. (2.3.6), while the moment about the mass center B^* of B of all forces exerted on B by $\bar{P}$ is approximated with $\tilde{\mathbf{M}}$ as defined in Eq. (2.6.3). Then a potential energy of B in A can be expressed as

$$W = -\tilde{V} \tag{3}$$

with $\widetilde{V}$ given by Eq. (2.12.13); but, if $\mathbf{F}$ is approximated with $\hat{\mathbf{F}}$ as defined in Eq. (2.3.5), while $\mathbf{M}$ is approximated with $\widetilde{\mathbf{M}}$, then there exists no potential energy, despite the fact that force functions for $\mathbf{F}$ and $\mathbf{M}$ do exist in the sense that $\mathbf{F}$ and $\widetilde{\mathbf{M}}$ can be expressed as $\hat{\mathbf{F}} = \nabla_{\mathbf{R}} V_1$ and $\widetilde{\mathbf{M}} = -\mathbf{R} \times \nabla_{\mathbf{R}} V_2$, where $\mathbf{R}$ is the position vector of B^* relative to $\bar{P}$ while $V_1 = G\bar{m}mR^{-1}$ and $V_2 = G\bar{m}mR^{-1}v^{(2)}$ with $v^{(2)}$ given in Eq. (2.12.2). The nonexistence of a potential energy under these circumstances may be inconvenient, but it does not invalidate approximating $\mathbf{F}$ with $\hat{\mathbf{F}}$ and $\mathbf{M}$ with $\widetilde{\mathbf{M}}$. On the contrary, precisely these approximations are particularly useful and are employed extensively in the analysis of motions of spacecraft.

When an expression for a potential energy W of S in A is readily available, generalized active forces for S in A usually can be found more expediently by differentiating W [see Eq. (1)] than by any other means. By the same token, when knowledge of generalized active forces is required for the formulation of W, the use of W cannot facilitate the determination of generalized active forces. However, even under these circumstances, a potential energy may be of interest, for instance, in connection with the construction of integrals of equations of motion. Finally, in the event that there exists no potential energy function, generalized active forces always can be evaluated by means of Eqs. (4.1.1) and (4.1.2).

Derivations The resultant of all contact and body forces acting on a single particle P of mass m moving under the action of the gravitational force exerted by a particle $\bar{P}$ of mass $\bar{m}$ is given by

$$\mathbf{F} \underset{(2.1.1)}{=} -G\bar{m}m\mathbf{p}(\mathbf{p}^2)^{-3/2} \tag{4}$$

From Prob. 1.29, the rth partial velocity of P in A can be expressed as

$$\mathbf{v}_r = \frac{\partial \mathbf{p}}{\partial q_r} \qquad (r = 1, 2, 3) \tag{5}$$

Consequently,

$$\begin{aligned} F_r &\underset{(4.1.1)}{=} \mathbf{v}_r \cdot \mathbf{F} \underset{(4,5)}{=} -G\bar{m}m \frac{\partial \mathbf{p}}{\partial q_r} \cdot \mathbf{p}(\mathbf{p}^2)^{-3/2} \\ &= \frac{\partial}{\partial q_r}\left[G\bar{m}m(\mathbf{p}^2)^{-1/2}\right] = \frac{\partial}{\partial q_r}(G\bar{m}mp^{-1}) \\ &\underset{(2.10.2)}{=} \frac{\partial V}{\partial q_r} \qquad (r = 1, 2, 3) \end{aligned} \tag{6}$$

Hence, if W is defined as

$$W \triangleq -V \tag{7}$$

then

$$F_r \underset{(6,7)}{=} -\frac{\partial W}{\partial q_r} \qquad (r = 1, 2, 3) \tag{8}$$

and W is a potential energy of P in A. Similarly, if S consists of both P and $\overline{P}$, the resultants $\mathbf{F}$ and $\overline{\mathbf{F}}$ of the forces acting on P and on $\overline{P}$ are given by

$$\mathbf{F} \underset{(2.1.1)}{=} -G\overline{m}m(\mathbf{p}_1 - \mathbf{p}_2)[(\mathbf{p}_1 - \mathbf{p}_2)^2]^{-3/2} \tag{9}$$

and by

$$\overline{\mathbf{F}} \underset{(2.1.2)}{=} -\mathbf{F} \tag{10}$$

respectively, where $\mathbf{p}_1$ and $\mathbf{p}_2$ are, respectively, the position vectors of P and $\overline{P}$ relative to a point fixed in A; and

$$\begin{aligned}
F_r &\underset{(4.1.1)}{=} \mathbf{v}_r^{P_1} \cdot \mathbf{F} + \mathbf{v}_r^{P_2} \cdot \overline{\mathbf{F}} \underset{(10)}{=} \left(\frac{\partial \mathbf{p}_1}{\partial q_r} - \frac{\partial \mathbf{p}_2}{\partial q_r}\right) \cdot \mathbf{F} \\
&= \left[\frac{\partial}{\partial q_r}(\mathbf{p}_1 - \mathbf{p}_2)\right] \cdot \mathbf{F} \\
&\underset{(9)}{=} -G\overline{m}m\left[\frac{\partial}{\partial q_r}(\mathbf{p}_1 - \mathbf{p}_2)\right] \cdot (\mathbf{p}_1 - \mathbf{p}_2)\left[(\mathbf{p}_1 - \mathbf{p}_2)^2\right]^{-3/2} \\
&= -G\overline{m}m\frac{\partial \mathbf{p}}{\partial q_r} \cdot \mathbf{p}(\mathbf{p}^2)^{-3/2} = \frac{\partial}{\partial q_r}(G\overline{m}mp^{-1}) \\
&\underset{(2.10.2)}{=} \frac{\partial V}{\partial q_r} \qquad (r = 1, \ldots, 6)
\end{aligned} \tag{11}$$

so that W as given in Eq. (7) is once again a potential energy of S in A.

When S consists of a rigid body of mass m and two particles, P_1 and P_2, of masses m_1 and m_2, the resultants $\mathbf{F}_1$ and $\mathbf{F}_2$ of all contact and body forces acting on P_1 and P_2, respectively, are [see Eqs. (2.1.1), (2.1.2), and (2.2.2)]

$$\mathbf{F}_1 = -Gm_1m_2\mathbf{p}_0(\mathbf{p}_0^2)^{-3/2} + Gm_1\int \mathbf{p}_1(\mathbf{p}_1^2)^{-3/2}\rho\, d\tau \tag{12}$$

and

$$\mathbf{F}_2 = Gm_1m_2\mathbf{p}_0(\mathbf{p}_0^2)^{-3/2} + Gm_2\int \mathbf{p}_2(\mathbf{p}_2^2)^{-3/2}\rho\, d\tau \tag{13}$$

where $\mathbf{p}_0$, $\mathbf{p}_1$, and $\mathbf{p}_2$ are position vectors directed as shown in Fig. 4.2.1. The contribution to F_r of forces acting on P_1 and P_2 is thus [refer to Eq. (4.1.1), see Fig. 4.2.1 for the position vectors $\mathbf{R}_1$ and $\mathbf{R}_2$, and note that $\mathbf{v}_r^{P_1} = \partial\mathbf{R}_1/\partial q_r$, $\mathbf{v}_r^{P_2} = \partial\mathbf{R}_2/\partial q_r$, and $\mathbf{R}_1 - \mathbf{R}_2 = \mathbf{p}_0$]

$$\begin{aligned}
\mathbf{v}_r^{P_1} \cdot \mathbf{F}_1 + \mathbf{v}_r^{P_2} \cdot \mathbf{F}_2 &\underset{(12,13)}{=} -Gm_1m_2\frac{\partial \mathbf{p}_0}{\partial q_r} \cdot \mathbf{p}_0(\mathbf{p}_0^2)^{-3/2} \\
&+ Gm_1\int \frac{\partial \mathbf{R}_1}{\partial q_r} \cdot \mathbf{p}_1(\mathbf{p}_1^2)^{-3/2}\rho\, d\tau + Gm_2\int \frac{\partial \mathbf{R}_2}{\partial q_r} \cdot \mathbf{p}_2(\mathbf{p}_2^2)^{-3/2}\rho\, d\tau \\
&\qquad (r = 1, \ldots, 12)
\end{aligned} \tag{14}$$

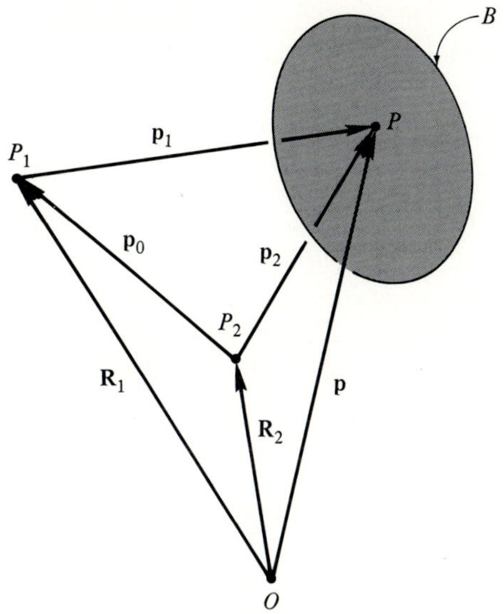

Figure 4.2.1

As for forces acting on the particles of B, these either are forces exerted by P_1 and P_2, or they are forces exerted by particles of B on each other. Since B is a *rigid* body, the latter forces make no contribution to F_r (see Sec. 4.1); and a typical force of the former kind can be expressed as

$$d\mathbf{F} \underset{(2.1.1)}{=} -G[m_1\mathbf{p}_1(\mathbf{p}_1{}^2)^{-3/2} + m_2\mathbf{p}_2(\mathbf{p}_2{}^2)^{-3/2}]\rho\, d\tau \tag{15}$$

The contribution from this force to F_r is thus given by (see Figure 4.2.1 for $\mathbf{p}$)

$$\begin{aligned}\mathbf{v}_r{}^P \cdot d\mathbf{F} &\underset{(15)}{=} -G\left[m_1\frac{\partial \mathbf{p}}{\partial q_r}\cdot \mathbf{p}_1(\mathbf{p}_1{}^2)^{-3/2} + m_2\frac{\partial \mathbf{p}}{\partial q_r}\cdot \mathbf{p}_2(\mathbf{p}_2{}^2)^{-3/2}\right]\rho\, d\tau \\ &= -G\left[m_1\frac{\partial \mathbf{R}_1}{\partial q_r}\cdot \mathbf{p}_1(\mathbf{p}_1{}^2)^{-3/2} + m_1\frac{\partial \mathbf{p}_1}{\partial q_r}\cdot \mathbf{p}_1(\mathbf{p}_1{}^2)^{-3/2}\right]\rho\, d\tau \\ &\quad -G\left[m_2\frac{\partial \mathbf{R}_2}{\partial q_r}\cdot \mathbf{p}_2(\mathbf{p}_2{}^2)^{-3/2} + m_2\frac{\partial \mathbf{p}_2}{\partial q_r}\cdot \mathbf{p}_2(\mathbf{p}_2{}^2)^{-3/2}\right]\rho\, d\tau\end{aligned}$$

$$(r = 1, \ldots, 12) \tag{16}$$

and F_r can now be formulated as

$$\begin{aligned}F_r &\underset{(4.1.1)}{=} \mathbf{v}_r{}^{P_1}\cdot \mathbf{F}_1 + \mathbf{v}_r{}^{P_2}\cdot \mathbf{F}_2 + \int \mathbf{v}_r{}^P \cdot d\mathbf{F} \\ &\underset{(14,16)}{=} -Gm_1m_2\frac{\partial \mathbf{p}_0}{\partial q_r}\cdot \mathbf{p}_0(\mathbf{p}_0{}^2)^{-3/2} - Gm_1\int \frac{\partial \mathbf{p}_1}{\partial q_r}\cdot \mathbf{p}_1(\mathbf{p}_1{}^2)^{-3/2}\rho\, d\tau \\ &\quad - Gm_2\int \frac{\partial \mathbf{p}_2}{\partial q_r}\cdot \mathbf{p}_2(\mathbf{p}_2{}^2)^{-3/2}\rho\, d\tau\end{aligned}$$

(*continued on next page*)

$$= \frac{\partial}{\partial q_r}\left[Gm_1m_2(\mathbf{p}_0^{\,2})^{-1/2} + Gm_1\int(\mathbf{p}_1^{\,2})^{-1/2}\rho\,d\tau + Gm_2\int(\mathbf{p}_2^{\,2})^{-1/2}\rho\,d\tau\right]$$

$$(r = 1, \ldots, 12) \quad (17)$$

or, with

$$p_j \triangleq (\mathbf{p}_j^{\,2})^{1/2} \qquad (j = 0, 1, 2) \tag{18}$$

as

$$F_r \underset{(17,18)}{=} \frac{\partial}{\partial q_r}\left(Gm_1m_2p_0^{-1} + Gm_1\int p_1^{-1}\rho\,d\tau + Gm_2\int p_2^{-1}\rho\,d\tau\right)$$

$$\underset{(2)}{=} -\frac{\partial W}{\partial q_r} \qquad (r = 1, \ldots, 12) \tag{19}$$

In accordance with Eq. (1), W as given by Eq. (2) is thus a potential energy of S.

Now consider a rigid body B of mass m moving under the action of gravitational forces exerted by a particle $\bar{P}$ of mass $\bar{m}$ fixed in a reference frame A, and let the resultant gravitational force exerted on B by $\bar{P}$ be approximated with $\tilde{\mathbf{F}}$ as defined in Eq. (2.3.6) while the moment about the mass center B^* of B of all forces exerted on B by $\bar{P}$ is approximated with $\tilde{\mathbf{M}}$ as defined in Eq. (2.6.3). Then

$$F_r \underset{(4.1.2)}{=} \boldsymbol{\omega}_r \cdot \tilde{\mathbf{M}} + \mathbf{v}_r \cdot \tilde{\mathbf{F}} \qquad (r = 1, 2, 3) \tag{20}$$

where $\mathbf{v}_r$, the rth partial velocity of B^* in A, can be expressed as (see Prob. 1.29)

$$\mathbf{v}_r = \frac{\partial}{\partial q_r}(R\mathbf{a}_1) \qquad (r = 1, 2, 3) \tag{21}$$

with R and $\mathbf{a}_1$ as defined in Sec. 2.13. Using Eqs. (2.3.6) and (2.6.13), one thus obtains

$$F_r = \frac{3G\bar{m}}{R^3}(\mathbf{a}_1 \times \mathbf{I} \cdot \mathbf{a}_1) \cdot \boldsymbol{\omega}_r$$

$$- \frac{G\bar{m}m}{R^2}[\mathbf{a}_1 + \mathbf{f}^{(2)}] \cdot \frac{\partial}{\partial q_r}(R\mathbf{a}_1) \qquad (r = 1, 2, 3) \tag{22}$$

or, equivalently,

$$F_r = \frac{3G\bar{m}}{R^3}(\boldsymbol{\omega}_r \times \mathbf{a}_1) \cdot (\mathbf{I} \cdot \mathbf{a}_1)$$

$$- \frac{G\bar{m}m}{R^2}\left[\mathbf{a}_1 \cdot \frac{\partial}{\partial q_r}(R\mathbf{a}_1) + \mathbf{f}^{(2)} \cdot \frac{\partial}{\partial q_r}(R\mathbf{a}_1)\right] \qquad (r = 1, 2, 3) \tag{23}$$

Now,

$$\frac{\partial}{\partial q_r}(R\mathbf{a}_1) = \frac{\partial R}{\partial q_r}\mathbf{a}_1 + R\frac{\partial \mathbf{a}_1}{\partial q_r} \qquad (r = 1, 2, 3) \tag{24}$$

and, using a superscript B to denote differentiation in B, one can write (see Prob. 1.29)

$$\frac{\partial \mathbf{a}_1}{\partial q_r} = \frac{{}^B\partial \mathbf{a}_1}{\partial q_r} + \boldsymbol{\omega}_r \times \mathbf{a}_1 \qquad (r = 1, 2, 3) \tag{25}$$

so that

$$(\boldsymbol{\omega}_r \times \mathbf{a}_1) \cdot (\mathbf{I} \cdot \mathbf{a}_1) \underset{(25)}{=} \frac{\partial \mathbf{a}_1}{\partial q_r} \cdot \mathbf{I} \cdot \mathbf{a}_1 - \frac{{}^B\partial \mathbf{a}_1}{\partial q_r} \cdot \mathbf{I} \cdot \mathbf{a}_1 \qquad (r = 1, 2, 3) \tag{26}$$

or, since $\mathbf{I}$ is independent of q_1, q_2, q_3 in B,

$$\begin{aligned}(\boldsymbol{\omega}_r \times \mathbf{a}_1) \cdot (\mathbf{I} \cdot \mathbf{a}_1) &= \frac{\partial \mathbf{a}_1}{\partial q_r} \cdot \mathbf{I} \cdot \mathbf{a}_1 - \frac{1}{2} \frac{{}^B\partial}{\partial q_r} (\mathbf{a}_1 \cdot \mathbf{I} \cdot \mathbf{a}_1) \\ &= \frac{\partial \mathbf{a}_1}{\partial q_r} \cdot \mathbf{I} \cdot \mathbf{a}_1 - \frac{1}{2} \frac{\partial I_{11}}{\partial q_r} \qquad (r = 1, 2, 3)\end{aligned} \tag{27}$$

where I_{11} is the moment of inertia of B about a line passing through B^* and parallel to $\mathbf{a}_1$. Substituting from Eqs. (24) and (27) into Eq. (23), one thus arrives at

$$\begin{aligned}F_r = {} & \frac{3G\overline{m}}{R^3} \left(\frac{\partial \mathbf{a}_1}{\partial q_r} \cdot \mathbf{I} \cdot \mathbf{a}_1 - \frac{1}{2} \frac{\partial I_{11}}{\partial q_r} \right) \\ & - \frac{G\overline{m}m}{R^2} \left[\frac{\partial R}{\partial q_r} + R\mathbf{a}_1 \cdot \frac{\partial \mathbf{a}_1}{\partial q_r} + \frac{\partial R}{\partial q_r} \mathbf{a}_1 \cdot \mathbf{f}^{(2)} + R \frac{\partial \mathbf{a}_1}{\partial q_r} \cdot \mathbf{f}^{(2)} \right] \\ & \qquad (r = 1, 2, 3)\end{aligned} \tag{28}$$

Since $\partial \mathbf{a}_1/\partial q_r$ is necessarily perpendicular to $\mathbf{a}_1$, the dot-product of $\partial \mathbf{a}_1/\partial q_r$ with $\mathbf{a}_1$ is equal to zero. Expressing $\mathbf{f}^{(2)}$ as in Eq. (2.3.3), one can, therefore, rewrite Eq. (28) as

$$F_r = -\frac{G\overline{m}m}{R^2} \frac{\partial R}{\partial q_r} - \frac{3G\overline{m}}{2R^3} \left\{ \frac{1}{R} \frac{\partial R}{\partial q_r} [\mathrm{tr}(\mathbf{I}) - 3I_{11}] + \frac{\partial I_{11}}{\partial q_r} \right\} \qquad (r = 1, 2, 3) \tag{29}$$

Differentiation of $\widetilde{V}$ as given in Eq. (2.12.13) gives

$$\begin{aligned}\frac{\partial \widetilde{V}}{\partial q_r} &= -\frac{G\overline{m}m}{R^2} \frac{\partial R}{\partial q_r} - \frac{G\overline{m}}{R^2} \frac{\partial R}{\partial q_r} v^{(2)} + \frac{G\overline{m}m}{R} \frac{\partial v^{(2)}}{\partial q_r} \\ &\underset{(2.12.2)}{=} -\frac{G\overline{m}m}{R^2} \frac{\partial R}{\partial q_r} - \frac{G\overline{m}}{2R^4} \frac{\partial R}{\partial q_r} [\mathrm{tr}(\mathbf{I}) - 3I_{11}] \\ &\quad + \frac{G\overline{m}m}{R} \left\{ -\frac{1}{mR^3} [\mathrm{tr}(\mathbf{I}) - 3I_{11}] \frac{\partial R}{\partial q_r} - \frac{3}{2mR^2} \frac{\partial I_{11}}{\partial q_r} \right\} \\ &\underset{(29)}{=} F_r \qquad (r = 1, 2, 3)\end{aligned} \tag{30}$$

In accordance with Eq. (1), W as given in Eq. (3) is thus a potential energy of B in A.

Example Figure 4.2.2 shows a uniform rod B of mass m and length L moving in a plane fixed in a reference frame N under the action of gravitational forces exerted by $\bar{P}$, a particle of mass $\bar{m}$, fixed in N; q_1, q_2, and q_3 are generalized coordinates characterizing the configuration of B in N.

If the resultant force exerted by $\bar{P}$ on B is approximated with $\tilde{\mathbf{F}}$ as defined in Eq. (2.3.6) while the moment about B^* of all forces exerted by $\bar{P}$ on B is approximated with $\tilde{\mathbf{M}}$ as defined in Eq. (2.6.3), then W, a potential energy of B in N, constructed by referring to Eqs. (3) and (2.12.13), can be expressed as

$$W = -\frac{G\bar{m}m}{q_2}\left[1 + \frac{L^2}{24q_2^{\,2}}(2 - 3\sin^2 q_3)\right] \tag{31}$$

With the aid of this function and Eq. (1), one can formulate the generalized active forces corresponding to $u_r \overset{\Delta}{=} \dot{q}_r$ ($r = 1, 2, 3$) as

$$F_1 \underset{(1)}{=} -\frac{\partial W}{\partial q_1} \underset{(31)}{=} 0 \tag{32}$$

$$F_2 \underset{(1)}{=} -\frac{\partial W}{\partial q_2} \underset{(31)}{=} -\frac{G\bar{m}m}{q_2^{\,2}}\left[1 + \frac{L^2}{8q_2^{\,2}}(2 - 3\sin^2 q_3)\right] \tag{33}$$

$$F_3 \underset{(1)}{=} -\frac{\partial W}{\partial q_3} \underset{(31)}{=} -\frac{G\bar{m}mL^2}{8q_2^{\,3}}\sin 2q_3 \tag{34}$$

in agreement with Eqs. (4.1.27)–(4.1.29).

Suppose that the resultant force exerted by $\bar{P}$ on B is approximated with $\hat{\mathbf{F}}$ as given in Eq. (2.3.5), rather than with $\tilde{\mathbf{F}}$, but that the moment about B^* of all forces exerted by $\bar{P}$ on B is approximated as before. Then the generalized active forces become [use Eq. (4.1.2)]

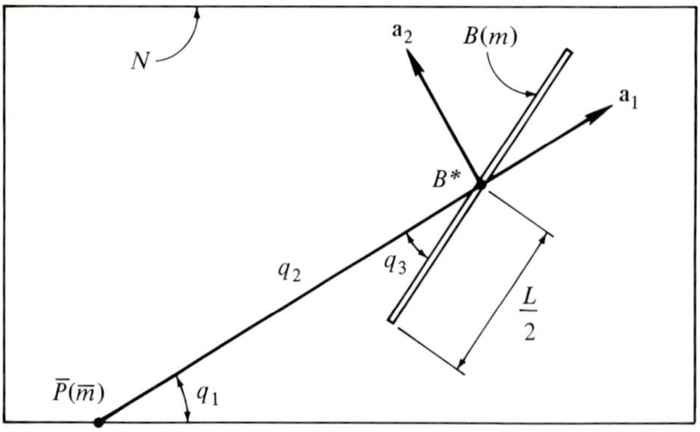

Figure 4.2.2

$$F_1 = -\frac{G\overline{m}mL^2}{8q_2^{\,3}}\sin 2q_3, \quad F_2 = -\frac{G\overline{m}m}{q_2^{\,2}}, \quad F_3 = -\frac{G\overline{m}mL^2}{8q_2^{\,3}}\sin 2q_3 \qquad (35)$$

Now, if there exists a function W that satisfies Eq. (1), then $\partial F_1/\partial q_3 = \partial F_3/\partial q_1$. Since here these derivatives are given by

$$\frac{\partial F_1}{\partial q_3} \underset{(35)}{=} -\frac{G\overline{m}mL^2}{4q_2^{\,3}}\cos 2q_3 \qquad \frac{\partial F_3}{\partial q_1} \underset{(35)}{=} 0 \qquad (36)$$

one is forced to conclude that a potential energy of B in N does not exist.

4.3 GENERALIZED INERTIA FORCES

Given a system S consisting of N particles $P_1, \ldots, P_N$, suppose that n generalized speeds are defined as in Eq. (1.21.1). Let $\mathbf{v}_r^{P_i}$ denote the rth partial velocity of P_i in reference frame A (see Sec. 1.21), and let $\mathbf{R}_i^*$ be the *inertia force* for P_i in A; that is, let

$$\mathbf{R}_i^* \triangleq -m_i\mathbf{a}_i \qquad (i = 1, \ldots, N) \qquad (1)$$

where m_i is the mass of P_i and $\mathbf{a}_i$ is the acceleration of P_i in A. Then $F_1^*, \ldots, F_n^*$ called *generalized inertia forces* for S in A, are defined as

$$F_r^* \triangleq \sum_{i=1}^{N} \mathbf{v}_r^{P_i} \cdot \mathbf{R}_i^* \qquad (i = 1, \ldots, n) \qquad (2)$$

$(F_r^*)_B$, the contribution to F_r^* of all inertia forces for the particles of a rigid body B belonging to S, can be expressed in terms of $\mathbf{R}^*$ and $\mathbf{T}^*$, defined as

$$\mathbf{R}^* \triangleq -M\mathbf{a}^* \qquad (3)$$

and

$$\mathbf{T}^* \triangleq -\sum_{i=1}^{\overline{N}} m_i\mathbf{r}_i \times \mathbf{a}_i \qquad (4)$$

where M is the mass of B, $\mathbf{a}^*$ is the acceleration of the mass center B^* of B in A, m_i is the mass of a generic particle P_i of B, $\overline{N}$ is the number of particles comprising B, $\mathbf{r}_i$ is the position vector from B^* to P_i, and $\mathbf{a}_i$ is the acceleration of P_i in A. $\mathbf{T}^*$ is called the *inertia torque* for B in A, and $(F_r^*)_B$ can be written

$$(F_r^*)_B = \boldsymbol{\omega}_r \cdot \mathbf{T}^* + \mathbf{v}_r \cdot \mathbf{R}^* \qquad (r = 1, \ldots, n) \qquad (5)$$

where $\mathbf{v}_r$ is the rth partial velocity of B^* in A, and $\boldsymbol{\omega}_r$ is the rth partial angular velocity of B in A (see Sec. 1.21). The utility of Eq. (5) derives from the fact that $\mathbf{T}^*$ can be expressed in a number of forms making it unnecessary to perform explicitly the summation indicated in Eq. (4). For example,

$$\mathbf{T}^* = -\boldsymbol{\alpha} \cdot \mathbf{I} - \boldsymbol{\omega} \times \mathbf{I} \cdot \boldsymbol{\omega} \qquad (6)$$

where $\boldsymbol{\alpha}$ and $\boldsymbol{\omega}$ are, respectively, the angular acceleration of B in A and the angular velocity of B in A, and $\mathbf{I}$ is the central inertia dyadic of B; and, if $\mathbf{c}_1$, $\mathbf{c}_2$, $\mathbf{c}_3$ form a dextral set of mutually perpendicular unit vectors, each parallel to a central principal axis of inertia of B, but not necessarily fixed in B, and α_j, ω_j, and I_j are defined as

$$\alpha_j \triangleq \boldsymbol{\alpha} \cdot \mathbf{c}_j \qquad \omega_j \triangleq \boldsymbol{\omega} \cdot \mathbf{c}_j \qquad I_j \triangleq \mathbf{c}_j \cdot \mathbf{I} \cdot \mathbf{c}_j \qquad (j = 1, 2, 3) \tag{7}$$

then Eq. (6) can be re-expressed as

$$\begin{aligned} \mathbf{T}^* = &- [\alpha_1 I_1 - \omega_2\omega_3(I_2 - I_3)]\mathbf{c}_1 \\ &- [\alpha_2 I_2 - \omega_3\omega_1(I_3 - I_1)]\mathbf{c}_2 \\ &- [\alpha_3 I_3 - \omega_1\omega_2(I_1 - I_2)]\mathbf{c}_3 \end{aligned} \tag{8}$$

When a subset of the particles of S forms a gyrostat G (see Fig. 4.3.1) consisting of a rigid body C that carries an axisymmetric rotor D whose mass center D^* and axis are fixed in C, then $(F_r{}^*)_G$, the contribution to $F_r{}^*$ of all inertia forces for the particles of G, can be expressed as the sum of two quantities, one associated with a (fictitious) rigid body R that has the same mass distribution as G, but moves like C, and the other accounting for rotation of D relative to C; that is, one can write

$$(F_r{}^*)_G = (F_r{}^*)_R + (F_r{}^*)_I \qquad (r = 1, \ldots, n) \tag{9}$$

where $(F_r{}^*)_R$ is the rth generalized inertia force associated with the motion of R in A, a quantity one can form most conveniently by using Eq. (5), and $(F_r{}^*)_I$ takes one of two forms, depending on whether the motion of D in C is prescribed as a function of time, in which event none of the generalized coordinates characterizing the motion of S in A is associated with rotation of D relative to C, or D is completely free to rotate relative to C. When the motion of D in C is prescribed (which implies that C exerts on D such forces as are required to produce the prescribed motion), then

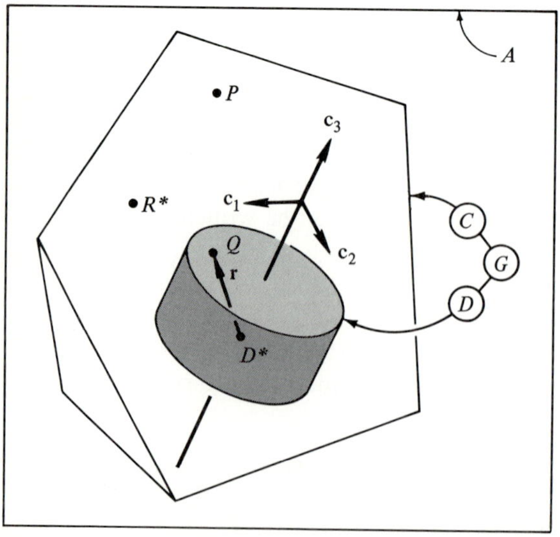

Figure 4.3.1

$$(F_r^*)_I = J[\Omega(-\omega_2\mathbf{c}_1 + \omega_1\mathbf{c}_2) - \dot{\Omega}\mathbf{c}_3] \cdot \boldsymbol{\omega}_r \qquad (r = 1, \ldots, n) \tag{10}$$

where $\mathbf{c}_1$, $\mathbf{c}_2$, $\mathbf{c}_3$ form a dextral set of mutually perpendicular unit vectors fixed in C, with $\mathbf{c}_3$ parallel to the axis of D (see Fig. 4.3.1); J is the moment of inertia of D about the axis of symmetry of D; $\Omega \triangleq {}^C\boldsymbol{\omega}^D \cdot \mathbf{c}_3$; $\omega_i \triangleq {}^A\boldsymbol{\omega}^C \cdot \mathbf{c}_i$ $(i = 1, 2, 3)$; and $\boldsymbol{\omega}_r$ is the rth partial angular velocity of C in A (see Sec. 1.21). Alternatively, when D is completely free to rotate relative to C, one can always define u_n, the nth generalized speed for S in A (see Sec. 1.21), as $u_n \triangleq {}^C\boldsymbol{\omega}^D \cdot \mathbf{c}_3$, and Eq. (10) then gives way to

$$(F_r^*)_I = J[u_n(-\omega_2\mathbf{c}_1 + \omega_1\mathbf{c}_2) - \dot{u}_n\mathbf{c}_3] \cdot \boldsymbol{\omega}_r \qquad (r = 1, \ldots, n-1) \tag{11}$$

together with

$$(F_n^*)_I = -J(\dot{\omega}_3 + \dot{u}_n) \tag{12}$$

Moreover, $(F_n^*)_R$ vanishes under these circumstances.

Derivations With m_i, $\mathbf{r}_i$, and $\mathbf{a}_i$ as defined,

$$\sum_{i=1}^{\bar{N}} m_i\mathbf{r}_i = 0 \tag{13}$$

and $\mathbf{a}_i$ can be expressed in terms of the acceleration $\mathbf{a}^*$ of the mass center of B in A, the angular acceleration $\boldsymbol{\alpha}$ of B in A, and the angular velocity $\boldsymbol{\omega}$ of B in A as

$$\mathbf{a}_i = \mathbf{a}^* + \boldsymbol{\alpha} \times \mathbf{r}_i + \boldsymbol{\omega} \times (\boldsymbol{\omega} \times \mathbf{r}_i) \qquad (i = 1, \ldots, \bar{N}) \tag{14}$$

Consequently,

$$\sum_{i=1}^{\bar{N}} m_i\mathbf{a}_i \underset{(14)}{=} \sum_{i=1}^{\bar{N}} m_i\mathbf{a}^* + \boldsymbol{\alpha} \times \sum_{i=1}^{\bar{N}} m_i\mathbf{r}_i + \boldsymbol{\omega} \times \left(\boldsymbol{\omega} \times \sum_{i=1}^{\bar{N}} m_i\mathbf{r}_i\right) \underset{(13)}{=} M\mathbf{a}^* \underset{(3)}{=} -\mathbf{R}^* \tag{15}$$

Now,

$$(F_r^*)_B \underset{(1,2)}{=} -\sum_{i=1}^{\bar{N}} m_i\mathbf{v}_r^{P_i} \cdot \mathbf{a}_i \qquad (r = 1, \ldots, n) \tag{16}$$

and (see Prob. 1.31)

$$\mathbf{v}_r^{P_i} = \mathbf{v}_r + \boldsymbol{\omega}_r \times \mathbf{r}_i \qquad (r = 1, \ldots, n;\ i = 1, \ldots, \bar{N}) \tag{17}$$

Hence,

$$\begin{aligned}(F_r^*)_B &\underset{(16,17)}{=} -\sum_{i=1}^{\bar{N}} m_i(\mathbf{v}_r + \boldsymbol{\omega}_r \times \mathbf{r}_i) \cdot \mathbf{a}_i \\ &= -\mathbf{v}_r \cdot \sum_{i=1}^{\bar{N}} m_i\mathbf{a}_i - \boldsymbol{\omega}_r \cdot \sum_{i=1}^{\bar{N}} m_i\mathbf{r}_i \times \mathbf{a}_i \\ &\underset{(15,4)}{=} \mathbf{v}_r \cdot \mathbf{R}^* + \boldsymbol{\omega}_r \cdot \mathbf{T}^* \qquad (r = 1, \ldots, n)\end{aligned} \tag{18}$$

which establishes the validity of Eq. (5).

With the aid of Eq. (14), one can express $\mathbf{T}^*$ as

$$\mathbf{T}^* \underset{(4)}{=} -\sum_{i=1}^{\bar{N}} m_i \mathbf{r}_i \times [\mathbf{a}^* + \boldsymbol{\alpha} \times \mathbf{r}_i + \boldsymbol{\omega} \times (\boldsymbol{\omega} \times \mathbf{r}_i)]$$

$$\underset{(13)}{=} -\sum_{i=1}^{\bar{N}} m_i \mathbf{r}_i \times (\boldsymbol{\alpha} \times \mathbf{r}_i) - \sum_{i=1}^{\bar{N}} m_i \mathbf{r}_i \times [\boldsymbol{\omega} \times (\boldsymbol{\omega} \times \mathbf{r}_i)] \tag{19}$$

Now,

$$\sum_{i=1}^{\bar{N}} m_i \mathbf{r}_i \times (\boldsymbol{\alpha} \times \mathbf{r}_i) = \sum_{i=1}^{\bar{N}} m_i (\mathbf{r}_i{}^2 \boldsymbol{\alpha} - \boldsymbol{\alpha} \cdot \mathbf{r}_i \mathbf{r}_i)$$

$$= \boldsymbol{\alpha} \cdot \sum_{i=1}^{\bar{N}} m_i (\mathbf{r}_i{}^2 \mathbf{U} - \mathbf{r}_i \mathbf{r}_i)$$

$$= \boldsymbol{\alpha} \cdot \mathbf{I} \tag{20}$$

and

$$\sum_{i=1}^{\bar{N}} m_i \mathbf{r}_i \times [\boldsymbol{\omega} \times (\boldsymbol{\omega} \times \mathbf{r}_i)] = -\sum_{i=1}^{\bar{N}} m_i \mathbf{r}_i \cdot \boldsymbol{\omega}\boldsymbol{\omega} \times \mathbf{r}_i$$

$$= -\boldsymbol{\omega} \times \left(\sum_{i=1}^{\bar{N}} m_i \mathbf{r}_i \mathbf{r}_i \right) \cdot \boldsymbol{\omega}$$

$$= -\boldsymbol{\omega} \times \left[\sum_{i=1}^{\bar{N}} m_i (\mathbf{r}_i \mathbf{r}_i - \mathbf{r}_i{}^2 \mathbf{U}) \right] \cdot \boldsymbol{\omega}$$

$$= \boldsymbol{\omega} \times \mathbf{I} \cdot \boldsymbol{\omega} \tag{21}$$

Substitution from Eqs. (20) and (21) into Eq. (19) produces Eq. (6).

With α_j, ω_j, and I_j as defined,

$$\mathbf{I} = I_1 \mathbf{c}_1 \mathbf{c}_1 + I_2 \mathbf{c}_2 \mathbf{c}_2 + I_3 \mathbf{c}_3 \mathbf{c}_3 \tag{22}$$

so that

$$\boldsymbol{\alpha} \cdot \mathbf{I} = \alpha_1 I_1 \mathbf{c}_1 + \alpha_2 I_2 \mathbf{c}_2 + \alpha_3 I_3 \mathbf{c}_3 \tag{23}$$

and

$$\boldsymbol{\omega} \times (\mathbf{I} \cdot \boldsymbol{\omega}) = -\omega_2 \omega_3 (I_2 - I_3) \mathbf{c}_1 - \omega_3 \omega_1 (I_3 - I_1) \mathbf{c}_2 - \omega_1 \omega_2 (I_1 - I_2) \mathbf{c}_3 \tag{24}$$

Equation (8) thus follows directly from Eqs. (6), (23), and (24).

Turning to the derivation of Eqs. (9)–(12), we let P and Q be generic points of C and D, respectively, and note that there corresponds to Q a certain point of R, the rigid body having the same mass distribution as G, but moving like C. Calling this point $\bar{Q}$, we now make a number of purely kinematical observations, regarding D as completely free to rotate relative to C for the time being and taking $u_n \triangleq {}^C\boldsymbol{\omega}^D \cdot \mathbf{c}_3$.

The partial velocities of $\overline{Q}$ and D^* in A are related to the partial angular velocities of C in A by (see Prob. 1.31)

$$\mathbf{v}_r^{\overline{Q}} = \mathbf{v}_r^{D^*} + \boldsymbol{\omega}_r \times \mathbf{r}^Q \qquad (r = 1, \ldots, n) \tag{25}$$

where $\mathbf{r}^Q$ is the position vector from D^* to Q; and $\mathbf{v}_n^{\overline{Q}} = 0$, because $\mathbf{v}_n^{D^*} = \boldsymbol{\omega}_n = 0$ since the velocity of D^* in A and the angular velocity of C in A are independent of u_n. To discover the relationship between the partial velocities of Q and D^* in A, we note first that the velocities of Q and D^* in A are related by

$$\mathbf{v}^Q = \mathbf{v}^{D^*} + ({}^A\boldsymbol{\omega}^C + u_n\mathbf{c}_3) \times \mathbf{r}^Q \tag{26}$$

from which it follows that

$$\mathbf{v}_r^Q = \mathbf{v}_r^{D^*} + \boldsymbol{\omega}_r \times \mathbf{r}^Q \qquad (r = 1, \ldots, n-1) \tag{27}$$

while

$$\mathbf{v}_n^Q = \mathbf{c}_3 \times \mathbf{r}^Q \tag{28}$$

The angular acceleration of D in A can be formulated as

$$\boldsymbol{\alpha}^D \underset{(1.20.1)}{=} \frac{{}^A d}{dt}({}^A\boldsymbol{\omega}^C + u_n\mathbf{c}_3) \underset{(1.16.1)}{=} \boldsymbol{\alpha}^C + \dot{u}_n\mathbf{c}_3 + u_n{}^A\boldsymbol{\omega}^C \underset{(1.11.9)}{\times} \mathbf{c}_3 \tag{29}$$

and this result may be used to express the acceleration of Q in A as

$$\begin{aligned}\mathbf{a}^Q &= \mathbf{a}^{D^*} + \boldsymbol{\alpha}^D \times \mathbf{r}^Q + ({}^A\boldsymbol{\omega}^C + u_n\mathbf{c}_3) \times [({}^A\boldsymbol{\omega}^C + u_n\mathbf{c}_3) \times \mathbf{r}^Q] \\ &\underset{(29)}{=} \mathbf{a}^{D^*} + \boldsymbol{\alpha}^C \times \mathbf{r}^Q + {}^A\boldsymbol{\omega}^C \times ({}^A\boldsymbol{\omega}^C \times \mathbf{r}^Q) + \dot{u}_n\mathbf{c}_3 \times \mathbf{r}^Q \\ &\quad + 2u_n{}^A\boldsymbol{\omega}^C \times (\mathbf{c}_3 \times \mathbf{r}^Q) + u_n^2\mathbf{c}_3 \times (\mathbf{c}_3 \times \mathbf{r}^Q)\end{aligned} \tag{30}$$

Similarly, the acceleration of $\overline{Q}$ in A is given by

$$\mathbf{a}^{\overline{Q}} = \mathbf{a}^{D^*} + \boldsymbol{\alpha}^C \times \mathbf{r}^Q + {}^A\boldsymbol{\omega}^C \times ({}^A\boldsymbol{\omega}^C \times \mathbf{r}^Q) \tag{31}$$

and, using Eqs. (25), (31), (27) and (30), one can write (after considerable rearranging)

$$\begin{aligned}\mathbf{v}_r^{\overline{Q}} \cdot \mathbf{a}^{\overline{Q}} - \mathbf{v}_r^Q \cdot \mathbf{a}^Q = &- \boldsymbol{\omega}_r \cdot [\dot{u}_n\mathbf{r}^Q \times (\mathbf{c}_3 \times \mathbf{r}^Q) + 2u_n{}^A\boldsymbol{\omega}^C \cdot \mathbf{r}^Q\mathbf{r}^Q \times \mathbf{c}_3 \\ &- u_n^2\mathbf{c}_3 \cdot \mathbf{r}^Q\mathbf{r}^Q \times \mathbf{c}_3] - \mathbf{v}_r^{D^*} \cdot [\dot{u}_n\mathbf{c}_3 \times \mathbf{r}^Q \\ &+ 2u_n{}^A\boldsymbol{\omega}^C \times (\mathbf{c}_3 \times \mathbf{r}^Q) + u_n^2\mathbf{c}_3 \times (\mathbf{c}_3 \times \mathbf{r}^Q)]\end{aligned}$$

$$(r = 1, \ldots, n-1) \tag{32}$$

To construct generalized inertia forces for G, we now let m^P and m^Q denote the masses of P and Q, respectively, and use the symbols Σ^C and Σ^D to denote summations extended over the particles of C and D, respectively. Referring to Eqs. (1) and (2), we can thus write for G

$$(F_r^*)_G = -\sum^C m^P \mathbf{v}_r^P \cdot \mathbf{a}^P - \sum^D m^Q \mathbf{v}_r^Q \cdot \mathbf{a}^Q \qquad (r = 1, \ldots, n) \quad (33)$$

and for R, the rigid body that has the same mass distribution as G but moves like C,

$$(F_r^*)_R = -\sum^C m^P \mathbf{v}_r^P \cdot \mathbf{a}^P - \sum^D m^Q \mathbf{v}_r^{\overline{Q}} \cdot \mathbf{a}^{\overline{Q}} \qquad (r = 1, \ldots, n) \quad (34)$$

Consequently,

$$(F_r^*)_G \underset{(33,34)}{=} (F_r^*)_R + \sum^D m^Q(\mathbf{v}_r^{\overline{Q}} \cdot \mathbf{a}^{\overline{Q}} - \mathbf{v}_r^Q \cdot \mathbf{a}^Q) \qquad (r = 1, \ldots, n) \quad (35)$$

from which Eq. (9) follows immediately if $(F_r^*)_I$ is defined as

$$(F_r^*)_I \triangleq \sum^D m^Q(\mathbf{v}_r^{\overline{Q}} \cdot \mathbf{a}^{\overline{Q}} - \mathbf{v}_r^Q \cdot \mathbf{a}^Q) \qquad (r = 1, \ldots, n) \quad (36)$$

We shall presently substitute from Eq. (32) into Eq. (36). Before doing so, it is worth noting that each member of the second bracketed term of Eq. (32) involves $\mathbf{r}^Q$, so that each such term will give rise to the sum $\Sigma^D m^Q \mathbf{r}^Q$, which is equal to zero because $\mathbf{r}^Q$ is the position vector from the mass center of D to a generic particle of D. By way of contrast, the first bracketed term of Eq. (32) leads to sums of the forms $\Sigma^D m^Q \mathbf{r}^Q \times (\mathbf{c}_3 \times \mathbf{r}^Q)$ and $\Sigma^D m^Q \mathbf{r}^Q \mathbf{r}^Q \times \mathbf{c}_3$, and these can be expressed as follows:

$$\sum^D m^Q \mathbf{r}^Q \times (\mathbf{c}_3 \times \mathbf{r}^Q) = \mathbf{c}_3 \cdot \sum^D m^Q[(\mathbf{r}^Q)^2 \mathbf{U} - \mathbf{r}^Q \mathbf{r}^Q]$$
$$= \mathbf{c}_3 \cdot \mathbf{J} \quad (37)$$

where $\mathbf{J}$ is the central inertia dyadic of D. Since $\mathbf{c}_3$ is parallel to a central principal axis of inertia of D, $\mathbf{c}_3 \cdot \mathbf{J}$ is parallel to $\mathbf{c}_3$, and

$$\mathbf{c}_3 \cdot \mathbf{J} = J\mathbf{c}_3 \quad (38)$$

where J is the moment of inertia of D about its axis of symmetry. As for the second sum, one can write

$$\sum^D m^Q \mathbf{r}^Q \mathbf{r}^Q \times \mathbf{c}_3 = \left[\sum^D m^Q (\mathbf{r}^Q)^2 \mathbf{U} - \mathbf{J}\right] \times \mathbf{c}_3$$
$$= \frac{J}{2}(-\mathbf{c}_1\mathbf{c}_2 + \mathbf{c}_2\mathbf{c}_1) \quad (39)$$

Substitution from Eq. (32) into Eq. (36) thus produces

$$(F_r^*)_I \underset{(37,38)}{=} -\boldsymbol{\omega}_r \cdot [\dot{u}_n J\mathbf{c}_3 + u_n J(-\omega_1\mathbf{c}_2 + \omega_2\mathbf{c}_1)] \qquad (r = 1, \ldots, n-1) \quad (40)$$

in agreement with Eq. (11). As for Eq. (12), this can be obtained by proceeding similarly, but using Eq. (28) in place of Eq. (27). Finally, if the motion of D in C is prescribed, the validity of Eq. (10) can be established by proceeding as in the derivation of Eq. (11), but replacing u_n with Ω.

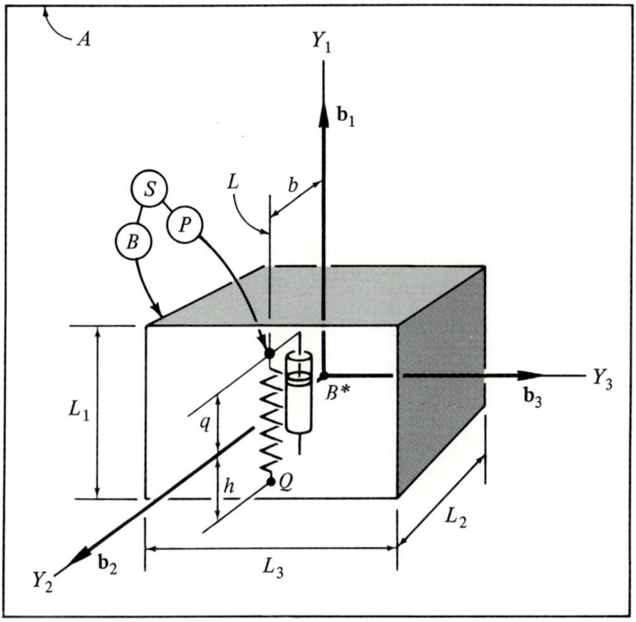

Figure 4.3.2

Example Figure 4.3.2 shows a system S composed of a particle P of mass m and a uniform rectangular block B that has a mass M and sides of lengths L_1, L_2, L_3. P is constrained to remain on a line L fixed in B and is attached to B by means of a spring and a damper. Y_1, Y_2, Y_3 are lines passing through B^*, the mass center of B, and parallel to the edges of B; L is parallel to Y_1 at a distance b from Y_1; and the spring is presumed to be undeformed when P lies on Y_2. Thus q, the distance from P to Y_2, measures the deformation of the spring.

Generalized speeds $u_1, \ldots, u_7$ for S in a reference frame A can be defined in terms of $\boldsymbol{\omega}$, the angular velocity of B in A, and $\mathbf{v}^*$, the velocity of S^*, the mass center of S, in A:

$$u_i \triangleq \begin{cases} \boldsymbol{\omega} \cdot \mathbf{b}_i & (i = 1, 2, 3) \\ \dot{q} & (i = 4) \\ \mathbf{v}^* \cdot \mathbf{b}_{i-4} & (i = 5, 6, 7) \end{cases} \tag{41}$$

where $\mathbf{b}_1$, $\mathbf{b}_2$, $\mathbf{b}_3$ are unit vectors respectively parallel to Y_1, Y_2, Y_3. The angular velocity and angular acceleration of B in A then can be expressed as

$$\boldsymbol{\omega} = u_1\mathbf{b}_1 + u_2\mathbf{b}_2 + u_3\mathbf{b}_3 \tag{42}$$

and

$$\boldsymbol{\alpha} = \dot{u}_1\mathbf{b}_1 + \dot{u}_2\mathbf{b}_2 + \dot{u}_3\mathbf{b}_3 \tag{43}$$

while the velocities of B^* and P in A are given by

$$\mathbf{v}^{B^*} = v_1^{B^*}\mathbf{b}_1 + v_2^{B^*}\mathbf{b}_2 + v_3^{B^*}\mathbf{b}_3 \tag{44}$$

and

$$\mathbf{v}^{P} = v_1^{P}\mathbf{b}_1 + v_2^{P}\mathbf{b}_2 + v_3^{P}\mathbf{b}_3 \tag{45}$$

where

$$v_1^{B^*} \triangleq u_5 + \frac{m}{m+M}(bu_3 - u_4) \tag{46}$$

$$v_2^{B^*} \triangleq u_6 - \frac{m}{m+M}qu_3 \tag{47}$$

$$v_3^{B^*} \triangleq u_7 + \frac{m}{m+M}(qu_2 - bu_1) \tag{48}$$

$$v_1^{P} \triangleq u_5 - \frac{M}{m+M}(bu_3 - u_4) \tag{49}$$

$$v_2^{P} \triangleq u_6 + \frac{M}{m+M}qu_3 \tag{50}$$

$$v_3^{P} \triangleq u_7 - \frac{M}{m+M}(qu_2 - bu_1) \tag{51}$$

The accelerations of B^* and P in A, found by differentiating Eqs. (44) and (45) with respect to t in A, are

$$\begin{aligned}\mathbf{a}^{B^*} = {}& \left[\dot{u}_5 + \frac{m}{m+M}(b\dot{u}_3 - \dot{u}_4) + u_2 v_3^{B^*} - u_3 v_2^{B^*}\right]\mathbf{b}_1 \\ &+ \left[\dot{u}_6 - \frac{m}{m+M}(u_4 u_3 + q\dot{u}_3) + u_3 v_1^{B^*} - u_1 v_3^{B^*}\right]\mathbf{b}_2 \\ &+ \left[\dot{u}_7 + \frac{m}{m+M}(u_4 u_2 + q\dot{u}_2 - b\dot{u}_1) + u_1 v_2^{B^*} - u_2 v_1^{B^*}\right]\mathbf{b}_3\end{aligned} \tag{52}$$

and

$$\begin{aligned}\mathbf{a}^{P} = {}& \left[\dot{u}_5 - \frac{M}{m+M}(b\dot{u}_3 - \dot{u}_4) + u_2 v_3^{P} - u_3 v_2^{P}\right]\mathbf{b}_1 \\ &+ \left[\dot{u}_6 + \frac{M}{m+M}(u_4 u_3 + q\dot{u}_3) + u_3 v_1^{P} - u_1 v_3^{P}\right]\mathbf{b}_2 \\ &+ \left[\dot{u}_7 - \frac{M}{m+M}(u_4 u_2 + q\dot{u}_2 - b\dot{u}_1) + u_1 v_2^{P} - u_2 v_1^{P}\right]\mathbf{b}_3\end{aligned} \tag{53}$$

The partial angular velocities of B in A and the partial velocities of B^* and P in A, formed by reference to Eqs. (42) and (44)–(51), are recorded in Table

Table 4.3.1

r	$\boldsymbol{\omega}_r$	$\mathbf{v}_r^{B*}$	$\mathbf{v}_r^{P}$
1	$\mathbf{b}_1$	$-\dfrac{m}{m+M}\,b\,\mathbf{b}_3$	$\dfrac{M}{m+M}\,b\,\mathbf{b}_3$
2	$\mathbf{b}_2$	$\dfrac{m}{m+M}\,q\,\mathbf{b}_3$	$-\dfrac{M}{m+M}\,q\,\mathbf{b}_3$
3	$\mathbf{b}_3$	$\dfrac{m}{m+M}(b\,\mathbf{b}_1 - q\,\mathbf{b}_2)$	$-\dfrac{M}{m+M}(b\,\mathbf{b}_1 - q\,\mathbf{b}_2)$
4	0	$-\dfrac{m}{m+M}\mathbf{b}_1$	$\dfrac{M}{m+M}\mathbf{b}_1$
5	0	$\mathbf{b}_1$	$\mathbf{b}_1$
6	0	$\mathbf{b}_2$	$\mathbf{b}_2$
7	0	$\mathbf{b}_3$	$\mathbf{b}_3$

4.3.1, and these are used to construct $(F_r{}^*)_B$ and $(F_r{}^*)_P$, the contributions of B and of P to the generalized inertia force $F_r{}^*$ for S in A. To this end, the inertia torque $\mathbf{T}^*$ for B in A is expressed as [see Eqs. (42) and (43)]

$$\begin{aligned}\mathbf{T}^* \underset{(8)}{=} &- [\dot{u}_1 B_1 - u_2 u_3 (B_2 - B_3)]\mathbf{b}_1 \\ &- [\dot{u}_2 B_2 - u_3 u_1 (B_3 - B_1)]\mathbf{b}_2 \\ &- [\dot{u}_3 B_3 - u_1 u_2 (B_1 - B_2)]\mathbf{b}_3 \end{aligned} \tag{54}$$

where

$$B_1 \triangleq \frac{M}{12}(L_2{}^2 + L_3{}^2) \qquad B_2 \triangleq \frac{M}{12}(L_3{}^2 + L_1{}^2) \qquad B_3 \triangleq \frac{M}{12}(L_1{}^2 + L_2{}^2) \tag{55}$$

and Eq. (5) then yields [see also Eqs. (3) and (52)]

$$\begin{aligned}(F_1{}^*)_B = &- \dot{u}_1 B_1 - u_2 u_3 (B_2 - B_3) \\ &+ \frac{mM}{m+M}\, b \left[\dot{u}_7 + \frac{m}{m+M}(u_4 u_2 + q\dot{u}_2 - b\dot{u}_1) + u_1 v_2{}^{B*} - u_2 v_1{}^{B*} \right] \end{aligned} \tag{56}$$

$$\begin{aligned}(F_2{}^*)_B = &- \dot{u}_2 B_2 - u_3 u_1 (B_3 - B_1) \\ &- \frac{mM}{m+M}\, q \left[\dot{u}_7 + \frac{m}{m+M}(u_4 u_2 + q\dot{u}_2 - b\dot{u}_1) + u_1 v_2{}^{B*} - u_2 v_1{}^{B*} \right] \end{aligned} \tag{57}$$

and so forth, while Eqs. (1), (2) and (53) lead to

$$(F_1^*)_P = -\frac{mM}{m+M}b\left[\dot{u}_7 - \frac{M}{m+M}(u_4u_2 + q\dot{u}_2 - b\dot{u}_1) + u_1v_2^P - u_2v_1^P\right] \tag{58}$$

and so forth. The generalized inertia forces for S in A now can be written by simply adding the contributions of B and P. Thus, for example,

$$F_1^* = -\dot{u}_1B_1 - u_2u_3(B_2 - B_3) + \frac{mM}{m+M}b(2u_2u_4 + q\dot{u}_2 - b\dot{u}_1 - qu_1u_3 - bu_2u_3) \tag{59}$$

Now consider the system $\widetilde{S}$ obtained by modifying S as follows: let D be a uniform axisymmetric rotor made of the same material as B; remove from B a cylindrical portion just big enough to accommodate D, letting the axis of the cylindrical cavity thus created be parallel to Y_3 and calling C what now remains of B; and place D into the cavity. Then C and D form a gyrostat G (see Fig. 4.3.3) with a mass distribution identical to that of B, and the generalized inertia forces, $\widetilde{F}_1^*, \ldots, \widetilde{F}_7^*$, for $\widetilde{S}$ in A can be obtained simply by adding to those for S in A terms generated with the aid of Eq. (10) or Eqs. (11) and (12). Suppose, for example, that D is presumed to be driven relative to C, say by means of a torque motor, in such a way that Ω, defined as $\Omega \triangleq {}^C\boldsymbol{\omega}^D \cdot \mathbf{b}_3$, remains constant. Then, if J is the moment of inertia of D about its axis,

$$\widetilde{F}_1^* = F_1^* - J\Omega u_2 \tag{60}$$

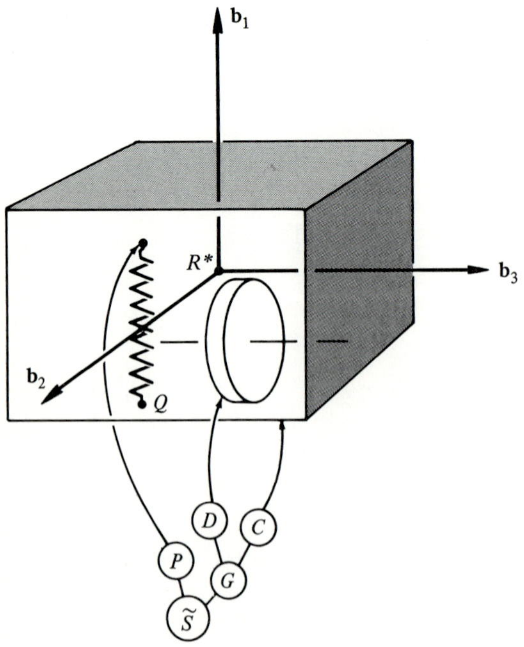

Figure 4.3.3

and so forth. If D is completely free to rotate relative to C, then a generalized speed u_8, defined as $u_8 \triangleq {}^C\boldsymbol{\omega}^D \cdot \mathbf{b}_3$, must be added to those introduced in Eq. (41); Eq. (60) is replaced with [see Eq. (11)]

$$\widetilde{F}_1{}^* = F_1{}^* - Ju_8u_2 \tag{61}$$

and Eqs. (9) and (12) furnish one relationship in addition to those available before, namely,

$$\widetilde{F}_8{}^* = -J(\dot{u}_3 + \dot{u}_8) \tag{62}$$

4.4 KINETIC ENERGY

If the configuration of a system S in a reference frame A is characterized by n generalized coordinates $q_1, \ldots, q_n$, and if n generalized speeds $u_1, \ldots, u_n$ are defined as in Eq. (1.21.1), then the kinetic energy K of S in A can be expressed as

$$K = K_0 + K_1 + K_2 \tag{1}$$

where K_j is homogeneous and of degree j in the variables $u_1, \ldots, u_n$. In particular, K_2 is given by

$$K_2 = \frac{1}{2}\sum_{r=1}^{n}\sum_{s=1}^{n} M_{rs}u_ru_s \tag{2}$$

where M_{rs}, called an *inertia coefficient* for S in A, is a function of $q_1, \ldots, q_n$, and t that is defined in terms of the masses and partial velocities of the particles of S in A (see Sec. 1.21) as

$$M_{rs} \triangleq \sum_{i=1}^{N} m_i \mathbf{v}_r{}^{P_i} \cdot \mathbf{v}_s{}^{P_i} \qquad (r, s = 1, \ldots, n) \tag{3}$$

where N is the number of particles of S.

The inertia coefficients furnish a measure of *dynamic coupling*: the generalized speeds u_r and u_s are said to be coupled dynamically if $M_{rs} \neq 0$, and uncoupled if $M_{rs} = 0$.

If B is a rigid body belonging to S, then $(M_{rs})_B$, the contribution of B to M_{rs}, is given by

$$(M_{rs})_B = m\mathbf{v}_r \cdot \mathbf{v}_s + \boldsymbol{\omega}_r \cdot \mathbf{I} \cdot \boldsymbol{\omega}_s \qquad (r, s = 1, \ldots, n) \tag{4}$$

where m is the mass of B, $\mathbf{v}_r$ and $\mathbf{v}_s$ are, respectively, the rth and the sth partial velocity of the mass center of B in A, $\boldsymbol{\omega}_r$ and $\boldsymbol{\omega}_s$ are, respectively, the rth and the sth partial angular velocity of B in A, and $\mathbf{I}$ is the central inertia dyadic of B.

If the generalized speeds are chosen as

$$u_r = \dot{q}_r \qquad (r = 1, \ldots, n) \tag{5}$$

and the kinetic energy of S in A is regarded as a function of the $2n + 1$ independent variables $q_1, \ldots, q_n, \dot{q}_1, \ldots, \dot{q}_n$, and t, then the generalized inertia forces for S in A (see Sec. 4.4) can be expressed as

$$F_r^* = \frac{\partial K}{\partial q_r} - \frac{d}{dt}\frac{\partial K}{\partial \dot{q}_r} \qquad (r = 1, \ldots, n) \tag{6}$$

Derivations Regard S as composed of N particles $P_1, \ldots, P_N$, and let $\mathbf{v}^{P_i}$ and $\mathbf{v}_r^{P_i}$ denote the velocity of P_i in A and the rth partial velocity of P_i in A, respectively. Then

$$\mathbf{v}^{P_i} \underset{(1.21.3)}{=} \sum_{r=1}^{n} \mathbf{v}_r^{P_i} u_r + \mathbf{v}_t^{P_i} \qquad (i = 1, \ldots, N) \tag{7}$$

where $\mathbf{v}_t^{P_i}$ is a function of $q_1, \ldots, q_n$, and t; and

$$(\mathbf{v}^{P_i})^2 \underset{(7)}{=} \sum_{r=1}^{n}\sum_{s=1}^{n} \mathbf{v}_r^{P_i} \cdot \mathbf{v}_s^{P_i} u_r u_s + 2\sum_{r=1}^{n} \mathbf{v}_r^{P_i} \cdot \mathbf{v}_t^{P_i} u_r + (\mathbf{v}_t^{P_i})^2 \qquad (i = 1, \ldots, N) \tag{8}$$

By definition,

$$K \triangleq \frac{1}{2}\sum_{i=1}^{N} m_i (\mathbf{v}^{P_i})^2 \tag{9}$$

where m_i is the mass of P_i. Hence,

$$K \underset{(8,9)}{=} \frac{1}{2}\sum_{i=1}^{N}\sum_{r=1}^{n}\sum_{s=1}^{n} m_i \mathbf{v}_r^{P_i} \cdot \mathbf{v}_s^{P_i} u_r u_s + \sum_{i=1}^{N}\sum_{r=1}^{n} m_i \mathbf{v}_r^{P_i} \cdot \mathbf{v}_t^{P_i} u_r + \frac{1}{2}\sum_{i=1}^{N} m_i (\mathbf{v}_t^{P_i})^2 \tag{10}$$

and, if K_0, K_1, and K_2 are defined as

$$K_0 \triangleq \frac{1}{2}\sum_{i=1}^{N} m_i (\mathbf{v}_t^{P_i})^2 \tag{11}$$

$$K_1 \triangleq \sum_{i=1}^{N}\sum_{r=1}^{n} m_i \mathbf{v}_r^{P_i} \cdot \mathbf{v}_t^{P_i} u_r \tag{12}$$

and

$$K_2 \triangleq \frac{1}{2}\sum_{i=1}^{N}\sum_{r=1}^{n}\sum_{s=1}^{n} m_i \mathbf{v}_r^{P_i} \cdot \mathbf{v}_s^{P_i} u_r u_s \tag{13}$$

then Eq. (1) follows immediately. Moreover, since the order in which the summations in Eq. (13) are performed is immaterial, K_2 can be expressed as

$$K_2 = \frac{1}{2}\sum_{r=1}^{n}\sum_{s=1}^{n}\sum_{i=1}^{N} m_i \mathbf{v}_r^{P_i} \cdot \mathbf{v}_s^{P_i} u_r u_s \tag{14}$$

which, together with Eq. (3), leads directly to Eq. (2).

To establish the validity of Eq. (4), let the rigid body B consist of the first $\overline{N}$ particles of S, let $\mathbf{r}_i$ be the position vector from the mass center B^* of B to P_i, and note that (see Prob. 1.31)

$$\mathbf{v}_r^{P_i} = \mathbf{v}_r + \boldsymbol{\omega}_r \times \mathbf{r}_i \qquad (i = 1, \ldots, \overline{N}) \tag{15}$$

where $\mathbf{v}_r$ is the rth partial velocity of B^* in A. Hence,

$$\begin{aligned}(M_{rs})_B \underset{(3)}{=} \sum_{i=1}^{\overline{N}} m_i[&\mathbf{v}_r \cdot \mathbf{v}_s + \mathbf{v}_r \cdot \boldsymbol{\omega}_s \times \mathbf{r}_i + \boldsymbol{\omega}_r \times \mathbf{r}_i \cdot \mathbf{v}_s \\ &+ (\boldsymbol{\omega}_r \times \mathbf{r}_i) \cdot (\boldsymbol{\omega}_s \times \mathbf{r}_i)] \qquad (r, s = 1, \ldots, n) \\ = \left(\sum_{i=1}^{\overline{N}} m_i\right)&\mathbf{v}_r \cdot \mathbf{v}_s + \mathbf{v}_r \cdot \boldsymbol{\omega}_s \times \sum_{i=1}^{\overline{N}} m_i\mathbf{r}_i \\ &+ \boldsymbol{\omega}_r \times \left(\sum_{i=1}^{\overline{N}} m_i\mathbf{r}_i\right) \cdot \mathbf{v}_s + \boldsymbol{\omega}_r \cdot \left[\sum_{i=1}^{\overline{N}} m_i(\mathbf{r}_i{}^2\mathbf{U} - \mathbf{r}_i\mathbf{r}_i)\right] \cdot \boldsymbol{\omega}_s \\ &\qquad (r, s = 1, \ldots, n)\end{aligned} \tag{16}$$

But,

$$\sum_{i=1}^{\overline{N}} m_i = m \qquad \sum_{i=1}^{\overline{N}} m_i\mathbf{r}_i = 0 \qquad \sum_{i=1}^{\overline{N}} m_i(\mathbf{r}_i{}^2\mathbf{U} - \mathbf{r}_i\mathbf{r}_i) = \mathbf{I} \tag{17}$$

Hence, Eq. (4) is equivalent to Eq. (16).

Finally, let $\mathbf{a}_i$ be the acceleration of P_i in A; write (see Prob. 1.30)

$$\mathbf{v}_r^{P_i} \cdot \mathbf{a}_i = \frac{1}{2}\left[\frac{d}{dt}\frac{\partial(\mathbf{v}^{P_i})^2}{\partial \dot{q}_r} - \frac{\partial(\mathbf{v}^{P_i})^2}{\partial q_r}\right] \qquad (i = 1, \ldots, N) \tag{18}$$

and, recalling that [see Eqs. (4.3.1) and (4.3.2)]

$$F_r^* = -\sum_{i=1}^{N} m_i\mathbf{v}_r^{P_i} \cdot \mathbf{a}_i \qquad (r = 1, \ldots, n) \tag{19}$$

proceed to

$$\begin{aligned}F_r^* &\underset{(18,19)}{=} \frac{1}{2}\sum_{i=1}^{N} m_i\left[\frac{\partial(\mathbf{v}^{P_i})^2}{\partial q_r} - \frac{d}{dt}\frac{\partial(\mathbf{v}^{P_i})^2}{\partial \dot{q}_r}\right] \\ &= \frac{\partial}{\partial q_r}\left[\frac{1}{2}\sum_{i=1}^{N} m_i(\mathbf{v}^{P_i})^2\right] - \frac{d}{dt}\frac{\partial}{\partial \dot{q}_r}\left[\frac{1}{2}\sum_{i=1}^{N} m_i(\mathbf{v}^{P_i})^2\right] \\ &\underset{(9)}{=} \frac{\partial K}{\partial q_r} - \frac{d}{dt}\frac{\partial K}{\partial \dot{q}_r} \qquad (r = 1, \ldots, n)\end{aligned} \tag{20}$$

in agreement with Eq. (6).

Example Figure 4.4.1 shows a system S consisting of a thin rod B of length $2L$ and mass m and a particle P, also of mass m. P can slide on B, but is attached to one end of B by means of a spring. X_1 and X_2 are Cartesian coordinate axes fixed in a reference frame A, and $q_1, \ldots, q_4$ are generalized coordinates for S in A, B being constrained to move in the X_1–X_2 plane.

The velocities of the mass center B^* of B and of the particle P in A are given, respectively, by

$$\mathbf{v}^{B^*} = \dot{q}_1\mathbf{a}_1 + \dot{q}_2\mathbf{a}_2 + \dot{q}_3 L\mathbf{b}_2 \tag{21}$$

and

$$\mathbf{v}^{P} = \dot{q}_1\mathbf{a}_1 + \dot{q}_2\mathbf{a}_2 + \dot{q}_3 q_4\mathbf{b}_2 + \dot{q}_4\mathbf{b}_1 \tag{22}$$

where $\mathbf{a}_1$, $\mathbf{a}_2$, $\mathbf{b}_1$, and $\mathbf{b}_2$ are unit vectors directed as shown in Fig. 4.4.1, and the angular velocity of B is

$$\boldsymbol{\omega} = \dot{q}_3\mathbf{b}_1 \times \mathbf{b}_2 \tag{23}$$

Hence, the kinetic energy of S in A is

$$\begin{aligned} K &= \frac{1}{2}m(\mathbf{v}^{B^*})^2 + \frac{1}{2}\frac{mL^2}{3}\boldsymbol{\omega}^2 + \frac{1}{2}m(\mathbf{v}^P)^2 \\ &\underset{(21,22)}{=} m\{\dot{q}_1{}^2 + \dot{q}_2{}^2 + \tfrac{1}{2}(\tfrac{4}{3}L^2 + q_4{}^2)\dot{q}_3{}^2 + \tfrac{1}{2}\dot{q}_4{}^2 \\ &\quad + [\dot{q}_1\dot{q}_4 + \dot{q}_2\dot{q}_3(L + q_4)]\cos q_3 \\ &\quad + [\dot{q}_2\dot{q}_4 - \dot{q}_1\dot{q}_3(L + q_4)]\sin q_3\} \end{aligned} \tag{24}$$

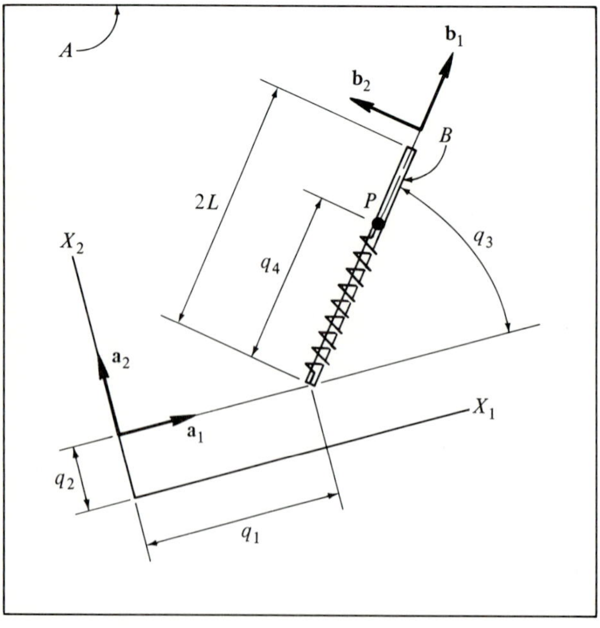

Figure 4.4.1

and, if $\dot{q}_1, \ldots, \dot{q}_4$ are chosen as generalized speeds, then the associated generalized inertia forces, found by using Eq. (6) together with

$$\frac{\partial K}{\partial q_1} = 0 \qquad \frac{\partial K}{\partial \dot{q}_1} = m[2\dot{q}_1 + \dot{q}_4 \cos q_3 - \dot{q}_3(L + q_4) \sin q_3] \tag{25}$$

$$\frac{\partial K}{\partial q_2} = 0 \qquad \frac{\partial K}{\partial \dot{q}_2} = m[2\dot{q}_2 + \dot{q}_3(L + q_4) \cos q_3 + \dot{q}_4 \sin q_3] \tag{26}$$

$$\frac{\partial K}{\partial q_3} = m\{[\dot{q}_1\dot{q}_4 - \dot{q}_2\dot{q}_3(1 + q_4)] \sin q_3 + [\dot{q}_2\dot{q}_4 - \dot{q}_1\dot{q}_3(L + q_4) \cos q_3]\} \tag{27}$$

$$\frac{\partial K}{\partial \dot{q}_3} = m[(\tfrac{4}{3}L^2 + q_4^2)\dot{q}_3 + \dot{q}_2(L + q_4) \cos q_3 - \dot{q}_1(L + q_4) \sin q_3] \tag{28}$$

$$\frac{\partial K}{\partial q_4} = m(q_4\dot{q}_3^2 + \dot{q}_2\dot{q}_3 \cos q_3 - \dot{q}_1\dot{q}_3 \sin q_3) \tag{29}$$

$$\frac{\partial K}{\partial \dot{q}_4} = m(\dot{q}_4 + \dot{q}_1 \cos q_3 + \dot{q}_2 \sin q_3) \tag{30}$$

are

$$F_1^* = -m[2\ddot{q}_1 + \ddot{q}_4 \cos q_3 - 2\dot{q}_4\dot{q}_3 \sin q_3 - (L + q_4)(\ddot{q}_3 \sin q_3 + \dot{q}_3^2 \cos q_3)] \tag{31}$$

$$F_2^* = -m[2\ddot{q}_2 + \ddot{q}_4 \sin q_3 + 2\dot{q}_4\dot{q}_3 \cos q_3 + (L + q_4)(\ddot{q}_3 \cos q_3 - \dot{q}_3^2 \sin q_3)] \tag{32}$$

$$F_3^* = -m[(L + q_4)(\ddot{q}_2 - \ddot{q}_1 \sin q_3) + (\tfrac{4}{3}L^2 + q_4^2)\ddot{q}_3 + 2\dot{q}_3\dot{q}_4q_4] \tag{33}$$

$$F_4^* = -m(\ddot{q}_4 + \ddot{q}_1 \cos q_3 + \ddot{q}_2 \sin q_3 - q_4\dot{q}_3^2) \tag{34}$$

It is worth noting that these results can be obtained with less effort as follows.

The accelerations of B^* and of P in A, found by differentiating $\mathbf{v}^{B^*}$ and $\mathbf{v}^P$, are

$$\mathbf{a}^{B^*} \underset{(21)}{=} \ddot{q}_1\mathbf{a}_1 + \ddot{q}_2\mathbf{a}_2 - \dot{q}_3^2L\mathbf{b}_1 + \ddot{q}_3L\mathbf{b}_2 \tag{35}$$

$$\mathbf{a}^P \underset{(22)}{=} \ddot{q}_1\mathbf{a}_1 + \ddot{q}_2\mathbf{a}_2 + (\ddot{q}_4 - \dot{q}_3^2q_4)\mathbf{b}_1 + (\ddot{q}_3q_4 + 2\dot{q}_3\dot{q}_4)\mathbf{b}_2 \tag{36}$$

and the angular acceleration of B in A is

$$\boldsymbol{\alpha} = \ddot{q}_3\mathbf{b}_1 \times \mathbf{b}_2 \tag{37}$$

so that the inertia torque for B in A can be expressed as

$$\mathbf{T}^* \underset{(4.3.6)}{=} -\frac{mL^2}{3}\ddot{q}_3\mathbf{b}_1 \times \mathbf{b}_2 \tag{38}$$

Hence [see Eqs. (4.3.2), (4.3.3), and (4.3.5)],

$$F_r^* = -m\mathbf{v}_r^P \cdot \mathbf{a}^P + \boldsymbol{\omega}_r \cdot \mathbf{T}^* - m\mathbf{v}_r^{B^*} \cdot \mathbf{a}^{B^*} \qquad (r = 1, \ldots, 4) \tag{39}$$

where the partial velocities $\mathbf{v}_r^P$ and $\mathbf{v}_r^{B^*}$ $(r = 1, \ldots, 4)$ can be written by inspection of Eqs. (21) and (22), and the partial angular velocities $\boldsymbol{\omega}_1, \ldots, \boldsymbol{\omega}_4$ all are equal to zero, except for $\boldsymbol{\omega}_3$, which is equal to $\mathbf{b}_1 \times \mathbf{b}_2$ [see Eq. (23)]. Substitution from Eqs. (35)–(38) into Eq. (39) leads directly to Eqs. (32)–(34). For example,

$$F_4^* \underset{(39)}{=} -m\mathbf{b}_1 \cdot \mathbf{a}^P + 0 \cdot \mathbf{T}^* - 0 \cdot \mathbf{a}^{B^*}$$

$$\underset{(36)}{=} -m(\ddot{q}_1 \cos q_3 + \ddot{q}_2 \sin q_3 + \ddot{q}_4 - \dot{q}_3^2 q_4) \tag{40}$$

in agreement with Eq. (34).

If, instead of using $\dot{q}_1, \ldots, \dot{q}_4$ as generalized speeds, one defines $u_1, \ldots, u_4$ as

$$u_1 \triangleq \dot{q}_1 \cos q_3 + \dot{q}_2 \sin q_3 \tag{41}$$

$$u_2 \triangleq -\dot{q}_1 \sin q_3 + \dot{q}_2 \cos q_3 + \dot{q}_3 L \tag{42}$$

$$u_3 \triangleq \dot{q}_3 \tag{43}$$

$$u_4 \triangleq \dot{q}_4 \tag{44}$$

then

$$\mathbf{v}^{B^*} = u_1\mathbf{b}_1 + u_2\mathbf{b}_2 \tag{45}$$

$$\mathbf{v}^P = (u_1 + u_4)\mathbf{b}_1 + [u_2 + (q_4 - L)u_3]\mathbf{b}_2 \tag{46}$$

$$\boldsymbol{\omega} = u_3\mathbf{b}_1 \times \mathbf{b}_2 \tag{47}$$

$$\mathbf{a}^{B^*} \underset{(45)}{=} (\dot{u}_1 - u_2u_3)\mathbf{b}_1 + (\dot{u}_2 + u_3u_1)\mathbf{b}_2 \tag{48}$$

$$\mathbf{a}^P \underset{(46)}{=} \{\dot{u}_1 + \dot{u}_4 - u_3[u_2 + (q_4 - L)u_3]\}\mathbf{b}_1$$

$$+ [\dot{u}_2 + 2u_4u_3 + (q_4 - L)\dot{u}_3 + u_3u_1]\mathbf{b}_2 \tag{49}$$

$$\boldsymbol{\alpha} \underset{(47)}{=} \dot{u}_3\mathbf{b}_1 \times \mathbf{b}_2 \tag{50}$$

and the generalized inertia forces $\widetilde{F}_1^*, \ldots, \widetilde{F}_4^*$ associated with $u_1, \ldots, u_4$ are [see Eq. (39)]

$$\widetilde{F}_1^* = -m\{2\dot{u}_1 + \dot{u}_4 - u_3[2u_2 + (q_4 - L)u_3]\} \tag{51}$$

$$\widetilde{F}_2^* = -m[2\dot{u}_2 + 2u_4u_3 + (q_4 - L)\dot{u}_3 - 2u_3u_1] \tag{52}$$

$$\widetilde{F}_3^* = -m(q_4 - L)[\dot{u}_2 + 2u_4u_3 + (q_4 - L)\dot{u}_3 + u_3u_1] - \frac{mL^2}{3}\dot{u}_3 \tag{53}$$

$$\widetilde{F}_4^* = -m\{\dot{u}_1 + \dot{u}_4 - u_3[u_2 + (q_4 - L)u_3]\} \tag{54}$$

These expressions have a number of advantages over their counterparts, Eqs. (31)–(34). Not only are Eqs. (51)–(54) free of trigonometric functions and shorter than Eqs. (31)–(34), but each contains only two of $\dot{u}_1, \ldots, \dot{u}_4$, whereas each of Eqs. (31)–(34) involves three of $\ddot{q}_1, \ldots, \ddot{q}_4$. The underlying reason for this is that $u_1, \ldots, u_4$ are less strongly coupled dynamically than are $\dot{q}_1, \ldots, \dot{q}_4$, a fact that can be ascertained by forming inertia coefficients [see Eqs. (3) and (4)] for both choices of generalized speeds. In the first case,

$$M_{12} = 0 \qquad M_{13} = -m(q_4 + L)\sin q_3 \qquad M_{14} = m\cos q_3 \tag{55}$$

$$M_{23} = m(q_4 + L)\cos q_3 \qquad M_{24} = m\sin q_3 \qquad M_{34} = 0 \tag{56}$$

while in the second case,

$$\widetilde{M}_{12} = 0 \qquad \widetilde{M}_{13} = 0 \qquad \widetilde{M}_{14} = m \tag{57}$$

$$\widetilde{M}_{23} = m(q_4 - L) \qquad \widetilde{M}_{24} = 0 \qquad \widetilde{M}_{34} = 0 \tag{58}$$

Thus, only two uncouplings occur when $\dot{q}_1, \ldots, \dot{q}_4$ are used as generalized speeds, whereas four take place when generalized speeds are defined in accordance with Eqs. (41)–(44).

4.5 DYNAMICAL EQUATIONS

Given a system S possessing n degrees of freedom in a Newtonian reference frame N, let $u_1, \ldots, u_n$ be generalized speeds for S in N (see Sec. 1.21), and form $F_1, \ldots, F_n$, the associated generalized active forces for S in N (see Sec. 4.1), and $F_1^*, \ldots, F_n^*$, the associated generalized inertia forces for S in N (see Sec. 4.3). Then all motions of S are governed by the equations

$$F_r + F_r^* = 0 \qquad (r = 1, \ldots, n) \tag{1}$$

These equations are called *Kane's dynamical equations*.

Derivation Regard S as composed of ν particles $P_1, \ldots, P_\nu$ having masses $m_1, \ldots, m_\nu$, respectively, and let $\mathbf{R}_i$ be the resultant of all contact and body forces acting on P_i. Then, in accordance with Newton's second law,

$$\mathbf{R}_i = m_i\mathbf{a}_i \qquad (i = 1, \ldots, \nu) \tag{2}$$

where $\mathbf{a}_i$ is the acceleration of P_i in N. Equivalently, if the inertia force $\mathbf{R}_i^*$ for P_i in N is defined as

$$\mathbf{R}_i^* \triangleq -m_i\mathbf{a}_i \qquad (i = 1, \ldots, \nu) \tag{3}$$

then

$$\mathbf{R}_i + \mathbf{R}_i^* \underset{(1,2)}{=} 0 \qquad (i = 1, \ldots, \nu) \tag{4}$$

and dot-multiplication with $\mathbf{v}_r^{P_i}$, the rth partial velocity of P_i in N (see Sec. 1.21), gives

$$\mathbf{v}_r^{P_i} \cdot \mathbf{R}_i + \mathbf{v}_r^{P_i} \cdot \mathbf{R}_i^* = 0 \qquad (r = 1, \ldots, n;\ i = 1, \ldots, \nu) \tag{5}$$

Summing over all particles of S, one can thus write

$$\sum_{i=1}^{\nu} \mathbf{v}_r^{P_i} \cdot \mathbf{R}_i + \sum_{i=1}^{\nu} \mathbf{v}_r^{P_i} \cdot \mathbf{R}_i^* \underset{(5)}{=} 0 \qquad (r = 1, \ldots, n) \tag{6}$$

and, using Eqs. (4.1.1) and (4.3.2), one arrives at Eq. (1).

Example Suppose that the system S described in the example in Sec. 4.3 moves in N in the absence of external forces, and let σ be a function of q and u_4 such that the force $\mathbf{R}$ exerted on P by the spring and damper is given by

$$\mathbf{R} = -\sigma \mathbf{b}_1 \tag{7}$$

Then the force exerted on B by the spring and damper is equal to $-\mathbf{R}$. The velocity in N of the point Q (see Fig. 4.3.2) at which the latter force is applied is

$$\mathbf{v}^Q = \mathbf{v}^{B^*} + \boldsymbol{\omega} \times (-h\mathbf{b}_1 + b\mathbf{b}_2) \tag{8}$$

or, in view of Eq. (4.3.42),

$$\mathbf{v}^Q = \mathbf{v}^{B^*} - bu_3\mathbf{b}_1 - hu_3\mathbf{b}_2 + (bu_1 + hu_2)\mathbf{b}_3 \tag{9}$$

Hence, the partial velocities of Q are

$$\mathbf{v}_1^Q = \mathbf{v}_1^{B^*} + b\mathbf{b}_3 \qquad \mathbf{v}_2^Q = \mathbf{v}_2^{B^*} + h\mathbf{b}_3 \qquad \mathbf{v}_3^Q = \mathbf{v}_3^{B^*} - b\mathbf{b}_1 \tag{10}$$

$$\mathbf{v}_r^Q = \mathbf{v}_r^{B^*} \qquad (r = 4, \ldots, 7) \tag{11}$$

and the generalized active forces for S in N, found by substituting into

$$F_r \underset{(4.1.1)}{=} \mathbf{v}_r^P \cdot \mathbf{R} + \mathbf{v}_r^Q \cdot (-\mathbf{R}) = (\mathbf{v}_r^P - \mathbf{v}_r^Q) \cdot \mathbf{R} \qquad (r = 1, \ldots, 7) \tag{12}$$

are [see Table 4.3.1 for $\mathbf{v}_r^P$ and $\mathbf{v}_r^{B^*}$ $(r = 1, \ldots, 7)$]

$$F_r = \begin{cases} 0 & (r = 1, 2, 3, 5, 6, 7) \\ -\sigma & (r = 4) \end{cases} \tag{13}$$

With $r = 1$, one thus obtains the dynamical equation [see Eq. (4.3.59) for F_1^*]

$$-\dot{u}_1 B_1 + (B_3 - B_2)u_2u_3 + \frac{mM}{m + M} b(2u_2u_4 + q\dot{u}_2 - b\dot{u}_1 - qu_1u_3 - bu_2u_3) \underset{(1,13)}{=} 0 \tag{14}$$

and similarly for $r = 2, \ldots, 7$. For instance, the dynamical equation corresponding to $r = 4$ is

$$-\sigma + \frac{mM}{m+M}[b(\dot{u}_3 - u_1u_2) - \dot{u}_4 + q(u_2^2 + u_3^2)] = 0 \tag{15}$$

4.6 LINEARIZED DYNAMICAL EQUATIONS

When the formulation of dynamical equations (see Sec. 4.5) for a system S is undertaken for the purpose of analyzing the stability of a particular motion of S or in connection with, say, designing a control system, it may occur that one is interested solely in linearized dynamical equations, that is, equations obtained from dynamical equations by dropping all terms of second or higher degree in some (or all) of $q_1, \ldots, q_n, u_1, \ldots, u_n$. Such equations can be formulated, without first writing exact dynamical equations, by proceeding as follows: Develop fully nonlinear expressions for angular velocities of rigid bodies belonging to S, for velocities of mass centers of such bodies, and for velocities of particles belonging to S, and use these to determine partial angular velocities and partial velocities (see Sec. 1.28) by inspection. Next, linearize all angular velocities, velocities, partial angular velocities, and partial velocities, and use the linear forms to construct generalized active forces (see Sec. 4.1) and generalized inertia forces (see Sec. 4.3), discarding all terms of second or higher degree. Finally, substitute into Eq. (4.5.1) to obtain the desired equations.

Example In the absence of external forces, the system S considered in the example in Sec. 4.3 can move in a Newtonian reference frame in such a way that $q = u_2 = \cdots = u_7 = 0$ (but $u_1 \neq 0$), provided the spring is unstretched when $q = 0$ and the damper exerts no force when $u_4 = 0$. To study the stability of such a motion, one may wish to construct dynamical equations linearized in q, $u_2, \ldots, u_7$. To obtain these, begin by expressing $\boldsymbol{\omega}$, $\mathbf{v}^{B^*}$, and $\mathbf{v}^P$ as in Eqs. (4.3.42)–(4.3.51), and record partial angular velocities and partial velocities as in Table 4.3.1. Next, linearize, obtaining for $\boldsymbol{\omega}$, $\mathbf{v}^{B^*}$, and $\mathbf{v}^P$,

$$\boldsymbol{\omega} = u_1\mathbf{b}_1 + u_2\mathbf{b}_2 + u_3\mathbf{b}_3 \tag{1}$$

$$\mathbf{v}^{B^*} = \left[u_5 + \frac{m}{m+M}(bu_3 - u_4)\right]\mathbf{b}_1 + u_6\mathbf{b}_2 + \left(u_7 - \frac{mbu_1}{m+M}\right)\mathbf{b}_3 \tag{2}$$

$$\mathbf{v}^P = \left[u_5 - \frac{M}{m+M}(bu_3 - u_4)\right]\mathbf{b}_1 + u_6\mathbf{b}_2 + \left(u_7 + \frac{Mbu_1}{m+M}\right)\mathbf{b}_3 \tag{3}$$

Note that Table 4.3.1 need not be altered since it is free of nonlinearities. Now differentiate Eqs. (1), (2), and (3) with respect to time in N to find

$$\boldsymbol{\alpha} = \dot{u}_1\mathbf{b}_1 + \dot{u}_2\mathbf{b}_2 + \dot{u}_3\mathbf{b}_3 \tag{4}$$

$$\mathbf{a}^{B^*} = \left[\dot{u}_5 + \frac{m}{m+M}(b\dot{u}_3 - \dot{u}_4) - \frac{mbu_1u_2}{m+M}\right]\mathbf{b}_1$$
$$+ \left[\dot{u}_6 - u_1\left(u_7 - \frac{mbu_1}{m+M}\right)\right]\mathbf{b}_2 + \left(\dot{u}_7 - \frac{mb\dot{u}_1}{m+M} + u_1u_6\right)\mathbf{b}_3 \tag{5}$$

$$\mathbf{a}^P = \left[\dot{u}_5 - \frac{M}{m+M}(b\dot{u}_3 - \dot{u}_4) + \frac{Mbu_1u_2}{m+M}\right]\mathbf{b}_1$$
$$+ \left[\dot{u}_6 - u_1\left(u_7 + \frac{Mbu_1}{m+M}\right)\right]\mathbf{b}_2 + \left(\dot{u}_7 + \frac{Mb\dot{u}_1}{m+M} + u_1u_6\right)\mathbf{b}_3 \quad (6)$$

Representing the action of the spring and damper as in the example in Sec. 4.5, and assuming that σ vanishes when $q = u_4 = 0$, one can expand σ in a Taylor series in q and u_4, retaining only terms linear in these variables, which gives

$$\sigma = \frac{\partial\sigma}{\partial q}q + \frac{\partial\sigma}{\partial u_4}u_4 \quad (7)$$

where the partial derivatives are evaluated at $q = u_4 = 0$. Proceeding as in the derivation of Eq. (4.5.13), one then obtains

$$F_r = \begin{cases} 0 & (r = 1, 2, 3, 5, 6, 7) \\ -\left(\dfrac{\partial\sigma}{\partial q}q + \dfrac{\partial\sigma}{\partial u_4}u_4\right) & (r = 4) \end{cases} \quad (8)$$

The linearized form of the inertia torque $\mathbf{T}^*$ for B in N is [see Eqs. (1) and (4)]

$$\mathbf{T}^* \underset{(4.3.8)}{=} -\dot{u}_1B_1\mathbf{b}_1 - [\dot{u}_2B_2 + (B_1 - B_3)u_3u_1]\mathbf{b}_2$$
$$- [\dot{u}_3B_3 + (B_2 - B_1)u_1u_2]\mathbf{b}_3 \quad (9)$$

where B_1, B_2, B_3 are given in Eqs. (4.3.55), and generalized inertia forces now can be formed as

$$F_r^* = \boldsymbol{\omega}_r \cdot \mathbf{T}^* + \mathbf{v}_r^{B^*} \cdot (-M\mathbf{a}^{B^*}) + \mathbf{v}_r^P \cdot (-m\mathbf{a}^P) \quad (r = 1, \ldots, 7) \quad (10)$$

so that, for $r = 1$ [see Table 4.3.1 and Eqs. (5) and (6)],

$$F_1^* = -\left(B_1 + \frac{mMb^2}{m+M}\right)\dot{u}_1 \quad (11)$$

and similarly for $r = 2, \ldots, 7$. Finally, substitution into Eq. (4.5.1) yields

$$\dot{u}_1 \underset{(8,11)}{=} 0 \quad (12)$$

and so forth. For instance, the linearized equation corresponding to $r = 4$ is

$$-\left(\frac{\partial\sigma}{\partial q}q + \frac{\partial\sigma}{\partial u_4}u_4\right) + \frac{mM}{m+M}(b\dot{u}_3 - \dot{u}_4 - bu_1u_2) \underset{(8,10)}{=} 0 \quad (13)$$

Equations (12) and (13) can be generated, alternatively, by linearizing Eqs. (4.5.14) and (4.5.15).

4.7 COMPUTERIZATION OF SYMBOL MANIPULATION

The labor required to formulate dynamical equations (see Sec. 4.5) can become very burdensome when a system consists of more than a few rigid bodies and/or particles. In that event, it is advantageous to resort to the use of a computer to perform such tasks as differentiations, multiplications, substitutions, and so forth, in *literal* form. Indeed, when one's ultimate goal is the numerical solution of differential equations of motion, one may be able to employ computerized symbol manipulation to bypass the explicit writing of equations of motion, that is, to proceed directly from a few handwritten equations to the creation of a computer program that yields simulation results. An additional advantage of this procedure over that of formulating explicit equations of motion by hand and then writing a program to solve them is that it offers one fewer opportunities to make mistakes.

Example Figure 4.7.1 is a schematic representation of a space shuttle A equipped with a manipulator arm consisting of n pin-connected bars $B_1, \ldots, B_n$. Simulations of motions of this system are to be performed in order to explore effects of arm movements on the motion of body A. Specifically, the angles $q_1, \ldots, q_n$ indicated in Fig. 4.7.1 will be specified as functions of time; control forces acting on A will be presumed to be equivalent to a couple of torque $\mathbf{T}$ together with a force $\mathbf{S}$ applied at the mass center A^* of A, where $\mathbf{T}$ and $\mathbf{S}$ are prescribed functions of generalized speeds characterizing the motion of A in a Newtonian reference frame N; and the time-history of $\boldsymbol{\omega}^A$, the angular velocity of A in N, and $\mathbf{v}^{A^*}$, the velocity of A^* in N, will then be found by

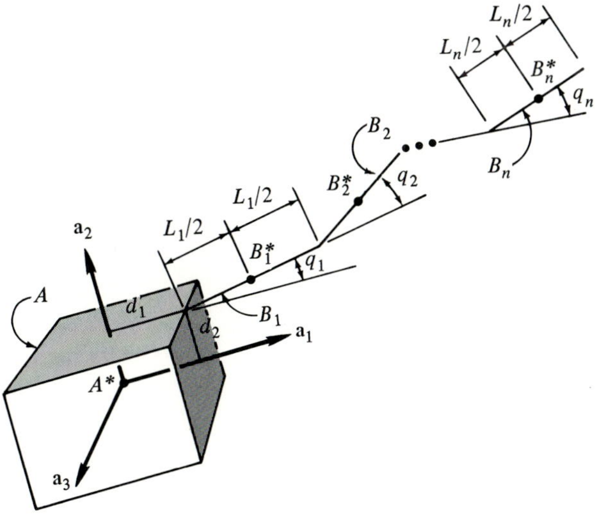

Figure 4.7.1

integrating numerically the differential equations governing the generalized speeds defined as

$$u_j \triangleq \begin{cases} \boldsymbol{\omega}^A \cdot \mathbf{a}_j & (j = 1, 2, 3) \\ \mathbf{v}^{A*} \cdot \mathbf{a}_{j-3} & (j = 4, 5, 6) \end{cases} \tag{1}$$

where $\mathbf{a}_1$, $\mathbf{a}_2$, $\mathbf{a}_3$ form a dextral set of mutually perpendicular unit vectors fixed in A, each parallel to a central principal axis of inertia of A.

To locate the mass center $B_i{}^*$ of B_i relative to A^*, we introduce $\mathbf{p}_i$ as

$$\mathbf{p}_i \triangleq \left[d_1 + \sum_{j=1}^{i} L_j \cos\left(\sum_{k=1}^{j} q_k\right) - \frac{L_j}{2}\cos\left(\sum_{k=1}^{i} q_k\right)\right]\mathbf{a}_1 + \left[d_2 + \sum_{j=1}^{i} L_j \sin\left(\sum_{k=1}^{j} q_k\right) - \frac{L_j}{2}\sin\left(\sum_{k=1}^{i} q_k\right)\right]\mathbf{a}_2 \qquad (i = 1, \ldots, n) \tag{2}$$

where d_1 and d_2 are distances used to select the point at which B_1 is attached to A, as shown in Fig. 4.7.1, and L_j is the length of B_j. Next, we express the angular velocities of A and $B_1, \ldots, B_n$ in N as

$$\boldsymbol{\omega}^A \underset{(1)}{=} u_1\mathbf{a}_1 + u_2\mathbf{a}_2 + u_3\mathbf{a}_3 \tag{3}$$

$$\boldsymbol{\omega}^{B_i} = u_1\mathbf{a}_1 + u_2\mathbf{a}_2 + \left(u_3 + \sum_{j=1}^{i} \dot{q}_j\right)\mathbf{a}_3 \qquad (r = 1, \ldots, n) \tag{4}$$

and write the velocities of A^* and $B_i{}^*$ in N as

$$\mathbf{v}^{A*} \underset{(1)}{=} u_4\mathbf{a}_1 + u_5\mathbf{a}_2 + u_6\mathbf{a}_3 \tag{5}$$

$$\mathbf{v}^{B_i*} = {}^N\mathbf{v}^{A*} + \frac{{}^A d\mathbf{p}_i}{dt} + \boldsymbol{\omega}^A \times \mathbf{p}_i \tag{6}$$

Furthermore, we note that the angular accelerations of A and $B_1, \ldots, B_n$ in N are given by

$$\boldsymbol{\alpha}^A = \frac{{}^A d\boldsymbol{\omega}^A}{dt} \tag{7}$$

$$\boldsymbol{\alpha}^{B_i} = \frac{{}^A d\boldsymbol{\omega}^{B_i}}{dt} + \boldsymbol{\omega}^A \times \boldsymbol{\omega}^{B_i} \qquad (i = 1, \ldots, n) \tag{8}$$

while the accelerations of A^* and $B_1{}^*, \ldots, B_n{}^*$ in N are

$$\mathbf{a}^{A*} = \frac{{}^A d\mathbf{v}^{A*}}{dt} + \boldsymbol{\omega}^A \times \mathbf{v}^{A*} \tag{9}$$

$$\mathbf{a}^{B_i*} = \frac{{}^A d\mathbf{v}^{B_i*}}{dt} + \boldsymbol{\omega}^A \times \mathbf{v}^{B_i*} \tag{10}$$

If A_1, A_2, A_3 denote the central principal moments of inertia of A, and J_i is the central transverse moment of inertia of B_i, then the central inertia dyadics of A and B_i are

$$\mathbf{I}_A = A_1\mathbf{a}_1\mathbf{a}_1 + A_2\mathbf{a}_2\mathbf{a}_2 + A_3\mathbf{a}_3\mathbf{a}_3 \tag{11}$$

and

$$\mathbf{I}_i = J_i\left[\sin^2\left(\sum_{j=1}^{i} q_j\right)\mathbf{a}_1\mathbf{a}_1 - \sin\left(\sum_{j=1}^{i} q_j\right)\cos\left(\sum_{j=1}^{i} q_j\right)(\mathbf{a}_1\mathbf{a}_2 + \mathbf{a}_2\mathbf{a}_1)\right.$$
$$\left. + \cos^2\left(\sum_{j=1}^{i} q_j\right)\mathbf{a}_2\mathbf{a}_2 + \mathbf{a}_3\mathbf{a}_3\right] \qquad (i = 1, \ldots, n) \tag{12}$$

and the inertia torques for A and B_i in N are given by

$$\mathbf{T}_A{}^* = -\boldsymbol{\alpha}^A \cdot \mathbf{I}_A - \boldsymbol{\omega}^A \times \mathbf{I}_A \cdot \boldsymbol{\omega}^A \tag{13}$$

$$\mathbf{T}_i{}^* = -\boldsymbol{\alpha}^{B_i} \cdot \mathbf{I}_i - \boldsymbol{\omega}^{B_i} \times \mathbf{I}_i \cdot \boldsymbol{\omega}^{B_i} \tag{14}$$

The generalized inertia forces (see Sec. 4.3) associated with $u_1, \ldots, u_6$ can be written

$$F_r{}^* = \boldsymbol{\omega}_r{}^A \cdot \mathbf{T}_A{}^* - m_A\mathbf{v}_r{}^{A*} \cdot \mathbf{a}^{A*} + \sum_{i=1}^{n} (\boldsymbol{\omega}_r{}^{B_i} \cdot \mathbf{T}_i{}^* - m_i\mathbf{v}_r{}^{B_i*} \cdot \mathbf{a}^{B_i*})$$
$$(r = 1, \ldots, 6) \tag{15}$$

where m_A is the mass of A, m_i is the mass of B_i, and $\boldsymbol{\omega}_r{}^A$, $\mathbf{v}_r{}^{A*}$, $\boldsymbol{\omega}_r{}^{B_i}$, $\mathbf{v}_r{}^{B_i*}$ are, respectively, the rth partial angular velocity of A in N, the rth partial velocity of A^* in N, and so forth; and, after expressing the aforementioned control torque **T** and control force **S** as

$$\mathbf{T} = T_1\mathbf{a}_1 + T_2\mathbf{a}_2 + T_3\mathbf{a}_3 \tag{16}$$

$$\mathbf{S} = S_1\mathbf{a}_1 + S_2\mathbf{a}_2 + S_3\mathbf{a}_3 \tag{17}$$

one can formulate the generalized active forces (see Sec. 4.1) associated with $u_1, \ldots, u_6$ as

$$F_r = \boldsymbol{\omega}_r{}^A \cdot \mathbf{T} + \mathbf{v}_r{}^{A*} \cdot \mathbf{S} \qquad (r = 1, \ldots, 6) \tag{18}$$

To generate the differential equations governing $u_1, \ldots, u_6$, what remains to be done is to perform the differentiations, dot multiplications, and so forth, indicated in Eqs. (6)–(18) and then to substitute the resulting expressions for $F_r{}^*$ and F_r $(r = 1, \ldots, 6)$ into Eq. (4.5.1). These are the tasks that can be relegated to a computer. For example, a FORMAC† program created for this purpose yields, for $r = 1$ and $n = 2$, the equations shown in Fig. 4.7.2, which, written in terms of symbols employed so far, are the equations

$$\begin{aligned}
F_1^* = &- m_1(d_2 + \tfrac{1}{2}L_1 \sin q_1)[\dot u_6 + \tfrac{1}{2}L_1 u_1 u_3 \cos q_1 + L_1 u_1 \dot q_1 \cos q_1 \\
&+ \tfrac{1}{2}L_1 \dot u_1 \sin q_1 + \tfrac{1}{2}L_1 u_2 u_3 \sin q_1 + L_1 u_2 \dot q_1 \sin q_1 - \tfrac{1}{2}L_1 \dot u_2 \cos q_1 \\
&+ d_1 u_1 u_3 - d_1 \dot u_2 + d_2 \dot u_1 + d_2 u_2 u_3 + u_5 u_1 - u_2 u_4] \\
&- m_2[d_2 + \tfrac{1}{2}L_2 \sin(q_1 + q_2) + L_1 \sin q_1][\dot u_6 + \tfrac{1}{2}L_2 u_1 u_3 \cos(q_1 + q_2) \\
&+ L_2 u_1 \dot q_1 \cos(q_1 + q_2) + L_2 u_1 \dot q_2 \cos(q_1 + q_2) + \tfrac{1}{2}L_2 \dot u_1 \sin(q_1 + q_2) \\
&+ \tfrac{1}{2}L_2 u_2 u_3 \sin(q_1 + q_2) + L_2 u_2 \dot q_1 \sin(q_1 + q_2) + L_2 u_2 \dot q_2 \sin(q_1 + q_2) \\
&- \tfrac{1}{2}L_2 \dot u_2 \cos(q_1 + q_2) + L_1 u_1 u_3 \cos q_1 + 2L_1 u_1 \dot q_1 \cos q_1 + L_1 \dot u_1 \sin q_1 \\
&+ L_1 u_2 u_3 \sin q_1 + 2L_1 u_2 \dot q_1 \sin q_1 - L_1 \dot u_2 \cos q_1 + d_1 u_1 u_3 - d_1 \dot u_2 \\
&+ d_2 \dot u_1 + d_2 u_2 u_3 + u_5 u_1 - u_2 u_4] - J_2 u_2(u_3 + \dot q_1 + \dot q_2) \\
&- J_2(\dot u_1 + u_2 \dot q_1 + u_2 \dot q_2) \sin^2(q_1 + q_2) + J_2(\dot u_2 - u_1 \dot q_1 \\
&- u_1 \dot q_2) \sin(q_1 + q_2) \cos(q_1 + q_2) - J_1 u_2(u_3 + \dot q_1) - J_1(\dot u_1 \\
&+ u_2 \dot q_1) \sin^2 q_1 + J_1(\dot u_2 - u_1 \dot q_1) \sin q_1 \cos q_1 - A_1 \dot u_1 + u_2 u_3(A_2 - A_3) \\
&+ (u_3 + \dot q_1)(- J_1 u_1 \sin q_1 \cos q_1 + J_1 u_2 \cos^2 q_1 + (u_3 + \dot q_1 \\
&+ \dot q_2)[- J_2 u_1 \sin(q_1 + q_2) \cos(q_1 + q_2) + J_2 u_2 \cos^2(q_1 + q_2)] \qquad (19)
\end{aligned}$$

$$F_1 = T_1 \qquad (20)$$

The generalized inertia forces and generalized active forces corresponding to $r = 2, \ldots, 6$ are obtained similarly. Now, since our goal is the numerical integration of the equations of motion, it is desirable to express the system of such equations in the form

$$X\dot u = Y \qquad (21)$$

where $\dot u$ is the 6×1 matrix having $\dot u_1, \ldots, \dot u_6$ as elements, X is a 6×6 matrix whose elements are functions of q_1 and q_2, and Y is a 6×1 matrix having as elements functions of $q_1, q_2, \dot q_1, \dot q_2, \ddot q_1, \ddot q_2, u_1, \ldots, u_6$. The matrices X and

† FORMAC is one of several symbol manipulation languages available at present (K. Bahr, "SHARE FORMAC/FORMAC73," SHARE Program Library Agency, Triangle Universities Computation Center, Research Triangle Park, N.C., Program 360D-03.3.013, July 1975). Some others are SYMBAL (M. E. Engeli, "User's Manual for the Formula Manipulation Language SYMBAL," University of Texas at Austin Computation Center, July 1969), MACSYMA ("MACSYMA Reference Manual," The Mathlab Laboratory for Computer Science, Version 9, M.I.T., 1977), REDUCE 2 (A. C. Hearn, "REDUCE 2 User's Manual," 2d ed., University of Utah Computational Physics Group Report No. UCP-19, 1973).

```
FSTAR(1) =  - M(1) ( D2 + 1/2 L(1) SIN ( Q(1) ) ) ( UD(6) + 1/2 L(1) U(1
-------------------------------------------------------------------------
) U(3) COS ( Q(1) ) + L(1) U(1) QD(1) COS ( Q(1) ) + 1/2 L(1) UD(1) SIN
-------------------------------------------------------------------------
 ( Q(1) ) + 1/2 L(1) U(2) U(3) SIN ( Q(1) ) + L(1) U(2) QD(1) SIN ( Q(1)
-------------------------------------------------------------------------
 ) - 1/2 L(1) UD(2) COS ( Q(1) ) + D1 U(1) U(3) - D1 UD(2) + D2 UD(1) +
-------------------------------------------------------------------------
D2 U(2) U(3) + U(5) U(1) - U(2) U(4) ) - M(2) ( D2 + 1/2 L(2) SIN ( Q(1)
-------------------------------------------------------------------------
 + Q(2) ) + L(1) SIN ( Q(1) ) ) ( UD(6) + 1/2 L(2) U(1) U(3) COS ( Q(1)
-------------------------------------------------------------------------
 + Q(2) ) + L(2) U(1) QD(1) COS ( Q(1) + Q(2) ) + L(2) U(1) QD(2) COS (
-------------------------------------------------------------------------
Q(1) + Q(2) ) + 1/2 L(2) UD(1) SIN ( Q(1) + Q(2) ) + 1/2 L(2) U(2) U(3)
-------------------------------------------------------------------------
SIN ( Q(1) + Q(2) ) + L(2) U(2) QD(1) SIN ( Q(1) + Q(2) ) + L(2) U(2) QD
-------------------------------------------------------------------------
(2) SIN ( Q(1) + Q(2) ) - 1/2 L(2) UD(2) COS ( Q(1) + Q(2) ) + L(1) U(1)
-------------------------------------------------------------------------
 U(3) COS ( Q(1) ) + 2 L(1) U(1) QD(1) COS ( Q(1) ) + L(1) UD(1) SIN ( Q
-------------------------------------------------------------------------
(1) ) + L(1) U(2) U(3) SIN ( Q(1) ) + 2 L(1) U(2) QD(1) SIN ( Q(1) ) - L
-------------------------------------------------------------------------
(1) UD(2) COS ( Q(1) ) + D1 U(1) U(3) - D1 UD(2) + D2 UD(1) + D2 U(2) U(
-------------------------------------------------------------------------
3) + U(5) U(1) - U(2) U(4) ) - J(2) U(2) ( U(3) + QD(1) + QD(2) ) - J(2)
-------------------------------------------------------------------------

                                            2
 ( UD(1) + U(2) QD(1) + U(2) QD(2) ) SIN   ( Q(1) + Q(2) ) + J(2) ( UD(2)
-------------------------------------------------------------------------
 - U(1) QD(1) - U(1) QD(2) ) SIN ( Q(1) + Q(2) ) COS ( Q(1) + Q(2) ) - J
-------------------------------------------------------------------------

                                                             2
(1) U(2) ( U(3) + QD(1) ) - J(1) ( UD(1) + U(2) QD(1) ) SIN   ( Q(1) ) +
-------------------------------------------------------------------------
J(1) ( UD(2) - U(1) QD(1) ) SIN ( Q(1) ) COS ( Q(1) ) - A1 UD(1) + U(2)
-------------------------------------------------------------------------
U(3) ( A2 - A3 ) + ( U(3) + QD(1) ) (  - J(1) U(1) SIN ( Q(1) ) COS ( Q(
-------------------------------------------------------------------------

                       2
1) ) + J(1) U(2) COS   ( Q(1) ) ) + ( U(3) + QD(1) + QD(2) ) (  - J(2) U(
-------------------------------------------------------------------------

                                                             2
1) SIN ( Q(1) + Q(2) ) COS ( Q(1) + Q(2) ) + J(2) U(2) COS   ( Q(1) + Q(2
-------------------------------------------------------------------------
) ) )
-----




F(1) = T(1)
-----------
```

Figure 4.7.2

Y can be formed readily with the aid of the computer. For example, the matrix elements X_{21} and Y_5 are displayed in Fig. 4.7.3. Here SQ(1) $\triangleq$ SIN (Q(1)), CQ(1) $\triangleq$ COS (Q(1)), SQ(2) $\triangleq$ SIN (Q(1) + Q(2)), CQ(2) $\triangleq$ COS (Q(1) + Q(2)); and QDD(1) and QDD(2) stand, respectively, for $\ddot{q}_1$ and $\ddot{q}_2$. Moreover, to produce a computer program for the numerical solution of Eq. (21), one does not need to write out and then code the elements of X and Y. Instead, one can proceed directly from the symbol manipulation program that produces the elements of X and Y in literal form to a punched deck of cards, a disk file, or a magnetic tape to be incorporated in a numerical integration program.

The use of a computer to formulate literal equations of motion is particularly advantageous when one must deal with several configurations of a system under consideration. For instance, to obtain equations of motion for the system

```
                                                                      2
X(2,1) = 1/2 CQ(2) SQ(1) M(2) L(2) L(1) + 1/4 CQ(2) SQ(2) M(2) L(2)    +
---------------------------------------------------------------------------
                                                                         2
CQ(2) SQ(2) J(2) + 1/2 CQ(2) M(2) L(2) D2 + 1/4 SQ(1) CQ(1) M(1) L(1)
---------------------------------------------------------------------------
                                   2
 + SQ(1) CQ(1) M(2) L(1)    + SQ(1) CQ(1) J(1) + 1/2 SQ(1) M(1) L(1) D1
---------------------------------------------------------------------------
 + SQ(1) M(2) L(1) D1 + 1/2 CQ(1) SQ(2) M(2) L(2) L(1) + 1/2 CQ(1) M(1)
---------------------------------------------------------------------------
L(1) D2 + CQ(1) M(2) L(1) D2 + 1/2 SQ(2) M(2) L(2) D1 + M(1) D1 D2 + M(2
---------------------------------------------------------------------------
) D1 D2
-------
```

```
Y(5) =  - S(2) + MA (  - U(6) U(1) + U(3) U(4) ) + M(1) (  - SQ(1) L(1)
---------------------------------------------------------------------------
                                   2                          2
U(3) QD(1) - 1/2 SQ(1) L(1) U(1)    - 1/2 SQ(1) L(1) U(3)    - 1/2 SQ(1) L
---------------------------------------------------------------------------
         2
(1) QD(1)    + 1/2 CQ(1) L(1) U(1) U(2) + 1/2 CQ(1) L(1) QDD(1) + D1 U(1)
---------------------------------------------------------------------------
               2            2
 U(2) - D2 U(1)    - D2 U(3)    - U(6) U(1) + U(3) U(4) ) + M(2) ( 1/2 CQ(
---------------------------------------------------------------------------
2) L(2) U(1) U(2) + 1/2 CQ(2) L(2) QDD(1) + 1/2 CQ(2) L(2) QDD(2) - 2 SQ
---------------------------------------------------------------------------
                                        2                    2
(1) L(1) U(3) QD(1) - SQ(1) L(1) U(1)    - SQ(1) L(1) U(3)    - SQ(1) L(1)
---------------------------------------------------------------------------
       2
 QD(1)    + CQ(1) L(1) U(1) U(2) + CQ(1) L(1) QDD(1) - SQ(2) L(2) U(3) QD
---------------------------------------------------------------------------
(1) - SQ(2) L(2) U(3) QD(2) - SQ(2) L(2) QD(1) QD(2) - 1/2 SQ(2) L(2) U(
---------------------------------------------------------------------------
  2                        2                          2
1)    - 1/2 SQ(2) L(2) U(3)    - 1/2 SQ(2) L(2) QD(1)    - 1/2 SQ(2) L(2)
---------------------------------------------------------------------------
      2                            2            2
QD(2)    + D1 U(1) U(2) - D2 U(1)    - D2 U(3)    - U(6) U(1) + U(3) U(4) )
---------------------------------------------------------------------------
```

Figure 4.7.3

in Fig. 4.7.1 when the arm consists of three bars, rather than two, one simply reruns the symbol manipulation program with $n = 3$ in place of $n = 2$, obtaining, for example, the expression for F_1^* shown in Fig. 4.7.4, where L(3), M(3), J(3), Q(3), QD(3) respectively denote L_3, m_3, J_3, q_3, $\dot{q}_3$. A comparison of Figs. 4.7.2 and 4.7.4 reveals that the addition of B_3 to the system is accompanied by a substantial increase in the number of terms in the equations of motion, the derivation of which by hand is a burdensome chore. Moreover, by using the computer to produce the augmented versions of the matrices X and Y in Eq. (21), one is freed from the task of extracting 42 matrix elements from the expressions for $F_1^*, \ldots, F_6^*, F_1, \ldots, F_6$.

4.8 DISCRETE MULTI-DEGREE-OF-FREEDOM SYSTEMS

When the number of degrees of freedom of a system is sufficiently large, the process of solving numerically the differential equations governing the motion of the

```
FSTAR(1) =  - M(1) ( D2 + 1/2 L(1) SIN ( Q(1) ) ) ( UD(6) + 1/2 L(1) U(1
) U(3) COS ( Q(1) ) + L(1) U(1) QD(1) COS ( Q(1) ) + 1/2 L(1) UD(1) SIN
 ( Q(1) ) + 1/2 L(1) U(2) U(3) SIN ( Q(1) ) + L(1) U(2) QD(1) SIN ( Q(1)
 ) - 1/2 L(1) UD(2) COS ( Q(1) ) + D1 U(1) U(3) - D1 UD(2) + D2 UD(1) +
D2 U(2) U(3) - U(4) U(2) + U(5) U(1) ) - M(2) ( D2 + 1/2 L(2) SIN ( Q(1)
 + Q(2) ) + L(1) SIN ( Q(1) ) ) ( UD(6) + 1/2 L(2) U(1) U(3) COS ( Q(1)
 + Q(2) ) + L(2) U(1) QD(1) COS ( Q(1) + Q(2) ) + L(2) U(1) QD(2) COS (
Q(1) + Q(2) ) + 1/2 L(2) UD(1) SIN ( Q(1) + Q(2) ) + 1/2 L(2) U(2) U(3)
SIN ( Q(1) + Q(2) ) + L(2) U(2) QD(1) SIN ( Q(1) + Q(2) ) + L(2) U(2) QD
(2) SIN ( Q(1) + Q(2) ) - 1/2 L(2) UD(2) COS ( Q(1) + Q(2) ) + L(1) U(1)
 U(3) COS ( Q(1) ) + 2 L(1) U(1) QD(1) COS ( Q(1) ) + L(1) UD(1) SIN ( Q
(1) ) + L(1) U(2) U(3) SIN ( Q(1) ) + 2 L(1) U(2) QD(1) SIN ( Q(1) ) - L
(1) UD(2) COS ( Q(1) ) + D1 U(1) U(3) - D1 UD(2) + D2 UD(1) + D2 U(2) U(
3) - U(4) U(2) + U(5) U(1) ) - M(3) ( D2 + L(2) SIN ( Q(1) + Q(2) ) + 1/
2 L(3) SIN ( Q(3) + Q(1) + Q(2) ) + L(1) SIN ( Q(1) ) ) ( UD(6) + L(2) U
(1) U(3) COS ( Q(1) + Q(2) ) + 2 L(2) U(1) QD(1) COS ( Q(1) + Q(2) ) + 2
 L(2) U(1) QD(2) COS ( Q(1) + Q(2) ) + L(2) UD(1) SIN ( Q(1) + Q(2) ) +
L(2) U(2) U(3) SIN ( Q(1) + Q(2) ) + 2 L(2) U(2) QD(1) SIN ( Q(1) + Q(2)
 ) + 2 L(2) U(2) QD(2) SIN ( Q(1) + Q(2) ) - L(2) UD(2) COS ( Q(1) + Q(2
) ) + L(3) QD(3) U(1) COS ( Q(3) + Q(1) + Q(2) ) + L(3) QD(3) U(2) SIN
 ( Q(3) + Q(1) + Q(2) ) + 1/2 L(3) U(1) U(3) COS ( Q(3) + Q(1) + Q(2) )
 + L(3) U(1) QD(1) COS ( Q(3) + Q(1) + Q(2) ) + L(3) U(1) QD(2) COS ( Q(
3) + Q(1) + Q(2) ) + 1/2 L(3) UD(1) SIN ( Q(3) + Q(1) + Q(2) ) + 1/2 L(3
) U(2) U(3) SIN ( Q(3) + Q(1) + Q(2) ) + L(3) U(2) QD(1) SIN ( Q(3) + Q(
1) + Q(2) ) + L(3) U(2) QD(2) SIN ( Q(3) + Q(1) + Q(2) ) - 1/2 L(3) UD(2
) COS ( Q(3) + Q(1) + Q(2) ) + L(1) U(1) U(3) COS ( Q(1) ) + 2 L(1) U(1)
 QD(1) COS ( Q(1) ) + L(1) UD(1) SIN ( Q(1) ) + L(1) U(2) U(3) SIN ( Q(1
) ) + 2 L(1) U(2) QD(1) SIN ( Q(1) ) - L(1) UD(2) COS ( Q(1) ) + D1 U(1)
 U(3) - D1 UD(2) + D2 UD(1) + D2 U(2) U(3) - U(4) U(2) + U(5) U(1) ) - J
(2) U(2) ( U(3) + QD(1) + QD(2) ) - J(2) ( UD(1) + U(2) QD(1) + U(2) QD(
        2
2) ) SIN  ( Q(1) + Q(2) ) + J(2) ( UD(2) - U(1) QD(1) - U(1) QD(2) ) SIN
 ( Q(1) + Q(2) ) COS ( Q(1) + Q(2) ) - J(3) U(2) ( QD(3) + U(3) + QD(1)
                                                                     2
 + QD(2) ) - J(3) ( UD(1) + QD(3) U(2) + U(2) QD(1) + U(2) QD(2) ) SIN
( Q(3) + Q(1) + Q(2) ) + J(3) ( UD(2) - QD(3) U(1) - U(1) QD(1) - U(1)
QD(2) ) SIN ( Q(3) + Q(1) + Q(2) ) COS ( Q(3) + Q(1) + Q(2) ) - J(1) U(2
                                                  2
) ( U(3) + QD(1) ) - J(1) ( UD(1) + U(2) QD(1) ) SIN  ( Q(1) ) + J(1) (
UD(2) - U(1) QD(1) ) SIN ( Q(1) ) COS ( Q(1) ) - A1 UD(1) + U(2) U(3) (
A2 - A3 ) + ( QD(3) + U(3) + QD(1) + QD(2) ) (  - J(3) U(1) SIN ( Q(3)
                                                         2
 + Q(1) + Q(2) ) COS ( Q(3) + Q(1) + Q(2) ) + J(3) U(2) COS  ( Q(3) + Q(
1) + Q(2) ) ) + ( U(3) + QD(1) ) (  - J(1) U(1) SIN ( Q(1) ) COS ( Q(1)
                      2
 ) + J(1) U(2) COS  ( Q(1) ) ) + ( U(3) + QD(1) + QD(2) ) (  - J(2) U(1)
                                                       2
 SIN ( Q(1) + Q(2) ) COS ( Q(1) + Q(2) ) + J(2) U(2) COS  ( Q(1) + Q(2)
 ) )
```

Figure 4.7.4

system can become prohibitively time-consuming. In some situations this difficulty can be surmounted by transforming from certain coordinates associated with vibratory motions of the system under consideration to normal coordinates arising in connection with a related classical vibrations problem, introducing corresponding generalized speeds, formulating equations of motion in terms of these, and then excluding all but a few of the equations when performing numerical integrations. This method rests upon certain results of classical vibrations theory,† which we review before proceeding to an illustrative example.

If M and S are constant $n \times n$ matrices such that the matrix $M^{-1}S$ possesses distinct, nonvanishing eigenvalues, and if $x(t)$ is an $n \times 1$ matrix that satisfies the differential equation

$$M\ddot{x} + Sx = 0 \tag{1}$$

then an approximation to x can be obtained by forming the product of an $n \times \nu$ ($\nu \leq n$) constant matrix A and a $\nu \times 1$ time-dependent matrix $q(t)$, that is, by taking

$$x = Aq \tag{2}$$

For $\nu = n$, Eq. (2) furnishes directly the exact, general solution of Eq. (1) when A and q are constructed as follows.

Let $\lambda_1, \ldots, \lambda_\nu$ be the first ν eigenvalues of $M^{-1}S$, and introduce the $n \times 1$ matrix B_i as an eigenvector of $M^{-1}S$ corresponding to λ_i ($i = 1, \ldots, \nu$). [When $M^{-1}S$ is unsymmetric, but M is symmetric, one can, nevertheless, work with a symmetric matrix by taking advantage of the facts that the eigenvalues of $M^{-1}S$ are the same as those of the symmetric $n \times n$ matrix $W \triangleq (V^{-1})^T SV^{-1}$, where V is an upper-triangular $n \times n$ matrix such that $M = V^T V$ and that, if C_i is an eigenvector of W corresponding to λ_i, then $V^{-1}C_i$ is an eigenvector of $M^{-1}S$ corresponding to λ_i. When M is diagonal, then $V = M^{1/2}$.] Form normalization constants $N_1, \ldots, N_\nu$ as

$$N_i \triangleq (B_i^T M B_i)^{1/2} \qquad (i = 1, \ldots, \nu) \tag{3}$$

and normalized $n \times 1$ eigenvectors $A_1, \ldots, A_\nu$ as

$$A_i \triangleq \frac{B_i}{N_i} \qquad (i = 1, \ldots, \nu) \tag{4}$$

Now construct the $n \times \nu$ modal matrix A in accordance with

$$A \triangleq [A_1 \quad A_2 \quad \ldots \quad A_\nu] \tag{5}$$

Next, denoting the initial values of x and $\dot{x}$ by $x(0)$ and $\dot{x}(0)$, respectively, express the initial values of q and $\dot{q}$ as

$$q(0) = A^T Mx(0) \qquad \dot{q}(0) = A^T M\dot{x}(0) \tag{6}$$

† See S. Timoshenko, D. H. Young, and W. Weaver Jr., *Vibration Problems in Engineering*, chap. 4, 4th ed., John Wiley & Sons, New York, 1974.

Also, introduce circular frequencies $p_1, \ldots, p_\nu$ as

$$p_i = \lambda_i^{1/2} \qquad (i = 1, \ldots, \nu) \tag{7}$$

and define diagonal $\nu \times \nu$ matrices f and g as

$$f \triangleq \begin{bmatrix} \cos p_1 t & 0 & \ldots & 0 \\ 0 & \cos p_2 t & \ldots & 0 \\ \ldots & \ldots & \ldots & \ldots \\ 0 & 0 & \ldots & \cos p_\nu t \end{bmatrix} \tag{8}$$

$$g \triangleq \begin{bmatrix} p_1^{-1} \sin p_1 t & 0 & \ldots & 0 \\ 0 & p_2^{-1} \sin p_2 t & \ldots & 0 \\ \ldots & \ldots & \ldots & \ldots \\ 0 & 0 & \ldots & p_\nu^{-1} \sin p_\nu t \end{bmatrix} \tag{9}$$

Finally, take

$$q \triangleq fq(0) + g\dot{q}(0) \tag{10}$$

(The elements of q are called *normal coordinates*.)

The matrix A defined in Eq. (5) satisfies two equations that will be of interest to us shortly. If U is the $\nu \times \nu$ unit matrix and λ is the $\nu \times \nu$ diagonal matrix having λ_i as the element in the ith row (and ith column), then, if M and S are symmetric,

$$A^T M A = U \qquad A^T S A = \lambda \tag{11}$$

Example Figure 4.8.1 shows n spring-connected particles $P_1, \ldots, P_n$ ($n > 2$) and a cylindrical body C. The particles are constrained to move on a line D that is fixed in C, and only motions during which D remains in a plane that is fixed in a Newtonian reference frame N are to be considered. The system formed by

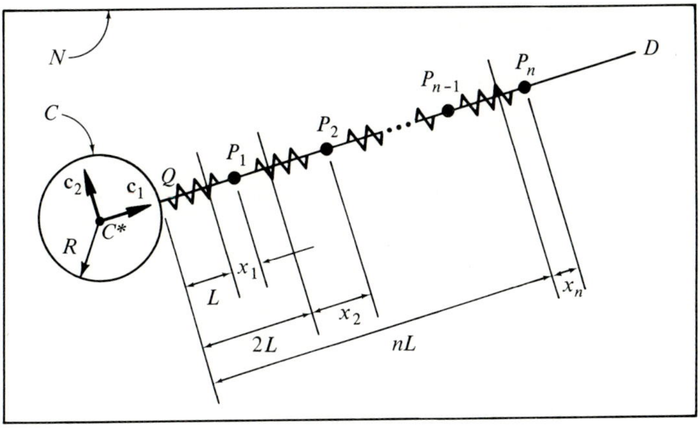

Figure 4.8.1

$P_1, \ldots, P_n$ and C is free of external forces. Its motion, presumed known for $t = 0$, is to be determined for $t > 0$.

To characterize the motion of C in N, introduce a dextral set of mutually perpendicular unit vectors $\mathbf{c}_1, \mathbf{c}_2, \mathbf{c}_3$ fixed in C as shown in Fig. 4.8.1 and define generalized speeds u_1, u_2, u_3 as

$$u_i \triangleq {}^N\mathbf{v}^{C^*} \cdot \mathbf{c}_i \qquad (i = 1, 2) \qquad u_3 \triangleq {}^N\boldsymbol{\omega}^C \cdot \mathbf{c}_3 \tag{12}$$

where ${}^N\mathbf{v}^{C^*}$ is the velocity in N of the mass center C^* of C and ${}^N\boldsymbol{\omega}^C$ is the angular velocity of C in N. Then it follows immediately that

$${}^N\mathbf{v}^{C^*} = u_1\mathbf{c}_1 + u_2\mathbf{c}_2 \qquad {}^N\boldsymbol{\omega}^C = u_3\mathbf{c}_3 \tag{13}$$

Next, after designating as L the common, natural length of all springs, let x_i be the distance from the point on D that P_i occupies when all springs are unstretched to the point on D with which P_i coincides at time t, regarding x_i as positive when the distance from point Q (Fig. 4.8.1) to P_i exceeds iL. Taking $\nu \le n$, express x_i as

$$x_i = \sum_{j=1}^{\nu} A_{ij}q_j \qquad (i = 1, \ldots, n) \tag{14}$$

where A_{ij} is an, as yet, totally unrestricted constant and q_j is an equally unrestricted function of t. Then the velocity of P_i in C is given by

$${}^C\mathbf{v}^{P_i} = \dot{x}_i\mathbf{c}_1 \underset{(14)}{=} \sum_{j=1}^{\nu} A_{ij}\dot{q}_j\mathbf{c}_1 \qquad (i = 1, \ldots, n) \tag{15}$$

which, if ν generalized speeds are defined [in addition to those introduced in Eqs. (12)] as

$$u_{3+k} \triangleq \dot{q}_k \qquad (k = 1, \ldots, \nu) \tag{16}$$

can be written

$${}^C\mathbf{v}^{P_i} \underset{(15,16)}{=} \sum_{j=1}^{\nu} A_{ij}u_{3+j}\mathbf{c}_1 \qquad (i = 1, \ldots, n) \tag{17}$$

The velocity of P_i in N can thus be expressed as

$$\begin{aligned} {}^N\mathbf{v}^{P_i} &= {}^N\mathbf{v}^{C^*} + {}^N\boldsymbol{\omega}^C \times [(R + iL + x_i)\mathbf{c}_1] + {}^C\mathbf{v}^{P_i} \\ &\underset{(13,14,17)}{=} \left(u_1 + \sum_{j=1}^{\nu} A_{ij}u_{3+j}\right)\mathbf{c}_1 + \left[u_2 + u_3\left(R + iL + \sum_{j=1}^{\nu} A_{ij}q_j\right)\right]\mathbf{c}_2 \\ &\qquad (i = 1, \ldots, n) \end{aligned} \tag{18}$$

where R is the radius of C, and partial angular velocities and partial velocities (see Sec. 1.2.1) of interest, formed by inspection of Eqs. (13) and (18), now can be recorded as in Table 4.8.1.

The angular acceleration of C in N is

$${}^N\boldsymbol{\alpha}^C \underset{(13)}{=} \dot{u}_3\mathbf{c}_3 \tag{19}$$

Table 4.8.1

r	$\boldsymbol{\omega}_r{}^C$	$\mathbf{v}_r{}^{C^*}$	$\mathbf{v}_r{}^{P_i}$ $(i = 1, \ldots, n)$
1	0	$\mathbf{c}_1$	$\mathbf{c}_1$
2	0	$\mathbf{c}_2$	$\mathbf{c}_2$
3	$\mathbf{c}_3$	0	$\left(R + iL + \sum_{j=1}^{\nu} A_{ij} q_j\right)\mathbf{c}_2$
$3 + k$ $(k = 1, \ldots, \nu)$	0	0	$A_{ik}\mathbf{c}_1$

while the accelerations of C^* and P_i in N are

$$
{}^N\mathbf{a}^{C^*} \underset{(13)}{=} (\dot{u}_1 - u_3 u_2)\mathbf{c}_1 + (\dot{u}_2 + u_3 u_1)\mathbf{c}_2 \tag{20}
$$

and

$$
\begin{aligned}
{}^N\mathbf{a}^{P_i} \underset{(18)}{=} & \left\{ \dot{u}_1 + \sum_{j=1}^{\nu} A_{ij}\dot{u}_{3+j} - u_3\left[u_2 + u_3\left(R + iL + \sum_{j=1}^{\nu} A_{ij} q_j \right)\right]\right\}\mathbf{c}_1 \\
& + \left[\dot{u}_2 + \dot{u}_3\left(R + iL + \sum_{j=1}^{\nu} A_{ij} q_j \right) + u_3 \sum_{j=1}^{\nu} A_{ij} u_{3+j} \right. \\
& \left. + u_3\left(u_1 + \sum_{j=1}^{\nu} A_{ij} u_{3+j} \right)\right]\mathbf{c}_2 \qquad (i = 1, \ldots, n)
\end{aligned} \tag{21}
$$

To express this in a more convenient form, one can introduce α_i, β_i, and γ_i as

$$
\alpha_i \triangleq R + iL + \sum_{j=1}^{\nu} A_{ij} q_j \qquad (i = 1, \ldots, n) \tag{22}
$$

$$
\beta_i \triangleq -u_3(u_2 + u_3\alpha_i), \quad \gamma_i \triangleq u_3\left(u_1 + 2\sum_{j=1}^{\nu} A_{ij} u_{3+j} \right) \quad (i = 1, \ldots, n) \tag{23}
$$

which permits one to write

$$
{}^N\mathbf{a}^{P_i} \underset{(21-23)}{=} \left(\dot{u}_1 + \sum_{j=1}^{\nu} A_{ij}\dot{u}_{3+j} + \beta_i \right)\mathbf{c}_1 + (\dot{u}_2 + \dot{u}_3\alpha_i + \gamma_i)\mathbf{c}_2 \qquad (i = 1, \ldots, n) \tag{24}
$$

The contribution of C to the generalized inertia force $F_r{}^*$ (see Sec. 4.3) is

$$
(F_r{}^*)_C \underset{(4.3.5)}{=} \boldsymbol{\omega}_r{}^C \cdot \mathbf{T}^* + \mathbf{v}_r{}^{C^*} \cdot \mathbf{R}^* \qquad (r = 1, \ldots, 3 + \nu) \tag{25}
$$

where the inertia torque $\mathbf{T}^*$ and inertia force $\mathbf{R}^*$, expressed in terms of J, the axial moment of inertia of C, and μ, the mass of C, are given by

$$
\mathbf{T}^* \underset{(4.3.6)}{=} -J\dot{u}_3\mathbf{c}_3 \qquad \mathbf{R}^* \underset{(4.3.3)}{=} -\mu[(\dot{u}_1 - u_3 u_2)\mathbf{c}_1 + (\dot{u}_2 + u_3 u_1)\mathbf{c}_2] \tag{26}
$$

Referring to Table 4.8.1, one thus finds

$$(F_1^*)_C = -\mu(\dot{u}_1 - u_3u_2), \quad (F_2^*)_C = -\mu(\dot{u}_2 + u_3u_1), \quad (F_3^*)_C = -J\dot{u}_3 \tag{27}$$

$$(F_{3+k}^*)_C = 0 \qquad (k = 1, \ldots, \nu) \tag{28}$$

Similarly, the contribution of P_i to F_r^* is

$$(F_r^*)_{P_i} \underset{(4.3.2)}{=} \mathbf{v}_r^{P_i} \cdot \mathbf{R}_i^* = -m\mathbf{v}_r^{P_i} \cdot {}^N\mathbf{a}^{P_i}$$

$$(r = 1, \ldots, 3 + \nu;\ i = 1, \ldots, n) \tag{29}$$

where m is the mass of each of $P_1, \ldots, P_n$. With the help of Table 4.8.1, one thus finds that

$$(F_1^*)_{P_i} = -m\left(\dot{u}_1 + \sum_{j=1}^{\nu} A_{ij}\dot{u}_{3+j} + \beta_i\right) \qquad (i = 1, \ldots, n) \tag{30}$$

$$(F_2^*)_{P_i} = -m(\dot{u}_2 + \dot{u}_3\alpha_i + \gamma_i) \qquad (i = 1, \ldots, n) \tag{31}$$

$$(F_3^*)_{P_i} = -m\alpha_i(\dot{u}_2 + \dot{u}_3\alpha_i + \gamma_i) \qquad (i = 1, \ldots, n) \tag{32}$$

$$(F_{3+k}^*)_{P_i} = -mA_{ik}\left(\dot{u}_1 + \sum_{j=1}^{\nu} A_{ij}\dot{u}_{3+j} + \beta_i\right) \qquad (k = 1, \ldots, \nu;\ i = 1, \ldots, n) \tag{33}$$

The complete generalized inertia forces, given by

$$F_r^* = (F_r^*)_C + \sum_{i=1}^{n} (F_r^*)_{P_i} \qquad (r = 1, \ldots, 3 + \nu) \tag{34}$$

are thus

$$F_1^* = -(\mu + nm)\dot{u}_1 + \mu u_3u_2 - m\left(\sum_{i=1}^{n}\sum_{j=1}^{\nu} A_{ij}\dot{u}_{3+j} + \sum_{i=1}^{n}\beta_i\right) \tag{35}$$

$$F_2^* = -(\mu + nm)\dot{u}_2 - \mu u_3u_1 - m\left(\dot{u}_3\sum_{i=1}^{n}\alpha_i + \sum_{i=1}^{n}\gamma_i\right) \tag{36}$$

$$F_3^* = -\left(J + m\sum_{i=1}^{n}\alpha_i^2\right)\dot{u}_3 - m\sum_{i=1}^{n}\alpha_i\dot{u}_2 - m\sum_{i=1}^{n}\gamma_i \tag{37}$$

$$F_{3+k}^* = -m\sum_{i=1}^{n} A_{ik}\dot{u}_1 - m\sum_{i=1}^{n}\sum_{j=1}^{\nu} A_{ij}A_{ik}\dot{u}_{3+j} - m\sum_{i=1}^{n} A_{ik}\beta_i$$

$$(k = 1, \ldots, \nu) \tag{38}$$

If all of the springs are linear and have a common spring constant σ, the resultants of all contact forces acting on C and P_i are, respectively,

$$\mathbf{F}^C = \sigma x_1 \mathbf{c}_1 \tag{39}$$

and

$$\mathbf{F}^{P_i} = \begin{cases} \sigma(x_2 - 2x_1)\mathbf{c}_1 & (i = 1) \\ \sigma(x_{i-1} - 2x_i + x_{i+1})\mathbf{c}_1 & (i = 2, \ldots, n-1) \\ \sigma(x_{n-1} - x_n)\mathbf{c}_1 & (i = n) \end{cases} \tag{40}$$

The velocity of point Q, the point of application of $\mathbf{F}^C$, is

$$^N\mathbf{v}^Q = {}^N\mathbf{v}^{C^*} + {}^N\boldsymbol{\omega}^C \times (R\mathbf{c}_1) \underset{(13)}{=} u_1\mathbf{c}_1 + (u_2 + Ru_3)\mathbf{c}_2 \tag{41}$$

so that the partial velocities of this point are

$$\mathbf{v}_1{}^Q = \mathbf{c}_1 \qquad \mathbf{v}_2{}^Q = \mathbf{c}_2 \qquad \mathbf{v}_3{}^Q = R\mathbf{c}_2 \qquad \mathbf{v}_{3+k}{}^Q = 0 \qquad (k = 1, \ldots, \nu) \tag{42}$$

Consequently, the contributions of C to the generalized active force F_r (see Sec. 4.1), which is given by,

$$(F_r)_C = \mathbf{v}_r{}^Q \cdot \mathbf{F}^C \qquad (r = 1, \ldots, 3 + \nu) \tag{43}$$

are

$$(F_1)_C = \sigma x_1 \qquad (F_2)_C = (F_3)_C = (F_{3+k})_C = 0 \qquad (k = 1, \ldots, \nu) \tag{44}$$

while the contribution of P_i to F_r is

$$(F_r)_{P_i} = \mathbf{v}_r{}^{P_i} \cdot \mathbf{F}^{P_i} \qquad (r = 1, \ldots, 3 + \nu;\ i = 1, \ldots, n) \tag{45}$$

or, in view of Table 4.8.1 and Eq. (40),

$$(F_1)_{P_i} = \begin{cases} \sigma(x_2 - 2x_1) & (i = 1) \\ \sigma(x_{i-1} - 2x_i + x_{i+1}) & (i = 2, \ldots, n - 1) \\ \sigma(x_{n-1} - x_n) & (i = n) \end{cases} \tag{46}$$

$$(F_2)_{P_i} = 0 \qquad (i = 1, \ldots, n) \tag{47}$$

$$(F_3)_{P_i} = 0 \qquad (i = 1, \ldots, n) \tag{48}$$

$$\begin{matrix}(F_{3+k})_{P_i} = \\ (k = 1, \ldots, \nu)\end{matrix} \begin{cases} \sigma A_{1k}(x_2 - 2x_1) & (i = 1) \\ \sigma A_{ik}(x_{i-1} - 2x_i + x_{i+1}) & (i = 2, \ldots, n - 1) \\ \sigma A_{nk}(x_{n-1} - x_n) & (i = n) \end{cases} \tag{49}$$

The complete generalized active forces, given by

$$F_r = (F_r)_C + \sum_{i=1}^{n} (F_r)_{P_i} \qquad (r = 1, \ldots, 3 + \nu) \tag{50}$$

are thus

$$F_1 = \sigma\left[x_1 + x_2 - 2x_1 + \sum_{i=2}^{n-1} (x_{i-1} - 2x_i + x_{i+1}) + x_{n-1} - x_n\right] = 0 \tag{51}$$

$$F_2 = F_3 = 0 \tag{52}$$

$$F_{3+k} = \sigma\left[A_{1k}(x_2 - 2x_1) + \sum_{i=2}^{n-1} A_{ik}(x_{i-1} - 2x_i + x_{i+1}) + A_{nk}(x_{n-1} - x_n)\right] \qquad (k = 1, \ldots, \nu) \tag{53}$$

and the following dynamical equations now can be written in accordance with Eqs. (4.5.1):

$$(\mu + nm)\dot{u}_1 + m \sum_{i=1}^{n} \sum_{j=1}^{\nu} A_{ij}\dot{u}_{3+j} = \mu u_3 u_2 - m \sum_{i=1}^{n} \beta_i \tag{54}$$

$$(\mu + nm)\dot{u}_2 + \sum_{i=1}^{n} \alpha_i \dot{u}_3 = - \mu u_3 u_1 - m \sum_{i=1}^{n} \gamma_i \tag{55}$$

$$\sum_{i=1}^{n} \alpha_i \dot{u}_2 + \left(J + m \sum_{i=1}^{n} \alpha_i^{\,2}\right)\dot{u}_3 = - m \sum_{i=1}^{n} \gamma_i \tag{56}$$

$$m \sum_{i=1}^{n} A_{ik}\dot{u}_1 + m \sum_{i=1}^{n} \sum_{j=1}^{\nu} A_{ij}A_{ik}\dot{u}_{3+j} = - m \sum_{i=1}^{n} A_{ik}\beta_i + \sigma\Bigg[A_{1k}(x_2 - 2x_1) + \sum_{i=2}^{n-1} A_{ik}(x_{i-1} - 2x_i + x_{i+1}) + A_{nk}(x_{n-1} - x_n)\Bigg] \qquad (k = 1, \ldots, \nu) \tag{57}$$

where $x_1, \ldots, x_n$ are given by Eqs. (14).

Equations (16) and (54)–(57) form a set of $3 + 2\nu$ equations that suffice [with the aid of Eqs. (14), (22), and (23)] for the determination, by numerical integration, of the $3 + 2\nu$ quantities $q_1, \ldots, q_\nu, u_1, \ldots, u_{3+\nu}$ whenever the initial values of these variables are known. Now, the initial values of u_1, u_2, u_3 [see Eqs. (12)] may be presumed to be given; but those of q_k and u_{3+k} $(k = 1, \ldots, \nu)$, introduced in Eqs. (14) and (16), are not so readily available, for it is not realistic to regard these as given. Rather, one must suppose that $x_i(0)$ and $\dot{x}_i(0)$, the initial values of x_i and $\dot{x}_i$ $(i = 1, \ldots, n)$, respectively, are specified. Consequently, $q_1(0), \ldots, q_\nu(0)$ must be found by solving the equations

$$x_i(0) \underset{(14)}{=} \sum_{j=1}^{\nu} A_{ij} q_j(0) \qquad (i = 1, \ldots, n) \tag{58}$$

and, similarly, $u_{3+j}(0)$ $(j = 1, \ldots, \nu)$ must satisfy the equations

$$\dot{x}_i(0) \underset{(14,15)}{=} \sum_{j=1}^{\nu} A_{ij} u_{3+j}(0) \qquad (i = 1, \ldots, n) \tag{59}$$

It is a straightforward matter to solve these two sets of linear equations if $\nu = n$. When $\nu < n$, so that each set of equations contains fewer unknowns than equations, one can determine $q_k(0)$ and $u_{3+k}(0)$ $(k = 1, \ldots, \nu)$ *uniquely* so as to satisfy all equations, provided one chooses A_{ij} $(i = 1, \ldots, n;\ j = 1, \ldots, \nu)$ properly. We shall pursue this matter presently. First, however, it is important to note that, if $\nu = n$, the simultaneous solution of Eqs. (16) and Eqs. (54)–(57) furnishes an *exact* description of the motion in question for *any* choice of A_{ij} $(i, j = 1, \ldots, n)$ such that the determinant having A_{ij} as the element in the ith row and jth column does not vanish, for Eqs. (14) then simply amount to a linear transformation from the dependent variables $x_1, \ldots, x_n$ to the new dependent variables $q_1, \ldots, q_n$. When $\nu < n$, on the other hand, solution of Eqs. (16) and (54)–(57) necessarily leads to an approximate description of the motion under consideration, no matter how A_{ij} $(i = 1, \ldots, n;$ $j = 1, \ldots, \nu)$ are chosen.

Returning to the matter of determining $q_k(0)$ and $u_{3+k}(0)$ $(k = 1, \ldots, \nu)$ so as to satisfy Eqs. (58) and (59) when $\nu < n$, we now focus attention on a classical vibrations problem intimately related to the problem at hand, namely, the free vibrations of the set of particles $P_1, \ldots, P_n$ when body C is held fixed in N. In that event, one can introduce generalized speeds $\overline{u}_1, \ldots, \overline{u}_n$ as

$$\overline{u}_r = \dot{x}_r \qquad (r = 1, \ldots, n) \tag{60}$$

so that the velocity of P_i in N is $\overline{u}_i\mathbf{c}_1$ and the partial velocities of P_i, denoted by $\overline{\mathbf{v}}_1{}^{P_i}, \ldots, \overline{\mathbf{v}}_n{}^{P_i}$, are given by

$$\overline{\mathbf{v}}_r{}^{P_i} = \delta_{ir}\mathbf{c}_1 \qquad (r, i = 1, \ldots, n) \tag{61}$$

where δ_{ir} is the Kronecker delta, while the acceleration of P_i in N is

$$^N\overline{\mathbf{a}}^{P_i} = \dot{\overline{u}}_i\mathbf{c}_1 \qquad (i = 1, \ldots, n) \tag{62}$$

The generalized inertia forces thus can be written

$$\overline{F}_r{}^* = -m\sum_{i=1}^{n}\overline{\mathbf{v}}_r{}^{P_i}\cdot\overline{\mathbf{a}}^{P_i} \underset{(61,62)}{=} -m\sum_{i=1}^{n}\delta_{ir}\dot{\overline{u}}_i = -m\dot{\overline{u}}_r \underset{(60)}{=} -m\ddot{x}_r$$

$$(r = 1, \ldots, n) \tag{63}$$

and the generalized active forces are

$$\overline{F}_r = \sum_{i-1}^{n}\overline{\mathbf{v}}_r{}^{P_i}\cdot\mathbf{F}^{P_i} = \overline{\mathbf{v}}_r{}^{P_1}\cdot\mathbf{F}^{P_1} + \sum_{i=2}^{n-1}\overline{\mathbf{v}}_r{}^{P_i}\cdot\mathbf{F}^{P_i} + \overline{\mathbf{v}}_r{}^{P_n}\cdot\mathbf{F}^{P_n}$$

$$\underset{(61,40)}{=} \sigma\left[\delta_{1r}(x_2 - 2x_1) + \sum_{i=2}^{n-1}\delta_{ir}(x_{i-1} - 2x_i + x_{i+1}) + \delta_{nr}(x_{n-1} - x_n)\right]$$

$$= \sigma\Bigg[(-2\delta_{1r} + \delta_{2r})x_1 + \sum_{i=2}^{n-1}(\delta_{i-1,r} - 2\delta_{ir} + \delta_{i+1,r})x_i$$

$$+ (\delta_{n-1,r} - \delta_{nr})x_n\Bigg] \qquad (r = 1, \ldots, n) \tag{64}$$

Consequently, the dynamical equations are now

$$m\ddot{x}_r + \sigma\left[(2\delta_{1r} - \delta_{2r})x_1 + \sum_{i=2}^{n-1}(2\delta_{ir} - \delta_{i-1,r} - \delta_{i+1,r})x_i + (\delta_{nr} - \delta_{n-1,r})x_n\right]$$

$$= 0 \qquad (r = 1, \ldots, n) \tag{65}$$

and these can be cast into the form of Eq. (1) if one defines x as the $n \times 1$ matrix having x_r as the element in the rth row, M as the $n \times n$ matrix

$$M \triangleq mU \tag{66}$$

where U is the $n \times n$ unit matrix, and S as the $n \times n$ matrix whose nonzero elements are

$$S_{ii} \triangleq 2\sigma \qquad (i = 1, \ldots, n-1) \tag{67}$$

$$S_{i,i+1} \triangleq S_{i+1,i} = -\sigma \qquad (i = 1, \ldots, n-1) \tag{68}$$

$$S_{nn} \triangleq \sigma \tag{69}$$

Now let A_{ij} in Eqs. (14) be the element in the ith row and jth column of the matrix A introduced in Eq. (5), and let q_j in Eqs. (14) be the jth element of the matrix q of Eq. (10). In other words, use Eqs. (14) to make a very special transformation from $x_1, \ldots, x_n$ to new variables, namely, a transformation involving the modal matrix A and the normal coordinates $q_1, \ldots, q_\nu$ associated with the vibration problem involving M and S as defined in Eqs. (66) and (67)–(69). Then Eqs. (2) and (14) are one and the same, and Eqs. (6) therefore furnish the solution of Eqs. (58) and (59) for $q_k(0)$ and $u_{3+k}(0)$; that is,

$$q_k(0) = \sum_{i=1}^{n} \sum_{j=1}^{n} A_{jk} M_{ji} x_i(0) \qquad (k = 1, \ldots, \nu) \tag{70}$$

$$u_{3+k}(0) = \sum_{i=1}^{n} \sum_{j=1}^{n} A_{jk} M_{ji} \dot{x}_i(0) \qquad (k = 1, \ldots, \nu) \tag{71}$$

Moreover, if A_{ij} is chosen as just described, Eq. (57) can be replaced with a simpler relationship. To this end, use Eqs. (14) to eliminate $x_1, \ldots, x_n$ from Eq. (53), and verify with the aid of Eqs. (67)–(69) that the resulting equation can be written

$$\begin{bmatrix} F_{3+1} \\ \vdots \\ F_{3+\nu} \end{bmatrix} = -A^T S A q \underset{(11)}{=} -\lambda q \tag{72}$$

Consequently,

$$F_{3+k} = -\lambda_k q_k \qquad (k = 1, \ldots, \nu) \tag{73}$$

and Eqs. (57) can be replaced with

$$m \sum_{i=1}^{n} A_{ik} \dot{u}_1 + m \sum_{i=1}^{n} \sum_{j=1}^{\nu} A_{ij} A_{ik} \dot{u}_{3+j} = -m \sum_{i=1}^{n} A_{ik} \beta_i - \lambda_k q_k$$

$$(k = 1, \ldots, \nu) \tag{74}$$

The appearance of λ_k in this equation imposes no burden since λ_k must be found, in any event, before A can be constructed.

To illustrate the use of the method under consideration, we turn to a specific case, taking $n = 10$, $\mu = 10$ kg, $J = 5$ kg $\cdot$ m^2, $R = 1$ m, $L = 1$ m, $\sigma = 1000$ N/m, $m = 10$ kg, $u_1(0) = u_2(0) = x_i(0) = \dot{x}_i(0) = 0$ $(i = 1, \ldots, 10)$, and $u_3(0) = 2$ rad/sec. In other words, we consider a motion that begins with C^* at rest in N, with $P_1, \ldots, P_{10}$ at rest in C, with all springs undeformed, and with C rotating at a rate of 2 rad/sec.

In preparation for a numerical integration of the equations of motion, M and S are formed in accordance with Eqs. (66) and Eqs. (67)–(69), respectively; the first ν eigenvalues and eigenvectors of $M^{-1}S$ are found; Eqs. (3)–(5) are used to construct A; and $q_k(0)$ and $u_{3+k}(0)$ $(k = 1, \ldots, \nu)$ are set equal to zero in view of Eqs. (70) and (71) and the given initial values of x_i and $\dot{x}_i$ $(i = 1, \ldots, n)$. With $\nu = 1$ and $\nu = 10$, numerical integration of the equations of motion then leads to the u_3 versus t plots shown in Fig. 4.8.2, where the solid curve represents $\nu = 10$, i.e., the exact solution of the equations of motion, and the dashed curve corresponds to $\nu = 1$, that is, to the exclusion of nine modes. Clearly, the one-mode approximation leaves something to be desired. The effect of taking one additional mode into account becomes apparent in Fig. 4.8.3.

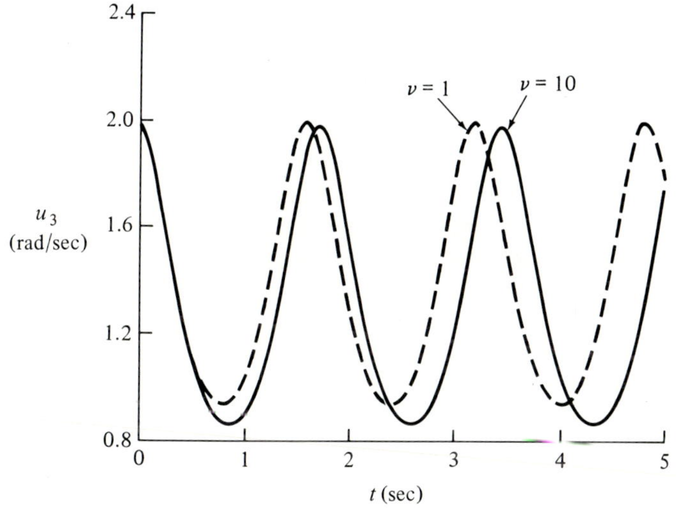

Figure 4.8.2

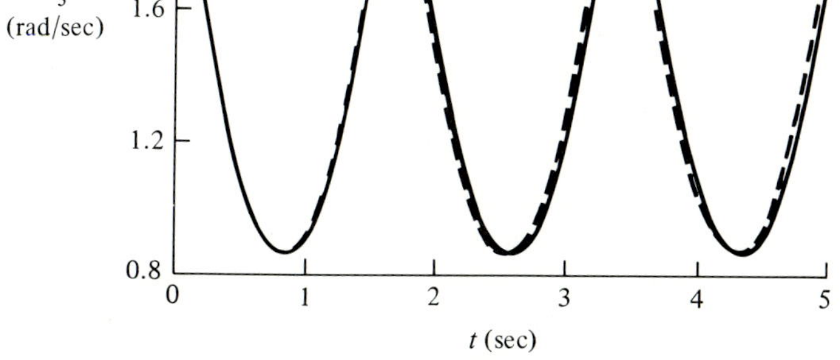

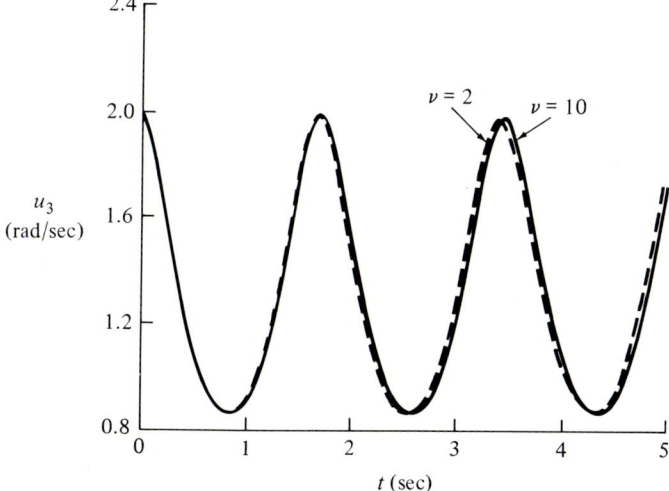

Figure 4.8.3

The value of ν that must be used to obtain satisfactory results depends on initial conditions and is not always so small as in the present example. Moreover, in the absence of the exact solution of a problem, one can never be completely certain that the omission of some modes will not lead to incorrect results.

4.9 LUMPED MASS MODELS OF SPACECRAFT

Certain spacecraft can be modeled as sets of elastically connected particles. For example, when a spacecraft is an unrestrained truss formed by a relatively large number of prismatic members, its motions may be analyzed by considering a set of particles placed at the joints of the structure, each particle having a mass equal to one-half the sum of the masses of all truss members meeting at the joint and the particles being connected to each other with massless springs whose stiffnesses reflect the elastic properties of the truss members. The differential equations governing the motions of such a set of particles can be formulated straightforwardly, and, in principle, can be integrated numerically to obtain descriptions of joint motions. However, when the number of structural elements is sufficiently large, the integration process can become prohibitively time-consuming. In this event, one can resort to the technique discussed in Sec. 4.8, which becomes particularly effective when one has access to a computer program capable of furnishing the elements of a structural stiffness matrix of the truss. (For a detailed discussion of this topic, see W. Weaver, Jr., and J. Gere, *Matrix Analysis of Framed Structures,* 2d ed., D. Van Nostrand Co., New York, 1980, pp. 254–263.) What follows is an initial-value problem solution algorithm that involves exclusion of modes.

Given a truss whose motion is to be simulated, number the members from 1 to m and the joints from 1 to n, and specify Young's modulus E_k, the cross-section area A_k, and the mass per unit of volume ρ_k of member k $(k = 1, \ldots, m)$. Introduce a Newtonian reference frame N, a point N^* fixed in N, and a dextral set of orthogonal unit vectors $\mathbf{n}_1$, $\mathbf{n}_2$, $\mathbf{n}_3$ fixed in N. Regarding the ith joint as a point P_i, and letting $\mathbf{p}_i$ be the position vector from N^* to P_i, specify the position of P_i in N for time $t = 0$ by assigning values to $x_{3i-3+j}(0)$, defined as the value of x_{3i-3+j} at $t = 0$ $(i = 1, \ldots, n; j = 1, 2, 3)$, where x_{3i-3+j} is, in turn, defined as $x_{3i-3+j} \triangleq \mathbf{p}_i \cdot \mathbf{n}_j$ $(i = 1, \ldots, n; j = 1, 2, 3)$. Similarly, to characterize the motion of the truss at $t = 0$, assign values to $\dot{x}_{3i-3+j}(0)$, defined as the value of $\dot{x}_{3i-3+j}$ at $t = 0$ $(i = 1, \ldots, n; j = 1, 2, 3)$, where $\dot{x}_{3i-3+j} \triangleq {}^N\mathbf{v}^{P_i} \cdot \mathbf{n}_j$ $(i = 1, \ldots, n; j = 1, 2, 3)$ and ${}^N\mathbf{v}^{P_i}$ denotes the velocity of P_i in N.

In preparation for performing a *modal analysis* of the truss, position the truss in its undeformed state in N in such a way that P_i occupies nearly or exactly the same position as it does at $t = 0$; and evaluate $\hat{x}_{3i-3+j}$, defined as $\hat{x}_{3i-3+j} \triangleq \hat{\mathbf{p}}_i \cdot \mathbf{n}_j$ $(i = 1, \ldots, n; j = 1, 2, 3)$, where $\hat{\mathbf{p}}_i$ is the position vector from N^* to P_i $(i = 1, \ldots, n)$ when the truss is positioned in N as just specified.

To account for external loads acting on the truss (e.g., control forces, gravitational forces, solar radiation pressure), let $\mathbf{Q}^{P_i}$ be the resultant of all external forces applied at P_i $(i = 1, \ldots, n)$. Form Q_{3i-3+j} as

$$Q_{3i-3+j} = \mathbf{Q}^{P_i} \cdot (C_{1j}\mathbf{n}_1 + C_{2j}\mathbf{n}_2 + C_{3j}\mathbf{n}_3) \qquad (i = 1, \ldots, n; j = 1, 2, 3) \quad (1)$$

and then, wherever x_{3i-3+j} or $\dot{x}_{3i-3+j}$ appear, replace them in accordance with

$$x_{3i-3+j} = z_j + \sum_{k=1}^{3} y_{3i-3+k} C_{jk} \qquad (i = 1, \ldots, n; j = 1, 2, 3) \qquad (2)$$

$$\dot{x}_{3i-3+j} = \dot{z}_j + \sum_{k=1}^{3} (\dot{y}_{3i-3+k} C_{jk} + y_{3i-3+k} \dot{C}_{jk}) \qquad (i = 1, \ldots, n; j = 1, 2, 3) \quad (3)$$

The stage is now set for the determination of $x_1, \ldots, x_{3n}$ for $t > 0$, a process that begins with the formation of a $3n \times 3n$ structural *stiffness matrix* S. Next, let m_i be half the total mass of all truss members meeting at P_i, and form the elements M_{rs} $(r, s = 1, \ldots, 3n)$ of a $3n \times 3n$ diagonal *mass matrix* M by setting

$$M_{3i-3+r,3l-3+r} = m_i \delta_{li} \qquad (i, l = 1, \ldots, n; r = 1, 2, 3) \qquad (4)$$

Now introduce a $3n \times 3n$ symmetric matrix W as

$$W = M^{-1/2} S M^{-1/2} \qquad (5)$$

and, after assigning a value to $\nu\,(1 \le \nu \le 3n - 6)$, discard the first six eigenvalues of W; arrange any ν of the rest in ascending order, calling these ν eigenvalues $\lambda_1, \ldots, \lambda_\nu$, and let $B_{1r}, \ldots, B_{3n,r}$ be the elements of the eigenvector of $M^{-1}S$ corresponding to λ_r, this eigenvector being obtained by premultiplying with $M^{-1/2}$ the eigenvector of W corresponding to λ_r $(r = 1, \ldots, \nu)$. Finally, form the elements A_{ij} of the *modal matrix* normalized with respect to the mass matrix as

$$A_{ij} = \left(\sum_{k=1}^{3n} M_{kk} B_{kj}^{\,2} \right)^{-1/2} B_{ij} \qquad (i = 1, \ldots, 3n; j = 1, \ldots, \nu) \qquad (6)$$

This completes the aforementioned modal analysis. The next task is that of formulating and integrating the differential equations that govern the motion of the truss in N.

The initial values of $x_1, \ldots, x_{3n}$ and $\dot{x}_1, \ldots, \dot{x}_{3n}$ that is, $x_1(0), \ldots, x_{3n}(0)$ and $\dot{x}_1(0), \ldots, \dot{x}_{3n}(0)$, have been assigned already. To determine associated initial values of C_{jk}, z_j $(j, k = 1, 2, 3)$, q_r $(r = 1, \ldots, \nu)$, and u_s $(s = 1, \ldots, 6 + \nu)$ (these are the dependent variables in the differential equations to be solved), form K_{rs} $(r = 1, \ldots, 3n; s = 1, \ldots, 6 + \nu)$ as

$$K_{3i-3+j,k} = -\frac{1}{2} \sum_{l=1}^{3} (j - k)(k - l)(l - j) x_{3i-3+l}(0)$$
$$(i = 1, \ldots, n; j, k = 1, 2, 3) \quad (7)$$

$$K_{3i-3+j,3+k} = \delta_{jk} \qquad (i = 1, \ldots, n; j, k = 1, 2, 3) \qquad (8)$$

$$K_{i,6+j} = A_{ij} \qquad (i = 1, \ldots, 3n; j = 1, \ldots, \nu) \qquad (9)$$

where δ_{ij} is the Kronecker delta; and let b be the $3n \times 1$ matrix whose elements are

$$b_i = x_i(0) - \hat{x}_i \qquad (i = 1, \ldots, 3n) \qquad (10)$$

Next, letting K be the matrix having K_{rs} as the element in the rth row and sth column, form a $(6 + \nu) \times 1$ matrix c as

$$c = (K^T K)^{-1} K^T b \tag{11}$$

where K^T is the transpose of K. Now, letting c_r be the element in the rth row of c, find $z_i(0)$, the initial value of z_i, from

$$z_i(0) = c_{3+i} \qquad (i = 1, 2, 3) \tag{12}$$

and, after introducing ϵ_{ij} $(i, j = 1, 2, 3)$ as

$$\epsilon_{ij} = \frac{1}{2} \sum_{k=1}^{3} (i - j)(j - k)(k - i) c_k \qquad (i, j = 1, 2, 3) \tag{13}$$

find $C_{ij}(0)$, the initial value of C_{ij}, from

$$C_{ij}(0) = \delta_{ij} + \epsilon_{ij} \qquad (i, j = 1, 2, 3) \tag{14}$$

Furthermore, find $q_k(0)$ and $y_{3i-3+l}(0)$ from

$$q_k(0) = c_{6+k} \qquad (k = 1, \ldots, \nu) \tag{15}$$

and

$$y_{3i-3+l}(0) = \hat{x}_{3i-3+l} + \sum_{k=1}^{\nu} A_{3i-3+l,k} q_k(0) \qquad (i = 1, \ldots, n;\ l = 1, 2, 3) \tag{16}$$

Finally, form L_{rs} $(r = 1, \ldots, 3n;\ s = 1, \ldots, 6 + \nu)$ and d_r $(r = 1, \ldots, 3n)$ as

$$L_{3i-3+j,k} = \frac{1}{2} \sum_{l=1}^{3} (j - k)(k - l)(l - j) y_{3i-3+l}(0)$$
$$(i = 1, \ldots, n;\ j, k = 1, 2, 3) \tag{17}$$

$$L_{3i-3+j,3+k} = \delta_{jk} \qquad (i = 1, \ldots, n;\ j, k = 1, 2, 3) \tag{18}$$

$$L_{i,6+j} = A_{ij} \qquad (i = 1, \ldots, 3n;\ j = 1, \ldots, \nu) \tag{19}$$

and as

$$d_{3i-2} = \dot{x}_{3i-2}(0) - \dot{x}_{3i-1}(0)\,\epsilon_{12} + \dot{x}_{3i}(0)\,\epsilon_{31} \qquad (i = 1, \ldots, n) \tag{20}$$

$$d_{3i-1} = \dot{x}_{3i-1}(0) - \dot{x}_{3i}(0)\,\epsilon_{23} + \dot{x}_{3i-2}(0)\,\epsilon_{12} \qquad (i = 1, \ldots, n) \tag{21}$$

$$d_{3i} = \dot{x}_{3i}(0) - \dot{x}_{3i-2}(0)\,\epsilon_{31} + \dot{x}_{3i-1}(0)\,\epsilon_{23} \qquad (i = 1, \ldots, n) \tag{22}$$

respectively; let d be the $3n \times 1$ matrix whose elements are $d_1, \ldots, d_{3n}$; let e_r be the element in the rth row of the $(6 + \nu) \times 1$ matrix e given by

$$e = (L^T L)^{-1} L^T d \tag{23}$$

and find $u_r(0)$, the initial value of u_r, from

$$u_r(0) = e_r \qquad (r = 1, \ldots, 6 + \nu) \tag{24}$$

The formulation of the differential equations to be solved is facilitated by first forming y_{3i-3+j}, $\dot{y}_{3i-3+j}$, v_{ij}, a_{ij}, I_{jk}, w_j $(i = 1, \ldots, n;\ j, k = 1, 2, 3)$, G_l $(l = 1, \ldots, \nu)$, and $\overline{m}$ as

$$y_{3i-3+j} = \hat{x}_{3i-3+j} + \sum_{k=1}^{\nu} A_{3i-3+j,k} q_k \qquad (i = 1, \ldots, n;\ j = 1, 2, 3) \tag{25}$$

$$\dot{y}_{3i-3+j} = \sum_{k=1}^{\nu} A_{3i-3+j,k} u_{6+k} \qquad (i = 1, \ldots, n;\ j = 1, 2, 3) \tag{26}$$

$$v_{i1} = u_4 + u_2 y_{3i} - u_3 y_{3i-1} + \dot{y}_{3i-2} \qquad (i = 1, \ldots, n) \tag{27}$$

$$v_{i2} = u_5 + u_3 y_{3i-2} - u_1 y_{3i} + \dot{y}_{3i-1} \qquad (i = 1, \ldots, n) \tag{28}$$

$$v_{i3} = u_6 + u_1 y_{3i-1} - u_2 y_{3i-2} + \dot{y}_{3i} \qquad (i = 1, \ldots, n) \tag{29}$$

$$a_{i1} = u_2(\dot{y}_{3i} + v_{i3}) - u_3(\dot{y}_{3i-1} + v_{i2}) \qquad (i = 1, \ldots, n) \tag{30}$$

$$a_{i2} = u_3(\dot{y}_{3i-2} + v_{i1}) - u_1(\dot{y}_{3i} + v_{i3}) \qquad (i = 1, \ldots, n) \tag{31}$$

$$a_{i3} = u_1(\dot{y}_{3i-1} + v_{i2}) - u_2(\dot{y}_{3i-2} + v_{i1}) \qquad (i = 1, \ldots, n) \tag{32}$$

$$I_{jk} = \sum_{i=1}^{n} m_i \left[\delta_{jk} \left(\sum_{l=1}^{3} y_{3i-3+l}{}^2 \right) - y_{3i-3+j} y_{3i-3+k} \right] \qquad (j, k = 1, 2, 3) \tag{33}$$

$$w_j = \sum_{i=1}^{n} m_i y_{3i-3+j} \qquad (j = 1, 2, 3) \tag{34}$$

$$G_l = \sum_{i=1}^{n} \sum_{j=1}^{3} A_{3i-3+j,l} Q_{3i-3+j} \qquad (l = 1, \ldots, \nu) \tag{35}$$

$$\overline{m} = \sum_{i=1}^{n} m_i \tag{36}$$

Having formed these quantities, solve simultaneously the differential equations

$$\dot{C}_{jk} = \frac{1}{2} \sum_{g=1}^{3} \sum_{h=1}^{3} C_{jg} u_h (g - h)(h - k)(k - g) \qquad (j, k = 1, 2, 3) \tag{37}$$

$$\dot{z}_j = \sum_{k=1}^{3} u_{3+j} C_{jk} \qquad (j = 1, 2, 3) \tag{38}$$

$$\dot{q}_k = u_{6+k} \qquad (k = 1, \ldots, \nu) \tag{39}$$

$$\begin{aligned} I_{11}\dot{u}_1 + I_{12}\dot{u}_2 + I_{13}\dot{u}_3 - w_3\dot{u}_5 + w_2\dot{u}_6 & \\ + \sum_{j=1}^{\nu} \sum_{i=1}^{n} m_i (y_{3i-1} A_{3i,j} - y_{3i} A_{3i-1,j}) \dot{u}_{6+j} & \\ = \sum_{i=1}^{n} \Big[y_{3i-1}(Q_{3i} - m_i a_{i3}) - y_{3i}(Q_{3i-1} - m_i a_{i2}) \Big] & \end{aligned} \tag{40}$$

$$\begin{aligned} I_{21}\dot{u}_1 + I_{22}\dot{u}_2 + I_{23}\dot{u}_3 - w_1\dot{u}_6 + w_3\dot{u}_4 & \\ + \sum_{j=1}^{\nu} \sum_{i=1}^{n} m_i (y_{3i} A_{3i-2,j} - y_{3i-2} A_{3i,j}) \dot{u}_{6+j} & \\ = \sum_{i=1}^{n} \Big[y_{3i}(Q_{3i-2} - m_i a_{i1}) - y_{3i-2} Q_{3i} - m_i a_{i3}) \Big] & \end{aligned} \tag{41}$$

$$I_{31}\dot{u}_1 + I_{32}\dot{u}_2 + I_{33}\dot{u}_3 - w_2\dot{u}_4 + w_1\dot{u}_5 + \sum_{j=1}^{\nu}\sum_{i=1}^{n} m_i(y_{3i-2}A_{3i-1,j} - y_{3i-1}A_{3i-2,j})\dot{u}_{6+j} = \sum_{i=1}^{n}\left[y_{3i-2}(Q_{3i-1} - m_i a_{i2}) - y_{3i-1}(Q_{3i-2} - m_i a_{i1})\right] \tag{42}$$

$$w_3\dot{u}_2 - w_2\dot{u}_3 + \overline{m}\dot{u}_4 = \sum_{i=1}^{n}(Q_{3i-2} - m_i a_{i1}) \tag{43}$$

$$w_1\dot{u}_3 - w_3\dot{u}_1 + \overline{m}\dot{u}_5 = \sum_{i=1}^{n}(Q_{3i-1} - m_i a_{i2}) \tag{44}$$

$$w_2\dot{u}_1 - w_1\dot{u}_2 + \overline{m}\dot{u}_6 = \sum_{i=1}^{n}(Q_{3i} - m_i a_{i3}) \tag{45}$$

$$\dot{u}_1\sum_{i=1}^{n} m_i(A_{3i-1,k}y_{3i} - A_{3i,k}y_{3i-1}) + \dot{u}_2\sum_{i=1}^{n} m_i(A_{3i,k}y_{3i-2} - A_{3i-2,k}y_{3i}) + \dot{u}_3\sum_{i=1}^{n} m_i(A_{3i-2,k}y_{3i-1} - A_{3i-1,k}y_{3i-2}) - \dot{u}_{6+k} = -G_k + \lambda_k q_k + \sum_{i=1}^{n} m_i(A_{3i-2,k}a_{i1} + A_{3i-1,k}a_{i2} + A_{3i,k}a_{i3}) \qquad (k = 1, \ldots, \nu) \tag{46}$$

Finally, form $y_1, \ldots, y_{3n}$ with the aid of Eq. (25), and find $x_1, \ldots, x_{3n}$ by substituting into Eq. (2). (The foregoing algorithm involves a number of procedures that can be replaced with other, computationally more efficient ones. In arriving at our choice of these, we have assigned a higher priority to clarity of presentation than to computational efficiency.)

Derivations Considering a truss having n joints and m members, let N^* be a point fixed in a Newtonian reference frame N; designate as P_i a typical joint of the truss; let $\hat{\mathbf{p}}_i$ and $\mathbf{p}_i$ be the position vectors from N^* to P_i at an instant at which the truss is undeformed and at a general time t, respectively, noting that $\hat{\mathbf{p}}_1, \ldots, \hat{\mathbf{p}}_n$ can be chosen in infinitely many ways, because the truss is unrestrained; and let ${}^N\mathbf{v}^{P_i}$ be the velocity of P_i in N at time t. Then, if E_k, A_k, and ρ_k denote Young's modulus, the cross-section area, and the mass per unit of volume of the kth member of the truss, the problem to be solved is this: given $\mathbf{p}_i$ and ${}^N\mathbf{v}^{P_i}$ for $t = 0$, $\hat{\mathbf{p}}_i$ $(i = 1, \ldots, n)$, and E_k, A_k, ρ_k $(k = 1, \ldots, m)$, find $\mathbf{p}_i$ for $t > 0$ $(i = 1, \ldots, n)$.

The first step in the solution is that of "discretizing" the truss by placing at joint P_i a particle of mass m_i equal to half the total mass of all truss members meeting at P_i $(i = 1, \ldots, n)$. (The mass of member k is $\rho_k A_k L_k$, where L_k is the length of member k.) Once this step has been taken, one can proceed to the formulation of differential equations of motion of a set of particles situated at $P_1, \ldots, P_n$, having masses $m_1, \ldots, m_n$, respectively, and connected to each other by massless, linear springs, each such spring representing a truss member, and the kth member having a spring constant equal to $A_k E_k / L_k$.

For the formulation of differential equations of motion, it is helpful to introduce an auxiliary reference frame R and two dextral sets of orthogonal unit vectors, namely $\mathbf{n}_1, \mathbf{n}_2, \mathbf{n}_3$, fixed in N, and $\mathbf{r}_1, \mathbf{r}_2, \mathbf{r}_3$, fixed in R. Furthermore, one may let R^* be a point fixed in R, denote the position vector of R^* relative to N^* by $\mathbf{p}^*$, and define $\hat{x}_{3i-3+j}$, x_{3i-3+j}, y_{3i-3+j}, z_j, and C_{jk} $(i = 1, \ldots, n; j, k = 1, 2, 3)$ as

$$\hat{x}_{3i-3+j} \triangleq \hat{\mathbf{p}}_i \cdot \mathbf{n}_j \qquad (i = 1, \ldots, n; j = 1, 2, 3) \tag{47}$$

$$x_{3i-3+j} \triangleq \mathbf{p}_i \cdot \mathbf{n}_j \qquad (i = 1, \ldots, n; j = 1, 2, 3) \tag{48}$$

$$y_{3i-3+j} \triangleq (\mathbf{p}_i - \mathbf{p}^*) \cdot \mathbf{r}_j \qquad (i = 1, \ldots, n; j = 1, 2, 3) \tag{49}$$

$$z_j \triangleq \mathbf{p}^* \cdot \mathbf{n}_j \qquad (j = 1, 2, 3) \tag{50}$$

and

$$C_{jk} \triangleq \mathbf{n}_j \cdot \mathbf{r}_k \qquad (j, k = 1, 2, 3) \tag{51}$$

(Note that $\hat{x}_i$, being defined in terms of $\hat{\mathbf{p}}_i$, is not unique.)

Six kinematical differential equations governing the motion of R in N may be generated as follows. In view of Eqs. (48)–(50), one can write

$$y_{3i-2}\mathbf{r}_1 + y_{3i-1}\mathbf{r}_2 + y_{3i}\mathbf{r}_3 = (x_{3i-2} - z_1)\mathbf{n}_1 + (x_{3i-1} - z_2)\mathbf{n}_2 + (x_{3i} - z_3)\mathbf{n}_3 \qquad (i = 1, \ldots, n) \tag{52}$$

which, upon successive scalar multiplication with $\mathbf{r}_1, \mathbf{r}_2, \mathbf{r}_3$ and subsequent use of Eq. (51), yields

$$y_{3i-3+j} = \sum_{k=1}^{3} (x_{3i-3+k} - z_k) C_{kj} \qquad (i = 1, \ldots, n; j = 1, 2, 3) \tag{53}$$

Next, generalized speeds $u_1, \ldots, u_6$ are defined in terms of the angular velocity of R in N, denoted by ${}^N\boldsymbol{\omega}^R$, and the velocity of R^* in N, written ${}^N\mathbf{v}^{R^*}$:

$$u_i \triangleq \begin{cases} {}^N\boldsymbol{\omega}^R \cdot \mathbf{r}_i & (i = 1, 2, 3) \\ {}^N\mathbf{v}^{R^*} \cdot \mathbf{r}_{i-3} & (i = 4, 5, 6) \end{cases} \tag{54}$$

Poisson's kinematical equations [see Eq. (1.10.10)] then can be stated as

$$\dot{C}_{jk} = \frac{1}{2} \sum_{g=1}^{3} \sum_{h=1}^{3} C_{jg} u_h (g - h)(h - k)(k - g) \qquad (j, k = 1, 2, 3) \tag{55}$$

and three additional kinematical differential equations result from the differentiation of Eq. (50) with respect to time, which yields

$$\dot{z}_j \underset{(50)}{=} {}^N\mathbf{v}^{R^*} \cdot \mathbf{n}_j \underset{(54)}{=} (u_4\mathbf{r}_1 + u_5\mathbf{r}_2 + u_6\mathbf{r}_3) \cdot \mathbf{n}_j$$

$$\underset{(51)}{=} u_4 C_{j1} + u_5 C_{j2} + u_6 C_{j3} \qquad (j = 1, 2, 3) \tag{56}$$

or equivalently,

$$\dot{z}_j = \sum_{k=1}^{3} u_{3+k} C_{jk} \qquad (j = 1, 2, 3) \tag{57}$$

In preparation for the formation of partial velocities of P_i, one can express ${}^N\mathbf{v}^{P_i}$ as

$${}^N\mathbf{v}^{P_i} = {}^R\mathbf{v}^{P_i} + {}^N\mathbf{v}^{\overline{R}_i} \qquad (i = 1, \ldots, n) \tag{58}$$

where ${}^N\mathbf{v}^{P_i}$ is the velocity of P_i in R and ${}^N\mathbf{v}^{\overline{R}_i}$ is the velocity in N of the point $\overline{R}_i$ of R that coincides with P_i. From Eq. (49),

$${}^R\mathbf{v}^{P_i} = \dot{y}_{3i-2}\mathbf{r}_1 + \dot{y}_{3i-1}\mathbf{r}_2 + \dot{y}_{3i}\mathbf{r}_3 \qquad (i = 1, \ldots, n) \tag{59}$$

As for ${}^N\mathbf{v}^{\overline{R}_i}$, this is given by

$$\begin{aligned} {}^N\mathbf{v}^{\overline{R}_i} &= {}^N\mathbf{v}^{R^*} + {}^N\boldsymbol{\omega}^R \times (\mathbf{p}_i - \mathbf{p}^*) \underset{(54,49)}{=} (u_4 + u_2 y_{3i} - u_3 y_{3i-1})\mathbf{r}_1 \\ &+ (u_5 + u_3 y_{3i-2} - u_1 y_{3i})\mathbf{r}_2 + (u_6 + u_1 y_{3i-1} - u_2 y_{3i-2})\mathbf{r}_3 \\ &\qquad (i = 1, \ldots, n) \end{aligned} \tag{60}$$

Substituting from Eqs. (59) and (60) into Eq. (58), one thus finds

$$\begin{aligned} {}^N\mathbf{v}^{P_i} &= (\dot{y}_{3i-2} + u_4 + u_2 y_{3i} - u_3 y_{3i-1})\mathbf{r}_1 \\ &+ (\dot{y}_{3i-1} + u_5 + u_3 y_{3i-2} - u_1 y_{3i})\mathbf{r}_2 \\ &+ (\dot{y}_{3i} + u_6 + u_1 y_{3i-1} - u_2 y_{3i-2})\mathbf{r}_3 \qquad (i = 1, \ldots, n) \end{aligned} \tag{61}$$

We shall return to this equation presently. First, however, we fix in R an "image" of the undeformed structure by letting $\widetilde{P}_i$ be a point fixed in R in such a way that the position vector from R^* to $\widetilde{P}_i$ is given by [see Eq. (47) for the definition of $\hat{x}_i$]

$$\widetilde{\mathbf{p}}_i \triangleq \hat{x}_{3i-2}\mathbf{r}_1 + \hat{x}_{3i-1}\mathbf{r}_2 + \hat{x}_{3i}\mathbf{r}_3 \qquad (i = 1, \ldots, n) \tag{62}$$

and we use the position vector from $\widetilde{P}_i$ to P_i, that is, $\mathbf{p}_i - \mathbf{p}^* - \widetilde{\mathbf{p}}_i$, to define "displacements" $\Delta_1, \ldots, \Delta_{3n}$ as

$$\Delta_{3i-3+j} \triangleq (\mathbf{p}_i - \mathbf{p}^* - \widetilde{\mathbf{p}}_i) \cdot \mathbf{r}_j \qquad (i = 1, \ldots, n;\ j = 1, 2, 3) \tag{63}$$

It then follows directly from Eqs. (49) and (62) that

$$\Delta_i = y_i - \hat{x}_i \qquad (i = 1, \ldots, 3n) \tag{64}$$

and, that Δ_i, like $\hat{x}_i$, is not unique. Next, we express the displacements in terms "generalized coordinates" $q_1, \ldots, q_\nu$ $(1 \le \nu \le 3n - 6)$ by setting

$$\Delta_i = \sum_{j=1}^{\nu} A_{ij} q_j \quad (i = 1, \ldots, 3n) \tag{65}$$

where A_{ij} $(i = 1, \ldots, 3n; j = 1, \ldots, \nu)$ are, as yet, any constants, whatsoever. Furthermore, we define generalized speeds $u_7, \ldots, u_{6+\nu}$ as

$$u_{6+k} \triangleq \dot{q}_k \qquad (k = 1, \ldots, \nu) \tag{66}$$

From these definitions it follows that

$$\dot{y}_i \underset{(64)}{=} \dot{\Delta}_i \underset{(65)}{=} \sum_{j=1}^{\nu} A_{ij}\dot{q}_j \underset{(66)}{=} \sum_{j=1}^{\nu} A_{ij}u_{6+j} \qquad (i = 1, \ldots, 3n) \tag{67}$$

and thus that

$$\begin{aligned} {}^N\mathbf{v}^{P_i} \underset{(61)}{=} & \left(u_4 + u_2 y_{3i} - u_3 y_{3i-1} + \sum_{j=1}^{\nu} A_{3i-2,j}u_{6+j}\right)\mathbf{r}_1 \\ & + \left(u_5 + u_3 y_{3i-2} - u_1 y_{3i} + \sum_{j=1}^{\nu} A_{3i-1,j}u_{6+j}\right)\mathbf{r}_2 \\ & + \left(u_6 + u_1 y_{3i-1} - u_2 y_{3i-2} + \sum_{j=1}^{\nu} A_{3i,j}u_{6+j}\right)\mathbf{r}_3 \\ & \qquad (i = 1, \ldots, n) \end{aligned} \tag{68}$$

Now, referring to Eqs. (48) and (51), one can express ${}^N\mathbf{v}^{P_i}$ also as

$${}^N\mathbf{v}^{P_i} = \sum_{j=1}^{3}\sum_{k=1}^{3} \dot{x}_{3i-3+j}C_{jk}\mathbf{r}_k \qquad (i = 1, \ldots, n) \tag{69}$$

Hence, equating the right-hand members of Eqs. (68) and (69), one arrives at the following equations, which will prove useful in the determination of initial values of $u_1, \ldots, u_{6+\nu}$:

$$u_4 + u_2 y_{3i} - u_3 y_{3i-1} + \sum_{j=1}^{\nu} A_{3i-2,j}u_{6+j} = \dot{x}_{3i-2}C_{11} + \dot{x}_{3i-1}C_{21} + \dot{x}_{3i}C_{31} \qquad (i = 1, \ldots, n) \tag{70}$$

$$u_5 + u_3 y_{3i-2} - u_1 y_{3i} + \sum_{j=1}^{\nu} A_{3i-1,j}u_{6+j} = \dot{x}_{3i-2}C_{12} + \dot{x}_{3i-1}C_{22} + \dot{x}_{3i}C_{32} \qquad (i = 1, \ldots, n) \tag{71}$$

$$u_6 + u_1 y_{3i-1} - u_2 y_{3i-2} + \sum_{j=1}^{\nu} A_{3i,j}u_{6+j} = \dot{x}_{3i-2}C_{13} + \dot{x}_{3i-1}C_{23} + \dot{x}_{3i}C_{33} \qquad (i = 1, \ldots, n) \tag{72}$$

Partial velocities of P_i, denoted by $\mathbf{v}_j^{P_i}$ $(j = 1, \ldots, 6 + \nu)$, can be written down by inspection of Eq. (68):

$$\mathbf{v}_1^{P_i} = -y_{3i}\mathbf{r}_2 + y_{3i-1}\mathbf{r}_3 \qquad (i = 1, \ldots, n) \tag{73}$$

$$\mathbf{v}_2^{P_i} = y_{3i}\mathbf{r}_1 - y_{3i-2}\mathbf{r}_3 \qquad (i = 1, \ldots, n) \tag{74}$$

$$\mathbf{v}_3^{P_i} = -y_{3i-1}\mathbf{r}_1 + y_{3i-2}\mathbf{r}_2 \qquad (i = 1, \ldots, n) \tag{75}$$

$$\mathbf{v}_4^{P_i} = \mathbf{r}_1 \qquad (i = 1, \ldots, n) \tag{76}$$

$$\mathbf{v}_5^{P_i} = \mathbf{r}_2 \qquad (i = 1, \ldots, n) \tag{77}$$

$$\mathbf{v}_6^{P_i} = \mathbf{r}_3 \qquad (i = 1, \ldots, n) \tag{78}$$

$$\mathbf{v}_{6+k}{}^{P_i} = A_{3i-2,k}\mathbf{r}_1 + A_{3i-1,k}\mathbf{r}_2 + A_{3i,k}\mathbf{r}_3 \qquad (i = 1, \ldots, n;\ k = 1, \ldots, \nu) \tag{79}$$

Equation (68) serves, in addition, as the point of departure for the determination of $\mathbf{a}^{P_i}$, the acceleration of P_i in N, which is found by differentiating Eq. (68) with respect to t in N. Before doing so, however, we define v_{ij} and a_{ij} ($i = 1, \ldots, n$; $j = 1, 2, 3$) as

$$v_{i1} \triangleq u_4 + u_2 y_{3i} - u_3 y_{3i-1} + \dot{y}_{3i-2} \qquad (i = 1, \ldots, n) \tag{80}$$

$$v_{i2} \triangleq u_5 + u_3 y_{3i-2} - u_1 y_{3i} + \dot{y}_{3i-1} \qquad (i = 1, \ldots, n) \tag{81}$$

$$v_{i3} \triangleq u_6 + u_1 y_{3i-1} - u_2 y_{3i-2} + \dot{y}_{3i} \qquad (i = 1, \ldots, n) \tag{82}$$

and

$$a_{i1} \triangleq u_2(\dot{y}_{3i} + v_{i3}) - u_3(\dot{y}_{3i-1} + v_{i2}) \qquad (i = 1, \ldots, n) \tag{83}$$

$$a_{i2} \triangleq u_3(\dot{y}_{3i-2} + v_{i1}) - u_1(\dot{y}_{3i} + v_{i3}) \qquad (i = 1, \ldots, n) \tag{84}$$

$$a_{i3} \triangleq u_1(\dot{y}_{3i-1} + v_{i2}) - u_2(\dot{y}_{3i-2} + v_{i1}) \qquad (i = 1, \ldots, n) \tag{85}$$

which makes it possible to express $\mathbf{a}^{P_i}$ as

$$\begin{aligned}\mathbf{a}^{P_i} = &\left(\dot{u}_4 + \dot{u}_2 y_{3i} - \dot{u}_3 y_{3i-1} + \sum_{j=1}^{\nu} A_{3i-2,j}\dot{u}_{6+j} + a_{i1}\right)\mathbf{r}_1 \\ &+ \left(\dot{u}_5 + \dot{u}_3 y_{3i-2} - \dot{u}_1 y_{3i} + \sum_{j=1}^{\nu} A_{3i-1,j}\dot{u}_{6+j} + a_{i2}\right)\mathbf{r}_2 \\ &+ \left(\dot{u}_6 + \dot{u}_1 y_{3i-1} - \dot{u}_2 y_{3i-2} + \sum_{j=1}^{\nu} A_{3i,j}\dot{u}_{6+j} + a_{i3}\right)\mathbf{r}_3 \\ &\qquad (i = 1, \ldots, n)\end{aligned} \tag{86}$$

The generalized inertia force $F_r{}^*$ associated with u_r, defined as

$$F_r{}^* \triangleq -\sum_{i=1}^{n} m_i \mathbf{v}_r{}^{P_i} \cdot \mathbf{a}^{P_i} \qquad (r = 1, \ldots, 6 + \nu) \tag{87}$$

now can be found by substitution from Eqs. (73)–(79) and (86) into Eq. (87), which leads to

$$\begin{aligned}F_1{}^* = -\sum_{i=1}^{n} m_i\Bigg[&-y_{3i}\left(\dot{u}_5 + \dot{u}_3 y_{3i-2} - \dot{u}_1 y_{3i} + \sum_{j=1}^{\nu} A_{3i-1,j}\dot{u}_{6+j} + a_{i2}\right) \\ &+ y_{3i-1}\left(\dot{u}_6 + \dot{u}_1 y_{3i-1} - \dot{u}_2 y_{3i-2} + \sum_{j=1}^{\nu} A_{3i,j}\dot{u}_{6+j} + a_{i3}\right)\Bigg]\end{aligned} \tag{88}$$

$$\begin{aligned}F_2{}^* = -\sum_{i=1}^{n} m_i\Bigg[&y_{3i}\left(\dot{u}_4 + \dot{u}_2 y_{3i} - \dot{u}_3 y_{3i-1} + \sum_{j=1}^{\nu} A_{3i-2,j}\dot{u}_{6+j} + a_{i1}\right) \\ &- y_{3i-2}\left(\dot{u}_6 + \dot{u}_1 y_{3i-1} - \dot{u}_2 y_{3i-2} + \sum_{j=1}^{\nu} A_{3i,j}\dot{u}_{6+j} + a_{i3}\right)\Bigg]\end{aligned} \tag{89}$$

$$F_3^* = -\sum_{i=1}^{n} m_i\left[-y_{3i-1}\left(\dot{u}_4 + \dot{u}_2 y_{3i} - \dot{u}_3 y_{3i-1} + \sum_{j=1}^{\nu} A_{3i-2,j}\dot{u}_{6+j} + a_{i1}\right)\right.$$
$$\left. + y_{3i-2}\left(\dot{u}_5 + \dot{u}_3 y_{3i-2} - \dot{u}_1 y_{3i} + \sum_{j=1}^{\nu} A_{3i-1,j}\dot{u}_{6+j} + a_{i2}\right)\right] \qquad (90)$$

$$F_4^* = -\sum_{i=1}^{n} m_i\left(\dot{u}_4 + \dot{u}_2 y_{3i} - \dot{u}_3 y_{3i-1} + \sum_{j=1}^{\nu} A_{3i-2,j}\dot{u}_{6+j} + a_{i1}\right) \qquad (91)$$

$$F_5^* = -\sum_{i=1}^{n} m_i\left(\dot{u}_5 + \dot{u}_3 y_{3i-2} - \dot{u}_1 y_{3i} + \sum_{j=1}^{\nu} A_{3i-1,j}\dot{u}_{6+j} + a_{i2}\right) \qquad (92)$$

$$F_6^* = -\sum_{i=1}^{n} m_i\left(\dot{u}_6 + \dot{u}_1 y_{3i-1} - \dot{u}_2 y_{3i-2} + \sum_{j=1}^{\nu} A_{3i,j}\dot{u}_{6+j} + a_{i3}\right) \qquad (93)$$

$$F_{6+k}^* = -\sum_{i=1}^{n} m_i\left[A_{3i-2,k}\left(\dot{u}_4 + \dot{u}_2 y_{3i} - \dot{u}_3 y_{3i-1} + \sum_{j=1}^{\nu} A_{3i-2,j}\dot{u}_{6+j} + a_{i1}\right)\right.$$
$$+ A_{3i-1,k}\left(\dot{u}_5 + \dot{u}_3 y_{3i-2} - \dot{u}_1 y_{3i} + \sum_{j=1}^{\nu} A_{3i-1,j}\dot{u}_{6+j} + a_{i2}\right)$$
$$\left. + A_{3i,k}\left(\dot{u}_6 + \dot{u}_1 y_{3i-1} - \dot{u}_2 y_{3i-2} + \sum_{j=1}^{\nu} A_{3i,j}\dot{u}_{6+j} + a_{i3}\right)\right]$$
$$(k = 1, \ldots, \nu) \qquad (94)$$

The next task is that of formulating expressions for the generalized active forces associated with $u_1, \ldots, u_{6+\nu}$. Here, two kinds of forces acting on P_i must be taken into account, namely, $\mathbf{S}^{P_i}$, the resultant of all forces exerted on P_i by truss members, and $\mathbf{Q}^{P_i}$, the resultant of all external forces acting on P_i (that is, contact and body forces exerted on P_i by agencies other than truss members). $\mathbf{Q}^{P_i}$ must be specified by the analyst as a function of, in general, $x_1, \ldots, x_{3n}, \dot{x}_1, \ldots, \dot{x}_{3n}$, and t, but can always be written

$$\mathbf{Q}^{P_i} = Q_{3i-2}\mathbf{r}_1 + Q_{3i-1}\mathbf{r}_2 + Q_{3i}\mathbf{r}_3 \qquad (i = 1, \ldots, n) \qquad (95)$$

As for $\mathbf{S}^{P_i}$, this can be expressed in terms of the elements of a $3n \times 3n$ structural stiffness matrix S and the displacements $\triangle_1, \ldots, \triangle_{3n}$ introduced in Eq. (63), provided that the following requirements be met when S is formed: the joints are numbered as heretofore; $\mathbf{n}_1$, $\mathbf{n}_2$, $\mathbf{n}_3$ are oriented in N as heretofore, and, for the performance of the structural analysis leading to S, the truss is placed in N in such a way that the position vector of P_i relative to N^* is the vector $\hat{\mathbf{p}}_i$ $(i = 1, \ldots, n)$ used heretofore. Under these circumstances,

$$\mathbf{S}^{P_i} = -\sum_{j=1}^{3n} (S_{3i-2,j}\mathbf{r}_1 + S_{3i-1,j}\mathbf{r}_2 + S_{3i,j}\mathbf{r}_3)\triangle_j \qquad (i = 1, \ldots, n) \qquad (96)$$

where S_{jk} is the element in the jth row and kth column of S.

F_r, the generalized active force associated with u_r, is given by

$$F_r = \sum_{i=1}^{n} \mathbf{v}_r^{P_i} \cdot (\mathbf{S}^{P_i} + \mathbf{Q}^{P_i}) \qquad (r = 1, \ldots, 6 + \nu) \tag{97}$$

Referring to Eqs. (73)–(79) for $\mathbf{v}_1^{P_i}, \ldots, \mathbf{v}_{\nu+6}^{P_i}$, and to Eqs. (95) and (96) for $\mathbf{Q}^{P_i}$ and $\mathbf{S}^{P_i}$, one thus finds that

$$F_1 = \sum_{i=1}^{n} \left[-y_{3i}Q_{3i-1} + y_{3i-1}Q_{3i} + \sum_{j=1}^{3n} \left(y_{3i}S_{3i-1,j} - y_{3i-1}S_{3i,j} \right) \triangle_j \right] \tag{98}$$

$$F_2 = \sum_{i=1}^{n} \left[y_{3i}Q_{3i-2} - y_{3i-2}Q_{3i} + \sum_{j=1}^{3n} \left(-y_{3i}S_{3i-2,j} + y_{3i-2}S_{3i,j} \right) \triangle_j \right] \tag{99}$$

$$F_3 = \sum_{i=1}^{n} \left[-y_{3i-1}Q_{3i-2} + y_{3i-2}Q_{3i-1} + \sum_{j=1}^{3n} \left(y_{3i-1}S_{3i-2,j} - y_{3i-2}S_{3i-1,j} \right) \triangle_j \right] \tag{100}$$

$$F_4 = \sum_{i=1}^{n} \left(Q_{3i-2} - \sum_{j=1}^{3n} S_{3i-2,j} \triangle_j \right) \tag{101}$$

$$F_5 = \sum_{i=1}^{n} \left(Q_{3i-1} - \sum_{j=1}^{3n} S_{3i-1,j} \triangle_j \right) \tag{102}$$

$$F_6 = \sum_{i=1}^{n} \left(Q_{3i} - \sum_{j=1}^{3n} S_{3i,j} \triangle_j \right) \tag{103}$$

$$F_{6+k} = \sum_{i=1}^{n} \Bigg[A_{3i-2,k}Q_{3i-2} + A_{3i-1,k}Q_{3i-1} + A_{3i,k}Q_{3i} - \sum_{j=1}^{3n} \left(A_{3i-2,k}S_{3i-2,j} + A_{3i-1,k}S_{3i-1,j} + A_{3i,k}S_{3i,j} \right) \triangle_j \Bigg] \qquad (k = 1, \ldots, \nu) \tag{104}$$

Now, as an immediate consequence of the definition of S_{ij},

$$\sum_{i=1}^{n} (y_{3i}S_{3i-1,j} - y_{3i-1}S_{3i,j}) = 0 \qquad (j = 1, \ldots, 3n) \tag{105}$$

$$\sum_{i=1}^{n} (y_{3i-2}S_{3i,j} - y_{3i}S_{3i-2,j}) = 0 \qquad (j = 1, \ldots, 3n) \tag{106}$$

$$\sum_{i=1}^{n} (y_{3i-1}S_{3i-2,j} - y_{3i-2}S_{3i-1,j}) = 0 \qquad (j = 1, \ldots, 3n) \tag{107}$$

$$\sum_{i=1}^{n} S_{3i-2,j} = 0 \qquad (j = 1, \ldots, 3n) \tag{108}$$

$$\sum_{i=1}^{n} S_{3i-1,j} = 0 \qquad (j = 1, \ldots, 3n) \tag{109}$$

$$\sum_{i=1}^{n} S_{3i,j} = 0 \qquad (j = 1, \ldots, 3n) \tag{110}$$

Hence, Eqs. (98)–(103) reduce to

$$F_1 = \sum_{i=1}^{n} (-\, y_{3i} Q_{3i-1} + y_{3i-1} Q_{3i}) \tag{111}$$

$$F_2 = \sum_{i=1}^{n} (-\, y_{3i-2} Q_{3i} + y_{3i} Q_{3i-2}) \tag{112}$$

$$F_3 = \sum_{i=1}^{n} (-\, y_{3i-1} Q_{3i-2} + y_{3i-2} Q_{3i-1}) \tag{113}$$

$$F_4 = \sum_{i=1}^{n} Q_{3i-2} \tag{114}$$

$$F_5 = \sum_{i=1}^{n} Q_{3i-1} \tag{115}$$

$$F_6 = \sum_{i=1}^{n} Q_{3i} \tag{116}$$

To accomplish a corresponding simplification of Eq. (104), we define G_l as

$$G_l \triangleq \sum_{i=1}^{n} \sum_{j=1}^{3} A_{3i-3+j,l} Q_{3i-3+j} \qquad (l = 1, \ldots, \nu) \tag{117}$$

and eliminate $\triangle_j$ by reference to Eq. (65), obtaining

$$F_{6+k} = G_k - \sum_{l=1}^{\nu} \left[\sum_{i=1}^{n} \sum_{j=1}^{3n} (A_{3i-2,k} S_{3i-2,j} A_{jl} + A_{3i-1,k} S_{3i-1,j} A_{jl} + A_{3i,k} S_{3i,j} A_{jl}) \right] q_l \qquad (k = 1, \ldots, \nu) \tag{118}$$

or, equivalently,

$$F_{6+k} = G_k - \sum_{l=1}^{\nu} \sum_{i=1}^{3n} \sum_{j=1}^{3n} A_{ik} S_{ij} A_{jl} q_l \qquad (k = 1, \ldots, \nu) \tag{119}$$

The dynamical differential equations governing all motions of the truss now can be formulated by substituting from Eqs. (88)–(94), (111)–(116), and (119) into

$$F_r + F_r{}^* = 0 \qquad (r = 1, \ldots, 6 + \nu) \tag{120}$$

which for $r = 1, \ldots, 6$, leads to Eqs. (40)–(45) when I_{jk}, w_j, and $\overline{m}$ are introduced as per Eqs. (33), (34), and (36), respectively. Before writing the equations corresponding to $r = 6 + k$ $(k = 1, \ldots, \nu)$, we note that, if $\nu = 3n - 6$, then the number of differential equations furnished by Eq. (120) is $3n$, which is precisely the number of degrees of freedom of the set of n particles under consideration.

Hence, if $\nu = 3n - 6$ and the constants A_{ij} $(i = 1, \ldots, 3n; j = 1, \ldots, \nu)$ in Eqs. (65) are such that these equations can be solved uniquely for $q_1, \ldots, q_{3n-6}$, then the solution of the dynamical equations leads to an *exact* description of the motion under consideration. But what if one takes $\nu < 3n - 6$, which one is well motivated to do in order to reduce the number of differential equations one must solve simultaneously? In that event, the solution of the dynamical equations leads to an approximate solution; and, if the constants A_{ij} $(i = 1, \ldots, 3n; j = 1, \ldots, \nu)$ are chosen "properly," one may obtain very good approximations to the exact solution, even with ν much smaller than $3n - 6$. How, then, can one assign "suitable" values to these constants? One way, it turns out, is to make use of the eigenvectors arising in connection with free vibrations of the unrestrained structure, which may be found as follows.

Construct a $3n \times 3n$ diagonal mass matrix M by taking

$$M_{3i-3+r,3l-3+r} = m_i \delta_{li} \qquad (i, l = 1, \ldots, n; r = 1, 2, 3) \tag{121}$$

Next, consider the eigenvalues of the matrix $M^{-1/2} SM^{-1/2}$, where S is the stiffness matrix used previously. (The eigenvalues of $M^{-1/2} SM^{-1/2}$ are the same as those of $M^{-1}S$, but they can be found more easily because $M^{-1/2} S^{-1/2}$ is a symmetric matrix, whereas $M^{-1}S$, in general, is not symmetric.) If the eigenvalues are arranged in ascending order, the first six (corresponding to "rigid-body" modes) will necessarily be equal to zero (or they will differ very slightly from zero, due to truncation errors, but they will be noticeably smaller than all remaining eigenvalues). Designate as $\lambda_1, \ldots, \lambda_\nu$ any† ν nonzero eigenvalues arranged in ascending order; form the corresponding eigenvectors; and premultiply each such eigenvector with $M^{-1/2}$, thus obtaining associated eigenvectors $B_{i1}, \ldots, B_{i\nu}$ $(i = 1, \ldots, 3n)$ of $M^{-1}S$. Now introduce normalization constants

$$\mu_j = \left(\sum_{k=1}^{3n} M_{kk} B_{kj}{}^2 \right)^{-1/2} \qquad (j = 1, \ldots, \nu) \tag{122}$$

and use these to express A_{ij} as

$$A_{ij} \triangleq \mu_j B_{ij} \qquad (i = 1, \ldots, 3n; j = 1, \ldots, \nu) \tag{123}$$

When A_{ij} is defined in this way, it can be shown that‡

$$\sum_{j=1}^{3n} \sum_{j=1}^{3n} A_{ik} S_{ij} A_{jl} = \lambda_l \delta_{lk} \qquad (k, l = 1, \ldots, \nu) \tag{124}$$

Consequently, Eq. (119) then can be replaced with

$$F_{6+k} = G_k - \lambda_k q_k \qquad (k = 1, \ldots, \nu) \tag{125}$$

† The most common practice is to use the *first* ν nonzero eigenvalues. The selection of eigenvalues other than these should be considered when there is reason to believe that particular vibration modes may become excited, say, due to the action of certain control forces.

‡ S. Timoshenko, D. H. Young, and W. Weaver, Jr., *Vibration Problems in Engineering*, 4th ed., John Wiley & Sons, New York, 1974, pp. 298–300.

The definition of A_{ij} as per Eq. (123) also leads to a simplification of Eqs. (91)–(94). Specifically, as will be shown presently,

$$\sum_{i=1}^{n} m_i A_{3i-3+j,k} = 0 \qquad (j = 1, 2, 3;\ k = 1, \ldots, \nu) \tag{126}$$

and

$$\sum_{i=1}^{n} \sum_{l=1}^{3} m_i A_{3i-3+l,j} A_{3i-3+l,k} = \delta_{jk} \qquad (j, k = 1, \ldots, \nu) \tag{127}$$

so that in Eqs. (91)–(93) the terms involving $\dot{u}_{6+j}$ $(j = 1, \ldots, \nu)$ vanish while Eq. (94) can be replaced with

$$\begin{aligned} F_{6+k}{}^* = -\sum_{i=1}^{n} m_i[&A_{3i-2,k}(\dot{u}_2 y_{3i} - \dot{u}_3 y_{3i-1} + a_{i1}) \\ &+ A_{3i-1,k}(\dot{u}_3 y_{3i-2} - \dot{u}_1 y_{3i} + a_{i2}) \\ &+ A_{3i,k}(\dot{u}_1 y_{3i-1} - \dot{u}_2 y_{3i-2} + a_{i3})] - \dot{u}_{6+k} \end{aligned}$$
$$(k = 1, \ldots, \nu) \tag{128}$$

To establish the validity of Eq. (126), we begin by noting that since m_i and A_{ij} are, respectively, a generic mass and a generic element of an eigenvector associated with an unrestrained structure, we can discover certain facts regarding these quantities by considering any particular motion that the unrestrained structure can perform. For instance, an unrestrained structure is capable of moving in such a way that its inertial linear momentum is initially equal to zero, in which event

$$\sum_{i=1}^{n} \sum_{j=1}^{3} m_i \dot{x}_{3i-3+j}(0)\mathbf{n}_j = 0 \tag{129}$$

so that, when x_{3i-3+j} is expressed in terms of $A_{3i-3+j,k}$ $(k = 1, \ldots, \nu)$ and ν functions of time, say $\bar{q}_1, \ldots, \bar{q}_\nu$, as

$$x_{3i-3+j} = \sum_{k=1}^{\nu} A_{3i-3+j,k}\bar{q}_k \qquad (i = 1, \ldots, n;\ j = 1, 2, 3) \tag{130}$$

one has

$$\sum_{i=1}^{n} \sum_{j=1}^{3} \sum_{k=1}^{\nu} m_i A_{3i-3+j,k}\dot{\bar{q}}_k(0)\mathbf{n}_j = 0 \tag{131}$$

which implies that

$$\sum_{k=1}^{\nu} \sum_{i=1}^{n} m_i A_{3i-3+j,k}\dot{\bar{q}}_k(0) = 0 \qquad (j = 1, 2, 3) \tag{132}$$

Moreover, since one always can choose $\dot{x}_i(0)$ $(i = 1, \ldots, 3n)$ such that a particular one of $\dot{\bar{q}}_1(0), \ldots, \dot{\bar{q}}_\nu(0)$ differs from zero while all the rest vanish, the coefficient of $\dot{\bar{q}}_k(0)$ in Eq. (132) must vanish; that is, Eq. (126) must be satisfied.

As for Eq. (127), this is simply an alternative way of expressing the orthogonality relationship†

$$\sum_{i=1}^{3n} \sum_{l=1}^{3n} A_{ij} M_{il} A_{lk} = \delta_{jk} \qquad (j, k = 1, \ldots, \nu) \tag{133}$$

since this equation can be rewritten as

$$\sum_{i=1}^{n} \sum_{l=1}^{n} \sum_{r=1}^{3} A_{3i-3+r,j} M_{3i-3+r,3l-3+r} A_{3l-3+r,k} = \delta_{jk} \qquad (j, k = 1, \ldots, \nu) \tag{134}$$

which reduces to Eq. (127) when Eq. (121) is taken into account.

The equation to which Eq. (120) leads for $r = 6 + k \ (k = 1, \ldots, \nu)$ now can be written by making use of Eqs. (125) and (128), which brings one directly to Eq. (46). Thus, all of the differential equations required for the solution of the problem initially posed are in hand. They are Eqs. (55), (57), (66) and the differential equations corresponding to Eq. (120), that is, Eqs. (40)–(46); and this set of first-order differential equations governs the unknowns C_{jk} $(j, k = 1, 2, 3)$, z_j $(j = 1, 2, 3)$, q_j $(j = 1, \ldots, \nu)$, and u_r $(r = 1, \ldots, 6 + \nu)$, all of whose initial values must be known before a solution of the equations can be undertaken. Now, since $\mathbf{p}_i$ at $t = 0$ and ${}^N\mathbf{v}^{P_i}$ at $t = 0$ are regarded as given, the only scalar variables whose initial values can be presumed to be readily available are [see Eq. (48)] the $6n$ quantities x_i $(i = 1, \ldots, 3n)$ and $\dot{x}_i$ $(i = 1, \ldots, 3n)$. Hence, one must express the $9 + 3 + \nu + 6 + \nu = 18 + 2\nu$ needed quantities in terms of the $6n$ available ones before one can hope to solve a realistically posed initial-value problem. We shall now show how to do this.

Throughout the sequel, the value of a function f of t at $t = 0$ is denoted by $f(0)$. Thus, referring to Eq. (53), one may write,

$$y_{3i-3+j}(0) = \sum_{k=1}^{3} [x_{3i-3+k}(0) - z_k(0)] C_{kj}(0)$$

$$(i = 1, \ldots, n; j = 1, 2, 3) \tag{135}$$

Similarly, after eliminating Δ_i from Eqs. (64) and (65), one has

$$y_{3i-3+j} = \hat{x}_{3i-3+j} + \sum_{k=1}^{\nu} A_{3i-3+j,k} q_k \qquad (i = 1, \ldots, n; j = 1, 2, 3) \tag{136}$$

and, upon setting $t = 0$,

$$y_{3i-3+j}(0) = \hat{x}_{3i-3+j} + \sum_{k=1}^{\nu} A_{3i-3+j,k} q_k(0) \qquad (i = 1, \ldots, n; j = 1, 2, 3) \tag{137}$$

Hence, one can equate the right-hand members of Eqs. (135) and (137), obtaining

$$\sum_{k=1}^{3} [x_{3i-3+k}(0) - z_k(0)] C_{kj}(0) = \hat{x}_{3i-3+j} + \sum_{k=1}^{\nu} A_{3i-3+j,k} q_k(0)$$

$$(i = 1, \ldots, n; j = 1, 2, 3) \tag{138}$$

† S. Timoshenko, D. H. Young, and W. Weaver, Jr., *Vibration Problems in Engineering*, 4th ed., John Wiley & Sons, New York, 1974, p. 229.

To this set of $3n$ equations in the $12+\nu$ unknowns $z_i(0)$, $C_{ij}(0)$, $q_k(0)$ ($i, j=1, 2, 3$; $k=1, \ldots, \nu$ one must add six independent equations expressing the fact that $\mathbf{r}_1$, $\mathbf{r}_2$, $\mathbf{r}_3$ are mutually perpendicular unit vectors, namely, any six of the nine equations

$$\sum_{k=1}^{3} C_{ki}(0)C_{kj}(0) = \delta_{ij} \qquad (i, j = 1, 2, 3) \tag{139}$$

Thus, we have here $3n + 6$ independent equations governing $12 + \nu$ unknowns. Now, $3n + 6$ exceeds $12 + \nu$, except when $\nu = 3n - 6$. Hence, precisely in the situations of greatest interest to us, that is, when $\nu < 3n - 6$, we have more equations than unknowns, which means, in general, that there exists no set of values of the unknowns that satisfies all of the equations. Moreover, the equations are nonlinear, so that solving them is not a simple matter, in any event. However, these difficulties can be surmounted by taking advantage of the fact that $\hat{x}_1, \ldots, \hat{x}_{3n}$ are not unique, which permits one to reason as follows.

If the truss were undeformed at $t = 0$, then $\hat{x}_1, \ldots, \hat{x}_{3n}$ could be assigned values such that

$$\hat{x}_i = x_i(0) \qquad (i = 1, \ldots, 3n) \tag{140}$$

and Eqs. (138) and (139) then would be satisfied identically by

$$z_i(0) = 0 \qquad C_{ij}(0) = \delta_{ij} \qquad q_k(0) = 0 \qquad (i, j = 1, 2, 3;\ k = 1, \ldots, \nu) \tag{141}$$

Now, in actuality, the truss is deformed only *slightly,* at most, at $t = 0$, for we are concerned solely with motions satisfying the requirements of linear structural theory. Hence, it is always possible to position the truss in N in its undeformed state in such a way that $\hat{x}_i$ differs only slightly from $x_i(0)$ ($i = 1, \ldots, 3n$) and that, consequently, $z_i(0)$, $C_{ij}(0)$, and $q_k(0)$ differ only slightly from 0, δ_{ij}, and 0, respectively; and, after introducing ϵ_{ij} by writing

$$C_{ij}(0) = \delta_{ij} + \epsilon_{ij} \qquad (i, j = 1, 2, 3) \tag{142}$$

one then may linearize Eqs. (138) and (139) in $z_i(0)$, ϵ_{ij}, and $q_k(0)$ ($i, j = 1, 2, 3$; $k = 1, \ldots, \nu$), obtaining from Eq. (138),

$$\sum_{k=1}^{3} \{[x_{3i-3+k}(0) - z_k(0)]\delta_{kj} + x_{3i-3+k}(0)\epsilon_{kj}\} = \hat{x}_{3i-3+j} + \sum_{k=1}^{\nu} A_{3i-3+j,k}\, q_k(0)$$
$$(i = 1, \ldots, n;\ j = 1, 2, 3) \tag{143}$$

and, from Eq. (139),

$$\sum_{k=1}^{3} (\delta_{ki}\delta_{kj} + \epsilon_{ki}\delta_{kj} + \delta_{ki}\epsilon_{kj}) = \delta_{ij} \qquad (i, j = 1, 2, 3) \tag{144}$$

which gives rise to the six independent equations

$$\epsilon_{11} = \epsilon_{22} = \epsilon_{33} = \epsilon_{12} + \epsilon_{21} = \epsilon_{23} + \epsilon_{32} = \epsilon_{31} + \epsilon_{13} = 0 \tag{145}$$

with the aid of which Eqs. (143) can be replaced with

$$x_{3i-2}(0) - z_1(0) - x_{3i-1}(0)\epsilon_{12} + x_{3i}(0)\epsilon_{31} = \hat{x}_{3i-2} + \sum_{k=1}^{\nu} A_{3i-2,k}q_k(0) \qquad (i = 1, \ldots, n) \qquad (146)$$

$$x_{3i-1}(0) - z_2(0) - x_{3i}(0)\epsilon_{23} + x_{3i-2}(0)\epsilon_{12} = \hat{x}_{3i-1} + \sum_{k=1}^{\nu} A_{3i-1,k}q_k(0) \qquad (i = 1, \ldots, n) \qquad (147)$$

$$x_{3i}(0) - z_3(0) - x_{3i-2}(0)\epsilon_{31} + x_{3i-1}(0)\epsilon_{23} = \hat{x}_{3i} + \sum_{k=1}^{\nu} A_{3i,k}q_k(0) \qquad (i = 1, \ldots, n) \qquad (148)$$

Now we are dealing with a set of linear equations, but we must still come to grips with the fact that the number of these equations, that is, $3n$, exceeds the number of unknowns, which is $6 + \nu$, the unknowns being $z_1(0)$, $z_2(0)$, $z_3(0)$, ϵ_{12}, ϵ_{23}, ϵ_{31}, and $q_1(0), \ldots, q_\nu(0)$. The resolution of this dilemma is to seek a "best approximate solution" by proceeding as follows:† Let b, c, and K be, respectively, the $3n \times 1$, $(6 + \nu) \times 1$, and $3n \times (6 + \nu)$ matrices whose elements are defined as

$$b_i \triangleq x_i(0) - \hat{x}_i \qquad (i = 1, \ldots, 3n) \qquad (149)$$

$$c_1 \triangleq \epsilon_{23} \qquad c_2 \triangleq \epsilon_{31} \qquad c_3 \triangleq \epsilon_{12} \qquad (150)$$

$$c_4 \triangleq z_1(0) \qquad c_5 \triangleq z_2(0) \qquad c_6 \triangleq z_3(0) \qquad (151)$$

$$c_{6+k} \triangleq q_k(0) \qquad (k = 1, \ldots, \nu) \qquad (152)$$

$$K_{3i-3+j,k} \triangleq -\frac{1}{2}\sum_{l=1}^{3} (j-k)(k-l)(l-j)x_{3i-3+l}(0) \qquad (i = 1, \ldots, n;\ j, k = 1, 2, 3) \qquad (153)$$

$$K_{3i-3+j,3+k} \triangleq \delta_{jk} \qquad (i = 1, \ldots, n;\ j, k = 1, 2, 3) \qquad (154)$$

$$K_{i,6+j} \triangleq A_{ij} \qquad (i = 1, \ldots, 3n;\ j = 1, \ldots, \nu) \qquad (155)$$

Then Eqs. (146)–(148) are equivalent to the matrix equation

$$Kc = b \qquad (156)$$

Since K is not a square matrix, and hence cannot be inverted, this equation cannot be solved for c. But K^TK, where K^T denotes the transpose of K, is a square matrix (nonsingular, in general). Hence, premultiply Eq. (156) with K^T and then solve for c, obtaining

$$c = (K^TK)^{-1}K^Tb \qquad (157)$$

(This "pseudo-inversion" of K can be performed even if K^TK is singular.)

†R. Penrose, "On Best Approximate Solutions of Linear Matrix Equations," *Proc. Cambridge Phil. Soc.*, vol. 52, part 1, 1956, pp. 17–19; E. H. Moore, "General Analysis, part 1," *Mem. Amer. Phil. Soc.*, vol. 1, 1935, pp. 197–209; R. Penrose, "A General Inverse for Matrices," *Proc. Cambridge Phil. Soc.*, vol. 51, part 1, 1955, pp. 406–413.

Equation (157) furnishes the desired "best approximate solution" of Eqs. (146)–(148) and, hence, a satisfactory solution of Eqs. (138) and (139). That is, once c has been determined, one can form $z_i(0)$ $(i = 1, 2, 3)$ by using Eqs. (151), $C_{ij}(0)$ $(i, j = 1, 2, 3)$ by reference to Eqs. (142), (145) and (150), and $q_k(0)$ $(k = 1, \ldots, \nu)$ with the aid of Eq. (152); and $y_i(0)$ $(i = 1, \ldots, 3n)$ then can be found by substitution of the now known values of $q_k(0)$ $(k = 1, \ldots, \nu)$ into Eq. (137). What remains to be done is to determine the initial values of u_r $(r = 1, \ldots, 6 + \nu)$, which can be accomplished by using a pseudo-inverse once more, this time in conjunction with Eqs. (70)–(72). That is, after introducing d, e, and L as $3n \times 1$, $(6 + \nu) \times 1$, and $3n \times (6 + \nu)$ matrices, respectively, defining their elements as

$$d_{3i-2} \triangleq \dot{x}_{3i-2}(0) - \dot{x}_{3i-1}(0)\,\epsilon_{12} + \dot{x}_{3i}(0)\,\epsilon_{31} \qquad (i = 1, \ldots, n) \tag{158}$$

$$d_{3i-1} \triangleq \dot{x}_{3i-1}(0) - \dot{x}_{3i}(0)\,\epsilon_{23} + \dot{x}_{3i-2}(0)\,\epsilon_{12} \qquad (i = 1, \ldots, n) \tag{159}$$

$$d_{3i} \triangleq \dot{x}_{3i}(0) - \dot{x}_{3i-2}(0)\,\epsilon_{31} + \dot{x}_{3i-1}(0)\,\epsilon_{23} \qquad (i = 1, \ldots, n) \tag{160}$$

$$e_r \triangleq u_r(0) \qquad (r = 1, \ldots, 6 + \nu) \tag{161}$$

$$L_{3i-3+j,k} \triangleq \frac{1}{2}\sum_{l=1}^{3} (j - k)(k - l)(l - j)\, y_{3i-3+l}(0) \qquad (i = 1, \ldots, n;\ j, k = 1, 2, 3) \tag{162}$$

$$L_{3i-3+j,\,3+k} \triangleq \delta_{jk} \qquad (i = 1, \ldots, n;\ j, k = 1, 2, 3) \tag{163}$$

$$L_{i,6+j} \triangleq A_{ij} \qquad (i = 1, \ldots, 3n;\ j = 1, \ldots, \nu) \tag{164}$$

one can write Eqs. (70)–(72) for $t = 0$ as

$$Le = d \tag{165}$$

and then "solve" for e as

$$e = (L^T L)^{-1} L^T d \tag{166}$$

whereupon Eq. (161) furnishes $u_r(0)$ $(r = 1, \ldots, 6 + \nu)$.

Once the differential equations of motion have been solved, that is, once C_{jk} $(j, k = 1, 2, 3)$, z_j $(j = 1, 2, 3)$, q_j $(j = 1, \ldots, \nu)$, and u_r $(r = 1, \ldots, 6 + \nu)$ have been determined for $t > 0$, one final, important task remains, namely, that of evaluating $x_1, \ldots, x_{3n}$, these being the quantities one needs to form $\mathbf{p}_i$ in accordance with

$$\mathbf{p}_i \underset{(48)}{=} \sum_{j=1}^{3} x_{3i-3+j}\,\mathbf{n}_j \qquad (i = 1, \ldots, n) \tag{167}$$

To find $x_1, \ldots, x_{3n}$, one needs only to perform successive scalar multiplications of Eq. (52) with $\mathbf{n}_1$, $\mathbf{n}_2$, $\mathbf{n}_3$, which yields, with the aid of Eq. (51),

$$x_{3i-3+j} = z_j + \sum_{k=1}^{3} y_{3i-3+k}\,C_{jk} \qquad (i = 1, \ldots, n;\ j = 1, 2, 3) \tag{168}$$

Example Figure 4.9.1 shows a truss consisting of members numbered from 1 to 57, these members meeting variously at joints numbered from 1 to 21. Members 1, 2, and 3 form a right-isosceles triangle, the right angle being at joint 1; and members 1 and 3 each have a length of 2 m. Members 4, 5 and 6 form a triangle congruent to that formed by members 1, 2, and 3, and similarly for

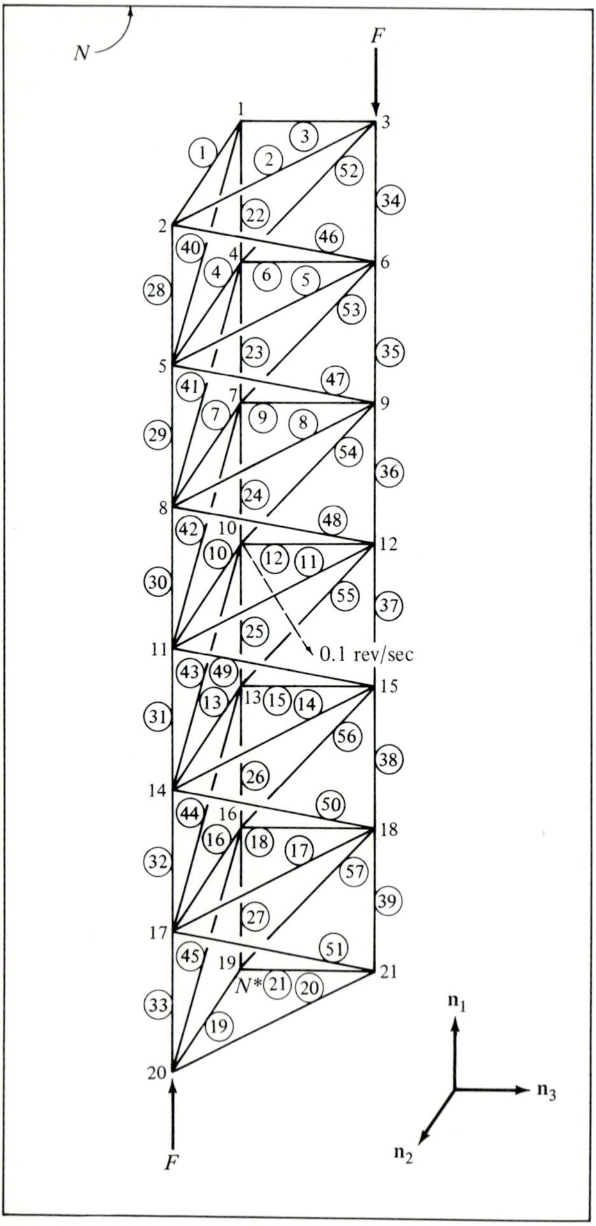

Figure 4.9.1

7, 8, and 9, . . . , 19, 20 and 21. Members 22, . . . , 39 all are 2 m long. Young's modulus for all members has the value 10^9 N/m^2; the cross-section area of each member is equal to 10^{-4} m^2; and each member has a mass density of 10^4 kg/m^3.

Also shown in Fig. 4.9.1 is a dextral set of mutually perpendicular unit vectors $\mathbf{n}_1$, $\mathbf{n}_2$, $\mathbf{n}_3$, presumed fixed in a Newtonian reference frame N. The motion of the truss in N is to be simulated for time $t > 0$, assuming that at $t = 0$ the truss is oriented relative to $\mathbf{n}_1$, $\mathbf{n}_2$, $\mathbf{n}_3$ as shown, that joint 19 coincides at $t = 0$ with a point N^* fixed in N, that the truss is undeformed at $t = 0$, that the truss is moving at $t = 0$ as if it were a rigid body rotating with an angular speed of 0.1 rev/sec about a line that passes through joint 10 and bisects the angle between members 10 and 12, the associated angular velocity vector being directed as shown in Fig. 4.9.1, and that two external forces, each of magnitude F, act on the truss, one applied at joint 3 and directed toward joint 6, the other applied at joint 20 and directed toward joint 17.

The quantities $x_i(0)$ $(i = 1, \ldots, 63)$ are found by reference to Fig. 4.9.1. For example, since

$$\mathbf{p}_5 = 10\mathbf{n}_1 + 2\mathbf{n}_2 \tag{169}$$

it follows that

$$x_{13}(0) = x_{15-3+1}(0) = \mathbf{p}_5 \cdot \mathbf{n}_1 = 10 \text{ m} \tag{170}$$

$$x_{14}(0) = x_{15-3+2}(0) = \mathbf{p}_5 \cdot \mathbf{n}_2 = 2 \text{ m} \tag{171}$$

$$x_{15}(0) = x_{15-3+3}(0) = \mathbf{p}_5 \cdot \mathbf{n}_3 = 0 \tag{172}$$

To determine $\dot{x}_i(0)$ $(i = 1, \ldots, 63)$, note that, for $t = 0$,

$$^N\mathbf{v}^{P_i} = 0.2\pi \frac{\mathbf{n}_2 + \mathbf{n}_3}{\sqrt{2}} \times (\mathbf{p}_i - \mathbf{p}_{10}) \qquad (i = 1, \ldots, 21) \tag{173}$$

and that

$$\mathbf{p}_i - \mathbf{p}_{10} = (x_{3i-2} - x_{28})\mathbf{n}_1 + (x_{3i-1} - x_{29})\mathbf{n}_2 + (x_{3i} - x_{30})\mathbf{n}_3 \qquad (i = 1, \ldots, 21) \tag{174}$$

so that

$$^N\mathbf{v}^{P_i} = \frac{0.2\pi}{\sqrt{2}} [(x_{3i} - x_{30} - x_{3i-1} + x_{29})\mathbf{n}_1 + (x_{3i-2} - x_{28})\mathbf{n}_2 + (x_{28} - x_{3i-2})\mathbf{n}_3] \qquad (i = 1, \ldots, 21) \tag{175}$$

Consequently, for example,

$$\dot{x}_{14}(0) = \dot{x}_{15-3+2}(0) = {}^N\mathbf{v}^{P_5} \cdot \mathbf{n}_2 = \frac{0.2\pi}{\sqrt{2}} [x_{13}(0) - x_{28}(0)] = \frac{0.2\pi}{\sqrt{2}} (10 - 6) = \frac{0.8\pi}{\sqrt{2}} \text{ m/sec} \tag{176}$$

Since the truss is undeformed at $t = 0$, we take

$$\hat{x}_i = x_i(0) \qquad (i = 1, \ldots, 63) \tag{177}$$

in order to satisfy the requirement that, for the purpose of performing a modal analysis, the undeformed truss be positioned in N in such a way that $P_1, \ldots, P_{21}$ occupy nearly or exactly the same positions as they do at $t = 0$. [If the initial conditions were altered slightly, by requiring, say, that at $t = 0$ joint 5 be displaced 0.01 m in the direction of $\mathbf{n}_3$, then Eq. (172) would be replaced with $x_{15}(0) = 0.01$ m; Eq. (177) would apply for all values of i except 15; and, for $i = 15$, we would have $\hat{x}_{15} = 0.00$ m.]

The forces of magnitude F applied to joints 3 and 20 give rise to six nonzero quantities of the kind previously designated Q_i ($i = 1, \ldots, 63$). Consider, for example, the force applied at joint 3. This can be expressed as

$$\mathbf{Q}^{P_3} = F\mathbf{n} \tag{178}$$

where $\mathbf{n}$ is a unit vector directed from joint 3 toward joint 6, that is,

$$\mathbf{n} = \frac{\mathbf{p}_6 - \mathbf{p}_3}{|\mathbf{p}_6 - \mathbf{p}_3|} \tag{179}$$

Now,

$$\mathbf{p}_6 = x_{16}\mathbf{n}_1 + x_{17}\mathbf{n}_2 + x_{18}\mathbf{n}_3 \qquad \mathbf{p}_3 = x_7\mathbf{n}_1 + x_8\mathbf{n}_2 + x_9\mathbf{n}_3 \tag{180}$$

Hence,

$$\mathbf{Q}^{P_3} = F\frac{(x_{16} - x_7)\mathbf{n}_1 + (x_{17} - x_8)\mathbf{n}_2 + (x_{18} - x_9)\mathbf{n}_3}{[(x_{16} - x_7)^2 + (x_{17} - x_8)^2 + (x_{18} - x_9)^2]^{1/2}} \tag{181}$$

and, in accordance with Eq. (1),

$$Q_7 = Q_{9-3+1} = \frac{F[(x_{16} - x_7)C_{11} + (x_{17} - x_8)C_{21} + (x_{18} - x_9)C_{31}]}{[(x_{16} - x_7)^2 + (x_{17} - x_8)^2 + (x_{18} - x_9)^2]^{1/2}} \tag{182}$$

$$Q_8 = Q_{9-3+2} = \frac{F[(x_{16} - x_7)C_{12} + (x_{17} - x_8)C_{22} + (x_{18} - x_9)C_{32}]}{[(x_{16} - x_7)^2 + (x_{17} - x_8)^2 + (x_{18} - x_9)^2]^{1/2}} \tag{183}$$

$$Q_9 = Q_{9-3+3} = \frac{F[(x_{16} - x_7)C_{13} + (x_{17} - x_8)C_{23} + (x_{18} - x_9)C_{33}]}{[(x_{16} - x_7)^2 + (x_{17} - x_8)^2 + (x_{18} - x_9)^2]^{1/2}} \tag{184}$$

Finally, $x_7, \ldots, x_{18}$ must be expressed in terms of y_i, z_j, C_{jk} ($i = 1, \ldots, 63$; $j, k = 1, 2, 3$) in accordance with Eq. (2). For example,

$$x_{16} = x_{18-3+1} = z_1 + y_{16}C_{11} + y_{17}C_{12} + y_{18}C_{13} \tag{185}$$

Using the numbering of members and joints indicated in Fig. 4.9.1, one obtains the following as representative elements of the structural stiffness matrix of the truss: $S_{4,6} = S_{6,4} = -9.62 \times 10^3$, $S_{7,11} = S_{11,7} = 0$, $S_{37,50} = S_{50,37} = 1.77 \times 10^4$. As for the mass matrix M, consider, for example, joint 7. The members that meet there are 7, 9, 23, 24, 42, and 53. The first four each have a length of 2 m, whereas each of the last two is $2\sqrt{2}$ m long. The total mass of the members meeting at the joint is thus $[(4)(2) + (2)(2\sqrt{2})](10^{-4})(10^4)$ kg, so that $m_7 = 6.83$ kg. Consequently, the elements $M_{19,19}$, $M_{20,20}$, and $M_{21,21}$ each have the value 6.83.

Suppose that ν is assigned the value 3 and that all but the first three "elastic" modes are to be excluded from the simulation to be performed. Then the first nine eigenvalues of the matrix W defined in Eq. (5) are found to have the values -1.06×10^{-15}, -3.04×10^{-15}, -5.69×10^{-15}, -7.28×10^{-15}, -1.45×10^{-14}, -1.50×10^{-14}, 65.3, 145.2, 232.7. Hence, $\lambda_1 = 65.3$, $\lambda_2 = 145.2$, $\lambda_3 = 232.7$; and Eq. (6) yields, for example, $A_{4,2} = -1.38 \times 10^{-2}$ and $A_{42,1} = 6.17 \times 10^{-2}$.

The initial values of C_{jk}, z_j $(j, k = 1, 2, 3)$, q_r $(r = 1, 2, 3)$, and u_s $(s = 1, \ldots, 9)$, determined by means of Eqs. (7)–(24), are $C_{ij}(0) = \delta_{ij}$ $(i, j = 1, 2, 3)$, $z_j(0) = 0$ $(j = 1, 2, 3)$, $q_r(0) = 0$ $(r = 1, 2, 3)$, $u_i(0) = 0$ $(i = 1, 4, 7, 8, 9)$, $u_2(0) = u_3(0) = 4.44 \times 10^{-1}$ rad/sec, $u_5(0) = -u_6(0) = -2.67$ m/sec; and the simultaneous solution of Eqs. (37)–(46), carried out by using a computer program that incorporates Eqs. (25)–(36), can be performed once F, the magnitude of the forces applied at joints 3 and 20, has been specified. Setting $F = 300$ N for the first 2 sec of the motion, and equal to zero thereafter, we thus find with the aid of Eq. (2) that, during the first 4 sec of the motion, $\sqrt{2}x_{29}$ and $\sqrt{2}x_{30}$ each have the values plotted as "Displacement of P" in Fig. 4.9.2, while x_{28} is equal to 6 m. In other words, joint 10 (that is, point P of Fig. 4.9.1)

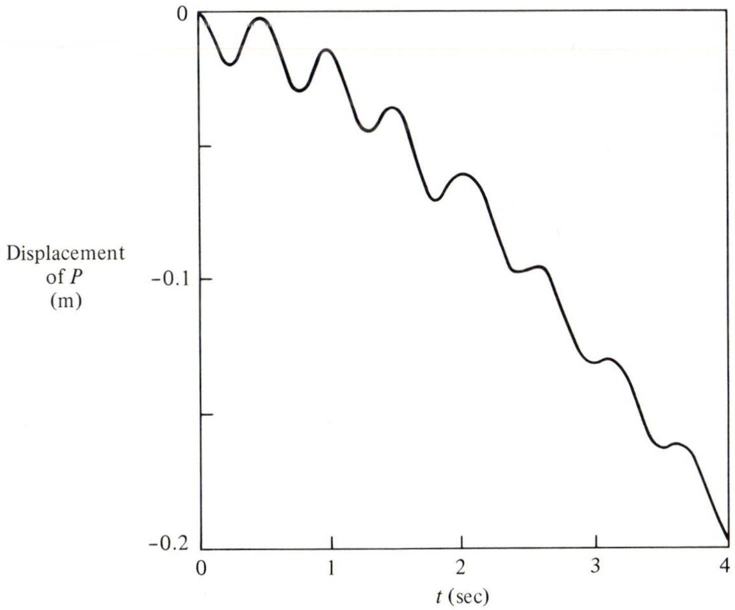

Figure 4.9.2

moves in a straight line, in the direction of the vector $-(\mathbf{n}_2 + \mathbf{n}_3)$.

Our reason for reporting results corresponding to $\nu = 3$ is that these results agree to five significant figures with those corresponding to $\nu = 2$. In other words, "convergence" seems to have occurred. Of course, this does not guarantee that modes higher than the third would not contribute significantly if they were taken into account.

4.10 SPACECRAFT WITH CONTINUOUS ELASTIC COMPONENTS

When a spacecraft consists in part of continuous elastic components supported by rigid bodies, small motions of the components relative to the rigid bodies generally are governed by partial differential equations that cannot be solved by the method of separation of variables. But it can occur that the partial differential equations governing *certain* motions of the components *can* be solved by this method, as is the case, for example, when a component is a uniform cantilever beam and one considers motions during which the supporting rigid body is fixed in a Newtonian reference frame. Under these circumstances, one can formulate equations governing motions of the spacecraft approximately by employing generalized speeds intimately related to vibration modes associated with the behavior governed by the separable equations. We shall illustrate this with an example after reviewing those portions of beam vibrations theory needed in the sequel.

Figure 4.10.1 shows a cantilever beam B of length L, constant flexural rigidity EI and constant mass per unit of length ρ. When B is supported by a rigid body fixed in a Newtonian reference frame, small flexural vibrations of B are governed by the equation†

$$EI \frac{\partial^4 y}{\partial x^4} + \rho \frac{\partial^2 y}{\partial t^2} = 0 \tag{1}$$

† W. Flugge, *Handbook of Engineering Mechanics,* McGraw-Hill, New York, 1962, pp. **61**-6–**61**-11.

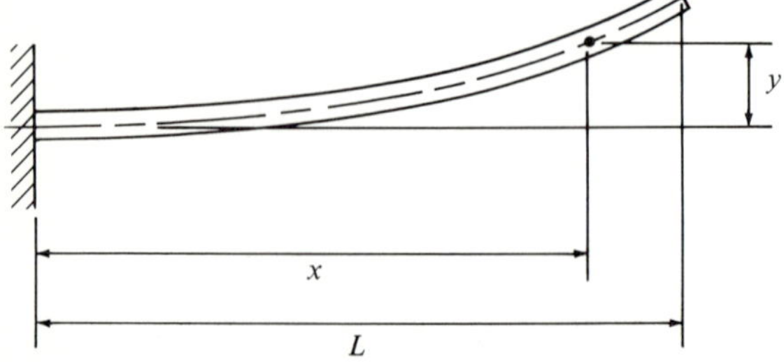

Figure 4.10.1

and by the boundary conditions

$$y(0, t) = y'(0, t) = y''(L, t) = y'''(L, t) = 0 \tag{2}$$

The general solution of Eq. (1) satisfying Eqs. (2) can be expressed as

$$y = \sum_{i=1}^{\infty} \overline{\phi}_i \overline{q}_i \tag{3}$$

where $\overline{\phi}_i$ and $\overline{q}_i$ are functions of x and t, respectively, defined as

$$\overline{\phi}_i \triangleq \cosh \frac{\lambda_i x}{L} - \cos \frac{\lambda_i x}{L} - \frac{\cosh \lambda_i + \cos \lambda_i}{\sinh \lambda_i + \sin \lambda_i} \left(\sinh \frac{\lambda_i x}{L} - \sin \frac{\lambda_i x}{L} \right) \tag{4}$$

$$\overline{q}_i \triangleq \alpha_i \cos p_i t + \beta_i \sin p_i t \tag{5}$$

and λ_i $(i = 1, \ldots, \infty)$ are the consecutive roots of the transcendental equation

$$\cos \lambda \cosh \lambda + 1 = 0 \tag{6}$$

while

$$p_i \triangleq \left(\frac{\lambda_i}{L} \right)^2 \left(\frac{EI}{\rho} \right)^{1/2} \tag{7}$$

and α_i and β_i are constants that depend upon initial conditions. The first 30 positive roots of Eq. (6) are listed in Table 4.10.1, as are those of the equation $\cos \lambda \cosh \lambda - 1 = 0$, which replaces Eq. (6) when B is an unrestrained beam.

The functions $\overline{\phi}_i$ satisfy the orthogonality relations

$$\int_0^L \overline{\phi}_i \overline{\phi}_j \rho \, dx = m_B \delta_{ij} \qquad (i, j = 1, \ldots, \infty) \tag{8}$$

and

$$EI \int_0^L \overline{\phi}_i'' \overline{\phi}_j'' \, dx = p_i^2 m_B \delta_{ij} \qquad (i, j = 1, \ldots, \infty) \tag{9}$$

where m_B is the mass of the beam and δ_{ij} is the Kronecker delta.

Example In Fig. 4.10.2, a schematic representation of a spacecraft S formed by a rigid body A that supports a uniform cantilever beam B of length L, flexural rigidity EI, and mass per unit of length ρ, O designates a particle of mass m_O that is fixed in a Newtonian reference frame N. In addition to gravitational forces exerted on S by O, certain control forces, to be described in more detail later, act on S. Only planar motions of S are to be considered.

Generalized speeds u_1, u_2, and u_3, used to characterize the motion of A in N, are defined as

$$u_i \triangleq {}^N\mathbf{v}^{A^*} \cdot \mathbf{a}_i \qquad (i = 1, 2) \qquad u_3 \triangleq {}^N\boldsymbol{\omega}^A \cdot \mathbf{a}_3 \tag{10}$$

where ${}^N\mathbf{v}^{A^*}$ is the velocity in N of the mass center A^* of A, ${}^N\boldsymbol{\omega}^A$ is the angular velocity of A in N, and $\mathbf{a}_1$, $\mathbf{a}_2$, $\mathbf{a}_3$ form a dextral set of mutually perpendicular

Table 4.10.1 Roots of characteristic equations for continuous beams

Root	$\cos\lambda \cosh\lambda + 1 = 0$ (cantilever beam)	$\cos\lambda \cosh\lambda - 1 = 0$ (free-free beam)
1	1.8751040687120	4.7300407448627
2	4.6940911329742	7.8532046240958
3	7.8547574382376	10.995607838002
4	10.995540734875	14.137165491257
5	14.137168391046	17.278759657399
6	17.278759532088	20.420352245626
7	20.420352251041	23.561944902040
8	23.561944901806	26.703537555508
9	26.703537555518	29.845130209103
10	29.845130209103	32.986722862693
11	32.986722862693	36.128315516283
12	36.128315516283	39.269908169872
13	39.269908169872	42.411500823462
14	42.411500823462	45.553093477052
15	45.553093477052	48.694686130642
16	48.694686130642	51.836278784232
17	51.836278784232	54.977871437821
18	54.977871437821	58.119464091411
19	58.119464091411	61.261056745001
20	61.261056745001	64.402649398591
21	64.402649398591	67.544242052181
22	67.544242052181	70.685834705770
23	70.685834705770	73.827427359360
24	73.827427359360	76.969020012950
25	76.969020012950	80.110612666540
26	80.110612666540	83.252205320130
27	83.252205320130	86.393797973719
28	86.393797973719	89.535390627309
29	89.535390627309	92.676983280899
30	92.676983280899	95.818575934489

unit vectors fixed in A and directed as shown in Fig. 4.10.2. It follows immediately that

$$^N\mathbf{v}^{A^*} = u_1\mathbf{a}_1 + u_2\mathbf{a}_2 \qquad ^N\boldsymbol{\omega}^A = u_3\mathbf{a}_3 \tag{11}$$

Deformations of B can be discussed in terms of the displacement y of a generic point P of B situated at a distance x from point Q, the point at which B is attached to A, and y is expressed as

$$y = \sum_{i=1}^{\nu} \phi_i q_i \tag{12}$$

where ϕ_i is an, as yet, totally unrestricted function of x, q_i is an equally unrestricted function of t, and ν is any positive integer. Generalized speeds u_{3+i} $(i = 1, \ldots, \nu)$ are introduced as

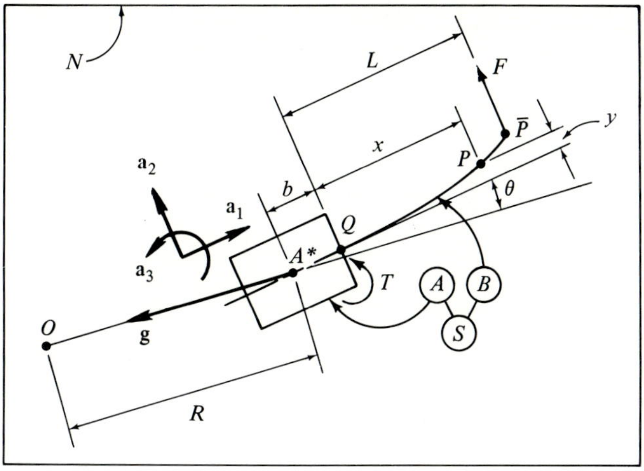

Figure 4.10.2

$$u_{3+i} \triangleq \dot{q}_i \qquad (i = 1, \ldots, \nu) \tag{13}$$

which makes it possible to write the velocities of P in Q in N as

$$^N\mathbf{v}^P = \left(u_1 - u_3 \sum_{i=1}^{\nu} \phi_i q_i\right) \mathbf{a}_1 + \left[u_2 + (b + x)u_3 + \sum_{i=1}^{\nu} \phi_i u_{3+i}\right] \mathbf{a}_2 \tag{14}$$

$$^N\mathbf{v}^Q = u_1 \mathbf{a}_1 + (u_2 + bu_3)\mathbf{a}_2 \tag{15}$$

where b is the distance from A^* to Q. Partial velocities and partial angular velocities of interest, formed by inspection of Eqs. (11), (14), and (15), then may be recorded as in Table 4.10.2.

The angular acceleration of A and the accelerations of A^* and P in N, needed for the formation of generalized inertia forces, are

$$^N\boldsymbol{\alpha}^A \underset{(11)}{=} \dot{u}_3 \mathbf{a}_3 \tag{16}$$

$$^N\mathbf{a}^{A^*} \underset{(11)}{=} (\dot{u}_1 - u_2 u_3)\mathbf{a}_1 + (\dot{u}_2 + u_3 u_1)\mathbf{a}_2 \tag{17}$$

$$^N\mathbf{a}^P \underset{(14)}{=} \left[\dot{u}_1 - u_2 u_3 - (b + x)u_3^2 - \sum_{i=1}^{\nu} \phi_i(\dot{u}_3 q_i + 2u_3 u_{3+i})\right] \mathbf{a}_1 + \left[\dot{u}_2 + u_3 u_1 + (b + x)\dot{u}_3 + \sum_{i=1}^{\nu} \phi_i(\dot{u}_{3+i} - u_3^2 q_i)\right] \mathbf{a}_2 \tag{18}$$

and, if m_A and J_3 are the mass of A and the moment of inertia of A about a line passing through A^* and parallel to $\mathbf{a}_3$, then the generalized inertia force F_r^*, formed in accordance with Eqs. (4.3.1)–(4.3.8), is given by

$$F_r^* = -J_3 \boldsymbol{\omega}_r^A \cdot {}^N\boldsymbol{\alpha}^A - m_A \mathbf{v}_r^{A^*} \cdot {}^N\mathbf{a}^{A^*} - \int_0^L \mathbf{v}_r^P \cdot {}^N\mathbf{a}^P \rho \, dx \qquad (r = 1, \ldots, 3 + \nu) \tag{19}$$

Table 4.10.2

r	$\mathbf{v}_r^{A^*}$	$\boldsymbol{\omega}_r^{A}$	$\mathbf{v}_r^{P}$	$\mathbf{v}_r^{Q}$
1	$\mathbf{a}_1$	0	$\mathbf{a}_1$	$\mathbf{a}_1$
2	$\mathbf{a}_2$	0	$\mathbf{a}_2$	$\mathbf{a}_2$
3	0	$\mathbf{a}_3$	$-\sum_{i=1}^{\nu} \phi_i q_i \mathbf{a}_1 + (b + x)\mathbf{a}_2$	$b\mathbf{a}_2$
$3 + j\ (j = 1, \ldots, \nu)$	0	0	$\phi_j \mathbf{a}_2$	0

Before substituting into this equation from Table 4.10.2 and Eqs. (16)–(18), we define constants m_B, e_B, I_B, E_i, F_i, and G_{ij} as

$$m_B \triangleq \int_0^L \rho\, dx \qquad e_B \triangleq \int_0^L x\rho\, dx \qquad I_B \triangleq \int_0^L x^2 \rho\, dx \tag{20}$$

$$E_i \triangleq \int_0^L \phi_i\, \rho\, dx \qquad F_i \triangleq \int_0^L x\, \phi_i\, \rho\, dx \qquad G_{ij} \triangleq \int_0^L \phi_i \phi_j\, \rho\, dx$$
$$(i, j = 1, \ldots, \nu) \tag{21}$$

Equation (19) then leads to

$$F_1^* = -(m_A + m_B)(\dot{u}_1 - u_2 u_3) + \dot{u}_3 \sum_{i=1}^{\nu} E_i q_i + 2u_3 \sum_{i=1}^{\nu} E_i u_{3+i}$$
$$+ u_3^2 (b m_B + e_B) \tag{22}$$

$$F_2^* = -(m_A + m_B)(\dot{u}_2 + u_3 u_1) - \sum_{i=1}^{\nu} E_i \dot{u}_{3+i} - \dot{u}_3 (b m_B + e_B)$$
$$+ u_3^2 \sum_{i=1}^{\nu} E_i q_i \tag{23}$$

$$F_3^* = (\dot{u}_1 - u_2 u_3) \sum_{i=1}^{\nu} E_i q_i - (\dot{u}_2 + u_3 u_1)(b m_B + e_B)$$
$$- \dot{u}_3 (b^2 m_B + 2b e_B + I_B + J_3) - \sum_{i=1}^{\nu} \dot{u}_{3+i} (b E_i + F_i) \tag{24}$$

$$F_{3+j}^* = -\dot{u}_2 E_j - \sum_{i=1}^{\nu} G_{ij} \dot{u}_{3+i} - \dot{u}_3 (b E_j + F_j) - u_3 u_1 E_j$$
$$+ u_3^2 \sum_{i=1}^{\nu} G_{ij} q_i \qquad (j = 1, \ldots, \nu) \tag{25}$$

As for generalized active forces, contributions to these are made by internal forces, by gravitational forces exerted on S by the particle at point O, and by control forces. Considering the internal forces first, we begin by noting that $d\mathbf{F}$, the net force exerted on a generic differential element of B by contiguous elements, is given by

$$d\mathbf{F} = -\frac{\partial V}{\partial x}\, dx\, \mathbf{a}_2 \tag{26}$$

where $V(x, t)$ is the shear at point P. Moreover, if rotatory inertia is left out of account, then V may be expressed in terms of the bending moment $M(x, t)$ as

$$V = \frac{\partial M}{\partial x} \tag{27}$$

Since

$$M = EI \frac{\partial^2 y}{\partial x^2} \tag{28}$$

Eqs. (26)–(28) thus yield

$$d\mathbf{F} = -\frac{\partial^2}{\partial x^2}\left(EI \frac{\partial^2 y}{\partial x^2}\right) dx\, \mathbf{a}_2 \tag{29}$$

The system of forces exerted on A by B is equivalent to a couple of torque $M(0, t)\mathbf{a}_3$ together with a force $-V(0, t)\mathbf{a}_2$ applied at point Q. Hence, $(F_r)_I$, the contribution of internal forces to the generalized active force F_r, is given by

$$(F_r)_I = \left[EI\left(\boldsymbol{\omega}_r{}^A \cdot \mathbf{a}_3 \frac{\partial^2 y}{\partial x^2} - \mathbf{v}_r{}^Q \cdot \mathbf{a}_2 \frac{\partial^3 y}{\partial x^3}\right)\right]_{x=0} - \int_0^L \mathbf{v}_r{}^P \cdot \mathbf{a}_2 \frac{\partial^2}{\partial x^2}\left(EI \frac{\partial^2 y}{\partial x^2}\right) dx \qquad (j = 1, \ldots, 3 + \nu) \tag{30}$$

which, with the aid of Table 4.10.2 and Eq. (12), leads to

$$(F_1)_I = 0 \qquad (F_2)_I = -\sum_{i=1}^{\nu} q_i \left[(EI\phi_i''')_{x=0} + \int_0^L (EI\phi_i'')'' \, dx\right] \tag{31}$$

$$(F_3)_I = b(F_2)_I + \sum_{i=1}^{\nu} q_i \left[(EI\phi_i'')_{x=0} - \int_0^L x(EI\phi_i'')'' \, dx\right] \tag{32}$$

$$(F_{3+j})_I = -\sum_{i=1}^{\nu} q_i \int_0^L \phi_j (EI\phi_i'')'' \, dx \qquad (j = 1, \ldots, \nu) \tag{33}$$

We now impose restrictions on ϕ_i so as to ensure that y and y' vanish at $x = 0$ while M and V vanish at $x = L$; that is, we require [see Eq. (12) and Eqs. (27) and (28)]

$$\phi_i(0) = \phi_i{}'(0) = \phi_i''(L) = \phi_i'''(L) = 0 \qquad (i = 1, \ldots, \nu) \tag{34}$$

When the integrations indicated in Eqs. (31)–(33) are carried out [integration by parts, in the case of Eqs. (32) and (33)] and Eqs. (34) are taken into account, the following expressions result for the contribution of internal forces to generalized active forces:

$$(F_1)_I = (F_2)_I = (F_3)_I = 0 \tag{35}$$

$$(F_{3+j})_I = -\sum_{i=1}^{\nu} H_{ij} q_i \qquad (j = 1, \ldots, \nu) \tag{36}$$

where H_{ij} is defined as

$$H_{ij} \triangleq \int_0^L EI\phi_i''\phi_j''\,dx \qquad (i, j = 1, \ldots, \nu) \tag{37}$$

To determine $(F_r)_G$, the contribution to F_r of gravitational forces exerted on S by the particle O, we assume that the maximum dimension of A is so small in comparison with the distance R between O and A^* that the system of forces exerted by O on A may be replaced with a couple of torque $\mathbf{T}_A$ given by

$$\mathbf{T}_A \underset{(2.6.3)}{=} \left(\frac{3Gm_O}{R^3}\right)\mathbf{g} \times \mathbf{I}_A \cdot \mathbf{g} \tag{38}$$

together with a force $\mathbf{F}_A$ applied at A^* and expressed as

$$\mathbf{F}_A \underset{(2.3.5)}{=} \frac{Gm_O m_A}{R^2}\mathbf{g} \tag{39}$$

where G is the universal gravitational constant, $\mathbf{g}$ is a unit vector directed from A^* toward O, and $\mathbf{I}_A$ is the central inertia dyadic of A. Letting θ denote the angle between the line joining O to A^* and the neutral axis of the undeformed beam, as shown in Fig. 4.10.2, defining γ as

$$\gamma \triangleq Gm_O \tag{40}$$

and introducing J_1 and J_2 as

$$J_i \triangleq \mathbf{a}_i \cdot \mathbf{I}_A \cdot \mathbf{a}_i \qquad (i = 1, 2) \tag{41}$$

one can replace Eqs. (38) and (39) with

$$\mathbf{T}_A = -\frac{3\gamma}{R^3}(J_2 - J_1)\sin\theta\cos\theta\,\mathbf{a}_3 \tag{42}$$

$$\mathbf{F}_A = -\frac{\gamma m_A}{R^2}(\cos\theta\,\mathbf{a}_1 - \sin\theta\,\mathbf{a}_2) \tag{43}$$

The particle O exerts gravitational forces also on B, the force acting on a differential element of B being

$$d\mathbf{F}_B \underset{(2.1.1)}{=} \gamma\rho(\mathbf{p}^2)^{-3/2}\mathbf{p}\,dx \tag{44}$$

where $\mathbf{p}$, the position vector from P to O, is given by

$$\mathbf{p} = R\mathbf{g} - (b + x)\mathbf{a}_1 - y\mathbf{a}_2 \tag{45}$$

Substituting from Eq. (45) into Eq. (44), and dropping all terms of second or higher degree in $|(b + x)/R|$ and $|y/R|$, one arrives at

$$d\mathbf{F}_B = \frac{\gamma\rho}{R^2}\left\{\left[-\cos\theta + \frac{(b + x)(3\cos^2\theta - 1) - 3y\sin\theta\cos\theta}{R}\right]\mathbf{a}_1 + \left[\sin\theta - \frac{3(b + x)\sin\theta\cos\theta + y(1 - 3\sin^2\theta)}{R}\right]\mathbf{a}_2\right\}dx \tag{46}$$

and $(F_r)_G$ can now be formed by substituting from Eqs. (42), (43), and (46) into

$$(F_r)_G = \boldsymbol{\omega}_r{}^A \cdot \mathbf{T}_A + \mathbf{v}_r{}^{A*} \cdot \mathbf{F}_A + \int_0^L \mathbf{v}_r{}^P \cdot d\mathbf{F}_B \qquad (r = 1, \ldots, 3 + \nu) \quad (47)$$

and using Table 4.10.2, which gives

$$(F_1)_G = \frac{\gamma}{R^2}\left[-(m_A + m_B)\cos\theta + \frac{(3\cos^2\theta - 1)(bm_B + e_B)}{R} - \frac{3\sin\theta\cos\theta}{R}\sum_{i=1}^{\nu} E_i q_i\right] \quad (48)$$

$$(F_2)_G = \frac{\gamma}{R^2}\left[(m_A + m_B)\sin\theta - \frac{3\sin\theta\cos\theta\,(bm_B + e_B)}{R} + \frac{3\sin^2\theta - 1}{R}\sum_{i=1}^{\nu} E_i q_i\right] \quad (49)$$

$$(F_3)_G = \frac{\gamma}{R^2}\left\{(bm_B + e_B)\sin\theta + \frac{3(J_1 - J_2 - b^2 m_B - 2be_B - I_B)\sin\theta\cos\theta}{R} + \sum_{i=1}^{\nu}\left[E_i\cos\theta + \frac{3(bE_i + F_i)(1 - 2\cos^2\theta)}{R}\right]q_i\right\} \quad (50)$$

$$(F_{3+j})_G = \frac{\gamma}{R^2}\left[E_j\sin\theta - \frac{3(bE_j + F_j)\sin\theta\cos\theta}{R} + \frac{3\sin^2\theta - 1}{R}\sum_{i=1}^{\nu} G_{ij} q_i\right] \qquad (j = 1, \ldots, \nu) \quad (51)$$

Finally, we must deal with $(F_r)_C$, the contribution to F_r of control forces acting on S. Supposing that the system of all such forces is equivalent to a couple of torque $\mathbf{T}_C$ applied to body A together with a force $\mathbf{F}_C$ applied at $\bar{P}$, the free end of B (see Fig. 4.10.2), and taking

$$\mathbf{T}_C = T\mathbf{a}_3 \qquad \mathbf{F}_C = F\mathbf{a}_2 \quad (52)$$

where F and T are unrestricted, we have

$$(F_r)_C \underset{(4.1.2)}{=} \boldsymbol{\omega}_r{}^A \cdot \mathbf{T}_C + \mathbf{v}_r{}^{\bar{P}} \cdot \mathbf{F}_C \qquad (r = 1, \ldots, 3 + \nu) \quad (53)$$

or, after using Eqs. (52) and Table 4.10.2,

$$(F_1)_C = 0 \qquad (F_2)_C = F \qquad (F_3)_C = (b + L)F + T \quad (54)$$

$$(F_{3+j})_C = F\phi_j(L) \qquad (j = 1, \ldots, \nu) \quad (55)$$

In accordance with Eqs. (4.5.1), dynamical equations now can be written by substituting from Eqs. (35), (36), (48)–(51), (54), (55), and (22)–(25) into

$$(F_r)_I + (F_r)_G + (F_r)_C + F_r^* = 0 \qquad (r = 1, \ldots, 3 + \nu) \tag{56}$$

which leads to

$$\begin{aligned} &\dot{u}_1(m_A + m_B) - \dot{u}_3 \sum_{i=1}^{\nu} E_i q_i - (m_A + m_B)u_2 u_3 - 2u_3 \sum_{i=1}^{\nu} E_i u_{3+i} \\ &\quad - u_3^2(bm_B + e_B) + \frac{\gamma}{R^2}\Bigg[(m_A + m_B)\cos\theta \\ &\quad + \frac{(1 - 3\cos^2\theta)(bm_B + e_B)}{R} + \frac{3\sin\theta\cos\theta}{R}\sum_{i=1}^{\nu} E_i q_i\Bigg] = 0 \end{aligned} \tag{57}$$

$$\begin{aligned} &\dot{u}_2(m_A + m_B) + \dot{u}_3(bm_B + e_B) + \sum_{i=1}^{\nu} E_i \dot{u}_{3+i} + (m_A + m_B)u_3 u_1 \\ &\quad - u_3^2 \sum_{i=1}^{\nu} E_i q_i + \frac{\gamma}{R^2}\Bigg[- (m_A + m_B)\sin\theta + \frac{3\sin\theta\cos\theta(bm_B + e_B)}{R} \\ &\quad + \frac{1 - 3\sin^2\theta}{R}\sum_{i=1}^{\nu} E_i q_i\Bigg] - F = 0 \end{aligned} \tag{58}$$

$$\begin{aligned} &\dot{u}_1 \sum_{i=1}^{\nu} E_i q_i - \dot{u}_2(bm_B + e_B) - \dot{u}_3(b^2 m_B + 2be_B + I_B + J_3) \\ &\quad - \sum_{i=1}^{\nu} (bE_i + F_i)\dot{u}_{3+i} - u_2 u_3 \sum_{i=1}^{\nu} E_i q_i - u_3 u_1(bm_B + e_B) \\ &\quad + \frac{\gamma}{R^2}\Bigg\{\Bigg[(bm_B + e_B)\sin\theta + \frac{3(J_1 - J_2 - b^2 m_B - 2be_B - I_B)\sin\theta\cos\theta}{R} \\ &\quad + \frac{1}{R}\sum_{i=1}^{\nu} [E_i\cos\theta + 3(bE_i + F_i)(1 - 2\cos^2\theta)]\, q_i\Bigg\} \\ &\quad + (b + L)F + T = 0 \end{aligned} \tag{59}$$

$$\begin{aligned} &\dot{u}_2 E_j + \dot{u}_3(bE_j + F_j) + \sum_{i=1}^{\nu} G_{ij}\dot{u}_{3+i} + u_3 u_1 E_j - u_3^2 \sum_{i=1}^{\nu} G_{ij} q_i \\ &\quad + \sum_{i=1}^{\nu} H_{ij} q_i + \frac{\gamma}{R^2}\Bigg[- E_j \sin\theta + \frac{3\sin\theta\cos\theta(bE_j + F_j)}{R} \\ &\quad + \frac{1 - 3\sin^2\theta}{R}\sum_{i=1}^{\nu} G_{ij} q_i\Bigg] - F\phi_j(L) = 0 \qquad (j = 1, \ldots, \nu) \end{aligned} \tag{60}$$

Before these equations can be employed for a numerical simulation of the motion of S, a number of additional relationships must be generated. Specifically, since Eqs. (57)–(60) involve the $5 + 2\nu$ dependent variables $u_1, \ldots, u_{3+\nu}, q_1, \ldots, q_\nu, R$, and θ, these $3 + \nu$ equations must be augmented by $2 + \nu$ kinematical equations; more must be said about the quantities

m_B, e_B, I_B, E_i, F_i, G_{ij}, H_{ij}, and $\phi_j(L)$, the last five of which depend on $\phi_1, \ldots, \phi_\nu$; and the functional forms of the quantities T and F introduced in Eqs. (52) must be specified.

To generate the requisite kinematical equations, we note that Eq. (13) furnishes ν of these. In addition, we express the velocity of A^* as

$$^N\mathbf{v}^{A^*} = [\dot{R}\cos\theta + R(u_3 - \dot{\theta})\sin\theta]\mathbf{a}_1 + [R(u_3 - \dot{\theta})\cos\theta - \dot{R}\sin\theta]\mathbf{a}_2 \tag{61}$$

and, after equating the right-hand member of this equation to that of the first of Eqs. (11), solve the associated two scalar equations for $\dot{R}$ and $\dot{\theta}$, obtaining

$$\dot{R} = u_1\cos\theta - u_2\sin\theta \tag{62}$$

$$\dot{\theta} = u_3 - \frac{u_1\sin\theta + u_2\cos\theta}{R} \tag{63}$$

The constants m_B, e_B, and I_B, defined in Eqs. (20), can be expressed directly in terms of system parameters because we are dealing with a uniform beam, so that ρ is independent of x, and the indicated integrations yield

$$m_B = \rho L \qquad e_B = \frac{\rho L^2}{2} \qquad I_B = \frac{\rho L^3}{3} \tag{64}$$

As for E_i, F_i, G_{ij}, defined in Eqs. (21), and H_{ij}, introduced in Eq. (37), the question that must be settled first is what to use for $\phi_1, \ldots, \phi_\nu$, which are, up to now, restricted only by Eqs. (34). Now, if the motion of B relative to A is not affected too profoundly by the motion of A in N, that is, if B moves in A almost as if A were fixed in a Newtonian reference frame, then ϕ_i should not differ greatly from $\overline{\phi}_i$ as given in Eq. (4). Hence, taking $\phi_i = \overline{\phi}_i$ may be physically sound. Moreover, this choice of ϕ_i brings with it the analytical advantage of facilitating the evaluation of G_{ij} and H_{ij} by reference to Eqs. (8) and (9), respectively. Accordingly, we take

$$\phi_i = \overline{\phi}_i \qquad (i = 1, \ldots, \nu) \tag{65}$$

which leads to

$$E_i \underset{(21,4)}{=} \frac{2m_B(1 + e^{-2\lambda_i} + 2e^{-\lambda_i}\cos\lambda_i)}{\lambda_i(1 - e^{-2\lambda_i} + 2e^{-\lambda_i}\sin\lambda_i)} \qquad (i = 1, \ldots, \nu) \tag{66}$$

$$F_i \underset{(21,4)}{=} \frac{2m_B L}{\lambda_i^2} \qquad (i = 1, \ldots, \nu) \tag{67}$$

$$G_{ij} \underset{(21,8)}{=} m_B\,\delta_{ij} \qquad H_{ij} \underset{(37,9,7)}{=} \frac{\lambda_i^4 EI}{L^3}\,\delta_{ij} \qquad (i, j = 1, \ldots, \nu) \tag{68}$$

$$\phi_j(L) \underset{(4)}{=} \frac{2[(1 + e^{-2\lambda_j})\sin\lambda_j - \cos\lambda_j(1 - e^{-2\lambda_j})]}{1 - e^{-2\lambda_j} + 2e^{-\lambda_j}\sin\lambda_j} \qquad (j = 1, \ldots, \nu) \tag{69}$$

[The reason for writing Eqs. (66) and (69) in forms involving solely negative exponents is that this precludes the appearance of large, computationally inconvenient numbers otherwise associated with the larger ones of $\lambda_1, \ldots, \lambda_{\nu}$.]

Finally, to dispose of the matter of specifying the control quantities T and F appearing in Eqs. (52), we note that the choices one makes in this connection depend on the control objective under consideration. For example, suppose one wished to have A^* move in a circular orbit centered at O while $\theta(t) = y(x, t) = 0$ for all t. We shall examine what happens when one chooses T and F in a manner that would help one accomplish this objective if B were a rigid body rather than an elastic beam. In this way we shall gain some insights into the effects of flexibility on system behavior.

In Fig. 4.10.3, the system S is shown in a general orientation, but with B undeformed. S^* designates the system mass center, R^* is the distance from O to S^*, and ψ denotes the angle between lines OS^* and A^*S^*. As before, a force of magnitude F is applied at $\overline{P}$ and a couple of torque T acts on A, as indicated in the sketch. Suppose now that B is a rigid body and that F and T vanish when $\dot{\psi} = 0$. Then the system can move in such a way that S^* and A^* trace out circles centered at O while ψ and θ remain equal to zero, and motions during which ψ is sufficiently small are governed by (see Sec. 3.5)

$$\ddot{\psi} + p^2\psi - \frac{T + sF}{I_3} = 0 \tag{70}$$

where p, I_3, and s, the distance from $\overline{P}$ to S^*, are given by

$$p \triangleq \Omega\left(3\,\frac{I_3 - J_1 + J_2 - J_3}{I_3}\right)^{1/2} \tag{71}$$

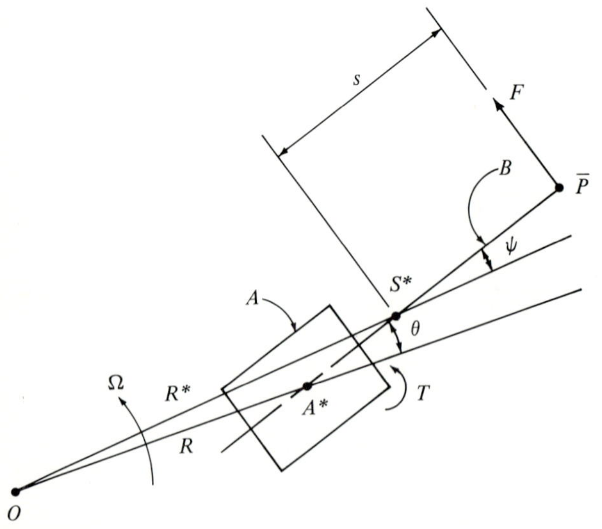

Figure 4.10.3

$$I_3 \triangleq J_3 + \frac{m_A m_B (b + L/2)^2}{m_A + m_B} + \frac{m_B L^2}{12} \tag{72}$$

$$s \triangleq \frac{m_A(b + L) + m_B L/2}{m_A + m_B} \tag{73}$$

and Ω is the spin rate in N of the line connecting O to S^*. To drive ψ to zero subsequent to an instant at which a disturbance has caused ψ to depart from zero, one can make T and F linear functions of the *attitude rate* ω or the *tip speed* v, defined as

$$\omega \triangleq \dot{\psi} \qquad v \triangleq s\dot{\psi} \tag{74}$$

That is, one can set

$$F = -(K_\omega \omega + K_v v) \qquad T = -(L_\omega \omega + L_v v) \tag{75}$$

where K_ω, K_v, L_ω, L_v are nonnegative constants, for Eq. (70) then leads to the equation of a damped oscillator, namely,

$$\ddot{\psi} + 2\mu p \dot{\psi} + p^2 \psi = 0 \tag{76}$$

with μ defined as

$$\mu \triangleq \frac{L_\omega + (L_v + K_\omega)s + K_v s^2}{2pI_3} \tag{77}$$

The oscillator is "critically" damped when $\mu = 1$.

Motivated by these considerations, we now investigate the desirability of employing, for the purpose of controlling S when B is deformable, the control law stated in Eqs. (75), but first we redefine ω and v [see Eqs. (74)] so as to express attitude rate and tip speed in ways that are appropriate under the new circumstances; that is, we let C be a reference frame in which A^* and line OA^* are fixed and replace Eqs. (74) with

$$\omega \triangleq {}^C\boldsymbol{\omega}^A \cdot \mathbf{a}_3 \qquad v \triangleq {}^C\mathbf{v}^{\bar{P}} \cdot \mathbf{a}_2 \tag{78}$$

so that, expressed in terms of variables appearing in the equations of motion, ω and v become

$$\omega = \dot{\theta} \qquad v = (b + L)\dot{\theta} + \sum_{i=1}^{\nu} \phi_i(L) u_{3+i} \tag{79}$$

Moreover, we shall confine attention to cases in which only one of the four constants in Eqs. (75) differs from zero. We note that under these circumstances the critical values of the constants, that is, the values for which $\mu = 1$, are

$$L_\omega{}^* = 2pI_3 \qquad K_v{}^* = \frac{2pI_3}{s^2} \qquad L_v{}^* = K_\omega{}^* = \frac{2pI_3}{s} \tag{80}$$

To investigate the effect of beam flexibility on the behavior of S, we let A be a solid, rectangular parallelepiped of mass 120 kg and having sides of lengths 2 m, 3 m, and 1 m, parallel to $\mathbf{a}_1$, $\mathbf{a}_2$, $\mathbf{a}_3$, respectively, so that [see Eq. (41)] $J_1 = 100$ kg · m², $J_2 = 50$ kg · m², $J_3 = 130$ kg · m²; let $L = 20$ m, $b = 1$ m, $EI = 5$ N · m², $\rho = 0.2$ kg/m, $\gamma = 3.9860 \times 10^{14}$ m³/sec² (corresponding to Earth); and set $R(0) = 7 \times 10^6$ m, $\dot{R}(0) = 0$, $\theta(0) = 45°$, $\dot{\theta}(0) = 0$, $u_3(0) = [\gamma/R^3(0)]^{1/2} = 1.0780 \times 10^{-3}$ rad/sec, $u_1(0) = \dot{R}(0)\cos\theta(0) + R(0)[u_3(0) - \dot{\theta}(0)]\sin\theta(0) = 5.3359 \times 10^3$ m/sec, $u_2(0) = -\dot{R}(0)\sin\theta(0) + R(0)[u_3(0) - \dot{\theta}(0)]\cos\theta(0) = 5.3359 \times 10^3$ m/sec [see Eqs. (11) and (61)]. This choice of initial values corresponds to placing A^* at a point 7×10^6 m from the center of the Earth (or 622 km above the Earth's surface), giving A^* an initial velocity in N equal to the velocity of a particle that occupies the same position as A^* and is moving on a circular orbit, and letting A and C have the same initial angular velocity in N, so that A is not rotating initially relative to the line joining O to A^*. Finally, we take $q_i(0) = u_{3+i}(0) = 0$ $(i = 1, \ldots, \nu)$, thus dealing with a beam that is initially undeformed and at rest in C.

The value of $L_\omega{}^*$ obtained from Eq. (80), with the aid of Eqs. (71) and (72) is 2.37 N · m · sec. Taking $L_\omega = 1.5$ N · m · sec (and $L_v = K_\omega = K_v = 0$), so that we are dealing with what may be expected to be a slightly underdamped system, we set $\nu = 3$ and perform a numerical integration of the equations of motion, Eqs. (57)–(60), (62), and (63), to generate the θ versus t plot dis-

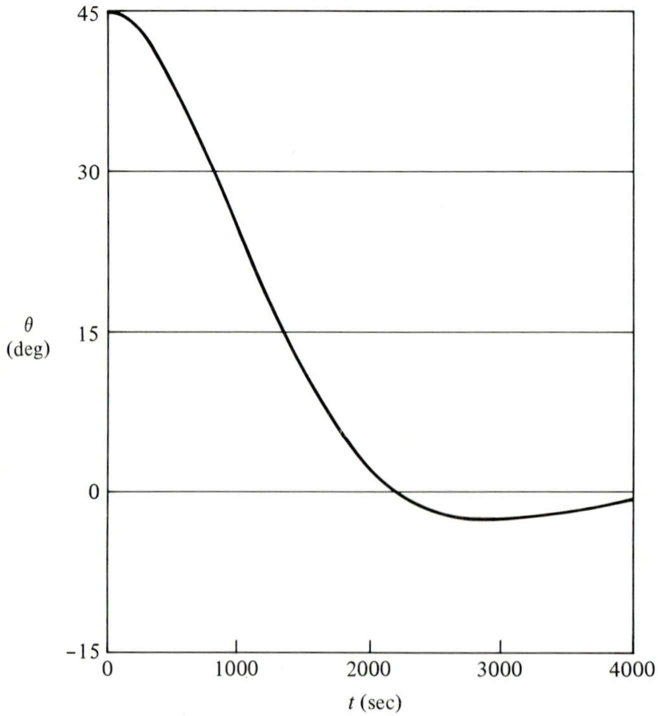

Figure 4.10.4

played in Fig. 4.10.4, which shows that θ decreases from its initial value of 45° to a value of less than 1° in 4000 sec (the orbital period is 5829 sec). Throughout this time interval, δ, the deflection of point $\overline{P}$, decays and never exceeds 0.02 m. Thus the anticipated damping is, in fact, occurring. Next, we repeat the integration with the same initial conditions but with $L_v = 0.07$ N · sec (and $L_\omega = K_\omega = K_v = 0$). Since $L_v{}^* = 0.115$ N · sec, one might once again expect θ to decrease and δ to decay. Instead, plots of θ and δ versus t appear as shown in Figs. 4.10.5 and 4.10.6, respectively, which reveal that the control system is, in fact, causing divergent behavior. Moreover, the instability here encountered is attributable to the flexibility of the beam, for no such instability arises when the beam is regarded as rigid, that is, when one uses Eq. (70) with $F = 0$ and $T = -0.07\, s\dot{\psi}$.

Responses similar to those just considered manifest themselves when $K_v = 3.5 \times 10^{-3}$ N · sec/m and when $K_\omega = 0.07$ N · sec (the associated values of $K_v{}^*$ and $K_\omega{}^*$ are 5.57×10^{-3} N · sec/m and 0.115 N · sec, respectively); that is, damping comes into evidence in the first case, and instability in the second. Again, the instability would be absent if the beam were rigid. These results can be summarized as follows. When the sensor and the actuator of the control system are *colocated,* that is, when the attitude rate of A is sensed and a control torque is applied to A, or when tip speed is sensed and a control force is

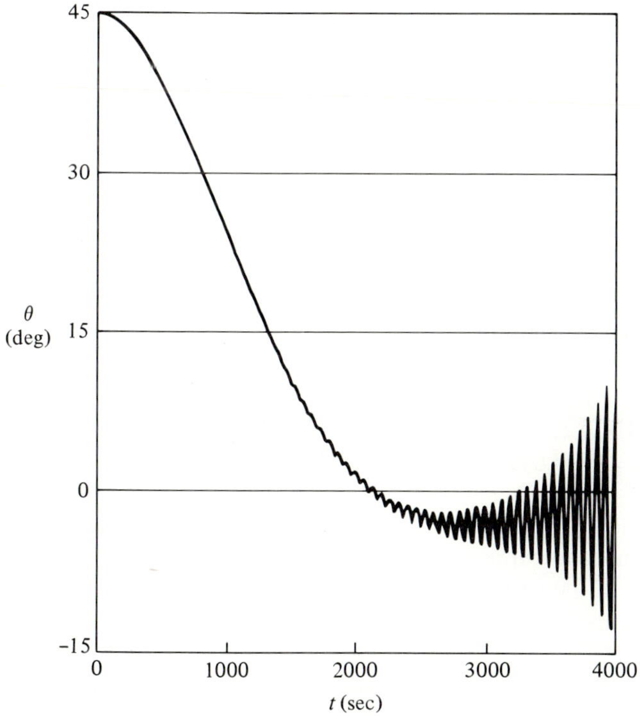

Figure 4.10.5

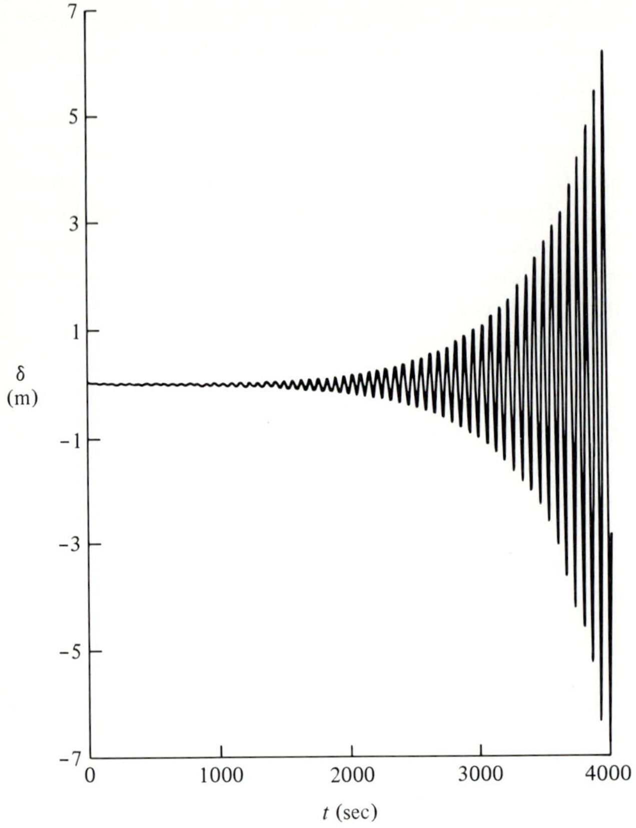

Figure 4.10.6

applied at $\bar{P}$, damping results; conversely, when the sensor and the actuator are *separated,* that is, when a control torque is applied to A in response to a tip-speed error signal, or when a control force is applied at $\bar{P}$ in response to an attitude-rate error signal, instability results.

4.11 USE OF THE FINITE ELEMENT METHOD FOR THE CONSTRUCTION OF MODAL FUNCTIONS

In Sec. 4.10, modal functions associated with certain free vibrations of a component of a spacecraft were used to express displacements of that component occurring during actual motions of the entire spacecraft as products of space-dependent and time-dependent functions [see Eqs. (4.10.12), (4.10.65), and (4.10.4)]. The modal functions $\bar{\phi}_1, \bar{\phi}_2, \ldots$ employed for this purpose were "exact" in the sense that they were obtained in the course of solving analytically (by the method of separation of variables) the partial differential equations governing certain vibrations

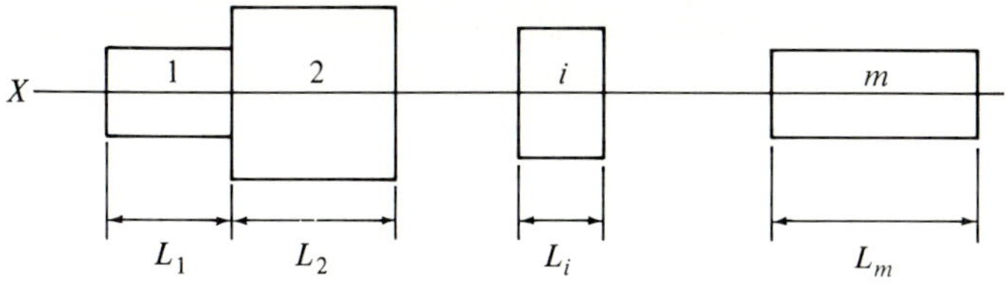

Figure 4.11.1

of the component under consideration. Alternatively, approximate modal functions can be constructed by means of a finite element analysis of the free vibrations. Since this technique is viable even when the partial differential equations governing the vibrations of interest cannot be solved analytically, as is the case, for example, when EI and/or ρ are discontinuous functions of x, one can formulate equations of motion of the spacecraft under these circumstances by proceeding as in Sec. 4.9, but use the finite element method to construct a substitute for Eq. (4.10.4). We shall illustrate this with an example after reviewing those portions of finite element theory† needed subsequently.

Figure 4.11.1 depicts a beam regarded as consisting of m prismatic elements, the ith element having a length L_i, flexural rigidity $(EI)_i$, and mass per unit of length ρ_i. The straight line X is the locus of cross-section centroids or the neutral axis of all elements when the beam is undeformed. In Fig. 4.11.2, the neutral axis X_i of the ith element is shown in a state of displacement relative to X, the latter now being regarded as fixed in a Newtonian reference frame N; Q_i and R_i designate, respectively, the left end and the right end of X_i; z_{2i-1} and z_{2i} are, respectively, the displacement of Q_i from X and the angle between X the tangent to X_i, at Q_i; and similarly for z_{2i+1} and z_{2i+2} at R_i. Finally, P_i is a generic point of X_i, and x_i and y_i are distances used to locate P_i.

Two fundamental assumptions are made regarding $y_i(x_i, t)$, namely, that this function can be expressed both as a linear function of the four time-dependent quan-

† See J. S. Przemieniecki, *Theory of Matrix Structural Analysis*, McGraw-Hill, New York, 1968.

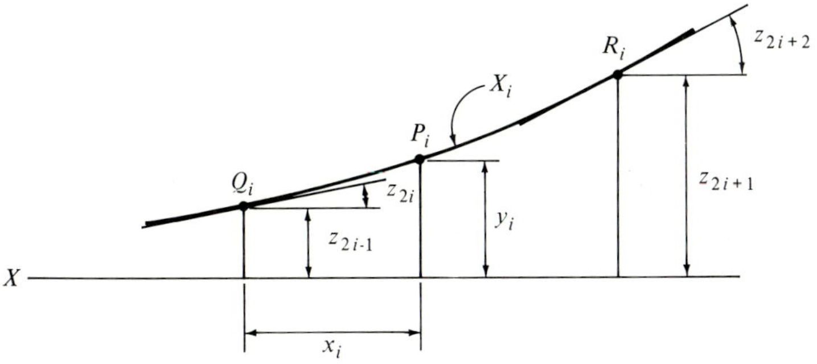

Figure 4.11.2

tities $z_{2i-1}(t)$, $z_{2i}(t)$, $z_{2i+1}(t)$, $z_{2i+2}(t)$ and as a cubic polynomial in x_i, that is, that there exist functions $\psi_{i1}(x_i), \ldots, \psi_{i4}(x_i)$ and functions $a_i(t)$, $b_i(t)$, $c_i(t)$, $d_i(t)$ such that

$$y_i = z_{2i-1}\psi_{i1} + z_{2i}\psi_{i2} + z_{2i+1}\psi_{i3} + z_{2i+2}\psi_{i4} \qquad (i = 1, \ldots, m) \tag{1}$$

and

$$y_i = a_i + b_i x_i + c_i x_i^2 + d_i x_i^3 \qquad (i = 1, \ldots, m) \tag{2}$$

Now, by definition (see Fig. 4.11.2),

$$z_{2i-1}(t) = y_i(0, t) \qquad z_{2i+1}(t) = y_i(L_i, t) \qquad (i = 1, \ldots, m) \tag{3}$$

and, if z_{2i} and z_{2i+2} are "small," which we presume, one may write (see Fig. 4.11.2)

$$z_{2i}(t) = \frac{\partial y_i(0, t)}{\partial x_i} \qquad z_{2i+2}(t) = \frac{\partial y_i(L_i, t)}{\partial x_i} \qquad (i = 1, \ldots, m) \tag{4}$$

It follows from these relationships that the "shape functions" $\psi_{i1}, \ldots, \psi_{i4}$ are given by

$$\psi_{i1}(x_i) = 1 - \frac{3x_i^2}{L_i^2} + \frac{2x_i^3}{L_i^3} \qquad (i = 1, \ldots, m) \tag{5}$$

$$\psi_{i2}(x_i) = x_i - \frac{2x_i^2}{L_i} + \frac{x_i^3}{L_i^2} \qquad (i = 1, \ldots, m) \tag{6}$$

$$\psi_{i3}(x_i) = \frac{3x_i^2}{L_i^2} - \frac{2x_i^3}{L_i^3} \qquad (i = 1, \ldots, m) \tag{7}$$

$$\psi_{i4}(x_i) = \frac{x_i^2}{L_i} + \frac{x_i^3}{L_i^2} \qquad (i = 1, \ldots, m) \tag{8}$$

Regarding $\dot{z}_1, \ldots, \dot{z}_{2m+2}$ as generalized speeds and letting z be the $(2m + 2) \times 1$ matrix having z_k as the element in the kth row, one can show by using Eq. (4.5.1) with $n = 2m + 2$ that free vibrations of the unrestrained beam are governed by the differential equation

$$M\ddot{z} + Sz = 0 \tag{9}$$

where M and S are $(2m + 2) \times (2m + 2)$ matrices, called, respectively, the *global mass matrix* and the *global stiffness matrix* of the unrestrained beam. The elements of M can be expressed in terms of the elements of 4×4 matrices $\hat{M}_1, \ldots, \hat{M}_m$, called *beam-element mass matrices*; similarly for S and the 4×4 matrices $\hat{S}_1, \ldots, \hat{S}_m$, called *beam-element stiffness matrices*. The elements of $\hat{M}_i$ and $\hat{S}_i$, denoted by $\hat{M}_{iab}$ and $\hat{S}_{iab}$, respectively, are defined as

$$\hat{M}_{iab} \triangleq \rho_i \int_0^{L_i} \psi_{ia}\psi_{ib}\, dx_i \qquad (i = 1, \ldots, m;\ a, b = 1, \ldots, 4) \tag{10}$$

and

$$\hat{S}_{iab} \triangleq (EI)_i \int_0^{L_i} \psi_{ia}''\psi_{ib}''\, dx_i \qquad (i = 1, \ldots, m;\ a, b = 1, \ldots, 4) \tag{11}$$

Consequently [see Eqs. (5)–(8)],

$$\hat{M}_i = \frac{\rho_i L_i}{420}\begin{bmatrix} 156 & 22L_i & 54 & -13L_i \\ 22L_i & 4L_i^2 & 13L_i & -3L_i^2 \\ 54 & 13L_i & 156 & -22L_i \\ -13L_i & -3L_i^2 & -22L_i & 4L_i^2 \end{bmatrix} \qquad (i = 1, \ldots, m) \qquad (12)$$

and

$$\hat{S}_i = \frac{2(EI)_i}{L_i^3}\begin{bmatrix} 6 & 3L_i & -6 & 3L_i \\ 3L_i & 2L_i^2 & -3L_i & L_i^2 \\ -6 & -3L_i & 6 & -3L_i \\ 3L_i & L_i^2 & -3L_i & 2L_i^2 \end{bmatrix} \qquad (i = 1, \ldots, m) \qquad (13)$$

To express the elements of M in terms of those of $\hat{M}_1, \ldots, \hat{M}_m$, and the elements of S in terms of those of $\hat{S}_1, \ldots, \hat{S}_m$, it is helpful to introduce A_j, B_j, C_j, D_j as

$$A_j \triangleq \frac{2j - 3 - (-1)^j}{4} \qquad B_j \triangleq A_j + 1 \qquad (j = 3, \ldots, 2m;\ m > 1) \qquad (14)$$

$$C_j \triangleq \frac{7 + (-1)^j}{2} \qquad D_j \triangleq C_j - 2 \qquad (j = 3, \ldots, 2m;\ m > 1) \qquad (15)$$

Letting M_{ab} and S_{ab} be the elements in the ath row and bth column of M and S, respectively, one then has

$$W_{ab} = \begin{cases} \hat{W}_{1ab} & (a = 1, 2;\ b = 1, 2, 3, 4) \\ \hat{W}_{A_jC_j1} & (a = j;\ b = 2A_j - 1;\ j = 3, \ldots, 2m;\ m > 1) \\ \hat{W}_{A_jC_j2} & (a = j;\ b = 2A_j;\ j = 3, \ldots, 2m;\ m > 1) \\ \hat{W}_{A_jC_j3} + \hat{W}_{B_jD_j1} & (a = j;\ b = 2A_j + 1;\ j = 3, \ldots, 2m;\ m > 1) \\ \hat{W}_{A_jC_j4} + \hat{W}_{B_jD_j2} & (a = j;\ b = 2A_j + 2;\ j = 3, \ldots, 2m;\ m > 1) \\ \hat{W}_{B_jD_j3} & (a = j;\ b = 2B_j + 1;\ j = 3, \ldots, 2m;\ m > 1) \\ \hat{W}_{B_jD_j4} & (a = j;\ b = 2B_j + 2;\ j = 3, \ldots, 2m;\ m > 1) \\ \hat{W}_{m,2+i,j} & (a = 2m + i;\ b = 2m - 2 + j; \\ & \qquad i = 1, 2;\ j = 1, 2, 3, 4) \\ 0 & \text{otherwise} \end{cases} \qquad (16)$$

where the symbol W is to be replaced with M and S, respectively, when M_{ab} and S_{ab} are being formed. For example,

$$M_{45} \underset{(16)}{=} \hat{M}_{B_4D_43} \underset{(14,15)}{=} \hat{M}_{223} \underset{(12)}{=} \frac{13\rho_2 L_2^2}{420} \qquad (17)$$

and

$$S_{24} \underset{(16)}{=} \hat{S}_{124} \underset{(13)}{=} \frac{2(EI)_1}{L_1} \tag{18}$$

When the beam is restrained, that is, when it is attached to a support fixed in N in such a way that r of $z_1, \ldots, z_{2m+2}$ vanish throughout the motion of the beam, then the remaining elements of z form a $(2m + 2 - r) \times 1$ matrix $\tilde{z}$ governed by the differential equation

$$\tilde{M}\ddot{\tilde{z}} + \tilde{S}\tilde{z} = 0 \tag{19}$$

where $\tilde{M}$ and $\tilde{S}$, called, respectively, the global mass matrix and the global stiffness matrix of the restrained beam, are $(2m + 2 - r) \times (2m + 2 - r)$ matrices formed by deleting from M and S the r rows and columns associated with vanishing elements of z. For instance, if $m = 2$ and the left end of the beam is clamped while the right end is simply-supported, so that $z_1 = z_2 = z_5 = 0$, then the first, second, and fifth rows and columns are deleted from M and S to form $\tilde{M}$ and $\tilde{S}$, respectively, and the matrices $\tilde{z}$, $\tilde{M}$, and $\tilde{S}$ are given by

$$\tilde{z} = \begin{bmatrix} z_3 \\ z_4 \\ z_6 \end{bmatrix} \qquad \tilde{M} = \begin{bmatrix} M_{33} & M_{34} & M_{36} \\ M_{43} & M_{44} & M_{46} \\ M_{63} & M_{64} & M_{66} \end{bmatrix} \qquad \tilde{S} = \begin{bmatrix} S_{33} & S_{34} & S_{36} \\ S_{43} & S_{44} & S_{46} \\ S_{63} & S_{64} & S_{66} \end{bmatrix} \tag{20}$$

Since Eqs. (9) and (19) each have the same form as Eq. (4.8.1), each implies the existence of a modal matrix defined as in Eq. (4.8.5). When $z_{2i-1}, z_{2i}, z_{2i+1}, z_{2i+2}$ in Eq. (1) are replaced with respective elements of the jth column of such a modal matrix, then the resulting equation characterizes the jth mode of vibration of the beam and can be used as a substitute for Eq. (4.10.4).

Example We consider the same system analyzed in the example in Sec. 4.10, but suppose that B, rather than being uniform, is a stepped beam consisting of portions C and D, as shown in Fig. 4.11.3. C has a length L_C, flexural rigidity $(EI)_C$, and mass per unit of length ρ_C, and the corresponding values for D are L_D, $(EI)_D$, and ρ_D.

The dynamical equations formulated previously, Eqs. (4.10.57)–(4.10.60), remain in force, but the constants m_B, e_B, I_B, E_i, F_i, G_{ij}, H_{ij}, and

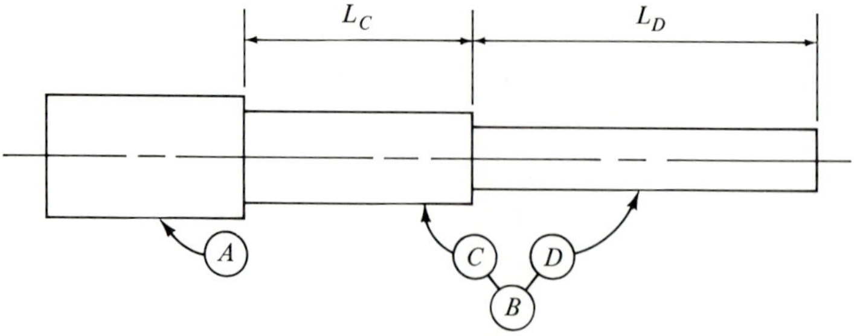

Figure 4.11.3

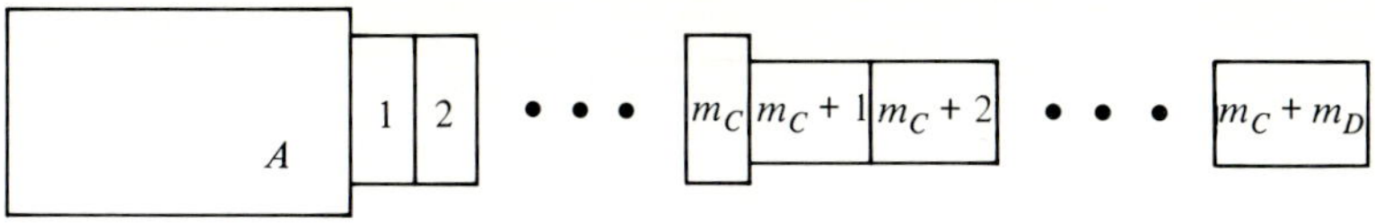

Figure 4.11.4

$\phi_i(L)$ $(i, j = 1, \ldots, \nu)$ appearing in these equations must be reevaluated since the expressions given for them in Eqs. (4.10.64) and (4.10.66)–(4.10.69) were found by using Eqs. (4.10.20), (4.10.21), (4.10.37), and (4.10.4), all of which involve x-dependent quantities, such as ρ, EI, and ϕ_i.

As a first step in the evaluation of $m_B, e_B, \ldots$, we divide B into elements in such a way that C contains m_C elements, each of length L_C/m_C, and D contains m_D elements, each of length L_D/m_D; number the elements as shown in Fig. 4.11.4; record the lengths, masses per unit of length, and flexural rigidities of the elements as in Table 4.11.1; set

$$m = m_C + m_D \tag{21}$$

and introduce x and s_i as shown in Fig. 4.11.5, so that

$$s_1 = 0 \qquad s_i = \sum_{j=1}^{i-1} L_j \qquad (i = 2, \ldots, m) \tag{22}$$

with the aid of which we form m_B, e_B, and I_B in accordance with Eqs. (4.10.20) as

$$m_B = \sum_{i=1}^{m} \int_0^{L_i} \rho_i \, dx_i = \sum_{i=1}^{m} \rho_i L_i \tag{23}$$

$$e_B = \sum_{i=1}^{m} \int_0^{L_i} (s_i + x_i)\rho_i \, dx_i = \sum_{i=1}^{m} \left(s_i L_i + \frac{L_i^2}{2}\right)\rho_i \tag{24}$$

$$I_B = \sum_{i=1}^{m} \int_0^{L_i} (s_i + x_i)^2 \rho_i \, dx_i = \sum_{i=1}^{m} \left[s_i L_i (s_i + L_i) + \frac{L_i^3}{3}\right]\rho_i \tag{25}$$

Turning to the construction of a substitute for Eq. (4.10.4), we begin by noting that, since B is clamped at A,

$$z_1 = z_2 = 0 \tag{26}$$

Next, after introducing $\tilde{z}_1, \ldots, \tilde{z}_{2m+2}$ such that

$$\tilde{z}_j = z_{j+2} \qquad (j = 1, \ldots, 2m) \tag{27}$$

Table 4.11.1

i	L_i	ρ_i	$(EI)_i$
$1, \ldots, m_C$	L_C/m_C	ρ_C	$(EI)_C$
$m_C + 1, \ldots, m_D$	L_D/m_D	ρ_D	$(EI)_D$

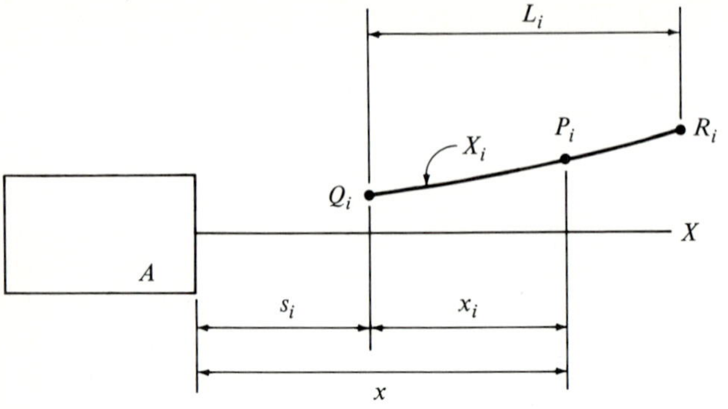

Figure 4.11.5

and defining $\widetilde{M}$ and $\widetilde{S}$ as

$$\widetilde{M} \triangleq \begin{bmatrix} M_{33} & \cdots & M_{3,2m+2} \\ \cdots & \cdots & \cdots \\ M_{2m+2,3} & \cdots & M_{2m+2,2m+2} \end{bmatrix} \qquad \widetilde{S} \triangleq \begin{bmatrix} S_{33} & \cdots & S_{3,2m+2} \\ \cdots & \cdots & \cdots \\ S_{2m+2,3} & \cdots & S_{2m+2,2m+2} \end{bmatrix} \tag{28}$$

where M_{ij} and S_{ij} $(i, j = 3, \ldots, 2m + 2)$ are given by Eq. (16), we use the procedure set forth in Sec. 4.8 to construct the first ν $(\nu \leq 2m)$ eigenvalues, $\widetilde{\lambda}_1, \ldots, \widetilde{\lambda}_\nu$, and the respective eigenvectors, $\widetilde{A}_1, \ldots, \widetilde{A}_\nu$, of $\widetilde{M}^{-1}\widetilde{S}$. Letting $\widetilde{A}_{ij}$ be the element in the ith row of $\widetilde{A}_j$, we then write by reference to Eqs. (1), (26), and (27)

$$\widetilde{\phi}_j(x) \triangleq \left\{ \begin{array}{l} \widetilde{A}_{1j}\,\psi_{13}(x) + \widetilde{A}_{2j}\,\psi_{14}(x) \qquad (0 \leq x \leq L_1) \\ \widetilde{A}_{2k-3,j}\,\psi_{k1}(x - s_k) + \widetilde{A}_{2k-2,j}\,\psi_{k2}(x - s_k) \\ \quad + \widetilde{A}_{2k-1,j}\,\psi_{k3}(x - s_k) + \widetilde{A}_{2k,j}\,\psi_{k4}(x - s_k) \\ (s_k \leq x \leq s_{k+1};\ k = 2, \ldots, m;\ s_{m+1} \triangleq L) \end{array} \right\} \qquad (j = 1, \ldots, \nu) \tag{29}$$

Proceeding as before [see Eq. (4.10.65)], we now set

$$\phi_i = \widetilde{\phi}_i \qquad (i = 1, \ldots, \nu) \tag{30}$$

after which we have from Eq. (4.10.21)

$$E_j = \int_0^{L_1} [\widetilde{A}_{1j}\,\psi_{13}(x) + \widetilde{A}_{2j}\,\psi_{14}(x)]\,\rho_1\,dx + \sum_{k=2}^{m} \int_{s_k}^{s_k+L_k} [\widetilde{A}_{2k-3,j}\,\psi_{k1}(x - s_k) + \widetilde{A}_{2k-2,j}\,\psi_{k2}(x - s_k) + \widetilde{A}_{2k-1,j}\,\psi_{k3}(x - s_k) + \widetilde{A}_{2k,j}\,\psi_{k4}(x - s_k)]\,\rho_k\,dx \qquad (j = 1, \ldots, \nu) \tag{31}$$

or, equivalently,

$$E_j = \int_0^{L_1} [\widetilde{A}_{1j}\,\psi_{13}(x_1) + \widetilde{A}_{2j}\,\psi_{14}(x_1)]\,\rho_1\,dx_1$$
$$+ \sum_{k=2}^{m} \int_0^{L_k} [\widetilde{A}_{2k-3,j}\,\psi_{k1}(x_k) + \widetilde{A}_{2k-2,j}\,\psi_{k2}(x_k)$$
$$+ \widetilde{A}_{2k-1,j}\,\psi_{k3}(x_k) + \widetilde{A}_{2k,j}\,\psi_{k4}(x_k)]\,\rho_k\,dx_k$$
$$(j = 1, \ldots, \nu) \quad (32)$$

Since ψ_{ij} $(i = 1, \ldots, m; j = 1, \ldots, 4)$ are available in Eqs. (5)–(8), the quadratures indicated in Eq. (32) can be performed explicitly. However, it is more convenient to note that, from Eqs. (5) and (7),

$$\psi_{k1} + \psi_{k3} = 1 \qquad (k = 1, \ldots, m) \quad (33)$$

so that multiplication with ψ_{ka} produces

$$\psi_{ka} = \psi_{ka}\,\psi_{k1} + \psi_{ka}\,\psi_{k3} \qquad (k = 1, \ldots, m;\ a = 1, \ldots, 4) \quad (34)$$

Integrating both sides of this equation from $x_k = 0$ to $x_k = L_k$ and multiplying with ρ_k, one has

$$\rho_k \int_0^{L_k} \psi_{ka}\,dx_k \underset{(10)}{=} \hat{M}_{ka1} + \hat{M}_{ka3} \qquad (k = 1, \ldots, m;\ a = 1, \ldots, 4) \quad (35)$$

and, upon substituting into Eq. (32),

$$E_j = \widetilde{A}_{1j}(\hat{M}_{131} + \hat{M}_{133}) + \widetilde{A}_{2j}(\hat{M}_{141} + \hat{M}_{143})$$
$$+ \sum_{k=2}^{m} [\widetilde{A}_{2k-3,j}(\hat{M}_{k11} + \hat{M}_{k13}) + \widetilde{A}_{2k-2,j}(\hat{M}_{k21} + \hat{M}_{k23})$$
$$+ \widetilde{A}_{2k-1,j}(\hat{M}_{k31} + \hat{M}_{k33}) + \widetilde{A}_{2k,j}(\hat{M}_{k41} + \hat{M}_{k43})]$$
$$(j = 1, \ldots, \nu) \quad (36)$$

or, after using Eq. (12),

$$E_j = \frac{\rho_1 L_1}{2}\left(\widetilde{A}_{1j} - \frac{\widetilde{A}_{2j} L_1}{6}\right)$$
$$+ \sum_{k=2}^{m} \frac{\rho_k L_k}{2}\left[\widetilde{A}_{2k-3,j} + \widetilde{A}_{2k-1,j} + \frac{(\widetilde{A}_{2k-2,j} - \widetilde{A}_{2k,j})L_k}{6}\right]$$
$$(j = 1, \ldots, \nu) \quad (37)$$

Similarly, substitution from Eq. (29) into the second of Eqs. (4.10.21) leads to

$$F_j = \int_0^{L_1} x_1[\widetilde{A}_{1j}\,\psi_{13}(x_1) + \widetilde{A}_{2j}\,\psi_{14}(x_1)]\,\rho_1\,dx_1$$

$$+ \sum_{k=2}^{m} \int_0^{L_k} (s_k + x_k)[\widetilde{A}_{2k-3,j}\,\psi_{k1}(x_k) + \widetilde{A}_{2k-2,j}\,\psi_{k2}(x_k)$$

$$+ \widetilde{A}_{2k-1,j}\,\psi_{k3}(x_k) + \widetilde{A}_{2k,j}\,\psi_{k4}(x_k)]\,\rho_k\,dx_k \qquad (j = 1, \ldots, \nu) \qquad (38)$$

and now one can make use of the fact that

$$x_k \underset{(6-8)}{=} \psi_{k2} + L_k\,\psi_{k3} + \psi_{k4} \qquad (k = 1, \ldots, m) \qquad (39)$$

so that

$$\psi_{ka}\,x_k = \psi_{ka}\,\psi_{k2} + L_k\,\psi_{ka}\,\psi_{k3} + \psi_{ka}\,\psi_{k4}$$

$$(k = 1, \ldots, m;\ a = 1, \ldots, 4) \qquad (40)$$

Integrating both sides of this equation from $x_k = 0$ to $x_k = L_k$, and using Eq. (10), one thus finds that

$$\int_0^{L_k} x_k\,\psi_{ka}\,\rho_k\,dx_k = \hat{M}_{ka2} + L_k\,\hat{M}_{ka3} + \hat{M}_{ka4}$$

$$(k = 1, \ldots, m;\ a = 1, \ldots, 4) \qquad (41)$$

and use of this result together with Eqs. (35) and (12) then yields

$$F_j \underset{(38)}{=} \frac{\rho_1 L_1{}^2}{20}\,(7\widetilde{A}_{1j} - \widetilde{A}_{2j}L_1)$$

$$+ \sum_{k=2}^{m} \left\{ \frac{s_k\,\rho_k\,L_k}{2} \left[\widetilde{A}_{2k-3,j} + \widetilde{A}_{2k-1,j} + \frac{(\widetilde{A}_{2k-2,j} - \widetilde{A}_{2k,j})\,L_k}{6} \right] \right.$$

$$\left. + \frac{\rho_k L_k{}^2}{20} \left[3\widetilde{A}_{2k-3,j} + 7\widetilde{A}_{2k-1,j} + \left(\frac{2\widetilde{A}_{2k-2,j}}{3} - \widetilde{A}_{2k,j} \right) L_k \right] \right\}$$

$$(j = 1, \ldots, \nu) \qquad (42)$$

As for G_{ij}, substitution from Eq. (29) into the third of Eqs. (4.10.21) and subsequent use of Eq. (10) leads to

$$G_{ij} = \widetilde{A}_{1i}\widetilde{A}_{1j}(\hat{M}_{133} + \hat{M}_{211}) + \widetilde{A}_{1i}\widetilde{A}_{2j}(\hat{M}_{134} + \hat{M}_{212})$$

$$+ \widetilde{A}_{1i}\widetilde{A}_{3j}\,\hat{M}_{213} + \widetilde{A}_{1i}\widetilde{A}_{4j}\,\hat{M}_{214} + \widetilde{A}_{2i}\widetilde{A}_{1j}(\hat{M}_{143} + \hat{M}_{221})$$

$$+ \widetilde{A}_{2i}\widetilde{A}_{2j}(\hat{M}_{144} + \hat{M}_{222}) + \widetilde{A}_{2i}\widetilde{A}_{3j}\,\hat{M}_{223} + \widetilde{A}_{2i}\widetilde{A}_{4j}\,\hat{M}_{224}$$

$$+ \cdots + \widetilde{A}_{2m,i}\widetilde{A}_{2m,j}\,\hat{M}_{m44} \qquad (i, j = 1, \ldots, \nu) \qquad (43)$$

and comparison of this equation with Eq. (16) enables one to write

$$G_{ij} = \sum_{r=1}^{2m} \sum_{s=1}^{2m} \widetilde{A}_{ri}\,\widetilde{M}_{rs}\,\widetilde{A}_{sj} \qquad (i, j = 1, \ldots, \nu) \qquad (44)$$

where $\widetilde{M}_{rs}$ is the element in the rth row and sth column of $\widetilde{M}$ as given in Eq. (28). Moreover, the first of Eqs. (4.8.11) permits one to conclude that

$$G_{ij} = \delta_{ij} \qquad (i, j = 1, \ldots, \nu) \tag{45}$$

where δ_{ij} is the Kronecker delta.

The evaluation of H_{ij} in accordance with Eq. (4.10.37) and Eq. (29) becomes a simple matter when Eq. (11) is taken into account, for this shows that

$$H_{ij} = \sum_{r=1}^{2m} \sum_{s=1}^{2m} \widetilde{A}_{ri} \widetilde{S}_{rs} \widetilde{A}_{sj} \qquad (i, j = 1, \ldots, \nu) \tag{46}$$

so that, with the aid of the second of Eqs. (4.8.11), one has

$$H_{ij} = \delta_{ij} \widetilde{\lambda}_j \qquad (i, j = 1, \ldots, \nu) \tag{47}$$

Finally, $\phi_j(L)$, needed for substitution into the last term of Eq. (4.10.60), is formed by using Eq. (29) with $k = m$, $x = L$, and $L - s_m = L_m$, which gives

$$\phi_j(L) = \widetilde{A}_{2m-3,j}\, \psi_{m1}(L_m) + \widetilde{A}_{2m-2,j}\, \psi_{m2}(L_m) + \widetilde{A}_{2m-1,j}\, \psi_{m3}(L_m) + \widetilde{A}_{2m,j}\, \psi_{m4}(L_m) \tag{48}$$

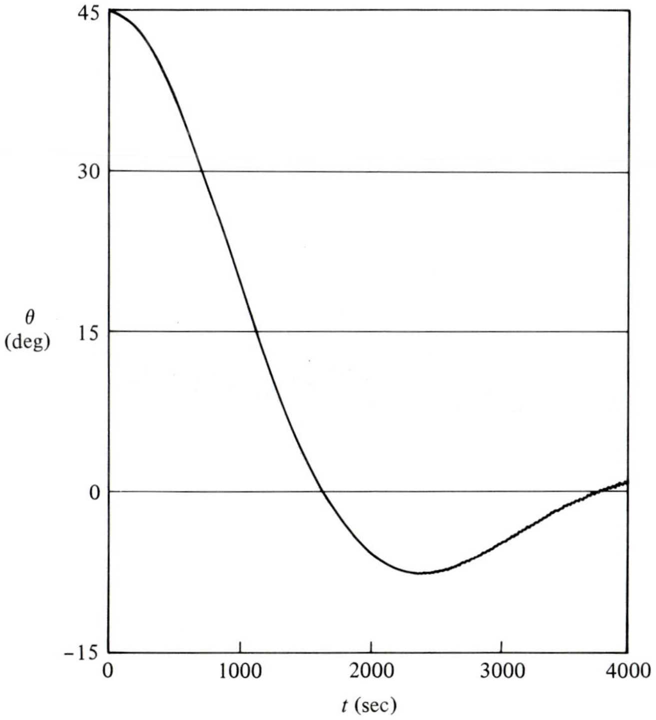

Figure 4.11.6

or, in view of Eqs. (5)–(8),

$$\phi_j(L) = \widetilde{A}_{2m-1,j} \qquad (j = 1, \ldots, \nu) \tag{49}$$

In summary, Eqs. (23)–(25), (37), (42), (44), (47), (49) replace Eqs. (4.10.64) and Eqs. (4.10.66)–(4.10.69).

To illustrate the use of these results, we turn to the following question concerning the system considered in the example in Sec. 4.10 (see Fig. 4.10.2): By stiffening a portion of B near the clamped end of B, to what extent can one bring about an attenuation of the effects of the instability manifested in Figs. 4.10.5 and 4.10.6? In what follows, it is shown how one can deal with such questions.

Let D (see Fig. 4.11.3) have the same mass per unit of length and flexural rigidity as the beam in the example in Sec. 4.10, that is, take (see Table 4.11.1) $\rho_D = 0.2$ kg/m and $(EI)_D = 5$ N $\cdot$ m^2, and set $\rho_C = 4\rho_D = 0.8$ kg/m and $(EI)_C = 16\ (EI)_D = 80$ N $\cdot$ m^2, which corresponds to regarding C and D as having circular cross sections such that C has twice the radius of D. Taking $L_C = L_D = 10$ m, divide C and D each into five elements of equal length, that

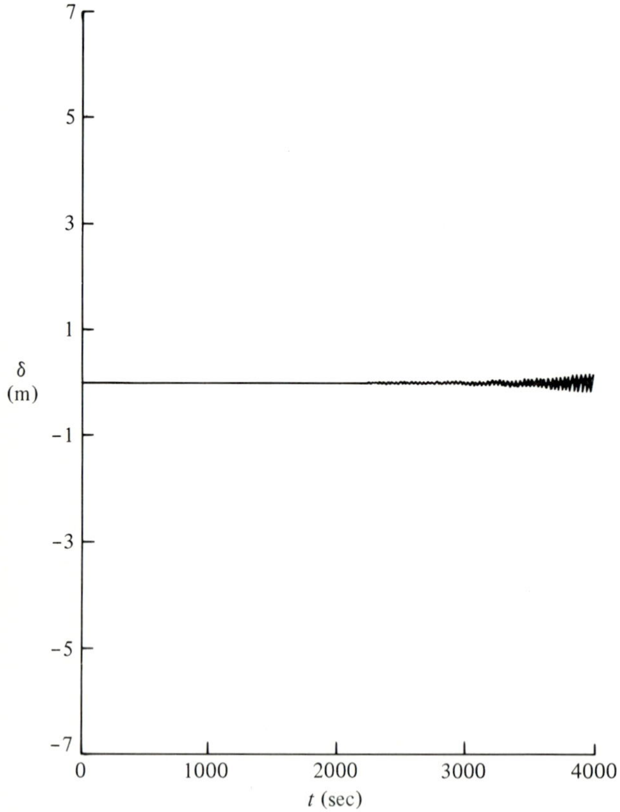

Figure 4.11.7

is, let (see Table 4.11.1) $m_C = m_D = 5$, and, after forming $\widetilde{M}$ and $\widetilde{S}$ in accordance with Eq. (28), determine $\widetilde{\lambda}_i$ and $\widetilde{A}_i$ ($i = 1, 2, 3$), the first three eigenvalues and eigenvectors, respectively, of $\widetilde{M}^{-1}\widetilde{S}$. (The reason for working with three modes is that Figs. 4.10.5 and 4.10.6 were generated with $\nu = 3$.) Evaluate m_B, e_B, I_B by reference to Eqs. (23)–(25), and form E_i, F_i, G_{ij}, H_{ij}, $\phi_i(L)$ ($i, j = 1, 2, 3$) in accordance with Eqs. (32), (42), (45), (47), (49), respectively. Finally, perform a numerical integration of Eqs. (4.10.57)–(4.10.60), using the same initial conditions and feedback gains previously employed in connection with Figs. 4.10.5 and 4.10.6. This leads to the plots shown in Figs. 4.11.6 and 4.11.7.

Comparison of Figs. 4.11.6 and 4.10.5 reveals that, so far as θ is concerned, the stiffening brought about by making $(EI)_C > (EI)_D$ is, indeed, effective, at least throughout the first 4000 seconds of the motion. Similarly, after comparing Fig. 4.11.7 with Fig. 4.10.6, it becomes clear that, although the tip of B performs an unstable motion in any event, one can reduce the associated amplitudes considerably by stiffening the beam.

PROBLEM SETS

PROBLEM SET 1

1.1 A dextral set of orthogonal unit vectors $\mathbf{a}_1$, $\mathbf{a}_2$, $\mathbf{a}_3$ is fixed in a reference frame A, and a similar set of unit vectors $\mathbf{b}_1$, $\mathbf{b}_2$, $\mathbf{b}_3$ is fixed in a rigid body B. Initially the orientation of B in A is such that

$$[\mathbf{b}_1 \quad \mathbf{b}_2 \quad \mathbf{b}_3] = [\mathbf{a}_1 \quad \mathbf{a}_2 \quad \mathbf{a}_3]M$$

where

$$M = \begin{bmatrix} 0.9363 & -0.2896 & 0.1987 \\ 0.3130 & 0.9447 & -0.0981 \\ -0.1593 & 0.1540 & 0.9751 \end{bmatrix}$$

Body B is then subjected to a 30° $\boldsymbol{\lambda}$-rotation relative to A, where

$$\boldsymbol{\lambda} = \frac{2\mathbf{a}_1 + 3\mathbf{a}_2 + 6\mathbf{a}_3}{7}$$

Find the matrix N such that, subsequent to the rotation,

$$[\mathbf{b}_1 \quad \mathbf{b}_2 \quad \mathbf{b}_3] = [\mathbf{a}_1 \quad \mathbf{a}_2 \quad \mathbf{a}_3]N$$

Result

$$N = \begin{bmatrix} 0.6527 & -0.6053 & 0.4556 \\ 0.7103 & 0.6981 & -0.0903 \\ -0.2634 & 0.3825 & 0.8856 \end{bmatrix}$$

1.2 By definition, the eigenvalues of a direction cosine matrix C are values of a scalar quantity μ that satisfy the *characteristic equation*

$$|C - \mu U| = 0$$

Show that this equation can be expressed as

$$(1 - \mu)(1 + \mu B + \mu^2) = 0$$

where

$$B = 1 - (C_{11} + C_{22} + C_{33})$$

1.3 In Fig. P1.3, $\mathbf{a}_1$, $\mathbf{a}_2$, $\mathbf{a}_3$ and $\mathbf{b}_1$, $\mathbf{b}_2$, $\mathbf{b}_3$ are dextral sets of orthogonal unit vectors, with $\mathbf{a}_1$ parallel to the line connecting a particle $\overline{P}$ and the mass center B^* of a rigid body B, and $\mathbf{b}_1$, $\mathbf{b}_2$, $\mathbf{b}_3$ each parallel to a principal axis of inertia of B for B^*.

The system of gravitational forces exerted by $\overline{P}$ on B produces a moment $\mathbf{M}$ about B^*. If the distance R between $\overline{P}$ and B^* exceeds the greatest distance from B^* to any point of B, $\mathbf{M}$ can be expressed approximately as (see Sec. 2.6)

$$\mathbf{M} \approx \widetilde{\mathbf{M}} \triangleq 3G\overline{m}R^{-3}\, \mathbf{a}_1 \times \mathbf{I} \cdot \mathbf{a}_1$$

where G is the universal gravitational constant, $\overline{m}$ is the mass of $\overline{P}$, and $\mathbf{I}$ is the inertia dyadic of B for B^*. $\widetilde{\mathbf{M}}$ thus depends on the orientation of $\mathbf{a}_1$ relative to $\mathbf{b}_1$, $\mathbf{b}_2$, $\mathbf{b}_3$.

Letting ϵ_1, ϵ_2, ϵ_3, ϵ_4 be Euler parameters characterizing the relative orientation of $\mathbf{a}_1$, $\mathbf{a}_2$, $\mathbf{a}_3$ and $\mathbf{b}_1$, $\mathbf{b}_2$, $\mathbf{b}_3$, express $\widetilde{\mathbf{M}}$ in terms of these parameters, the unit vectors $\mathbf{b}_1$, $\mathbf{b}_2$, $\mathbf{b}_3$, and the principal moments of inertia I_1, I_2, I_3 of B for B^*.

Result

$$\begin{aligned}\widetilde{\mathbf{M}} = 6G\overline{m}R^{-3}[&2(\epsilon_1\epsilon_2 - \epsilon_3\epsilon_4)(\epsilon_3\epsilon_1 + \epsilon_2\epsilon_4)(I_3 - I_2)\,\mathbf{b}_1 \\ &+ (\epsilon_3\epsilon_1 + \epsilon_2\epsilon_4)(1 - 2\epsilon_2^2 - 2\epsilon_3^2)(I_1 - I_3)\,\mathbf{b}_2 \\ &+ (1 - 2\epsilon_2^2 - 2\epsilon_3^2)(\epsilon_1\epsilon_2 - \epsilon_3\epsilon_4)(I_2 - I_1)\,\mathbf{b}_3]\end{aligned}$$

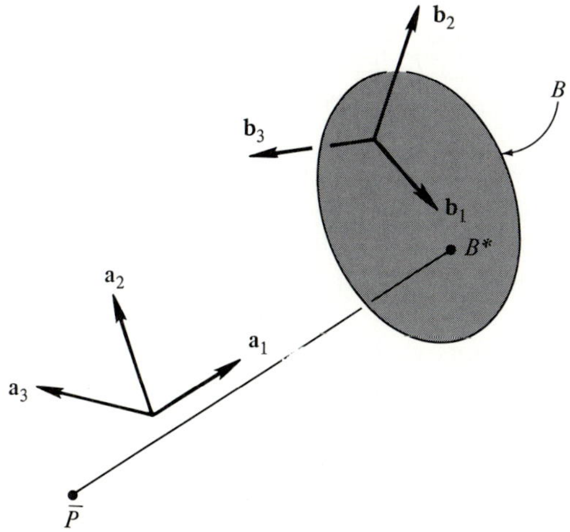

Figure P1.3

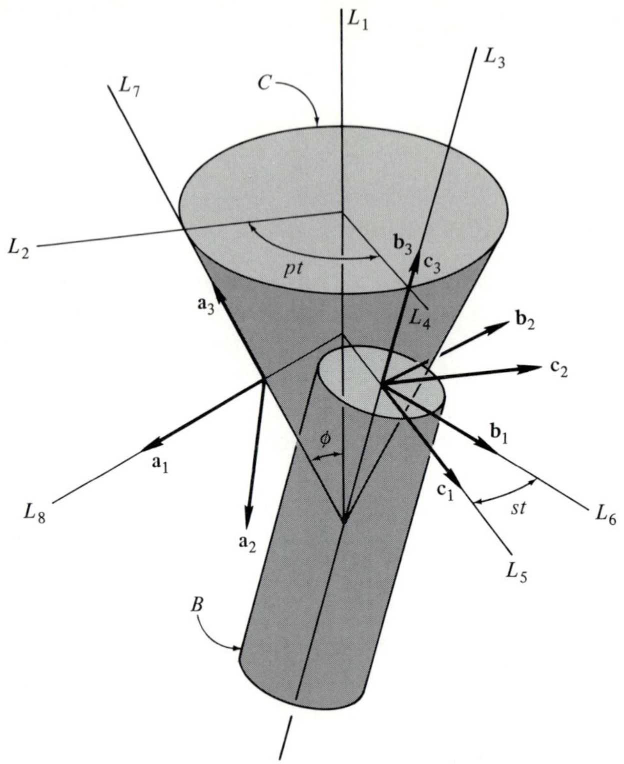

Figure P1.4

1.4 In Fig. P1.4, L_1 is the axis of a cone C of semivertex angle ϕ; L_2 is perpendicular to L_1 and fixed in C; L_3 is the axis of symmetry of a cylindrical body B which moves in such a way that L_3 always coincides with a generator of C; L_4 is perpendicular to L_1 and lies in the plane determined by L_1 and L_3; L_5 is perpendicular to L_3 and lies in the plane determined by L_1 and L_3; L_6 is perpendicular to L_3 and fixed in B; L_7 is a generator of C that intersects L_2; and L_8 is perpendicular to L_7 and lies in the plane determined by L_1 and L_7.

Assuming that B moves in such a way that the angle between L_2 and L_4 is equal to pt while the angle between L_5 and L_6 is equal to st, and letting $\mathbf{a}_1$, $\mathbf{a}_2$, $\mathbf{a}_3$ and $\mathbf{b}_1$, $\mathbf{b}_2$, $\mathbf{b}_3$ be dextral sets of orthogonal unit vectors directed as shown, express the Euler parameters $\epsilon_1, \ldots, \epsilon_4$ relating $\mathbf{a}_1$, $\mathbf{a}_2$, $\mathbf{a}_3$ to $\mathbf{b}_1$, $\mathbf{b}_2$, $\mathbf{b}_3$ as functions of ϕ, p, s, and t.

Suggestion: Introduce a set of unit vectors $\mathbf{c}_1$, $\mathbf{c}_2$, $\mathbf{c}_3$ as shown, and use Eqs. (1.2.23)–(1.2.31) to generate a direction cosine matrix L such that

$$[\mathbf{c}_1 \quad \mathbf{c}_2 \quad \mathbf{c}_3] = [\mathbf{a}_1 \quad \mathbf{a}_2 \quad \mathbf{a}_3]L$$

Next, use Eq. (1.2.37) to find a matrix M such that

$$[\mathbf{b}_1 \quad \mathbf{b}_2 \quad \mathbf{b}_3] = [\mathbf{c}_1 \quad \mathbf{c}_2 \quad \mathbf{c}_3]M$$

Then

$$[\mathbf{b}_1 \quad \mathbf{b}_2 \quad \mathbf{b}_3] = [\mathbf{a}_1 \quad \mathbf{a}_2 \quad \mathbf{a}_3]N$$

where

$$N = LM$$

and the elements of N can be used in Eqs. (1.3.15)–(1.3.18) to find $\epsilon_1, \ldots, \epsilon_4$.

Result

$$\epsilon_1 = -\sin\phi \sin\frac{pt}{2}\cos\frac{st}{2}$$

$$\epsilon_2 = \sin\phi \sin\frac{pt}{2}\sin\frac{st}{2}$$

$$\epsilon_3 = \cos\phi \sin\frac{pt}{2}\cos\frac{st}{2} + \cos\frac{pt}{2}\sin\frac{st}{2}$$

$$\epsilon_4 = -\cos\phi \sin\frac{pt}{2}\sin\frac{st}{2} + \cos\frac{pt}{2}\cos\frac{st}{2}$$

1.5 Dextral sets of orthogonal unit vectors $\mathbf{a}_1$, $\mathbf{a}_2$, $\mathbf{a}_3$ and $\mathbf{b}_1$, $\mathbf{b}_2$, $\mathbf{b}_3$ are fixed in bodies A and B, respectively. At $t = 0$, $\mathbf{a}_i = \mathbf{b}_i$ $(i = 1, 2, 3)$. B then moves relative to A in such a way that the first time-derivatives of $\mathbf{b}_1$ and $\mathbf{b}_2$ in A are given by

$$\frac{{}^A d\mathbf{b}_1}{dt} = -\cos pt\,\mathbf{b}_3 \qquad \frac{{}^A d\mathbf{b}_2}{dt} = \sin pt\,\mathbf{b}_3$$

During this motion, the Euler parameters associated with any relative orientation of $\mathbf{a}_1$, $\mathbf{a}_2$, $\mathbf{a}_3$ and $\mathbf{b}_1$, $\mathbf{b}_2$, $\mathbf{b}_3$ must satisfy a set of differential equations that can be expressed in the form

$$[\dot{\epsilon}_1 \quad \dot{\epsilon}_2 \quad \dot{\epsilon}_3 \quad \dot{\epsilon}_4] = \tfrac{1}{2}[f_1 \quad f_2 \quad f_3 \quad f_4]M$$

where $f_1, \ldots, f_4$ are functions of pt and where M is a 4×4 matrix whose elements depend solely on $\epsilon_1, \ldots, \epsilon_4$.

Find the set of differential equations governing the Euler parameters.

Suggestion: Use Eqs. (1.2.2) and (1.2.5) to formulate general expressions for ${}^A d\mathbf{b}_1/dt$ and ${}^A d\mathbf{b}_2/dt$ in terms of $\mathbf{b}_1$, $\mathbf{b}_2$, $\mathbf{b}_3$ and the elements C_{ij} $(i, j = 1, 2, 3)$ of the direction cosine matrix relating $\mathbf{a}_1$, $\mathbf{a}_2$, $\mathbf{a}_3$ to $\mathbf{b}_1$, $\mathbf{b}_2$, $\mathbf{b}_3$; and substitute the resulting expressions into the given equations for ${}^A d\mathbf{b}_1/dt$ and ${}^A d\mathbf{b}_2/dt$. Noting that only three of the six scalar equations thus obtained are linearly independent and not satisfied identically, use Eqs. (1.3.6)–(1.3.14) to express these three equations, together with the equation

$$\epsilon_1\dot{\epsilon}_1 + \epsilon_2\dot{\epsilon}_2 + \epsilon_3\dot{\epsilon}_3 + \epsilon_4\dot{\epsilon}_4 = 0$$

in the form

$$[\dot{\epsilon}_1 \quad \dot{\epsilon}_2 \quad \dot{\epsilon}_3 \quad \dot{\epsilon}_4]N = \tfrac{1}{2}[0 \quad 0 \quad \cos pt \quad -\sin pt]$$

Finally, verify that $NN^T = U$, and use this fact to find the desired expressions for $\dot{\epsilon}_1, \ldots, \dot{\epsilon}_4$.

Result

$$2\dot{\epsilon}_1 = -\epsilon_3 \cos pt + \epsilon_4 \sin pt$$

$$2\dot{\epsilon}_2 = \epsilon_3 \sin pt + \epsilon_4 \cos pt$$

$$2\dot{\epsilon}_3 = \epsilon_1 \cos pt - \epsilon_2 \sin pt$$

$$2\dot{\epsilon}_4 = -\epsilon_1 \sin pt - \epsilon_2 \cos pt$$

1.6 Dextral sets of orthogonal unit vectors $\mathbf{a}_1, \mathbf{a}_2, \mathbf{a}_3$ and $\mathbf{b}_1, \mathbf{b}_2, \mathbf{b}_3$ are fixed in rigid bodies A and B, respectively, and two vectors, $\mathbf{p}$ and $\mathbf{q}$, are observed simultaneously from A and B in order to determine the twelve quantities $\mathbf{a}_i \cdot \mathbf{p}$, $\mathbf{b}_i \cdot \mathbf{p}$, $\mathbf{a}_i \cdot \mathbf{q}$, $\mathbf{b}_i \cdot \mathbf{q}$ $(i = 1, 2, 3)$. For the values of these quantities recorded in Table P1.6, determine C such that Eq. (1.2.2) is satisfied.

Table P1.6

	$\mathbf{a}_i \cdot \mathbf{p}$	$\mathbf{b}_i \cdot \mathbf{p}$	$\mathbf{a}_i \cdot \mathbf{q}$	$\mathbf{b}_i \cdot \mathbf{q}$
$i = 1$	−0.092	0.928	0.554	0.004
$i = 2$	−0.208	−0.349	−0.803	−0.658
$i = 3$	−0.974	0.130	−0.221	0.753

Result

$$C = \begin{bmatrix} -0.138 & 0.253 & 0.957 \\ 0.168 & 0.958 & -0.229 \\ -0.976 & 0.130 & -0.175 \end{bmatrix}$$

1.7 A rigid body is subjected to successive rotations characterized by Rodrigues vectors $\boldsymbol{\rho}_1, \boldsymbol{\rho}_2, \boldsymbol{\rho}_3$. Show that the Rodrigues vector $\boldsymbol{\rho}$ characterizing a single equivalent rotation can be expressed as

$$\boldsymbol{\rho} = (\boldsymbol{\rho}_1 + \boldsymbol{\rho}_2 + \boldsymbol{\rho}_3 - \boldsymbol{\rho}_1 \times \boldsymbol{\rho}_2 - \boldsymbol{\rho}_2 \times \boldsymbol{\rho}_3 + \boldsymbol{\rho}_3 \times \boldsymbol{\rho}_1 - \boldsymbol{\rho}_2 \cdot \boldsymbol{\rho}_3 \boldsymbol{\rho}_1 + \boldsymbol{\rho}_3 \cdot \boldsymbol{\rho}_1 \boldsymbol{\rho}_2 - \boldsymbol{\rho}_1 \cdot \boldsymbol{\rho}_2 \boldsymbol{\rho}_3)/(1 - \boldsymbol{\rho}_1 \cdot \boldsymbol{\rho}_2 - \boldsymbol{\rho}_2 \cdot \boldsymbol{\rho}_3 - \boldsymbol{\rho}_3 \cdot \boldsymbol{\rho}_1 + \boldsymbol{\rho}_1 \cdot \boldsymbol{\rho}_2 \times \boldsymbol{\rho}_3)$$

1.8 A rigid body B is brought into a desired orientation in a reference frame A by being subjected successively to an $\mathbf{a}_1$-rotation of amount θ_1, an $\mathbf{a}_2$-rotation of amount θ_2, and an $\mathbf{a}_3$-rotation of amount θ_3, where $\mathbf{a}_1, \mathbf{a}_2, \mathbf{a}_3$ form a dextral set of orthogonal unit vectors fixed in A.

Show that the Rodrigues vector $\boldsymbol{\rho}$ associated with a single rotation by means of which B can be brought into the same orientation in A is given by

$$\boldsymbol{\rho} = \frac{1}{1 + t_1 t_2 t_3} \left[(t_1 - t_2 t_3)\mathbf{a}_1 + (t_2 + t_3 t_1)\mathbf{a}_2 + (t_3 - t_1 t_2)\mathbf{a}_3 \right]$$

where $t_i \triangleq \tan(\theta_i/2)$ $(i = 1, 2, 3)$.

1.9 Letting θ_1, θ_2, θ_3 be a set of body-two angles used to describe the orientation of a rigid body, express the associated Euler parameters in terms of these angles.

Result

$$\epsilon_1 = \cos(\theta_2/2)\sin[(\theta_1 + \theta_3)/2]$$

$$\epsilon_2 = \sin(\theta_2/2)\cos[(\theta_1 - \theta_3)/2]$$

$$\epsilon_3 = \sin(\theta_2/2)\sin[(\theta_1 - \theta_3)/2]$$

$$\epsilon_4 = \cos(\theta_2/2)\cos[(\theta_1 + \theta_3)/2]$$

1.10 A dextral set of orthogonal unit vectors $\mathbf{a}_1$, $\mathbf{a}_2$, $\mathbf{a}_3$ is fixed in a reference frame A; $\mathbf{b}_1$, $\mathbf{b}_2$, $\mathbf{b}_3$ is a similar set fixed in a rigid body B; and $\mathbf{c}_1$, $\mathbf{c}_2$, $\mathbf{c}_3$ is a third such set fixed in a rigid body C. Initially $\mathbf{a}_i = \mathbf{b}_i = \mathbf{c}_i$ $(i = 1, 2, 3)$, and B is then subjected to successive rotations $\theta_1\mathbf{a}_1$, $\theta_2\mathbf{a}_2$, $\theta_3\mathbf{a}_3$ while C undergoes successive rotations $\theta_3\mathbf{c}_3$, $\theta_2\mathbf{c}_2$, $\theta_1\mathbf{c}_1$.

Show that $\mathbf{b}_i = \mathbf{c}_i$ $(i = 1, 2, 3)$ when all rotations have been completed.

1.11 At a certain instant the angles ϕ, θ, and ψ shown in Fig. P1.11 have the

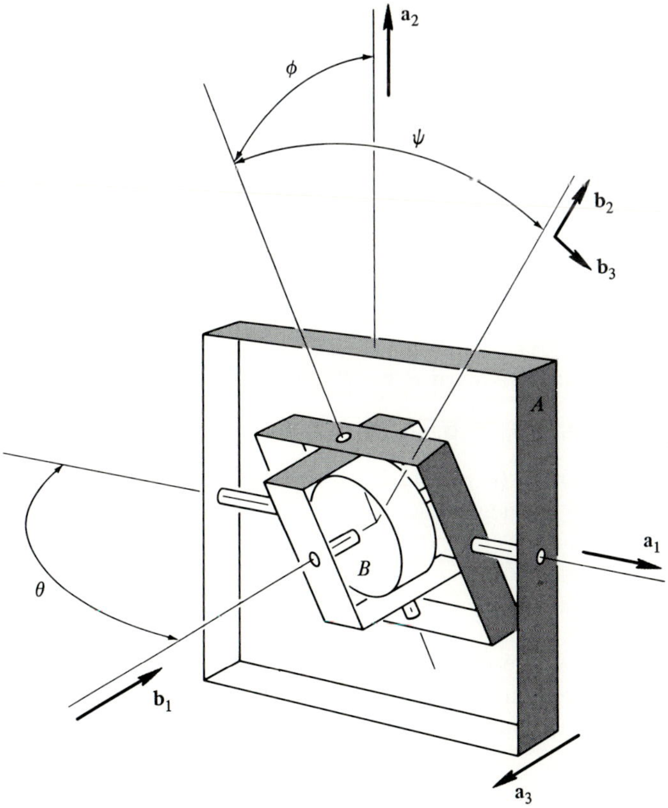

Figure P1.11

values $\phi = 63.03°$, $\theta = 20.56°$, and $\psi = -55.55°$. The angles are then changed in such a way that the rotor B experiences a 30° $\boldsymbol{\lambda}$-rotation relative to the frame A, where

$$\boldsymbol{\lambda} = \frac{2\mathbf{a}_1 + 3\mathbf{a}_2 + 6\mathbf{a}_3}{7}$$

Determine the changes $\Delta\phi$, $\Delta\theta$, and $\Delta\psi$ in the angles ϕ, θ, and ψ.

Suggestion: Use the results of Prob. 1.1 after verifying that, when ϕ, θ, and ψ have the given values,

$$[\mathbf{b}_1 \quad \mathbf{b}_2 \quad \mathbf{b}_3] = [\mathbf{a}_1 \quad \mathbf{a}_2 \quad \mathbf{a}_3]M$$

with M as in Prob. 1.1.

Results $\Delta\phi = 6.62°$, $\Delta\theta = 28.69°$, $\Delta\psi = 2.52°$

1.12 Dextral sets of orthogonal unit vectors $\mathbf{a}_1$, $\mathbf{a}_2$, $\mathbf{a}_3$ and $\mathbf{b}_1$, $\mathbf{b}_2$, $\mathbf{b}_3$ are fixed in a reference frame A and in a rigid body B, respectively. Initially $\mathbf{a}_i = \mathbf{b}_i$ $(i = 1, 2, 3)$, and B is then subjected to successive rotations $\phi\mathbf{a}_2$, $\psi\mathbf{a}_3$, and $\theta\mathbf{a}_2$.

Letting $s\phi \triangleq \sin\phi$, $c\phi \triangleq \cos\phi$, etc., determine M such that subsequent to the last rotation

$$[\mathbf{b}_1 \quad \mathbf{b}_2 \quad \mathbf{b}_3] = [\mathbf{a}_1 \quad \mathbf{a}_2 \quad \mathbf{a}_3]M$$

Result

$$M = \begin{bmatrix} c\phi c\psi c\theta - s\theta s\phi & -s\psi c\theta & s\phi c\psi c\theta + s\theta c\phi \\ c\phi s\psi & c\psi & s\phi s\psi \\ -c\phi c\psi s\theta - c\theta s\phi & s\psi s\theta & -s\phi c\psi s\theta + c\theta c\phi \end{bmatrix}$$

1.13 Two rigid bodies, A and B, are attached to each other at a point P, and dextral sets of orthogonal unit vectors $\mathbf{a}_1$, $\mathbf{a}_2$, $\mathbf{a}_3$ and $\mathbf{b}_1$, $\mathbf{b}_2$, $\mathbf{b}_3$ are fixed in A and B, respectively. Initially $\mathbf{a}_i = \mathbf{b}_i$ $(i = 1, 2, 3)$, and B is then subjected to successive rotations $\theta_1\mathbf{b}_1$, $\theta_2\mathbf{b}_2$, $\theta_3\mathbf{b}_3$.

Letting $\mathbf{I}$ denote the inertia dyadic of B for P, and defining A_{ij} and B_{ij} as

$$A_{ij} \triangleq \mathbf{a}_i \cdot \mathbf{I} \cdot \mathbf{a}_j \qquad B_{ij} \triangleq \mathbf{b}_i \cdot \mathbf{I} \cdot \mathbf{b}_j \qquad (i, j = 1, 2, 3)$$

express A_{11} and A_{23} in terms of B_{ij} $(i, j = 1, 2, 3)$, assuming that terms of second or higher degree in θ_1, θ_2, θ_3 are negligible.

Results

$$A_{11} = B_{11} + 2(\theta_2 B_{31} - \theta_3 B_{12})$$

$$A_{23} = B_{23} + \theta_1(B_{22} - B_{33}) - \theta_2 B_{12} + \theta_3 B_{31}$$

1.14 In Fig. P1.14, X_1, X_2, X_3 are mutually perpendicular lines, and $ABCD$ is a square plate which is to be brought from the X_1–X_2 plane into the X_1–X_3 plane by means of a translation followed by a rotation. Determine the minimum magnitude of the translation.

Result $2\sqrt{3}L$

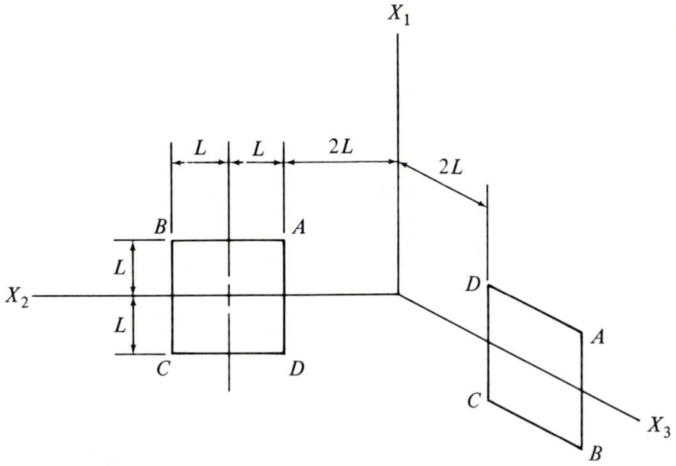

Figure P1.14

1.15 Referring to Prob. 1.11, and supposing that B moves in such a way that at a certain instant $\phi = \theta = \psi = \pi/4$ rad while the angular velocity $\boldsymbol{\omega}$ of B in A is given by

$$\boldsymbol{\omega} = \mathbf{a}_1 + \mathbf{a}_2 + \mathbf{a}_3 \text{ rad/sec}$$

determine $\dot{M}_{22}$ for this instant.

Result

$$\dot{M}_{22} = -\frac{\sqrt{2}}{4} \text{ sec}^{-1}$$

1.16 In Fig. P1.16, L represents the line of sight from an Earth satellite to a star; B_1, B_2, B_3 are lines fixed in the satellite; and M is the intersection of the plane determined by B_1 and B_2 with the plane determined by B_3 and L.

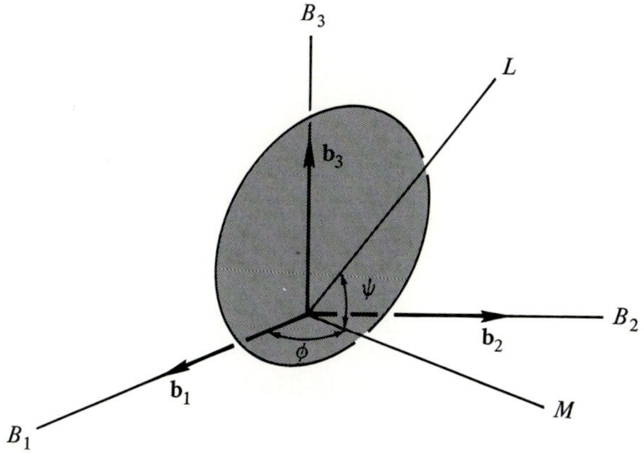

Figure P1.16

Letting ϕ and ψ be the angles shown, express $\dot{\phi}$ and $\dot{\psi}$ in terms of ϕ, ψ, and ω_i $(i = 1, 2, 3)$, where

$$\omega_i \triangleq \boldsymbol{\omega} \cdot \mathbf{b}_i \qquad (i = 1, 2, 3)$$

and $\boldsymbol{\omega}$ is the angular velocity of the satellite in a reference frame in which L remains fixed, while $\mathbf{b}_i$ is a unit vector parallel to line B_i.

Result
$$\dot{\phi} = (\omega_1 \cos\phi + \omega_2 \sin\phi)\tan\psi - \omega_3$$

$$\dot{\psi} = -\omega_1 \sin\phi + \omega_2 \cos\phi$$

1.17 The time-derivatives of a dyadic $\mathbf{D}$ in two reference frames A and B are defined as

$$\frac{{}^A d\mathbf{D}}{dt} \triangleq \mathbf{a}_i\mathbf{a}_j \frac{d}{dt}(\mathbf{a}_i \cdot \mathbf{D} \cdot \mathbf{a}_j)$$

and

$$\frac{{}^B d\mathbf{D}}{dt} \triangleq \mathbf{b}_i\mathbf{b}_j \frac{d}{dt}(\mathbf{b}_i \cdot \mathbf{D} \cdot \mathbf{b}_j)$$

where $\mathbf{a}_i$ and $\mathbf{b}_i$ $(i = 1, 2, 3)$ are sets of orthogonal unit vectors fixed, respectively, in A and B, and the summation convention is used for repeated subscripts. Show that these two derivatives are related to each other and to the angular velocity ${}^A\boldsymbol{\omega}^B$ of B in A as follows:

$$\frac{{}^A d\mathbf{D}}{dt} = \frac{{}^B d\mathbf{D}}{dt} + {}^A\boldsymbol{\omega}^B \times \mathbf{D} - \mathbf{D} \times {}^A\boldsymbol{\omega}^B$$

1.18 Solve Prob. 1.5 by using the given expressions for ${}^A d\mathbf{b}_1/dt$ and ${}^A d\mathbf{b}_2/dt$ together with Eqs. (1.11.1) and (1.11.4) to find ω_1, ω_2, ω_3 and then substituting into Eq. (1.13.8).

1.19 Defining ω_i $(i = 1, 2, 3)$ and ϵ as in Secs. 1.10 and 1.13, respectively, show that Eq. (1.13.8) is equivalent to

$$\dot{\epsilon} = \epsilon\Omega$$

where

$$\Omega = \frac{1}{2}\begin{bmatrix} 0 & -\omega_3 & \omega_2 & -\omega_1 \\ \omega_3 & 0 & -\omega_1 & -\omega_2 \\ -\omega_2 & \omega_1 & 0 & -\omega_3 \\ \omega_1 & \omega_2 & \omega_3 & 0 \end{bmatrix}$$

1.20 The angular velocity $\boldsymbol{\omega}$ of an axisymmetric rigid body B in an inertial reference frame A, when B is subjected to the action of a body-fixed, transverse torque of constant magnitude T, can be expressed as†

† Eugene Leimanis, *The General Problem of the Motion of Coupled Rigid Bodies About a Fixed Point*, Springer-Verlag, New York, 1965, p. 138.

$$\boldsymbol{\omega} = \omega_1 \mathbf{b}_1 + \omega_2 \mathbf{b}_2 + \omega_3 \mathbf{b}_3$$

and

$$\omega_1 = \overline{\omega}_1 \cos rt + (\overline{\omega}_2 + \mu) \sin rt$$

$$\omega_2 = -\mu - \overline{\omega}_1 \sin rt + (\overline{\omega}_2 + \mu) \cos rt$$

$$\omega_3 = \overline{\omega}_3$$

where $\mathbf{b}_1$, $\mathbf{b}_2$, $\mathbf{b}_3$ are mutually perpendicular unit vectors fixed in B and parallel to central principal axes of inertia of B; $\mathbf{b}_3$ is parallel to the symmetry axis of B; $\mathbf{b}_1$ is parallel to the applied torque; $\overline{\omega}_1$, $\overline{\omega}_2$, $\overline{\omega}_3$ are the initial values of ω_1, ω_2, ω_3; r and μ are defined as

$$r \triangleq \overline{\omega}_3 \left(1 - \frac{J}{I}\right) \qquad \mu \triangleq \frac{T}{\overline{\omega}_3(I - J)}$$

and I and J are, respectively, the transverse and the axial moment of inertia of B.

Letting ϕ be the angle between the symmetry axis of B and the line fixed in A with which the symmetry axis coincides initially, determine ϕ for $t = 1, 2, \ldots,$ 10 sec if $\overline{\omega}_1 = \overline{\omega}_2 = 1.0$ rad/sec, $\overline{\omega}_3 = 1.5$ rad/sec, $I = 60$ kg · m^2, $J = 40$ kg · m^2 and $T = 1.5$ N · m.

Result Table P1.20

Table P1.20

t (sec)	1	2	3	4	5	6	7	8	9	10
ϕ (deg)	77.5	108.0	46.7	37.3	105.2	82.9	3.6	76.1	108.3	51.2

1.21 Dextral sets of orthogonal unit vectors $\mathbf{a}_1$, $\mathbf{a}_2$, $\mathbf{a}_3$ and $\mathbf{b}_1$, $\mathbf{b}_2$, $\mathbf{b}_3$ are fixed in rigid bodies A and B, respectively, and two vectors, $\mathbf{p}$ and $\mathbf{q}$, are observed simultaneously from A and B in order to determine $\mathbf{a}_i \cdot \mathbf{p}$, $\mathbf{b}_i \cdot \mathbf{p}$, $\mathbf{a}_i \cdot \mathbf{q}$, $\mathbf{b}_i \cdot \mathbf{q}$, as well as $({}^A d\mathbf{p}/dt) \cdot \mathbf{a}_i$, $({}^B d\mathbf{p}/dt) \cdot \mathbf{b}_i$, $({}^A d\mathbf{q}/dt) \cdot \mathbf{a}_i$, $({}^B d\mathbf{q}/dt) \cdot \mathbf{b}_i$ $(i = 1, 2, 3)$.

Letting $\boldsymbol{\omega}$ be the angular velocity of B in A, refer to Tables P1.6 and P1.21 to determine $\boldsymbol{\omega} \cdot \mathbf{a}_i$ $(i = 1, 2, 3)$.

Table P1.21

	$({}^A d\mathbf{p}/dt) \cdot \mathbf{a}_i$	$({}^B d\mathbf{p}/dt) \cdot \mathbf{b}_i$	$({}^A d\mathbf{q}/dt) \cdot \mathbf{a}_i$	$({}^B d\mathbf{q}/dt) \cdot \mathbf{b}_i$
$i = 1$	0.166	1.522	8.142	4.622
$i = 2$	−5.838	4.025	5.485	−2.238
$i = 3$	1.231	−0.050	0.498	−1.981

Result −8.808, 3.827, 15.053 rad/sec

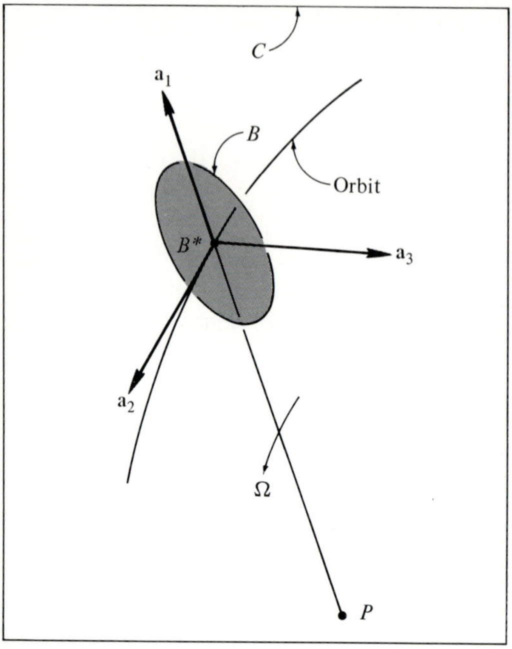

Figure P1.22

1.22 The mass center B^* of a rigid body B moves in a circular orbit fixed in a reference frame C and centered at a point P, as shown in Fig. P1.22, where $\mathbf{a}_1$, $\mathbf{a}_2$, $\mathbf{a}_3$ are mutually perpendicular unit vectors, $\mathbf{a}_1$ being parallel to line PB^*, $\mathbf{a}_2$ pointing in the direction of motion of B^* in C, and $\mathbf{a}_3$ thus being normal to the orbit plane. If θ_1, θ_2, θ_3 are space-three angles governing the orientation of B in a reference frame A in which $\mathbf{a}_1$, $\mathbf{a}_2$, $\mathbf{a}_3$ are fixed, the angular velocity of B in C can be expressed as $\boldsymbol{\omega} = \omega_1\mathbf{a}_1 + \omega_2\mathbf{a}_2 + \omega_3\mathbf{a}_3$; and $\dot{\theta}_i$ $(i = 1, 2, 3)$ can be expressed as a function f_i of θ_1, θ_2, θ_3, ω_1, ω_2, ω_3, and Ω, the angular speed of line PB^* in C.

Determine f_1, f_2, f_3, using the abbreviations $\mathrm{s}_i \triangleq \sin\theta_i$, $\mathrm{c}_i \triangleq \cos\theta_i$ $(i = 1, 2, 3)$.

Result

$$f_1 = (\mathrm{c}_3\omega_1 + \mathrm{s}_3\omega_2)/\mathrm{c}_2$$

$$f_2 = -\mathrm{s}_3\omega_1 + \mathrm{c}_3\omega_2$$

$$f_3 = (\mathrm{c}_3\omega_1 + \mathrm{s}_3\omega_2)(\mathrm{s}_2/\mathrm{c}_2) + \omega_3 - \Omega$$

1.23 Derive Eq. (1.17.5) by using Eq. (1.16.1), selecting auxiliary reference frames such that each term in Eq. (1.16.1) represents the angular velocity of a body performing a motion of simple rotation and can, therefore, be expressed as in Eq. (1.11.5). Noting that a similar derivation can be used to obtain Eq. (1.17.9), comment on the feasibility of using this approach to derive Eqs. (1.17.3) and (1.17.7).

1.24 The orientation of a rigid body B in a reference frame A is described in terms of the angles ϕ, ψ, θ of Prob. 1.12. Letting $\omega_i \triangleq \boldsymbol{\omega} \cdot \mathbf{b}_i$ $(i = 1, 2, 3)$, where $\boldsymbol{\omega}$

is the angular velocity of B in A, find the matrix L such that

$$[\omega_1 \quad \omega_2 \quad \omega_3] = [\dot{\phi} \quad \dot{\psi} \quad \dot{\theta}]L$$

Result

$$L = \begin{bmatrix} 0 & 1 & 0 \\ -\text{s}\phi & 0 & \text{c}\phi \\ \text{c}\phi\,\text{s}\psi & \text{c}\psi & \text{s}\phi\,\text{s}\psi \end{bmatrix}$$

1.25 Two rigid bodies, B and C, are connected to each other at one point. At a certain instant the velocity of this point and the angular velocities of the bodies in a reference frame A are given by

$$\mathbf{v} = V(\mathbf{a}_1 + \mathbf{a}_2)$$

$$\boldsymbol{\omega}^B = \Omega\mathbf{a}_1 \qquad \boldsymbol{\omega}^C = \Omega\mathbf{a}_2$$

where $\mathbf{a}_1$ and $\mathbf{a}_2$ are unit vectors which are perpendicular to each other. Determine the distance between the instantaneous axes of B and C in A at this instant.

Result $2V/\Omega$

1.26 A rigid body B performs slow, small rotational motions in a reference frame A while carrying a rotor C. The orientation of B in A is specified in terms of body-three or space-three angles θ_1, θ_2, θ_3 relating unit vectors $\mathbf{a}_1$, $\mathbf{a}_2$, $\mathbf{a}_3$ fixed in A to unit vectors $\mathbf{b}_1$, $\mathbf{b}_2$, $\mathbf{b}_3$ fixed in B, and the angular velocity of C in B is equal to $\Omega\mathbf{b}_1$, where Ω is time-dependent.

The angular acceleration of C in A can be expressed as

$$\boldsymbol{\alpha} = \alpha_1\mathbf{a}_1 + \alpha_2\mathbf{a}_2 + \alpha_3\mathbf{a}_3$$

Determine α_1, α_2, α_3.

Result

$$\alpha_1 = \ddot{\theta}_1 + \dot{\Omega}$$

$$\alpha_2 = \ddot{\theta}_2 + \dot{\Omega}\theta_3 + \Omega\dot{\theta}_3$$

$$\alpha_3 = \ddot{\theta}_3 - \dot{\Omega}\theta_2 - \Omega\dot{\theta}_2$$

1.27 In Fig. P1.27, A, B, and C designate the rotor, the inner gimbal, and the outer gimbal of a gyroscopic device carried by a spacecraft D; **a**, **b**, and **c** are unit vectors respectively parallel to the rotor axis, the inner gimbal axis, and the outer gimbal axis, with **a** perpendicular to **b**, and **b** perpendicular to **c**; and **d** is a unit vector fixed in D and perpendicular to **c**. By means of a motor attached to B, A is driven relative to B in such a way that Ω, defined as $\Omega \triangleq {}^B\boldsymbol{\omega}^A \cdot \mathbf{a}$, is a prescribed function of time t.

Denoting by q_1 and q_2 the radian measures of the angles between **a** and **c** and between **b** and **d**, respectively, letting q_1 and q_2 serve as generalized coordinates characterizing the configuration in D of the system formed by A, B, and C (q_1 and q_2 are positive for the configuration shown), and defining u_r as $u_r \triangleq \dot{q}_r$ ($r = 1, 2$), determine the partial angular velocities of A, B, and C in D.

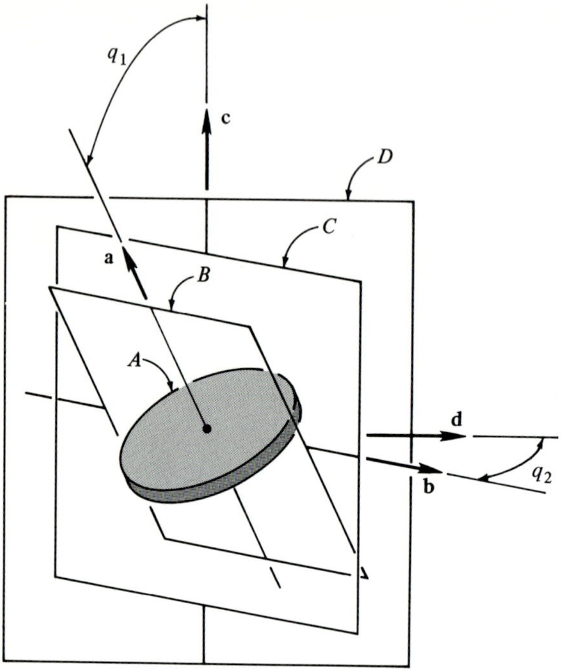

Figure P1.27

Result Table P1.27

Table P1.27

r	${}^D\boldsymbol{\omega}_r{}^A$	${}^D\boldsymbol{\omega}_r{}^B$	${}^D\boldsymbol{\omega}_r{}^C$
1	$\mathbf{b}$	$\mathbf{b}$	0
2	$-\mathbf{c}$	$-\mathbf{c}$	$-\mathbf{c}$

1.28 Figure P1.28 is a schematic representation of a spacecraft S modeled as a rigid body A that carries a planar linkage L_1–L_2–L_3–L_4 which, in turn, supports a particle P that can slide on L_2–L_3; $\mathbf{a}_1$, $\mathbf{a}_2$, $\mathbf{a}_3$ form a dextral set of mutually perpendicular unit vectors fixed in A, and the linkage lies in a plane perpendicular to $\mathbf{a}_3$.

With $u_1, \ldots, u_8$ defined as

$$u_i \triangleq \begin{cases} \boldsymbol{\omega} \cdot \mathbf{a}_i & (i = 1, 2, 3) \\ \mathbf{v}^{A^*} \cdot \mathbf{a}_{i-3} & (i = 4, 5, 6) \\ b\dot{\theta} & (i = 7) \\ b\dot{\theta} \sin \theta - \dot{r} & (i = 8) \end{cases}$$

where $\boldsymbol{\omega}$ is the angular velocity of A in N, $\mathbf{v}^{A^*}$ is the velocity of A^* in N, and b, θ, and r are shown in Fig. P1.28, determine the partial angular velocities of A in N, the partial velocities of A^* in N, and the partial velocities of P in N.

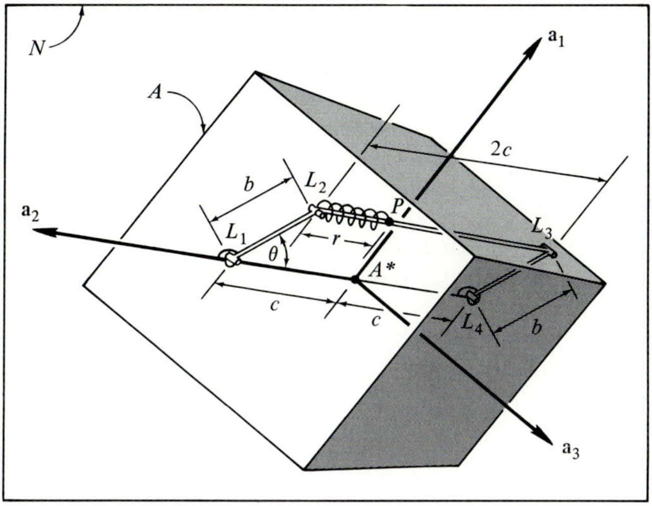

Figure P1.28

Result Table P1.28

Table P1.28

i	$\boldsymbol{\omega}_i$	$\mathbf{v}_i^{A^*}$	$\mathbf{v}_i^P$
1	$\mathbf{a}_1$	0	$(c - r - b\cos\theta)\mathbf{a}_3$
2	$\mathbf{a}_2$	0	$-b\sin\theta\,\mathbf{a}_3$
3	$\mathbf{a}_3$	0	$-(c - r - b\cos\theta)\mathbf{a}_1 + b\sin\theta\,\mathbf{a}_2$
4	0	$\mathbf{a}_1$	$\mathbf{a}_1$
5	0	$\mathbf{a}_2$	$\mathbf{a}_2$
6	0	$\mathbf{a}_3$	$\mathbf{a}_3$
7	0	0	$\cos\theta\,\mathbf{a}_1$
8	0	0	$\mathbf{a}_2$

1.29 The configuration of a system S in a reference frame A is characterized by n generalized coordinates $q_1, \ldots, q_n$. Taking $u_r = \dot{q}_r$ $(r = 1, \ldots, n)$, letting $\mathbf{v}_r$ be the rth partial velocity of a generic particle P of S, and letting $\boldsymbol{\omega}_r$ be the rth partial angular velocity of a rigid body B belonging to S, verify that

$$\frac{{}^A\partial \mathbf{v}_r}{\partial q_s} = \frac{{}^A\partial \mathbf{v}_s}{\partial q_r} \qquad (r, s = 1, \ldots, n)$$

$$\mathbf{v}_r = \frac{{}^A\partial \mathbf{p}}{\partial q_r} \qquad (r = 1, \ldots, n)$$

where $\mathbf{p}$ is the position vector of P relative to a point fixed in A; and show that, if $\mathbf{b}$ is a vector fixed in B, the partial derivatives of $\mathbf{b}$ with respect to q_r in A are given by

$$\frac{{}^A\partial \mathbf{b}}{\partial q_r} = \boldsymbol{\omega}_r \times \mathbf{b} \qquad (r = 1, \ldots, n)$$

1.30 Referring to Prob. 1.29, letting $\mathbf{v}$ and $\mathbf{a}$ denote the velocity of P in A and the acceleration of P in A, respectively, and regarding $\mathbf{v}$ as a function of $q_1, \ldots, q_n$, $\dot{q}_1, \ldots, \dot{q}_n$, and t in A, verify that

$$\mathbf{v}_r \cdot \mathbf{a} = \frac{1}{2}\left(\frac{d}{dt}\frac{\partial \mathbf{v}^2}{\partial \dot{q}_r} - \frac{\partial \mathbf{v}^2}{\partial q_r}\right) \qquad (r = 1, \ldots, n)$$

1.31 X and Y are points of a rigid body B moving in a reference frame A, and $\mathbf{v}_r^X$, $\mathbf{v}_r^Y$, and $\boldsymbol{\omega}_r$ are, respectively, the rth partial velocity of X in A, the rth partial velocity of Y in A, and the rth partial angular velocity of B in A. Show that

$$\mathbf{v}_r^Y = \mathbf{v}_r^X + \boldsymbol{\omega}_r \times \mathbf{z} \qquad (r = 1, \ldots, n)$$

where $\mathbf{z}$ is the position vector from X to Y.

PROBLEM SET 2

2.1 Letting P and $\bar{P}$ be particles of mass m and $\bar{m}$, respectively, and assuming that the law used in Sec. 2.1 to construct an expression for the force $\mathbf{F}$ experienced by P in the presence of $\bar{P}$ applies equally well to the force $\bar{\mathbf{F}}$ experienced by $\bar{P}$ in the presence of P, show that $\bar{\mathbf{F}}$ is given by Eq. (2.1.2).

2.2 The magnitude of the gravitational force exerted on a particle P of mass m by the Earth can be expressed as mg. Assuming that, for the purpose of calculating this force, the Earth may be replaced with a particle situated at the Earth's mass center, and taking the distance from this point to P to be 6.373×10^6 m, determine g.

Result 9.81 m/sec^2

2.3 Particles P_1, P_2, and Q of equal mass are placed as shown in Fig. P2.3. Letting

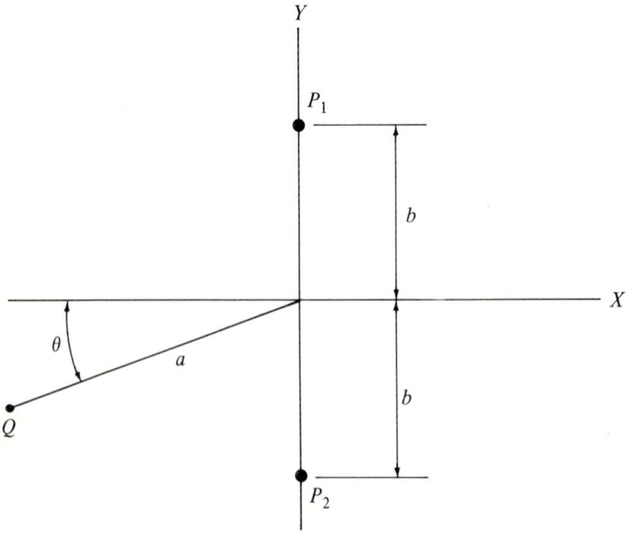

Figure P2.3

C be the center of gravity of P_1 and P_2 for Q, (*a*) determine the X and Y coordinates of C for $\theta = 45°$ and $a = 0.5b$, $a = b$, $a = 2b$; (*b*) letting d be the distance between Q and C, and taking $\theta = 0$, determine d/a when Q is placed so as to make d as small as possible.

Results (*a*) $(0.42b, -1.21b)$, $(0.34b, -0.94b)$, $(0.29b, -0.61b)$; (*b*) $3^{3/4}$

2.4 A particle $\overline{P}$ of mass $\overline{m}$ is placed in the gravitational field of an infinitely long, uniform rod R of mass ρ per unit of length. The distance from $\overline{P}$ to R is equal to r. Determine the magnitude of the gravitational force exerted by R on $\overline{P}$.

Result $2G\rho\overline{m}/r$

2.5 In Fig. P2.5, $\overline{P}$ designates a particle of mass $\overline{m}$, B is a uniform, thin rod of length L and mass m, $\mathbf{a}_1$ and $\mathbf{a}_2$ are unit vectors, and R denotes the distance from $\overline{P}$ to the midpoint of B.

(*a*) Verify that the gravitational force $\mathbf{F}$ exerted on B by $\overline{P}$ can be expressed (exactly) as

$$\mathbf{F} = -\frac{G\overline{m}m}{R^2}(\mathbf{a}_1 + \mathbf{f}_1 + \mathbf{f}_2)$$

where

$$\mathbf{f}_1 \triangleq \frac{L^2}{16R^2}(\mathbf{a}_1 - 2\mathbf{a}_2)$$

and that, for $\psi = \pi/4$ rad, Eq. (2.3.38) can be written

$$\widetilde{\mathbf{F}} = -\frac{G\overline{m}m}{R^2}(\mathbf{a}_1 + \mathbf{f}_1)$$

(*b*) To acquire a measure of the relative importance of $\mathbf{f}_1$ and $\mathbf{f}_2$, evaluate $|\mathbf{f}_2 \cdot \mathbf{a}_1/\mathbf{f}_1 \cdot \mathbf{a}_1|$ for $L/R = 0.1, 1, 3, 5, 10, 100, 1000$.

Results (*b*) 4.066×10^{-3}, 4.239×10^{-1}, 1.558, 1.377, 1.128, 1.002, 1.000

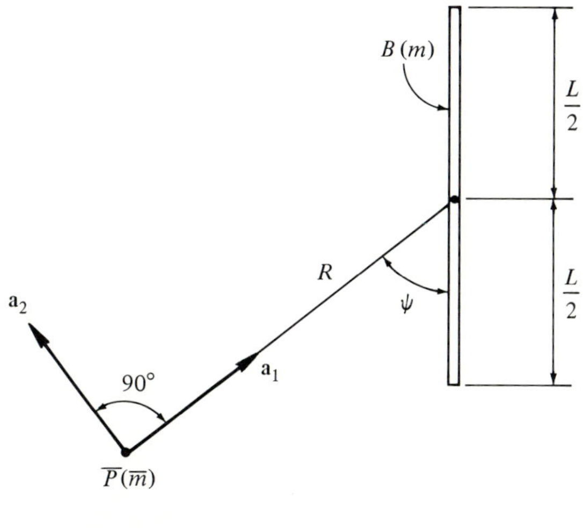

Figure P2.5

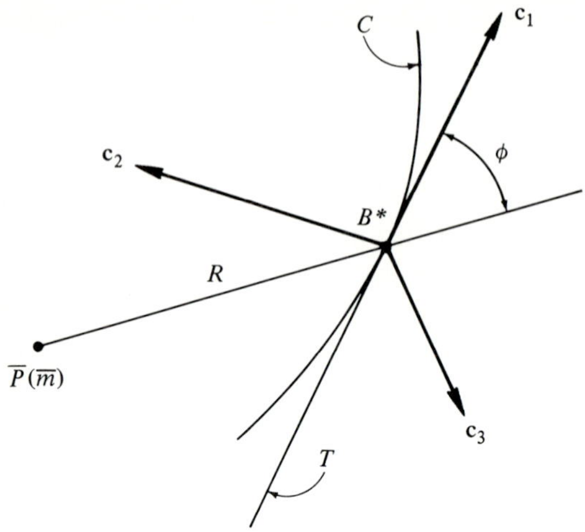

Figure P2.6(a)

2.6 A solid, uniform right-circular cylinder B of radius r, height h, and mass m moves in the gravitational field of a particle $\overline{P}$ of mass $\overline{m}$. At a certain instant, the distance between $\overline{P}$ and B^*, the mass center of B, is equal to R, and B^* is moving on a curve C, as indicated in Fig. P2.6(a), which shows also T, the tangent to C at B^*, and $\mathbf{c}_1$, $\mathbf{c}_2$, $\mathbf{c}_3$, a dextral set of mutually perpendicular unit vectors. The plane determined by lines $\overline{P}B^*$ and T is perpendicular to $\mathbf{c}_3$, and the orientation of the axis of B relative to $\mathbf{c}_1$, $\mathbf{c}_2$, $\mathbf{c}_3$ is shown in Fig. P2.6(b) where line L is perpendicular to $\mathbf{c}_2$.

The gravitational force $\mathbf{F}$ exerted on B by $\overline{P}$ can be expressed as

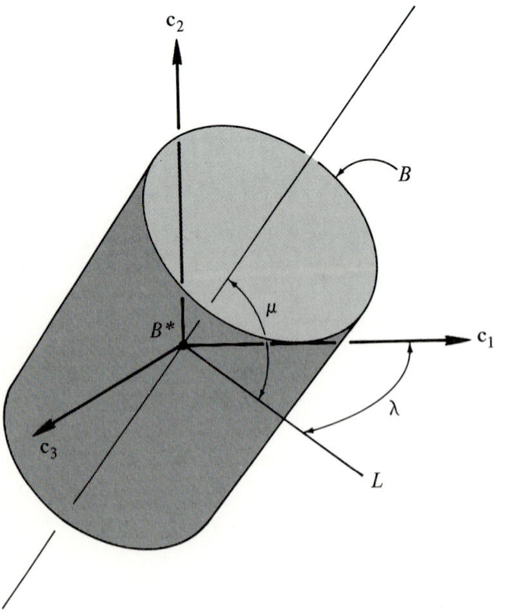

Figure P2.6(b)

$$\mathbf{F} = -\frac{G\overline{m}m}{R^2}(H_1\mathbf{c}_1 + H_2\mathbf{c}_2 + H_3\mathbf{c}_3)$$

Taking $\phi = 30°$, $\lambda = 40°$, $\mu = 50°$, and $r/R = h/R = 0.01$, determine H_1, H_2, H_3 approximately, (*a*) treating B as a particle and (*b*) taking into account the fact that B is an extended body.

Results (*a*) 0.866025, −0.500000, 0

(*b*) 0.866048, −0.500011, 0.896841×10^{-6}

2.7 At times it is convenient to evaluate integrals such as the one in Eq. (2.2.2) by means of numerical procedures (for example, Simpson's rule). Determine $F/(G\overline{m}M)$, where F is the magnitude of the gravitational force exerted on a particle $\overline{P}$ of mass $\overline{m}$ by a thin, circular ring of mass M, assuming that the ring has a radius of 1 m and that $\overline{P}$ is placed at the point (10 m, 10 m, 10 m) in a rectangular Cartesian coordinate system originating at the center of the ring and having the normal to the plane of the ring as one of the coordinate axes.

Result 3.33332×10^{-3} m^{-2}

2.8 Referring to Prob. 2.7, determine $F/(G\overline{m}M)$ approximately by using Eqs. (2.3.5) and (2.3.6).

Results 3.33333×10^{-3} m^{-2}, 3.33334×10^{-3} m^{-2}

2.9 A particle $\overline{P}$ is placed into the gravitational field produced by a particle P and a thin, circular ring R centered at P and having a radius r. Letting P and R have the same mass, determine the distance from P to a point Q such that the total gravitational force exerted on $\overline{P}$ by P and R vanishes when $\overline{P}$ is placed at Q.

Result $0.83r$

2.10 Figure P2.10 shows two identical, uniform, thin, circular disks, D and $\overline{D}$, each of mass m and radius r. Letting R be the distance between C and $\overline{C}$, the centers of the disks, and dropping terms of degree three or higher in r/R, determine the magnitude of the gravitational force exerted on D by $\overline{D}$.

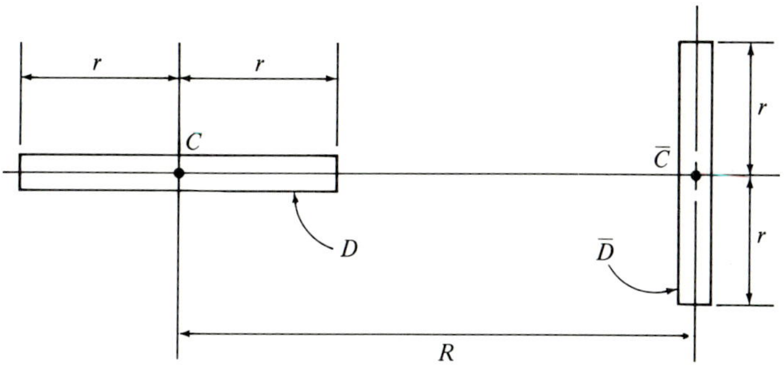

Figure P2.10

Result $$\frac{Gm^2}{R^2}\left[1 - \frac{3}{8}\left(\frac{r}{R}\right)^2\right]$$

2.11 One way to find the gravitational force **F** exerted on a body B by a particle $\bar{P}$ is to divide the figure (curve, surface, or solid) occupied by B into n elements, place within each element a particle having the same mass as the element, and calculate the gravitational force **F**′ exerted by $\bar{P}$ on the set of particles thus formed. **F**′ then furnishes an approximation to **F**, and this approximation improves with increasing n. For example, suppose B is a uniform rod of length L and mass m, and $\bar{P}$ is situated relative to B as shown in Fig. P2.11(a). If P_i, the ith of n particles, is placed at the midpoint of the ith of n elements, each of length L/n, as indicated in Fig. P2.11(a), and if **F** and **F**′ are expressed as $\mathbf{F} = F_1\mathbf{b}_1 + F_2\mathbf{b}_2$ and $\mathbf{F}' = F_1'\mathbf{b}_1 + F_2'\mathbf{b}_2$, respectively, then it may be verified with the aid of Eq. (2.2.7) that, for instance,

$$\frac{F_2'}{F_2} = \frac{100}{n}(101)^{1/2}\sum_{i=1}^{n}\left[100 - \left(\frac{i-0.5}{n}\right)^2\right]^{-3/2}$$

so that, for $n = 5, 10, 20$, F_2'/F_2 has the values 1.000049, 1.000012, 1.000003, respectively.

Formulate an approximate expression for $|\mathbf{F}|$, the magnitude of the gravitational force exerted by one of the two identical, uniform rods depicted in Fig. P2.11(b) on the other rod, using (*a*) the method just described (divide each rod into n elements of equal length), and (*b*) Eq. (2.4.14). Evaluate the ratio of the results of (*a*) and (*b*) for $n = 5$ and $n = 10$, and use these values to determine whether Eq. (2.4.14) furnishes an upper or a lower bound for F_2.

Results 1.00011, 1.00004, lower

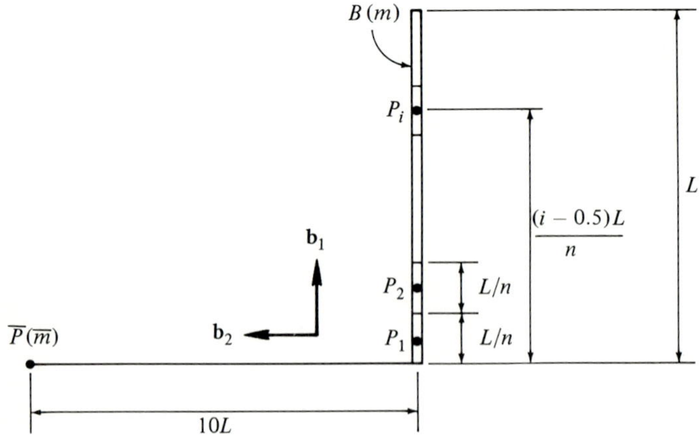

Figure P2.11(a)

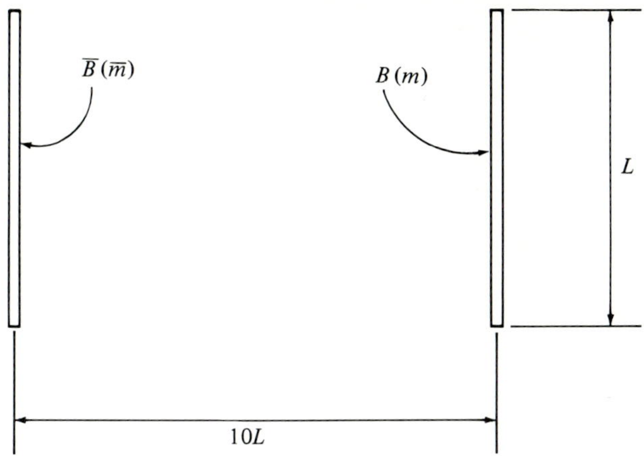

Figure P2.11(b)

2.12 Particles of masses M and M' are placed on the axis of a thin, uniform hemispherical shell, as shown in Fig. P2.12. Determine the value of M/M' for which the gravitational forces exerted on the shell by the particles are equal in magnitude.

Result 0.25

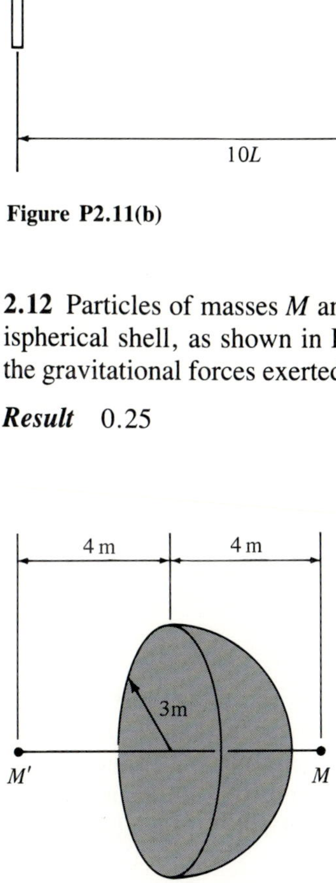

Figure P2.12

2.13 Noting that eight identical particles situated at the corners of a cube form a body that has a spherical central inertia ellipsoid, show that this body is not centrobaric.

2.14 Figure P2.14 shows a body B consisting of particles P_1, P_2, P_3 having masses of 2 kg, 1 kg, and 1 kg, respectively. The system of gravitational forces exerted

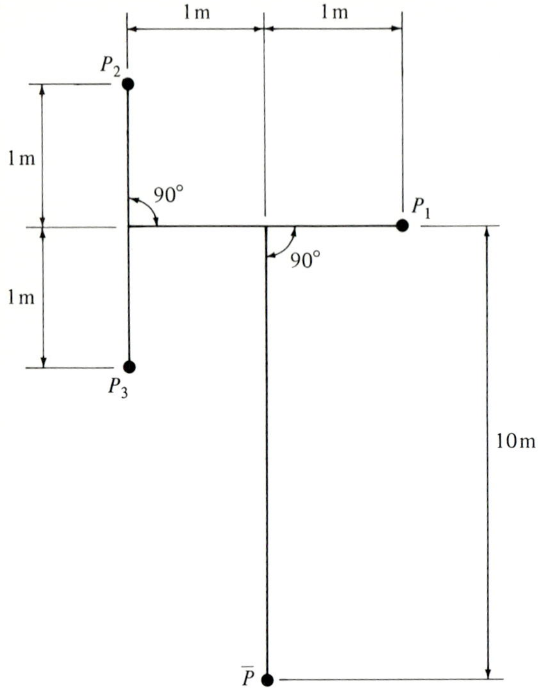

Figure P2.14

on B by a particle $\overline{P}$ that has a mass of 10 kg and is placed as indicated produces a moment about the mass center of B. Determine (*a*) the magnitude of this moment, and (*b*) the magnitude of the vector $\widetilde{\mathbf{M}}$ defined in Eq. (2.6.3).

Results (*a*) 7.9×10^{-13} N · m; (*b*) 0

2.15 A uniform, rectangular plate P having the dimensions shown in Fig. P2.15 is placed in the gravitational field of two uniform spheres, S_1 and S_2, of masses 3×10^{23} kg and 10^{24} kg, respectively. Determine the "zero-moment" orientation of the plate; that is, find θ $(0 \le \theta \le 180°)$ such that the total moment about the mass center of the plate of all gravitational forces exerted on the plate by S_1 and S_2 is equal to zero.

Result 20.44°

2.16 In Fig. P2.16, A represents a satellite modeled as a dumbbell consisting of two particles connected by a light, rigid rod of length $2a$. B is a damper boom modeled similarly and hinged to A at point O, the two particles forming B each having a mass β and being separated by a distance $2b$. E designates the Earth, regarded as a homogeneous sphere of mass ϵ.

To compare the moments about O of the gravitational forces exerted on A by B and by E, let z be the ratio of the magnitudes of these moments and determine the maximum value of z (by varying θ) for (*a*) $a = 11$ m, $b = 10$ m, and (*b*) $a =$

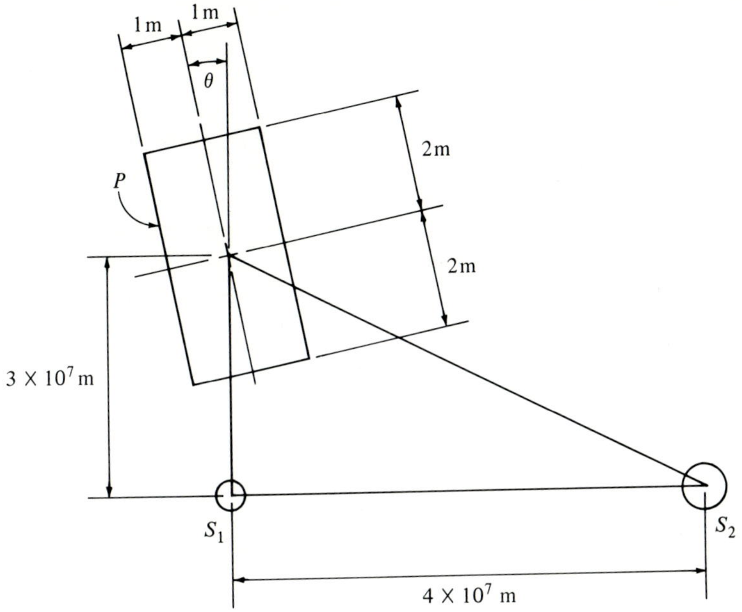

Figure P2.15

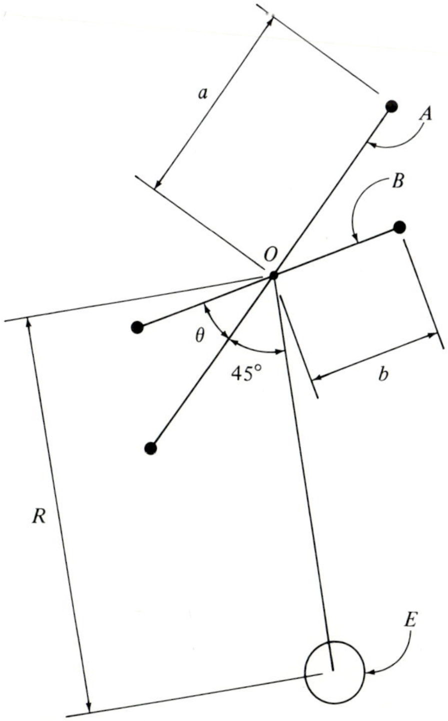

Figure P2.16

0.11 m, $b = 0.10$ m, taking $\beta = 10$ kg, $\epsilon = 6 \times 10^{24}$ kg, and $R = 7 \times 10^6$ m in both cases.

Results (*a*) 13×10^{-6}; (*b*) 13

2.17 A body B is placed in the gravitational field of an infinitely long, uniform rod R of mass ρ per unit of length. The largest dimension of B is small in comparison with the distance s from R to the mass center B^* of B.

Letting $\mathbf{M}$ be the moment about B^* of all gravitational forces exerted on B by R, express $\mathbf{M}$ in terms of ρ, s, $\mathbf{n}_1$, $\mathbf{n}_2$, $\mathbf{I}$, and G, where $\mathbf{n}_1$ is a unit vector parallel to R, $\mathbf{n}_2$ is a unit vector perpendicular to R and parallel to the plane determined by R and B^*, $\mathbf{I}$ is the central inertia dyadic of B, and G is the universal gravitational constant.

Result $2G\rho(\mathbf{n}_1 \times \mathbf{I} \cdot \mathbf{n}_1 + 2\mathbf{n}_2 \times \mathbf{I} \cdot \mathbf{n}_2)/s^2$

2.18 F and $\mathbf{G}$ are, respectively, a scalar function of a vector $\mathbf{v}$ and a vector function of $\mathbf{v}$. Show that

$$\nabla_{\mathbf{v}} F^n = nF^{n-1}\nabla_{\mathbf{v}} F$$

$$\nabla_{\mathbf{v}} (\mathbf{G}^2)^n = 2n(\mathbf{G}^2)^{n-1}\mathbf{G} \cdot \nabla_{\mathbf{v}} \mathbf{G}$$

$$\nabla_{\mathbf{v}} (F\mathbf{G}) = (\nabla_{\mathbf{v}} F)\mathbf{G} + F\nabla_{\mathbf{v}} \mathbf{G}$$

2.19 In Fig. P2.19, r, θ, and z are cylindrical coordinates of a point P and $\mathbf{b}_1$, $\mathbf{b}_2$, $\mathbf{b}_3$ are unit vectors pointing, respectively, in the directions in which P moves when r, θ, z are made to vary, one at a time. Letting $\mathbf{p}(r, \theta, z)$ be the position vector of P relative to O, $F(\mathbf{p})$ a function of $\mathbf{p}$, and $G(r, \theta, z) = F[\mathbf{p}(r, \theta, z)]$, express $\nabla^2 F$ in terms of partial derivatives of G.

Result
$$\frac{\partial^2 G}{\partial r^2} + \frac{1}{r}\left(\frac{\partial G}{\partial r} + \frac{1}{r}\frac{\partial^2 G}{\partial \theta^2}\right) + \frac{\partial^2 G}{\partial z^2}$$

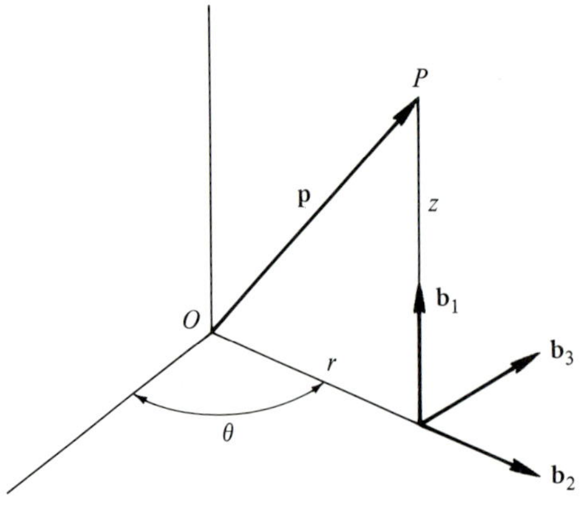

Figure P2.19

2.20 If $\mathbf{a}_1$, $\mathbf{a}_2$, $\mathbf{a}_3$ designate orthogonal unit vectors fixed in a reference frame A, O is a point fixed in A, and $\mathbf{p}$ is the position vector from O to a point P, then $\mathbf{p}$ can be expressed as $\mathbf{p} = p_1\mathbf{a}_1 + p_2\mathbf{a}_2 + p_3\mathbf{a}_3$; and, if p_r $(r = 1, 2, 3)$ is regarded as a function of three scalar quantities s_1, s_2, s_3, then these quantities are called *curvilinear coordinates* of P in A. For example, the quantities r, λ, β in the example in Sec. 2.9 are curvilinear coordinates.

The partial derivatives of $\mathbf{p}$ with respect to s_r $(r = 1, 2, 3)$ in A always can be expressed as

$$\frac{\partial \mathbf{p}}{\partial s_r} = h_r \mathbf{c}_r \qquad (r = 1, 2, 3)$$

where h_r is a function of s_1, s_2, s_3 and $\mathbf{c}_r$ is a unit vector. If $|\mathbf{c}_1 \cdot \mathbf{c}_2 \times \mathbf{c}_3| = 1$ for all s_1, s_2, s_3, then s_1, s_2, s_3 are called *orthogonal* curvilinear coordinates.

(*a*) Letting F denote a function of $\mathbf{p}$, and defining a function G of s_1, s_2, s_3 as $G(s_1, s_2, s_3) = F[\mathbf{p}(s_1, s_2, s_3)]$, show that, if s_1, s_2, s_3 are orthogonal curvilinear coordinates, then

$$\nabla_{\mathbf{p}} F = \frac{1}{h_1}\frac{\partial G}{\partial s_1}\mathbf{c}_1 + \frac{1}{h_2}\frac{\partial G}{\partial s_2}\mathbf{c}_2 + \frac{1}{h_3}\frac{\partial G}{\partial s_3}\mathbf{c}_3$$

(*b*) Verify that r, λ, β are orthogonal curvilinear coordinates and that the associated expressions for h_1, h_2, h_3 are $h_1 = \pm 1$, $h_2 = \pm rc\beta$, $h_3 = \pm r$.

(*c*) Show that, if s_1, s_2, s_3 are orthogonal curvilinear coordinates, then

$$\nabla_{\mathbf{p}}{}^2 F = \frac{1}{h_1 h_2 h_3}\left[\frac{\partial}{\partial s_1}\left(\frac{h_2 h_3}{h_1}\frac{\partial G}{\partial s_1}\right) + \frac{\partial}{\partial s_2}\left(\frac{h_3 h_1}{h_2}\frac{\partial G}{\partial s_2}\right) + \frac{\partial}{\partial s_3}\left(\frac{h_1 h_2}{h_3}\frac{\partial G}{\partial s_3}\right)\right]$$

Suggestion: Make use of the fact that, for example,

$$\frac{\partial^2 \mathbf{p}}{\partial s_1\,\partial s_2} = \frac{\partial^2 \mathbf{p}}{\partial s_2\,\partial s_1}$$

which implies

$$\frac{\partial h_1}{\partial s_2}\mathbf{c}_1 + h_1\frac{\partial \mathbf{c}_1}{\partial s_2} = \frac{\partial h_2}{\partial s_1}\mathbf{c}_2 + h_2\frac{\partial \mathbf{c}_2}{\partial s_1}$$

from which it follows that

$$h_1\mathbf{c}_2 \cdot \frac{\partial \mathbf{c}_1}{\partial s_2} = \frac{\partial \mathbf{c}_2}{\partial s_1}$$

2.21 A particle P of mass m and two particles P_1 and P_2, each of mass $\underline{m}$, are situated as shown in Fig. P2.21, where r, θ, z are cylindrical coordinates of P, and $\mathbf{c}_1$, $\mathbf{c}_2$, $\mathbf{c}_3$ are unit vectors pointing in the directions in which P moves when r, θ, z are made to increase, one at a time. Use a force function for the gravitational force $\mathbf{F}$ exerted on P by P_1 and P_2 to express $\mathbf{F}$ in terms of components parallel to $\mathbf{c}_1$, $\mathbf{c}_2$, $\mathbf{c}_3$. Compare this method for constructing $\mathbf{F}$ with the use of Eq. (2.2.1).

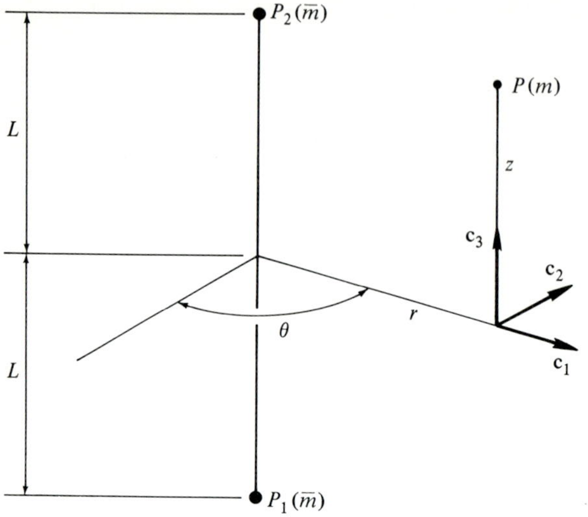

Figure P2.21

Result $-G\bar{m}m(\{r[r^2 + (z + L)^2]^{-3/2} + r[r^2 + (z - L)^2]^{-3/2}\}\mathbf{c}_1 + \{(z + L)[r^2 + (z + L)^2]^{-3/2} + (z - L)[r^2 + (z - L)^2]^{-3/2}\}\mathbf{c}_3)$

2.22 A particle $\bar{P}$ of mass $\bar{m}$ is situated on the axis of a thin circular disk D of radius r and mass m. Letting R be the distance from the center of D to $\bar{P}$, construct a force function for the gravitational force exerted on D by $\bar{P}$.

Result $2G\bar{m}M\{[1 + (R/r)^2]^{1/2} - R/r\}r^{-1}$

2.23 Use the result of Prob. 2.22 to find a force function for the gravitational force exerted on a uniform sphere S of mass m by a particle $\bar{P}$ of mass $\bar{m}$, assuming that $\bar{P}$ and the center of S are separated by a distance p that exceeds the radius of S.

Result $G\bar{m}m/p$

2.24 Show that any constant is a force function for the force exerted on a uniform spherical shell by a particle situated inside the shell.

2.25 Figure P2.25 shows a rectangular parallelepiped B of mass m and a particle $\bar{P}$ of mass $\bar{m}$ located a distance R from the mass center B^* of B. The lines X_1, X_2, X_3 are parallel to central principal axes of inertia of B, and $\bar{P}$ has coordinates x_1, x_2, x_3 with respect to these axes. The sides of B have lengths $2a$, $2b$, $2c$, as shown.

(*a*) Assuming that R exceeds $(a^2 + b^2 + c^2)^{1/2}$, construct an approximation $\tilde{V}$ to a force function for the gravitational force exerted on B by $\bar{P}$.

(*b*) It is known that, so long as $\bar{P}$ lies outside of B, a force function V for the gravitational force exerted on B by $\bar{P}$ can be expressed exactly as†

$$V = \frac{G\bar{m}m}{8abc}\left\{(x_1 + a)(x_2 + b)\ln\frac{r_1 + r_2 + 2c}{r_1 + r_2 - 2c}\right.$$

(*continued on next page*)

† W. D. Macmillan, *The Theory of the Potential*, Dover, New York, 1958, pp. 72–79.

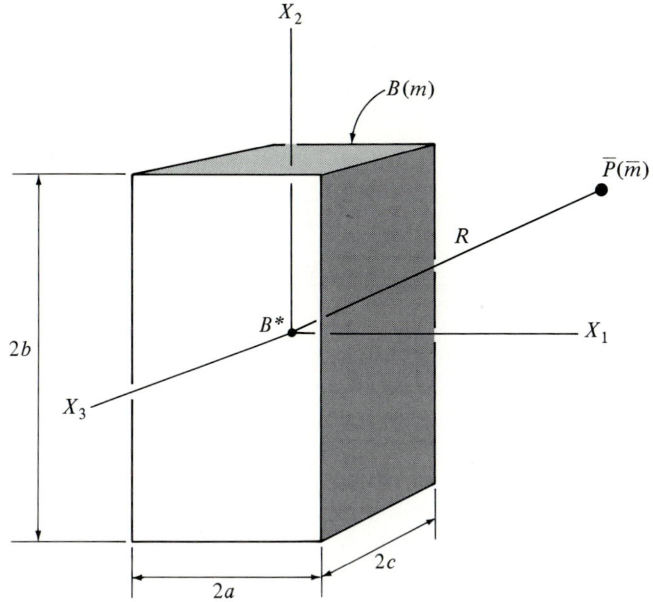

Figure P2.25

$$- (x_1 + a)(x_2 - b) \ln \frac{r_3 + r_4 + 2c}{r_3 + r_4 - 2c} + (x_1 + a)(x_3 + c) \ln \frac{r_1 + r_3 + 2b}{r_1 + r_3 - 2b}$$

$$- (x_1 + a)(x_3 - c) \ln \frac{r_2 + r_4 + 2b}{r_2 + r_4 - 2b} + (x_1 - a)(x_2 - b) \ln \frac{r_7 + r_8 + 2c}{r_7 + r_8 - 2c}$$

$$- (x_1 - a)(x_2 + b) \ln \frac{r_5 + r_6 + 2c}{r_5 + r_6 - 2c} + (x_1 - a)(x_3 - c) \ln \frac{r_6 + r_8 + 2b}{r_6 + r_8 - 2b}$$

$$- (x_1 - a)(x_3 + c) \ln \frac{r_5 + r_7 + 2b}{r_5 + r_7 - 2b} + (x_2 + b)(x_3 + c) \ln \frac{r_1 + r_5 + 2a}{r_1 + r_5 - 2a}$$

$$- (x_2 + b)(x_3 - c) \ln \frac{r_2 + r_6 + 2a}{r_2 + r_6 - 2a} + (x_2 - b)(x_3 - c) \ln \frac{r_4 + r_8 + 2a}{r_4 + r_8 - 2a}$$

$$- (x_2 - b)(x_3 + c) \ln \frac{r_3 + r_7 + 2a}{r_3 + r_7 - 2a}$$

$$+ \tfrac{1}{2}(x_1 + a)^2 \left[\tan^{-1} \frac{(x_2 + b)(x_3 - c)}{(x_1 + a)r_2} - \tan^{-1} \frac{(x_2 + b)(x_3 + c)}{(x_1 + a)r_1} \right.$$

$$\left. + \tan^{-1} \frac{(x_2 - b)(x_3 + c)}{(x_1 + a)r_3} - \tan^{-1} \frac{(x_2 - b)(x_3 - c)}{(x_1 + a)r_4} \right]$$

(continued on next page)

$$+ \tfrac{1}{2}(x_1 - a)^2 \left[\tan^{-1} \frac{(x_2 + b)(x_3 + c)}{(x_1 - a)r_5} - \tan^{-1} \frac{(x_2 + b)(x_3 - c)}{(x_1 - a)r_6} \right.$$
$$\left. + \tan^{-1} \frac{(x_2 - b)(x_3 - c)}{(x_1 - a)r_8} - \tan^{-1} \frac{(x_2 - b)(x_3 + c)}{(x_1 - a)r_7} \right]$$
$$+ \tfrac{1}{2}(x_2 + b)^2 \left[\tan^{-1} \frac{(x_1 - a)(x_3 + c)}{(x_2 + b)r_5} - \tan^{-1} \frac{(x_1 + a)(x_3 + c)}{(x_2 + b)r_1} \right.$$
$$\left. + \tan^{-1} \frac{(x_1 + a)(x_3 - c)}{(x_2 + b)r_2} - \tan^{-1} \frac{(x_1 - a)(x_3 - c)}{(x_2 + b)r_6} \right]$$
$$+ \tfrac{1}{2}(x_2 - b)^2 \left[\tan^{-1} \frac{(x_1 + a)(x_3 + c)}{(x_2 - b)r_3} - \tan^{-1} \frac{(x_1 - a)(x_3 + c)}{(x_2 - b)r_7} \right.$$
$$\left. + \tan^{-1} \frac{(x_1 - a)(x_3 - c)}{(x_2 - b)r_8} - \tan^{-1} \frac{(x_1 + a)(x_3 - c)}{(x_2 - b)r_4} \right]$$
$$+ \tfrac{1}{2}(x_3 + c)^2 \left[\tan^{-1} \frac{(x_1 + a)(x_2 - b)}{(x_3 + c)r_3} - \tan^{-1} \frac{(x_1 + a)(x_2 + b)}{(x_3 + c)r_1} \right.$$
$$\left. + \tan^{-1} \frac{(x_1 - a)(x_2 + b)}{(x_3 + c)r_5} - \tan^{-1} \frac{(x_1 - a)(x_2 - b)}{(x_3 + c)r_7} \right]$$
$$+ \tfrac{1}{2}(x_3 - c)^2 \left[\tan^{-1} \frac{(x_1 + a)(x_2 + b)}{(x_3 - c)r_2} - \tan^{-1} \frac{(x_1 + a)(x_2 - b)}{(x_3 - c)r_4} \right.$$
$$\left. \left. + \tan^{-1} \frac{(x_1 - a)(x_2 - b)}{(x_3 - c)r_8} - \tan^{-1} \frac{(x_1 - a)(x_2 + b)}{(x_3 - c)r_6} \right] \right\}$$

where

$$r_1 \triangleq [(x_1 + a)^2 + (x_2 + b)^2 + (x_3 + c)^2]^{1/2}$$
$$r_2 \triangleq [(x_1 + a)^2 + (x_2 + b)^2 + (x_3 - c)^2]^{1/2}$$
$$r_3 \triangleq [(x_1 + a)^2 + (x_2 - b)^2 + (x_3 + c)^2]^{1/2}$$
$$r_4 \triangleq [(x_1 + a)^2 + (x_2 - b)^2 + (x_3 - c)^2]^{1/2}$$
$$r_5 \triangleq [(x_1 - a)^2 + (x_2 + b)^2 + (x_3 + c)^2]^{1/2}$$
$$r_6 \triangleq [(x_1 - a)^2 + (x_2 + b)^2 + (x_3 - c)^2]^{1/2}$$
$$r_7 \triangleq [(x_1 - a)^2 + (x_2 - b)^2 + (x_3 + c)^2]^{1/2}$$
$$r_8 \triangleq [(x_1 - a)^2 + (x_2 - b)^2 + (x_3 - c)^2]^{1/2}$$

Taking $a = 1$ m, $b = 2$ m, $c = 9$ m, $x_1 = 1.01$ m, $x_2 = 3.01$ m, and $x_3 = 1$ m, 9.01 m, 100 m, determine $\widetilde{V}/V$ and comment briefly on the results.

(*c*) For $x_1 = x_2 = x_3 = 10^7$ m, evaluate $\widetilde{V}/(G\overline{m}m/R)$ and $V/(G\overline{m}m/R)$, then comment briefly on the utility of the "closed form" expression for the force function.

Results (*a*) $$\widetilde{V} = \frac{G\bar{m}m}{R}\left\{1 + \frac{1}{6R^2}\left[(b^2 + c^2)\left(1 - \frac{3x_1^{\,2}}{R^2}\right) + (c^2 + a^2)\left(1 - \frac{3x_2^{\,2}}{R^2}\right) + (a^2 + b^2)\left(1 - \frac{3x_3^{\,2}}{R^2}\right)\right]\right\}$$

(*b*) 0.43, 0.91, 0.999990

2.26 Figure P2.26 depicts a satellite S consisting of an axisymmetric body B of mass m that carries an axisymmetric rotor B' of mass m'. B^* is the common mass center of B and B', and B_1, B_2, B_3 are mutually perpendicular axes fixed in B. B_2 is fixed also in B', thus serving as the axis of rotation of B' relative to B. The moments of inertia of B and B' about B_1, B_2, B_3 have the values I, I, J and I', J', I', respectively.

S moves in the gravitational field of a particle $\bar{P}$ of mass $\bar{m}$, the distance R between $\bar{P}$ and B^* being large in comparison with the largest dimension of B. To describe the orientation of B_1, B_2, B_3 relative to mutually perpendicular axes A_1,

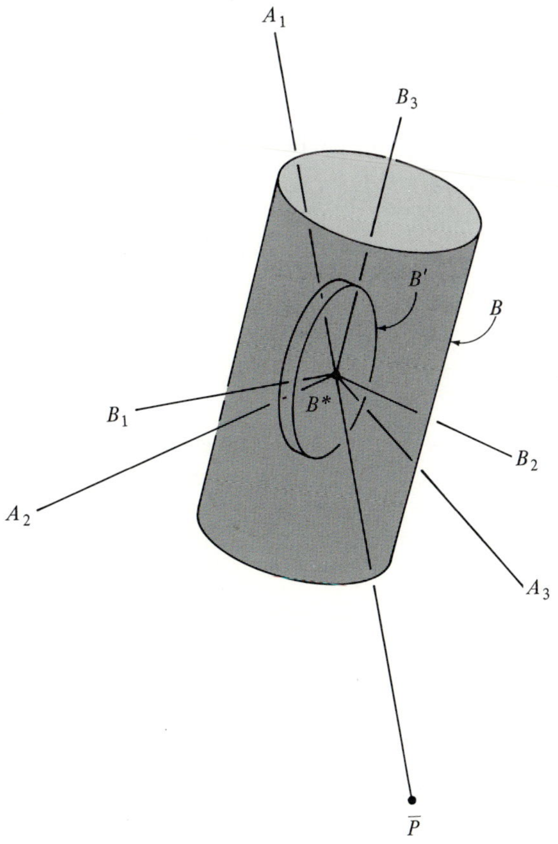

Figure P2.26

A_2, A_3, with A_1 the extension of line $\overline{P}B^*$, body-three orientation angles θ_1, θ_2, θ_3 (see Sec. 1.7) are employed.

Verify that a force function V for the gravitational force exerted on S by $\overline{P}$ can be expressed approximately as

$$V \approx \frac{G\overline{m}}{R}\left[m + m' + \frac{\alpha(J - I) + \beta(J' - I')}{2R^2}\right]$$

where α and β are functions of θ_1, θ_2, θ_3. Evaluate α and β for $\theta_1 = 10°$, $\theta_2 = 30°$, $\theta_3 = 60°$.

Result $\alpha = \frac{1}{4}$, $\beta = -\frac{11}{16}$

2.27 A circular ring of radius r is centered at the origin of a set of rectangular Cartesian coordinate axes X_1, X_2, X_3 and lies in the X_1–X_2 plane. A particle $\overline{P}$ has coordinates $\overline{x}_1$, $\overline{x}_2$, $\overline{x}_3$ relative to these axes. Compute the quantity $v^{(2)}$ by using (*a*) Eq. (2.13.1) and (*b*) Eq. (2.12.2).

Result
$$\frac{r^2(\overline{x}_1{}^2 + \overline{x}_2{}^2 - 2\overline{x}_3{}^2)}{4(\overline{x}_1{}^2 + \overline{x}_2{}^2 + \overline{x}_3{}^2)^2}$$

2.28 For $i = 2$, the quantities defined in Eqs. (2.13.8)–(2.13.10) can be expressed in terms of I'_{jk} defined as $I'_{jk} \triangleq \mathbf{b}'_j \cdot \mathbf{I} \cdot \mathbf{b}'_k$ $(j, k = 1, 2, 3)$, where $\mathbf{I}$ is the central inertia dyadic of B and $\mathbf{b}'_1$, $\mathbf{b}'_2$, $\mathbf{b}'_3$ are mutually perpendicular unit vectors directed as shown in Fig. 2.12.1.

(*a*) Refer to Eqs. (2.12.7) and (2.13.11) to verify that

$$C_{20} = \frac{I'_{11} + I'_{22} - 2I'_{33}}{2mR_B{}^2}$$

$$C_{21} = \frac{-I'_{13}}{mR_B{}^2} \qquad C_{22} = -\frac{I'_{11} - I'_{22}}{4mR_B{}^2}$$

$$S_{21} = \frac{-I'_{13}}{mR_B{}^2} \qquad S_{22} = \frac{-I'_{12}}{2mR_B{}^2}$$

(*b*) Measurements of gravitational forces exerted on objects at the Earth's surface and on artificial satellites show that, if R_B is taken to be the Earth's mean equatorial radius, then $C_{20} \approx -1.0827 \times 10^{-3}$, $C_{2j} \approx 0$, $S_{2j} \approx 0$ $(j = 1, 2)$. Assuming that the Earth is a homogeneous spheroid having a polar radius R_C, show that $R_C/R_B \approx 0.9973$.

2.29 Referring to Prob. 2.27, determine $v^{(2)}$ by using Eq. (2.13.11).

Result
$$\frac{r^2(\overline{x}_1{}^2 + \overline{x}_2{}^2 - 2\overline{x}_3{}^2)}{4(\overline{x}_1{}^2 + \overline{x}_2{}^2 + \overline{x}_3{}^2)^2}$$

2.30 Figure P2.30 shows a particle $\overline{P}$ and a rigid body B of mass density ρ. Mutually perpendicular axes B_1, B_2, B_3 are fixed in B and originate at the mass center B^* of B; $\overline{P}$ has coordinates $\overline{x}_1$, $\overline{x}_2$, $\overline{x}_3$ relative to B_1, B_2, B_3; and a point P of B has coordinates x_1, x_2, x_3.

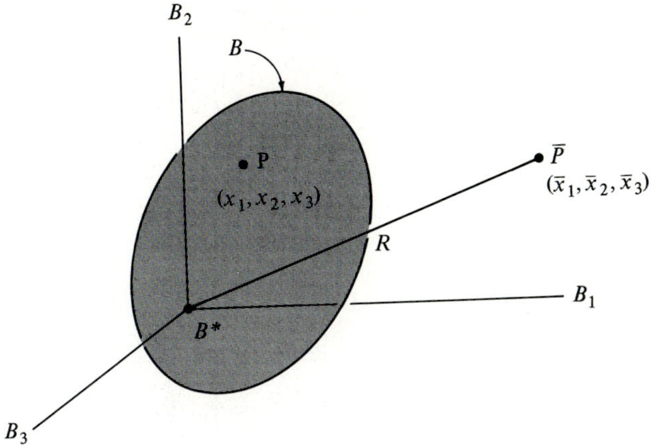

Figure P2.30

(*a*) If $R = (\bar{x}_1^2 + \bar{x}_2^2 + \bar{x}_3^2)^{1/2}$, m is the mass of B, and $d\tau$ is the volume of a differential element of B, verify with the aid of Eqs. (2.13.1) and (2.13.3) that $v^{(4)}$ of Eq. (2.12.1) can be expressed as

$$
\begin{aligned}
v^{(4)} = \frac{1}{8mR^4}\Bigg[& \frac{35}{R^4}(\bar{x}_1^4 I_{400} + \bar{x}_2^4 I_{040} + \bar{x}_3^4 I_{004} + 4\bar{x}_1^3\bar{x}_2 I_{310} + 4\bar{x}_1^3\bar{x}_3 I_{301} \\
& + 4\bar{x}_2^3\bar{x}_1 I_{130} + 4\bar{x}_2^3\bar{x}_3 I_{031} + 4\bar{x}_3^3\bar{x}_1 I_{103} + 4\bar{x}_3^3\bar{x}_2 I_{013} + 6\bar{x}_1^2\bar{x}_2^2 I_{220} \\
& + 6\bar{x}_2^2\bar{x}_3^2 I_{022} + 6\bar{x}_3^2\bar{x}_1^2 I_{202} + 12\bar{x}_1\bar{x}_2\bar{x}_3^2 I_{112} + 12\bar{x}_2\bar{x}_3\bar{x}_1^2 I_{211} \\
& + 12\bar{x}_3\bar{x}_1\bar{x}_2^2 I_{121}) - \frac{30}{R^2}(\bar{x}_1^2 I_{400} + \bar{x}_2^2 I_{040} + \bar{x}_3^2 I_{004} + \bar{x}_1^2 I_{220} + \bar{x}_1^2 I_{202} \\
& + \bar{x}_2^2 I_{022} + \bar{x}_2^2 I_{220} + \bar{x}_3^2 I_{202} + \bar{x}_3^2 I_{022} + 2\bar{x}_1\bar{x}_2 I_{310} + 2\bar{x}_1\bar{x}_2 I_{130} \\
& + 2\bar{x}_2\bar{x}_3 I_{031} + 2\bar{x}_2\bar{x}_3 I_{013} + 2\bar{x}_3\bar{x}_1 I_{103} + 2\bar{x}_3\bar{x}_1 I_{301} + 2\bar{x}_1\bar{x}_2 I_{112} \\
& + 2\bar{x}_2\bar{x}_3 I_{211} + 2\bar{x}_3\bar{x}_1 I_{121}) + 3(I_{400} + I_{040} + I_{004} + 2I_{220} + 2I_{022} + 2I_{202})\Bigg]
\end{aligned}
$$

where

$$I_{ijk} \triangleq \rho \int_B x_1^i x_2^j x_3^k \, d\tau \qquad (i, j, k = 0, 1, 2, 3, 4)$$

(*b*) Letting B be a rectangular parallelepiped having sides of lengths $2a$, $2b$, $2c$ parallel to B_1, B_2, B_3, respectively, verify that

$$I_{ijk} = \frac{ma^i b^j c^k [1 - (-1)^{i+1}][1 - (-1)^{j+1}][1 - (-1)^{k+1}]}{8(i+1)(j+1)(k+1)}$$
$$(i, j, k = 0, 1, 2, 3, 4)$$

2.31 A particle P of mass m is situated at the point x_1, x_2, x_3 of a rectangular Cartesian coordinate system whose axes X_1, X_2, X_3 originate at the center of a circular ring of radius r and mass $\bar{m}$. Assuming that the ring lies in the X_1–X_2 plane and that

$x_1^2 + x_2^2 + x_3^2 < r^2$, construct a force function for the gravitational force exerted on P by the ring. Express the result in terms of ascending powers of x_i/r ($i = 1, 2, 3$), displaying only terms of degree less than four in these quantities.

Result
$$\frac{Gm\overline{m}}{4r}\left[\left(\frac{x_1}{r}\right)^2 + \left(\frac{x_2}{r}\right)^2 - 2\left(\frac{x_3}{r}\right)^2\right]$$

2.32 In Fig. P2.32, S and $\overline{S}$ designate a set of n particles $P_1, \ldots, P_n$ of masses $m_1, \ldots, m_n$ and a set of $\overline{n}$ particles $\overline{P}_1, \ldots, \overline{P}_{\overline{n}}$ of masses $\overline{m}_1, \ldots, \overline{m}_{\overline{n}}$, respectively; p_{ij} denotes the distance from P_i to $\overline{P}_j$; $\mathbf{R}$ is the position vector from $\overline{Q}$ to Q, these being two arbitrary points; and $\mathbf{r}_i$ and $\overline{\mathbf{r}}_j$ are the position vectors from Q to P_i and from $\overline{Q}$ to $\overline{P}_j$, respectively.

Letting

$$V \triangleq G \sum_{i=1}^{n} \sum_{j=1}^{\overline{n}} \frac{m_i \overline{m}_j}{p_{ij}}$$

and considering changes in the positions of $P_1, \ldots, P_n, \overline{P}_1, \ldots, \overline{P}_{\overline{n}}$, Q, and $\overline{Q}$ such that $\mathbf{r}_1, \ldots, \mathbf{r}_n$ and $\overline{\mathbf{r}}_1, \ldots, \overline{\mathbf{r}}_{\overline{n}}$ remain unaltered, show that the gravitational force $\mathbf{F}$ exerted by $\overline{S}$ on S is given by

$$\mathbf{F} = \nabla_{\mathbf{R}} V$$

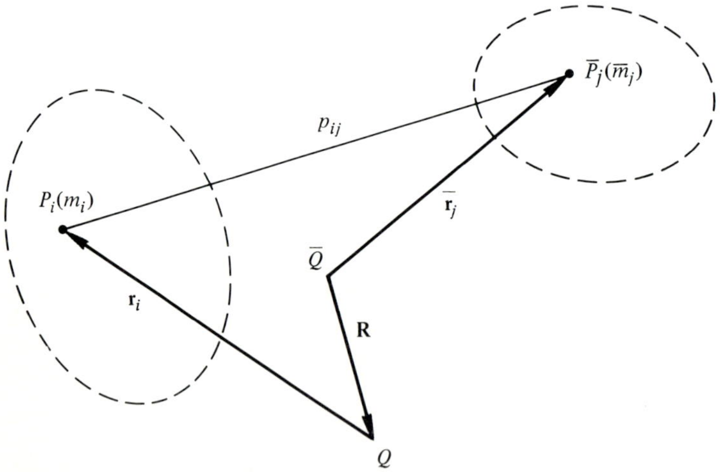

Figure P2.32

2.33 The distance between the mass centers B^* and $\overline{B}^*$ (see Fig. P2.33) of two bodies B and $\overline{B}$ exceeds the greatest distance from B^* to any point of B, and a force function $V(\mathbf{p})$ for the force exerted by B on a particle of unit mass at a point P of B is known in all detail. Moreover, $V(\mathbf{R})$ is available in the form of an explicit function V^* of the cylindrical coordinates r, θ, z shown in Fig. P2.33. Under these circumstances, there exists a function $\widetilde{V}(\mathbf{R})$, such that the resultant of all gravitational forces exerted on $\overline{B}$ by B can be approximated by $\widetilde{\mathbf{F}} \triangleq \nabla_{\mathbf{R}} \widetilde{V}(\mathbf{R})$, with $\widetilde{V}(\mathbf{R})$ ex-

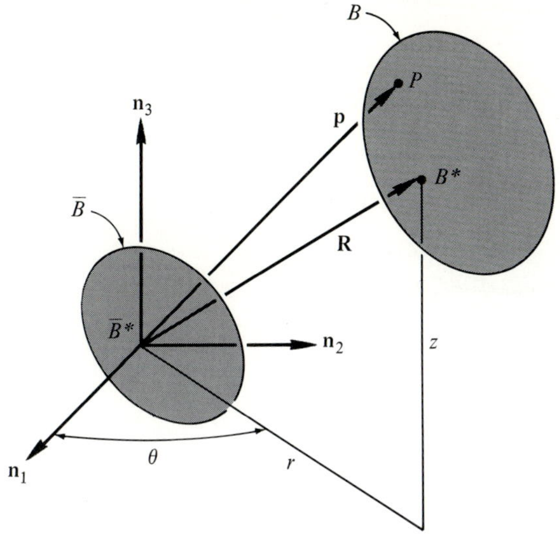

Figure P2.33

pressed as $\widetilde{V}(\mathbf{R}) = mV^* - \frac{1}{2}(I_{11}Q_{11} + I_{22}Q_{22} + I_{33}Q_{33}) - (I_{12}Q_{12} + I_{23}Q_{23} + I_{31}Q_{31})$, where $I_{ij} \overset{\Delta}{=} \mathbf{n}_i \cdot \mathbf{I} \cdot \mathbf{n}_j$ $(i, j = 1, 2, 3)$, $\mathbf{I}$ is the central inertia dyadic of B, and $\mathbf{n}_1$, $\mathbf{n}_2$, $\mathbf{n}_3$ are mutually perpendicular unit vectors (see Fig. P2.33). Determine $Q_{11}, Q_{22}, Q_{33}, Q_{12}, Q_{23}, Q_{31}$.

Results $\quad Q_{11} = \dfrac{\partial^2 V^*}{\partial r^2} \qquad Q_{22} = \dfrac{1}{r^2}\dfrac{\partial^2 V^*}{\partial \theta^2} - \dfrac{1}{r}\dfrac{\partial V^*}{\partial r} \qquad Q_{33} = \dfrac{\partial^2 V^*}{\partial z^2}$

$$Q_{12} = -\frac{1}{r^2}\frac{\partial V^*}{\partial \theta} + \frac{1}{r}\frac{\partial^2 V^*}{\partial r\,\partial \theta} \qquad Q_{23} = \frac{1}{r}\frac{\partial^2 V^*}{\partial \theta\,\partial z} \qquad Q_{31} = \frac{\partial^2 V^*}{\partial r\,\partial z}$$

2.34 Using Eq. (2.10.2) to form a force function V for two particles, determine $\nabla\nabla V$ with $m = 1$ and show that this result, together with Eq. (2.16.2) leads to Eq. (2.12.13) with $v^{(2)}$ given by Eq. (2.12.2).

Result $\quad -G\bar{m}(p^{-3}\mathbf{U} - 3p^{-5}\mathbf{pp})$

2.35 Figure P2.35 shows a uniform rod $\bar{B}$ of mass $\bar{M}$ and a uniform prolate spheroid B of mass M.

(*a*) Restricting attention to situations in which a and b are small in comparison with $(x_1^2 + x_2^2)^{1/2}$, but L is unrestricted, construct a force function for the gravitational interaction of B and $\bar{B}$.

(*b*) Referring to Eq. (2.14.3) to construct a force function that applies when a, b, and L all are small in comparison with $(x_1^2 + x_2^2)^{1/2}$, and setting the arbitrary constant in both this force function and the one found in (*a*) equal to zero, determine the ratio of the two functions for $L = 10{,}000$ m, $a = 20$ m, $b = 10$ m, and $x_1 = x_2 = 10^3$ m, 5×10^3 m, 25×10^3 m.

Results $\quad (a) \quad \dfrac{GM}{L}\left[M \ln \dfrac{A + L - 2x_1}{B - L - 2x_1} + \dfrac{1}{2}(J_1 Q_1 + J_2 Q_2)\right] + C$

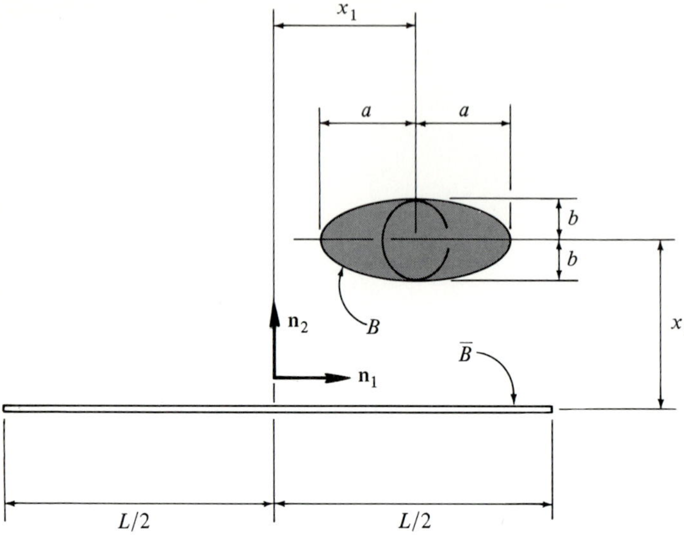

Figure P2.35

where

$$A = [(2x_1 - L)^2 + 4x_2^2]^{1/2} \qquad B = [(2x_1 + L)^2 + 4x_2^2]^{1/2}$$

$$J_1 = \frac{2Mb^2}{5} \qquad J_2 = \frac{M(a^2 + b^2)}{5}$$

$$Q_1 = 4(A + L - 2x_1)^{-2}[(2x_1 - L)A^{-1} - 1] + 4(A + L - 2x_1)^{-1}[A^{-1} - (2x_1 - L)^2 A^{-3}] - 4(B - L - 2x_1)^{-2}[(2x_1 + L)B^{-1} - 1] - 4(B - L - 2x_1)^{-1}[B^{-1} - (2x_1 + L)^2 B^{-3}]$$

$$Q_2 = 8(A + L - 2x_1)^{-2} x_2 A^{-1} + 4(A + L - 2x_1)^{-1}(A^{-1} - 4x_2^2 A^{-3}) - 8(B - L - 2x_1)^{-2} x_2 B^{-1} - 4(B - L - 2x_1)^{-1}(B^{-1} - 4x_2^2 B^{-3})$$

(*b*) 0.3, 0.98, 0.99997

2.36 The mass center B^* of a body B of mass m is situated at the point x_1^*, x_2^*, x_3^* of a rectangular Cartesian coordinate system whose axes X_1, X_2, X_3 originate at the center O of a circular ring of radius r and mass $\bar{m}$. Assuming that the ring lies in the X_1–X_2 plane and that the distance from O to any point of B is smaller than r, determine (*a*) the gravitational force exerted on B by the ring and (*b*) the moment about B^* of the gravitational forces exerted on B by the ring. Express the results in terms of unit vectors $\mathbf{n}_1$, $\mathbf{n}_2$, $\mathbf{n}_3$ pointing in the directions of the positive X_1, X_2, X_3 axes, respectively, displaying only terms of degree less than three in x_1^*/r, x_2^*/r, x_3^*/r.

Results (*a*) $(Gm\bar{m}/2r^3)(x_1^*\,\mathbf{n}_1 + x_2^*\,\mathbf{n}_2 - 2x_3^*\,\mathbf{n}_3)$

(*b*) $(3G\bar{m}/2r^3)(I_{23}\,\mathbf{n}_1 - I_{31}\,\mathbf{n}_2)$

where $I_{23} \triangleq \mathbf{n}_2 \cdot \mathbf{I} \cdot \mathbf{n}_3$ and $I_{31} \triangleq \mathbf{n}_3 \cdot \mathbf{I} \cdot \mathbf{n}_1$, with $\mathbf{I}$ the inertia dyadic of B for B^*.

2.37 In Fig. P2.37, B is a rigid body of mass m, $\overline{P}$ is a particle of mass $\overline{m}$, $\mathbf{a}_1$, $\mathbf{a}_2$, $\mathbf{a}_3$ form a dextral set of orthogonal unit vectors such that $\mathbf{R}$, the position vector from $\overline{P}$ to B^*, can be written $\mathbf{R} = R\mathbf{a}_1$, and $\mathbf{b}_1$, $\mathbf{b}_2$, $\mathbf{b}_3$ form a dextral set of orthogonal unit vectors fixed in B.

If the orientation of B in a reference frame in which $\mathbf{a}_1$, $\mathbf{a}_2$, $\mathbf{a}_3$ are fixed is described in terms of body-three orientation angles θ_1, θ_2, θ_3, then the existence of the force function $V(\mathbf{R})$ for the gravitational force exerted on B by $\overline{P}$ implies the existence of a function W of θ_1, θ_2, θ_3, and R such that the total moment $\mathbf{M}$ about B^* of all gravitational forces exerted on B by $\overline{P}$ can be expressed as

$$\mathbf{M} = \left(C\frac{\partial W}{\partial \theta_2} + D\frac{\partial W}{\partial \theta_3}\right)\mathbf{a}_2 + \left(E\frac{\partial W}{\partial \theta_2} + F\frac{\partial W}{\partial \theta_3}\right)\mathbf{a}_3$$

where C, D, E, F are functions of θ_1 and θ_2. Determine these functions.

Results c_1, $-s_1/c_2$, s_1, c_1/c_2

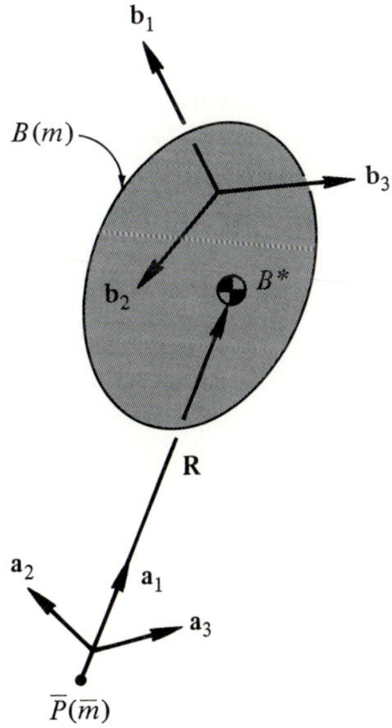

Figure P2.37

2.38 Referring to Prob. 2.35, and again restricting attention to situations in which a and b are small in comparison with $(x_1^2 + x_2^2)^{1/2}$, but L is unrestricted, construct an approximate expression for the magnitude of the moment about B^* of the gravitational forces exerted on B by $\overline{B}$.

Result
$$8G\overline{M}^{-1}\frac{M}{5}(a^2 - b^2)\Big|(A + L - 2x_1)^{-2}x_2A^{-1}[(L - 2x_1)A^{-1} + 1]$$
$$+ (A + L - 2x_1)^{-1}(L - 2x_1)x_2A^{-3} + (B - L - 2x_1)^{-2}x_2B^{-1}[(L$$
$$+ 2x_1)B^{-1} - 1] + (B - L - 2x_1)^{-1}(L + 2x_1)x_2B^{-3}\Big|$$

where

$$A \triangleq [(2x_1 - L)^2 + 4x_2^2]^{1/2} \qquad B \triangleq [(2x_1 + L)^2 + 4x_2^2]^{1/2}$$

PROBLEM SET 3

3.1 A uniform rectangular block having the dimensions shown in Fig. P3.1 performs a torque-free motion. Initially, $\boldsymbol{\omega}$, the angular velocity of the block, has a magnitude of 10 rad/sec and is directed as indicated. Letting L be an inertially fixed line that is parallel to the edge AB initially, determine the angle between L and AB for $t = 1, 5, 10$ sec.

Results 75.7°, 84.4°, 58.6°

3.2 Letting K denote the rotational kinetic energy of an axisymmetric rigid body, express K in terms of H, I, J, and ϕ, defined as in Sec. 3.1.

Result

$$K = \frac{H^2}{2I}\left[1 + \left(\frac{I}{J} - 1\right)\cos^2\phi\right]$$

3.3 An Earth satellite formed by a uniform circular disk D and a uniform rod R having the dimensions shown in Fig. P3.3 is intended to move in a circular orbit

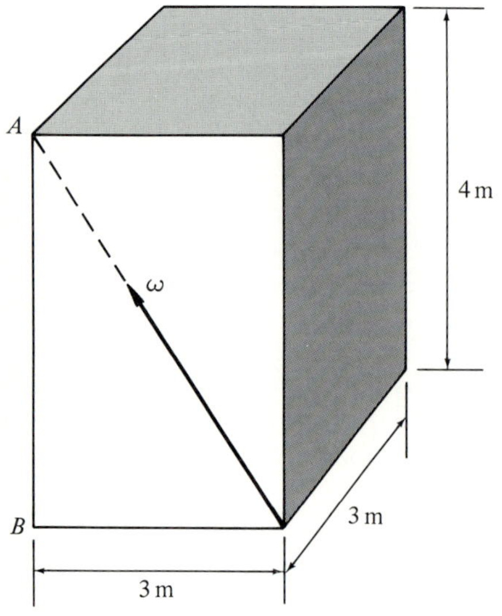

Figure P3.1

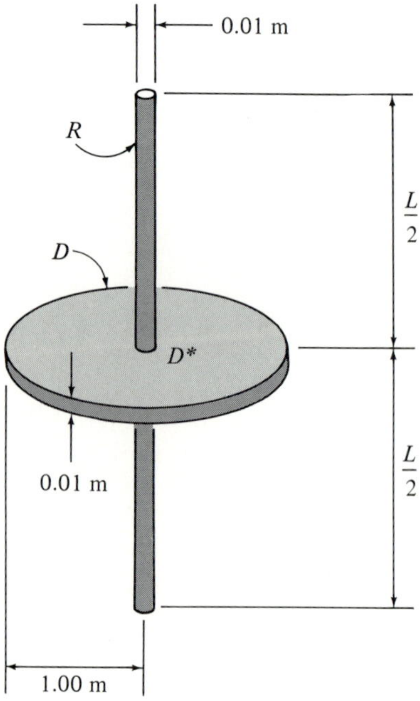

Figure P3.3

in such a way that R remains normal to the orbit plane at all times. D and R are made of the same material, and the satellite is to be spin-stabilized; that is, D is to perform a uniform rotational motion, completing Z rotations in a Newtonian reference frame N during each orbital revolution of the center of D. These rotations are to proceed in such a way that ${}^N\boldsymbol{\omega}^D \times {}^N\mathbf{v}^{D^*}$ points from D^* toward the center of the orbit.

(*a*) For $Z = 2$, determine the smallest value of L^* such that the motion is unstable if $L > L^*$.

(*b*) For $L = 20$ m, determine the largest value of Z^* such that the motion is unstable if $0 < Z < Z^*$.

Results (*a*) $L^* = 12.2$ m; (*b*) $Z^* = 12.3$

3.4 Referring to Prob. 3.3 and letting ψ be the angle between R and a unit vector $\mathbf{k}$ that is normal to the orbit plane, consider motions such that, at $t = 0$, ψ has a value $\hat{\psi}$, R lies in the plane determined by the orbit normal and the line joining the orbit center to D^*, the center of D, and the inertial angular velocity of D is parallel to R, as shown in Fig. P3.4(a), where the velocity of D^* is presumed to have the same direction as $\mathbf{k} \times {}^N\boldsymbol{\omega}^D$. Finally, let $\hat{Z}$ be the ratio of the initial magnitude of ${}^N\boldsymbol{\omega}^D$ to the orbital angular speed.

Taking $L = 20$ m, make two plots of ψ versus the number of orbits traversed by D^*, one plot applicable to $\hat{Z} = 11$, the other to $\hat{Z} = 14$, and each showing two curves, one corresponding to $\hat{\psi} = 0.1$ rad, the other to $\hat{\psi} = 0.05$ rad. Comment briefly on the results.

Results Figures P3.4(b), P3.4(c)

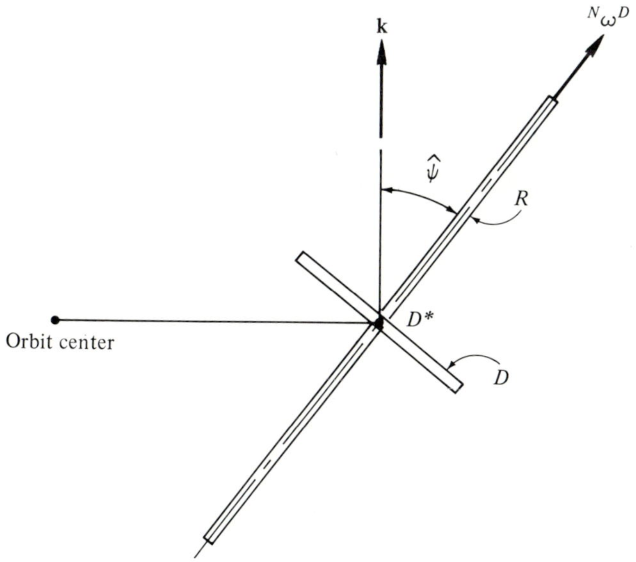

Figure P3.4(a)

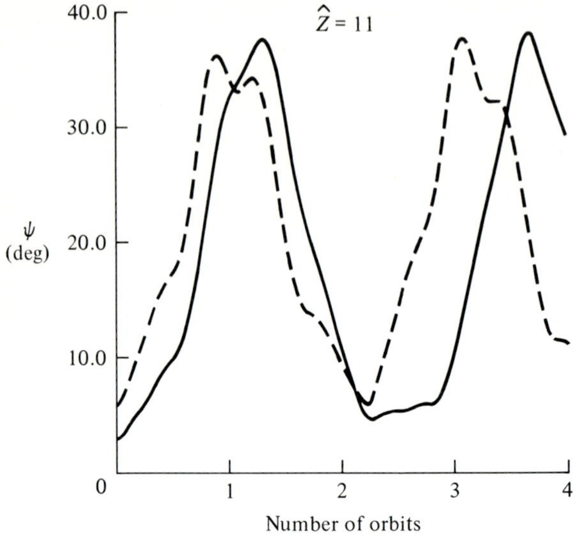

Figure P3.4(b)

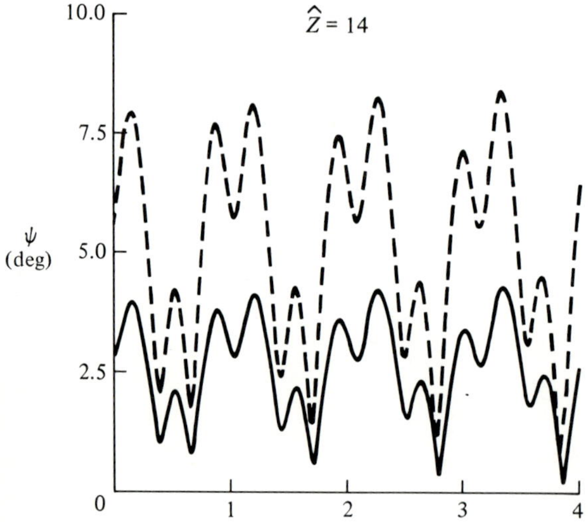

Figure P3.4(c)

3.5 To explore the utility of Eq. (3.2.22), compare the results of using this equation with those of numerical solutions of the equations of motion governing the behavior of the satellite considered in Prob. 3.3. Take $L = 20$ m; let ϕ be the angle between the plane P determined by R and the orbit normal, on the one hand, and the inertially fixed plane with which P coincides initially, on the other hand; and plot ϕ versus the number of orbits traversed by D^*, using the same initial conditions as in Prob. 3.4, with $\hat{\psi} = 30°$ and $\hat{Z} = 20, 40, 80$. Display results for two orbits.

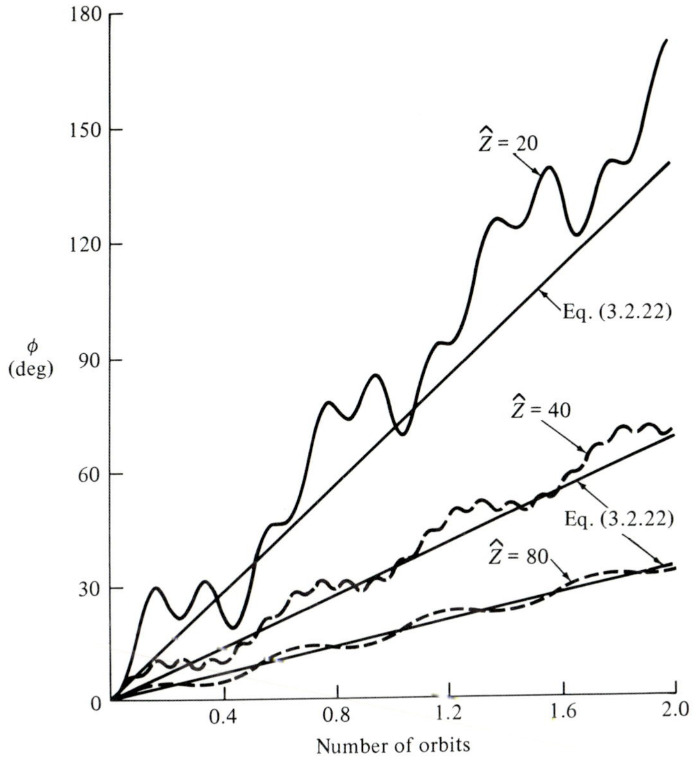

Figure P3.5

Also, determine in each case the maximum value of $|\psi - \hat{\psi}|$. Comment briefly on the results.

Results Figure P3.5; 11°, 4°, 2°

3.6 The satellite described in Prob. 3.3 is to be placed into an elliptic orbit having an eccentricity of 0.5. Once again R is to remain normal to the orbit plane at all times, and D is to perform a uniform rotational motion, completing Z rotations in an inertial reference frame during each orbital revolution of D^*.

(*a*) For $Z = 2$, determine the smallest value of L^* such that the motion is unstable if $L^* < L < 8.0$ m.

(*b*) Taking $L = 20$ m, and considering only motions satisfying the initial conditions stated in Prob. 3.4 (with D^* at the periapsis at $t = 0$), determine the smallest positive integer value of Z^* such that, if $Z > Z^*$, then ψ does not exceed 10° during the first five orbital revolutions of D^*.

Results (*a*) $L^* = 6.7$ m; (*b*) $Z^* = 32$

3.7 For $e = 0.3$, construct an instability chart such as those in Fig. 3.3.3.

Result Figure P3.7

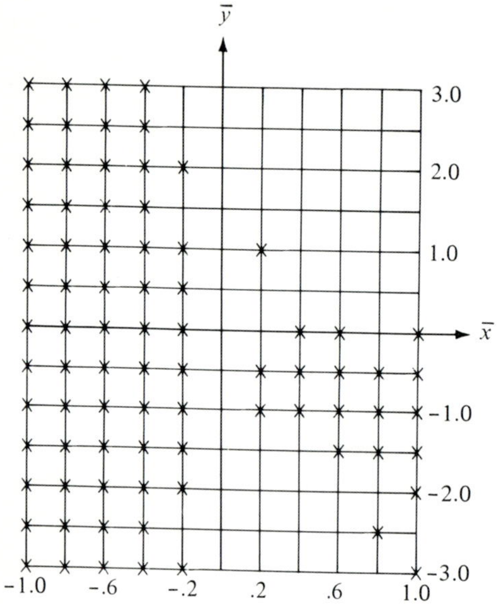

Figure P3.7

3.8 At time $t = 0$, the inertial angular velocity of a rectangular plate in torque-free motion has a magnitude Ω and is directed as shown in Fig. P3.8. Determine (*a*) the

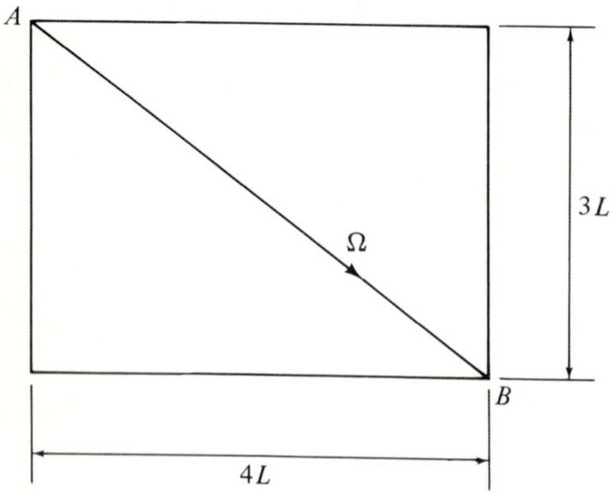

Figure P3.8

maximum magnitude of the angular velocity subsequent to $t = 0$ and (*b*) the time at which the maximum angular velocity is first acquired.

Results (*a*) 1.049Ω; (*b*) $3.3\Omega^{-1}$

3.9 Referring to Prob. 3.8, let $\mathbf{n}$ be a unit vector normal to the plate and suppose that at $t = 0$ the angular velocity $\boldsymbol{\omega}$ of the plate has a magnitude Ω, is perpendicular to $\mathbf{n}$, and makes a small angle with the side of length $3L$. Determine approximately the largest value acquired by $\boldsymbol{\omega} \cdot \mathbf{n}$ for $t > 0$ and relate the result to considerations of stability of simple rotational motions of the plate.

Result 0.53Ω

3.10 If I_1, I_2, I_3 are the central principal moments of inertia of a rigid body B, and K_1 and K_2 are defined as

$$K_1 \triangleq \frac{I_2 - I_3}{I_1} \qquad K_2 \triangleq \frac{I_3 - I_1}{I_2}$$

then every real body is represented by a point having coordinates $x = K_1$, $y = K_2$ in a rectangular Cartesian coordinate system; and all such points lie within a square bounded by the lines $x = -1$, $y = -1$, $x = 1$, $y = 1$. Draw this square and shade those of its portions corresponding to unstable simple rotational motions if B is torque-free and the inertial angular velocity of B is parallel to the 3-axis.

Result Figure P3.10

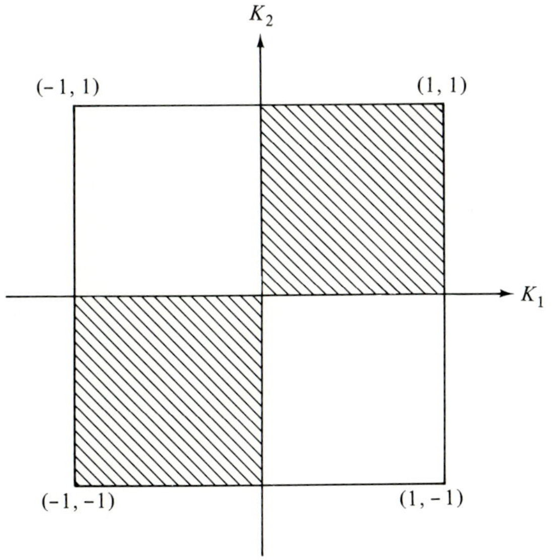

Figure P3.10

3.11 A uniform rectangular block B has the dimensions shown in Fig. P3.11(a). X_1, X_2, X_3 are the central principal axes of inertia of B. In Fig. P3.11(b), six regions of the K_1–K_2 space of Fig. 3.5.3 are labeled 1, . . . , 6.

Referring to Sec. 3.5, suppose that B remains at rest in A with each of X_1, X_2, X_3 parallel to one of $\mathbf{a}_1$, $\mathbf{a}_2$, $\mathbf{a}_3$. To each such alignment there corresponds one of the numbers in Fig. P3.11(b). For example, the arrangement indicated in Fig. P3.11(c) corresponds to region 6. Draw sketches similar to Fig. P3.11(c) for regions 1, . . . , 5.

Results Figure P3.11(d)

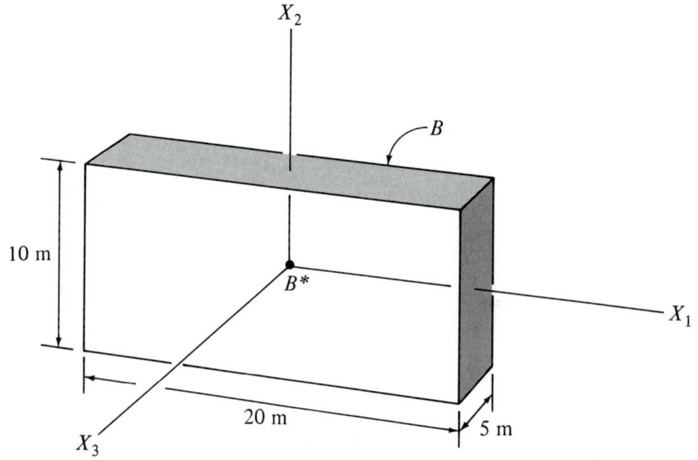

Figure P3.11(a)

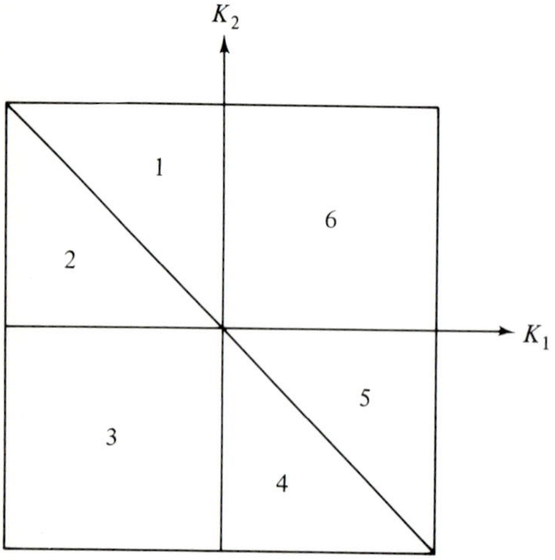

Figure P3.11(b)

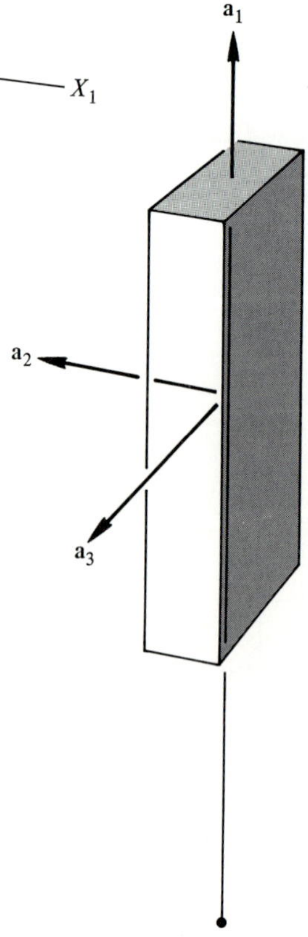

Figure P3.11(c)

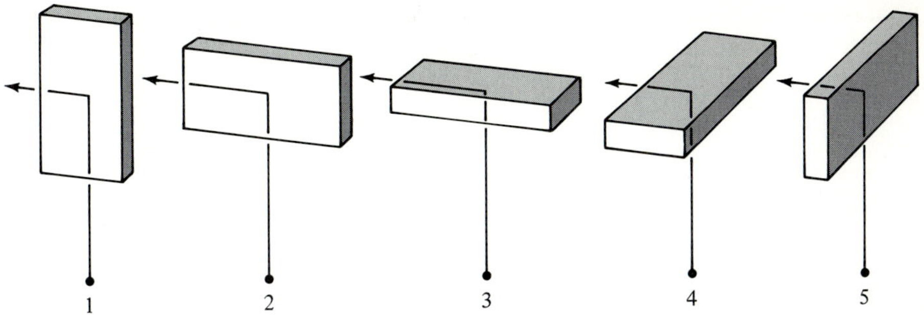

Figure P3.11(d)

3.12 Referring to Problem 3.11, consider the motions associated with regions 1, 2, and 4. Assume that, at $t = 0$, $\omega_3/\Omega = 1.0$, and X_1, X_2, X_3 are suitably aligned with $\mathbf{a}_1$, $\mathbf{a}_2$, $\mathbf{a}_3$, but, as the result of a disturbance, $\omega_1/\Omega = \omega_2/\Omega = 0.1$. Plot the angle ϕ between the orbit normal and that one of X_1, X_2, X_3 that is aligned with the orbit normal at $t = 0$, displaying results for 10 orbits and showing two curves in each case, one obtained by ignoring the gravitational moment $\mathbf{M}$, the other resulting

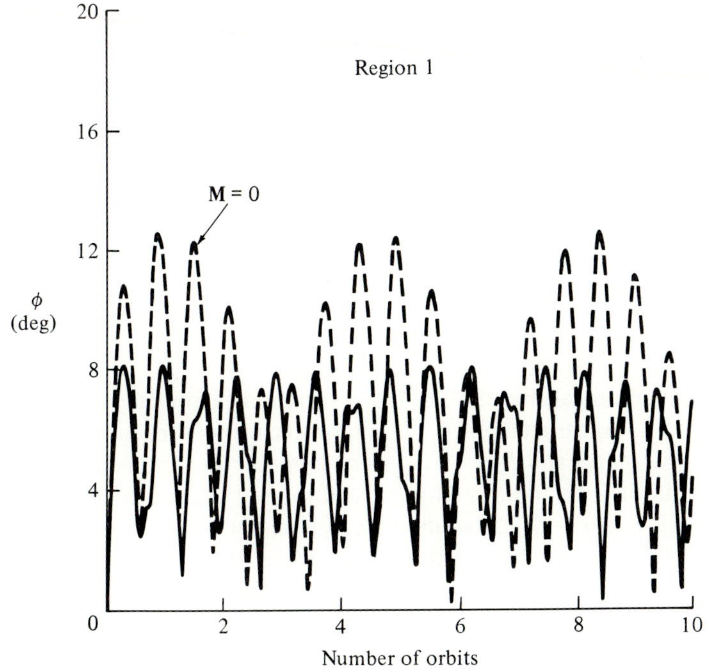

Figure P3.12(a)

from taking this moment into account. Comment on the relationship between these curves, Fig. P3.10, and Fig. 3.5.3.

Results Figures P3.12(a)–P3.12(c)

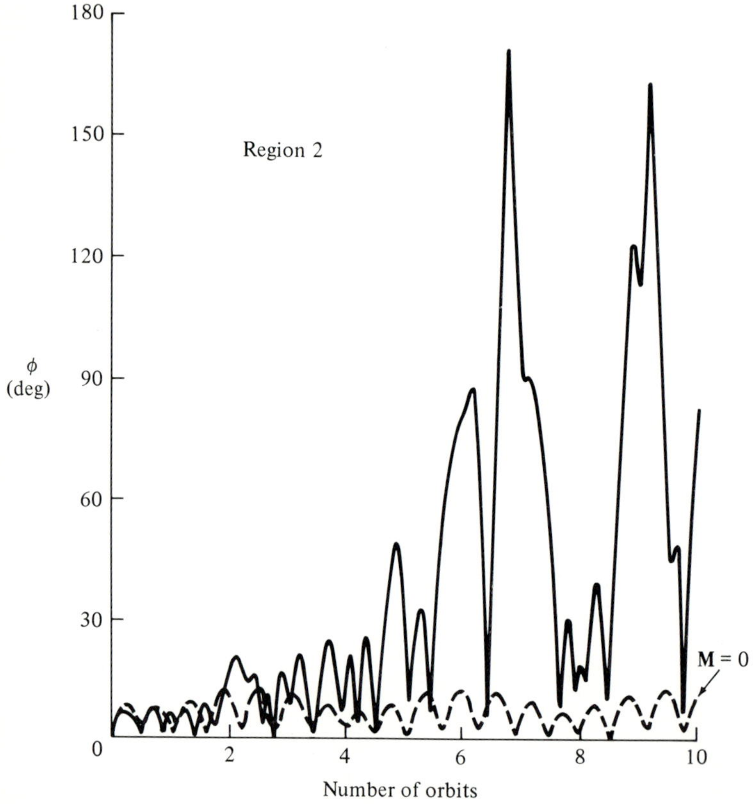

Figure P3.12(b)

3.13 To form a conjecture regarding the stability of motions associated with the black region in Fig. 3.5.3, construct a plot containing two curves of the kind shown in Fig. 3.5.4, taking $K_1 = 0.3$ and $K_2 = -0.1$. For both curves, let each central principal axis of inertia of B be initially aligned with one of $\mathbf{a}_1$, $\mathbf{a}_2$, $\mathbf{a}_3$, but take $\omega_1(0) = \omega_2(0) = 0.1\Omega$, $\omega_3(0) = \Omega$ in the first case, and $\omega_1(0) = \omega_2(0) = 0.05\Omega$, $\omega_3(0) = \Omega$ in the second case. Display results for ten orbits. Would the conjecture you form in this way be altered if only two orbits were considered?

Result Figure P3.13

3.14 Referring to Sec. 3.6, and letting $\mathbf{c}$ be the position vector from A^* to B^*, verify that $\mathbf{I}_G$ can be expressed as

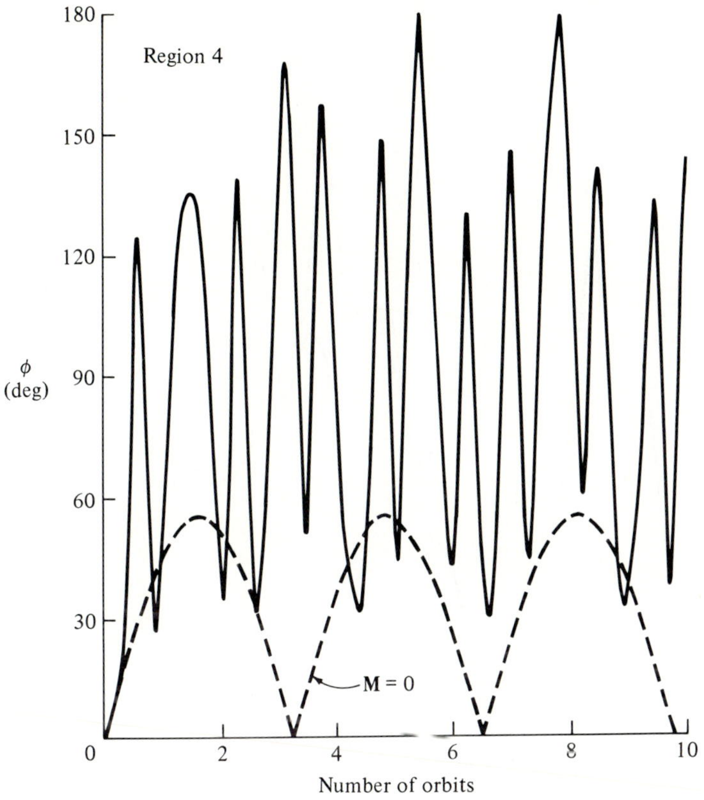

Figure P3.12(c)

$$\mathbf{I}_G = \mathbf{I}_A + \mathbf{I}_B + \frac{m_A m_B}{m_A + m_B} (\mathbf{U}\mathbf{c}^2 - \mathbf{c}\mathbf{c})$$

3.15 Figure P3.15 shows a simple gyrostat G formed by a uniform rectangular block A of mass 1200 kg and a thin circular disk of mass 100 kg. The sides of A have lengths 4 m, 6 m, 12 m, and B has a radius of 2 m. The center of B is situated at a corner of A, and the axis of B is parallel to the unit vector $\mathbf{a}_1$.

B is driven relative to A in such a way that the angular velocity of B in A is at all times equal to $500\mathbf{a}_1$ rad/sec. At a certain instant, the velocity of the mass center of A, the angular velocity of A, and the angular acceleration of A are given by $200\mathbf{a}_1$ m/sec, $20\mathbf{a}_2 + 10\mathbf{a}_3$ rad/sec, and $30\mathbf{a}_1$ rad/sec^2, respectively. For this instant, determine the central angular momentum of G, the inertia torque of G, and the kinetic energy of G.

Results

$$\mathbf{H}_G = (1.39\mathbf{a}_1 + 4.40\mathbf{a}_2 + 1.87\mathbf{a}_3) \times 10^5 \text{ kg} \cdot \text{m}^2/\text{sec}$$

$$\mathbf{T}_G = (0.460\mathbf{a}_1 - 1.421\mathbf{a}_2 + 2.73\mathbf{a}_3) \times 10^6 \text{ N} \cdot \text{m}$$

$$K_G = 5.56 \times 10^7 \text{ N} \cdot \text{m}$$

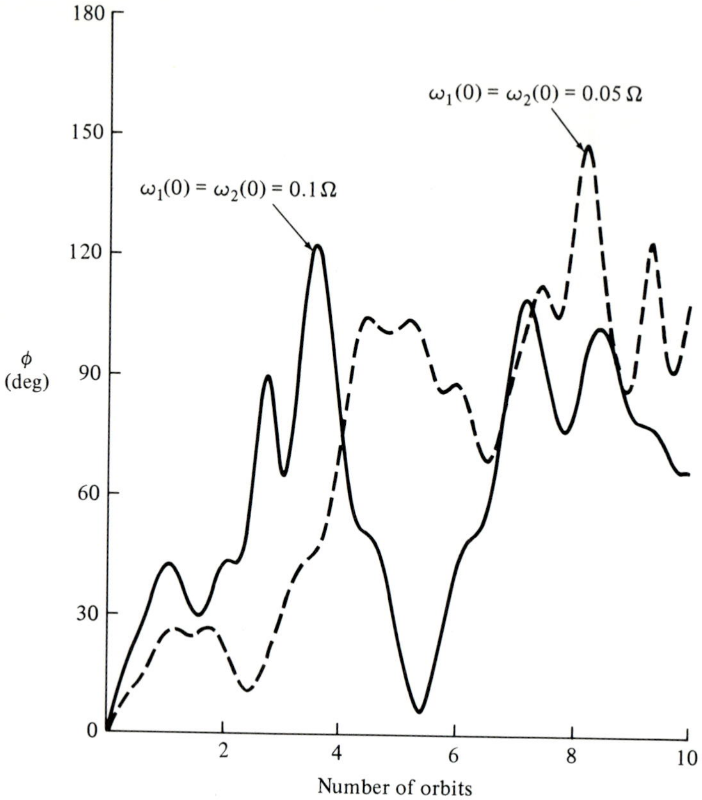

Figure P3.13

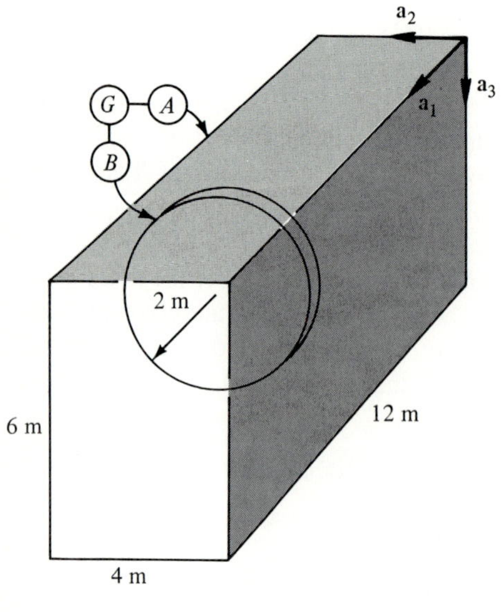

Figure P3.15

3.16 Referring to Prob. 3.15, suppose that A and B are at rest in a Newtonian reference frame N, that no forces act on A or B except those exerted by A on B and vice versa, and that, beginning at some instant, B is made to rotate relative to A by means of a motor that connects A to B and has negligible mass. Then A begins to rotate in N at this instant about a line L that passes through the mass center of G. Determine the angle between L and $\mathbf{a}_1$, and find the number of revolutions B must make relative to A in order for A to complete one revolution in N.

Results 5.73°, 32 rev

3.17 Assuming that the gyrostat G of Prob. 3.15 is moving in a Newtonian reference frame under the action of forces whose resultant moment about the mass center of G is equal to zero, determine the smallest and largest values that the rotational kinetic energy of G can possibly have during the motion under consideration.

Results 2.97×10^7 N · m, 4.38×10^7 N · m

3.18 The rotor B of a simple gyrostat G is driven with a constant angular speed ${}^A\hat{\omega}^B$ relative to the carrier A, and the system of all forces acting on G has zero moment about the mass center of G. Under these circumstances, ${}^N\boldsymbol{\omega}^A$, the angular velocity of A in a Newtonian reference frame N, can be permanently equal to zero; and, to explore the stability of this motion, one may construct the set of linearized differential equations governing ω_i $(i = 1, 2, 3)$ defined as $\omega_i \triangleq {}^N\boldsymbol{\omega}^A \cdot \mathbf{a}_i$, where $\mathbf{a}_i$ $(i = 1, 2, 3)$ are unit vectors fixed in A, each parallel to a central principal axis of inertia of G.

Formulate the characteristic equation for the linearized differential equations (that is, let $\omega_i = C_i e^{\lambda t}$, where C_i and λ are constants, and set the determinant of the coefficients of C_1, C_2, C_3 equal to zero), letting I_1, I_2, I_3 denote the central principal moments of inertia of G, J the axial moment of inertia of B, and $\beta_i \triangleq \boldsymbol{\beta} \cdot \mathbf{a}_i$ $(i = 1, 2, 3)$, where $\boldsymbol{\beta}$ is a unit vector parallel to the axis of B.

Result

$$\lambda^3 + \lambda(J\,{}^A\hat{\omega}^B)^2 \left(\frac{\beta_1{}^2}{I_2 I_3} + \frac{\beta_2{}^2}{I_3 I_1} + \frac{\beta_3{}^2}{I_1 I_2}\right) = 0$$

3.19 Letting $\mathbf{a}_1, \mathbf{a}_2, \mathbf{a}_3$ be unit vectors parallel to the central principal axes of inertia of a gyrostat G formed by a rigid body A and a rotor B, suppose that B is completely free to rotate relative to A, that the axis of B is parallel to $\mathbf{a}_3$, and that the system of all forces acting on G has zero moment about G^*, the mass center of G. Then one possible motion of G is described as follows: B rotates relative to A with a constant angular speed ${}^A\hat{\omega}^B$, and the angular velocity of A in a Newtonian reference frame is permanently parallel to $\mathbf{a}_3$ and has a constant magnitude $\hat{\omega}_3$.

Taking ${}^A\hat{\omega}^B$ and $\hat{\omega}_3$ positive when ${}^A\boldsymbol{\omega}^B$ and ${}^N\boldsymbol{\omega}^A$ have the same direction as $\mathbf{a}_3$, show that this motion is unstable if

$$\left(\frac{I_2 - I_3}{J} - \frac{{}^A\hat{\omega}^B}{\hat{\omega}_3}\right)\left(\frac{I_3 - I_1}{J} + \frac{{}^A\hat{\omega}^B}{\hat{\omega}_3}\right) > 0$$

where I_i is the moment of inertia of G about a line passing through G^* and parallel to $\mathbf{a}_i$ $(i = 1, 2, 3)$ and J is the axial moment of inertia of B.

Verify that this result remains unaltered if B is driven relative to A with a constant angular speed ${}^A\hat{\omega}^B$.

3.20 Figure P3.20 shows a gyrostat G formed by a uniform, rectangular block A that contains a cylindrical cavity which is completely filled by a uniform rotor B made of the same material as A. The system of all forces acting on G has zero moment about the mass center of G, and the angular velocity of A in a Newtonian reference frame N is parallel to the axis of B (which is parallel to the axis of maximum central moment of inertia of G).

As is shown in Sec. 3.4, this motion is stable if B is held fixed in A. To show that a rotor can have a destabilizing effect, assume that B is rotating relative to A with an angular velocity having a constant magnitude and directed oppositely to that of A in N, and verify that the motion of G is unstable if the ratio of the magnitude of the angular velocity of B in A to the magnitude of the angular velocity of A in N lies between 9.82 and 49.1.

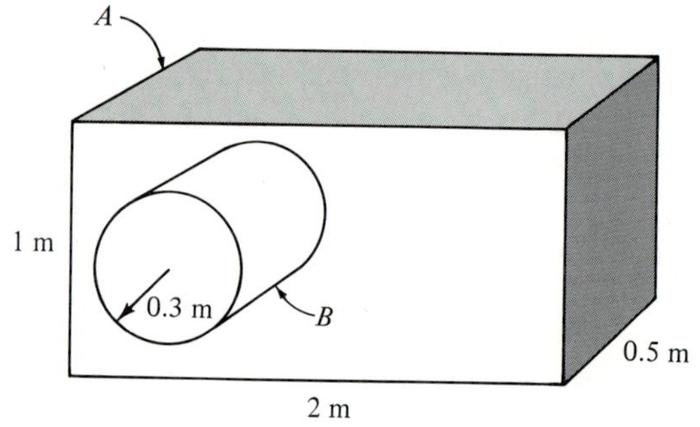

Figure P3.20

3.21 The gyrostat G shown in Fig. P3.21 consists of a thick, right-circular, cylindrical shell A and a uniform right-circular cylinder B whose axis coincides with that of A. The mass density of B is equal to that of A. G is to be set into motion in such a way that ${}^N\hat{\boldsymbol{\omega}}^A$, the initial angular velocity of A in a Newtonian reference frame N, is parallel to the axis of symmetry of G; but an "injection" error of as much as 1° (see Fig. P3.21) is anticipated, and the angle between the symmetry axis at time t and the (fixed) line with which the symmetry axis coincides at $t = 0$ may, therefore, differ from zero, acquiring a maximum value δ.

Letting $\hat{r}$ and $\hat{\omega}$ denote the initial values of ${}^A\boldsymbol{\omega}^B \cdot \mathbf{n}$ and ${}^N\boldsymbol{\omega}^A \cdot \mathbf{n}$, respectively, where $\mathbf{n}$ is a unit vector parallel to the axis of G, determine $\hat{r}/\hat{\omega}$ for $\delta = 0.1°$, 1°, and 10°, and find δ for $\hat{r}/\hat{\omega} = -15.9$.

Results 357, 21, −12; 146°

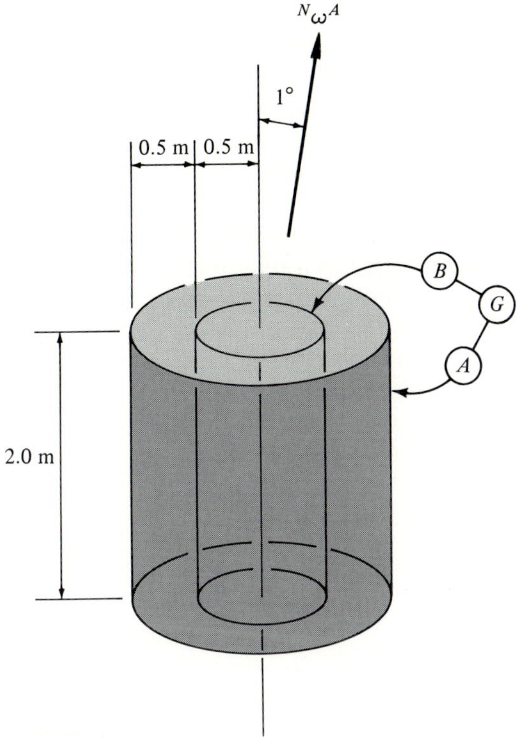

Figure P3.21

3.22 Referring to Prob. 3.15, suppose that A and B are at rest in a Newtonian reference frame N, that no forces act on A or B except those exerted by A on B and vice versa, and that, beginning at some instant, B is made to rotate relative to A by means of a motor that connects A to B and that has negligible mass. Then (see Prob. 3.16) A begins to rotate in N at this instant about a line that passes through the mass center of G and that makes an angle of 5.7° with $\mathbf{a}_1$. In order to make A rotate, instead, about a line parallel to $\mathbf{a}_1$, one can leave the center of B in place, but reorient the axis of B, making it parallel to a certain unit vector $\boldsymbol{\beta}$.

Determine the angle between $\boldsymbol{\beta}$ and $\mathbf{a}_1$, and find the number of revolutions B must make relative to A in order for A to complete one revolution in N.

Results 17.1°, 34.5 rev

3.23 A torque-free simple gyrostat that is initially at rest is to be reoriented by means of a rotation of the rotor relative to the carrier. The associated rotation of the carrier is to take place about a central principal axis of inertia of the gyrostat. Show that this is impossible unless the principal axis under consideration is parallel to the rotor axis.

3.24 An Earth satellite formed by a uniform circular disk D and a uniform rod R having the dimensions shown in Fig. P3.24(a) is intended to move in a circular orbit in such a way that R remains normal to the orbit plane at all times while D rotates uniformly, completing two rotations in a Newtonian reference frame N during each orbital revolution of the center of D. The rotations are to proceed in such a way that ${}^N\boldsymbol{\omega}^D \times {}^N\mathbf{v}^{D*}$ points from $D*$ toward the center of the orbit. D and R are made of the same material.

As was shown in Prob. 3.3, part (*b*), this motion is unstable. To overcome this difficulty, C, a central portion of D, is to be used as a rotor that is driven with a constant angular speed relative to the remaining peripheral part of D [see Fig. P3.24(b)]. If C has a radius of 0.8 m, what is the range of values of ${}^D\omega^C/\Omega$ that must be excluded in order to avoid instability?

Result $-23.9 < {}^D\omega^C/\Omega < 25.2$

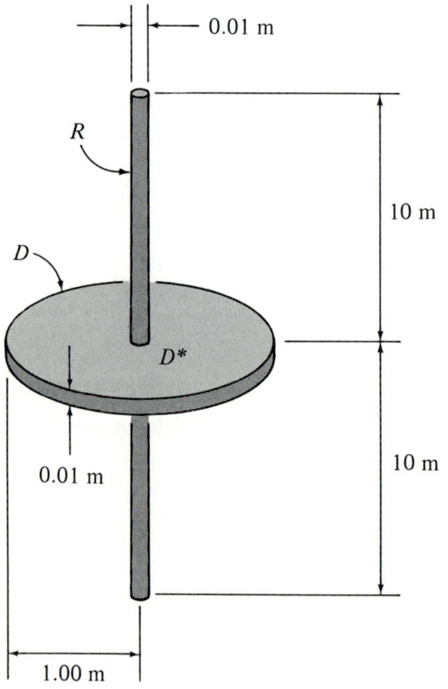

Figure P3.24(a)

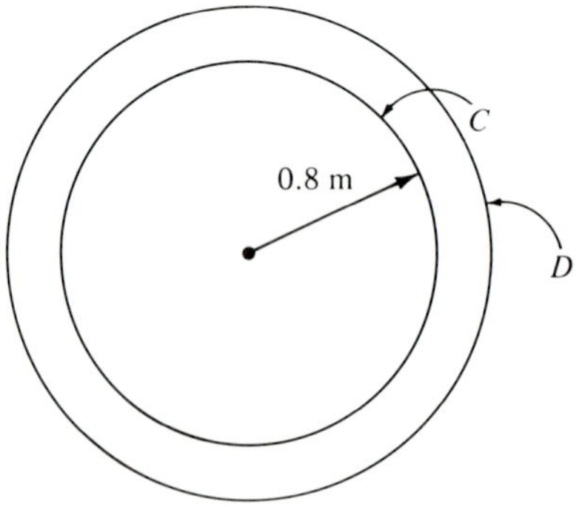

Figure P3.24(b)

3.25 In Fig. P3.25(a), which shows the satellite considered in Prob. 3.24, $\mathbf{e}_1$, $\mathbf{e}_2$, $\mathbf{e}_3$ form a dextral set of mutually perpendicular unit vectors; $\mathbf{e}_3$ is normal to the orbit plane; $\mathbf{e}_2$ points in the direction of motion of $D*$.

Suppose that, at $t = 0$, R is aligned with $\mathbf{e}_3$ and the angular velocity of D in a Newtonian reference frame N is given by

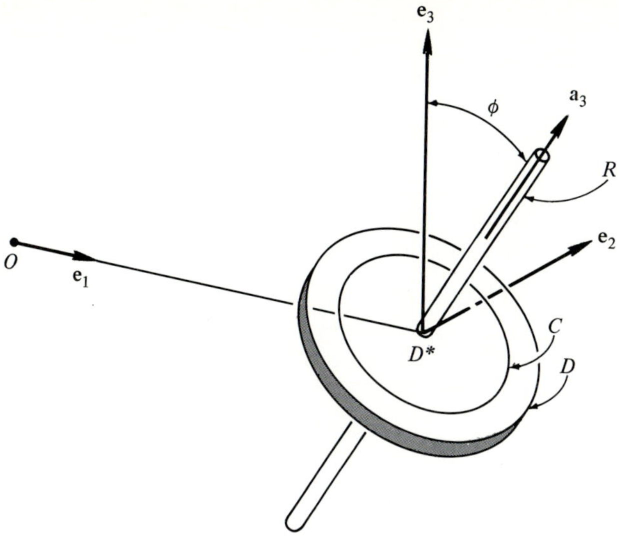

Figure P3.25(a)

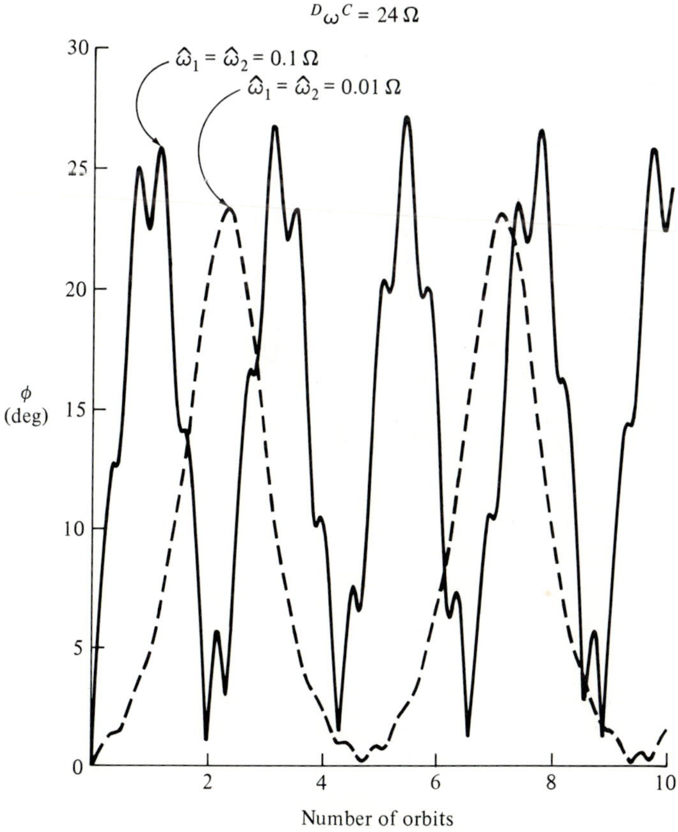

Figure P3.25(b)

$$^N\boldsymbol{\omega}^D \triangleq \hat{\omega}_1 \mathbf{e}_1 + \hat{\omega}_2 \mathbf{e}_2 + \hat{\omega}_3 \mathbf{e}_3$$

Letting $^D\hat{\omega}^C = {}^D\boldsymbol{\omega}^C \cdot \mathbf{a}_3$, where $\mathbf{a}_3$ is a unit vector directed as shown, make two plots of ϕ, the angle between R and $\mathbf{e}_3$, versus the number of orbits traversed by D^*, one plot applying to $^D\hat{\omega}^C = 24\Omega$, the other to $^D\hat{\omega}^C = -24\Omega$, and each containing two curves, one corresponding to $\hat{\omega}_1 = \hat{\omega}_2 = 0.1\Omega$, $\hat{\omega}_3 = 2\Omega$, the other to $\hat{\omega}_1 = \hat{\omega}_2 = 0.01\Omega$, $\hat{\omega}_3 = 2\Omega$. Discuss the results in relationship to those of Prob. 3.24.

Results Figures P3.25(b) and P3.25(c).

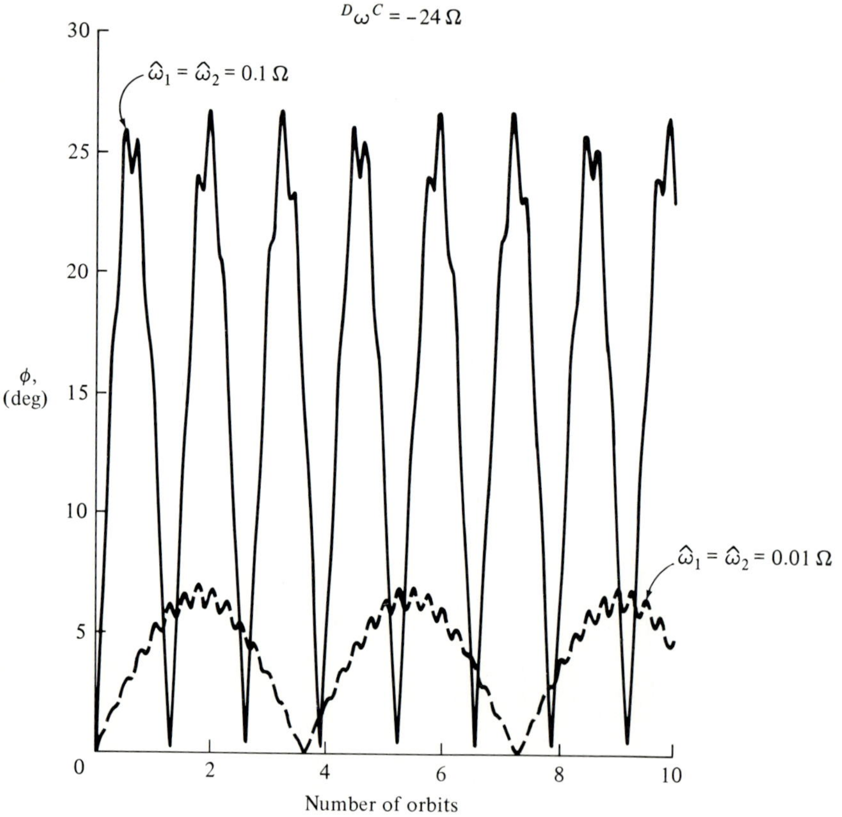

Figure P3.25(c)

3.26 Since neither K_3 as defined in Eq. (3.11.16) nor b and c as given by Eqs. (3.11.20) and (3.11.21) involve $\bar{I}_3$ or σ, the instability conditions (3.11.18) and (3.11.19) apply both when B is completely free to rotate relative to A and when B is made to rotate relative to A with a constant angular speed. However, the motion of G may proceed quite differently in the two cases. To verify this, plot two curves of the kind shown in Fig. 3.11.2, one applicable to the free rotor case, the other to the constant angular speed case. Use the same inertia properties as in the example in Sec. 3.11, and let C_{ij} $(i = 1, 3; j = 1, 2, 3)$, ω_i $(i = 1, 2, 3)$, and ${}^A\omega^B$ have the following initial values: $C_{ij} = \delta_{ij}$ $(i = 1, 3; j = 1, 2, 3)$, $\omega_1 = \omega_2 = 0.1\Omega$, $\omega_3 = \Omega$, ${}^A\omega^B = -10\Omega$. Display results for the two orbits. Also, determine the maximum and minimum values of ${}^A\omega^B$ during the first two orbits.

Results Figure P3.26; $({}^A\omega^B)_{\max} = -7.49\Omega$, $({}^A\omega^B)_{\min} = -10.12\Omega$

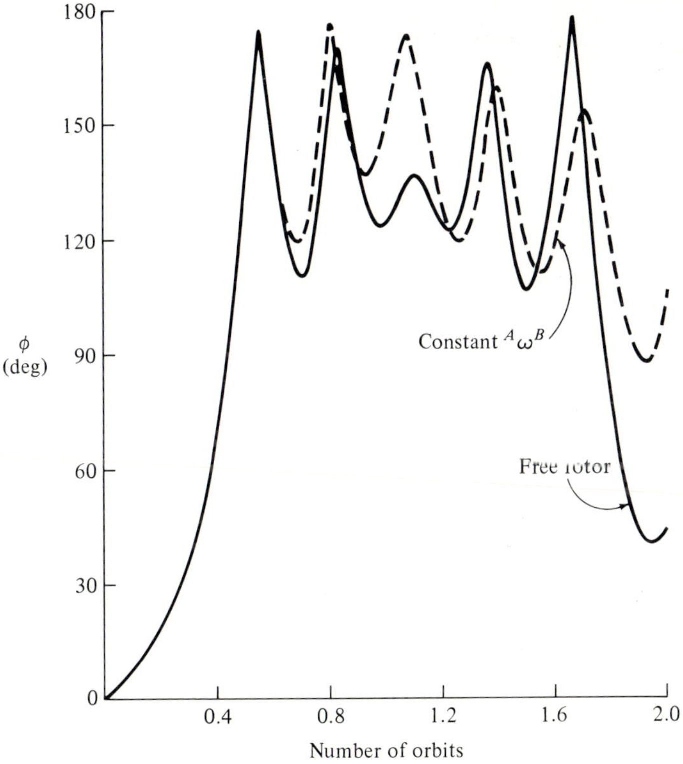

Figure P3.26

PROBLEM SET 4

4.1 Figure P4.1 shows a rigid body B carrying two particles, P_1 and P_2, constrained to move in a smooth tube T that is fixed in B. P_1 is connected to B and to P_2 by means of light linear springs having spring constants k_1 and k_2 and unstretched lengths L_1 and L_2. Thus s_1 and s_2 in Fig. P4.1 measure spring extensions.

The system S formed by B, P_1, and P_2 moves in a reference frame A in the absence of external forces. Letting

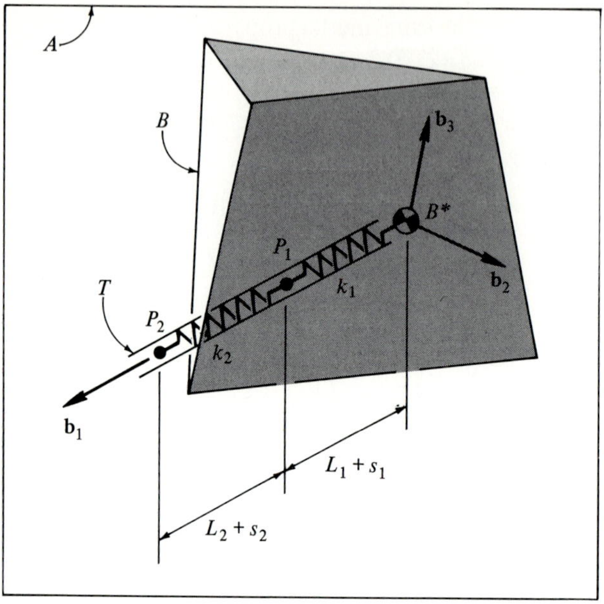

Figure P4.1

$$u_i \triangleq \begin{cases} \boldsymbol{\omega} \cdot \mathbf{b}_i & (i = 1, 2, 3) \\ \mathbf{v} \cdot \mathbf{b}_{i-3} & (i = 4, 5, 6) \\ \dot{s}_1 & (i = 7) \\ \dot{s}_1 + \dot{s}_2 & (i = 8) \end{cases}$$

where $\boldsymbol{\omega}$ is the angular velocity of B in A, $\mathbf{v}$ is the velocity of B^* in A, and $\mathbf{b}_1$, $\mathbf{b}_2$, $\mathbf{b}_3$ are mutually perpendicular unit vectors directed as shown, determine the generalized active forces $F_1, \ldots, F_8$ for S in A.

Results $F_1 = \cdots = F_6 = 0$, $F_7 = -k_1 s_1 + k_2 s_2$, $F_8 = -k_2 s_2$

4.2 Referring to Prob. 1.28, suppose that equal torsional springs and dampers at points L_1 and L_4 resist rotation of the linkage relative to A and that an extensional spring and damper connect P to point L_2 (dampers are not shown in Fig. P1.28). The resultant force exerted on P by the extensional spring and damper is given by $\sigma \mathbf{a}_2$, where σ is a function of r and $\dot{r}$, and the torque associated with the action of each torsional spring and damper on a link is $\tau \mathbf{a}_3$, with τ a function of θ and $\dot{\theta}$. Assume that the linkage is sufficiently light to be treated as massless, and let the set of all external body and contact forces acting on S be represented by a force $R_1\mathbf{a}_1 + R_2\mathbf{a}_2 + R_3\mathbf{a}_3$ applied to P, a force $S_1\mathbf{a}_1 + S_2\mathbf{a}_2 + S_3\mathbf{a}_3$ applied to A at A^*, and a couple of torque $T_1\mathbf{a}_1 + T_2\mathbf{a}_2 + T_3\mathbf{a}_3$ applied to A.

With $u_1, \ldots, u_8$ defined as in Prob. 1.28, determine the generalized active forces $F_1, \ldots, F_8$.

Results

$$F_1 = T_1 + (c - r - b\cos\theta)R_3$$
$$F_2 = T_2 - bR_3\sin\theta$$
$$F_3 = T_3 - (c - r - b\cos\theta)R_1 + bR_2\sin\theta$$
$$F_4 = R_1 + S_1$$
$$F_5 = R_2 + S_2$$
$$F_6 = R_3 + S_3$$
$$F_7 = 2\tau/b - \sigma\sin\theta + R_1\cos\theta$$
$$F_8 = \sigma + R_2$$

4.3 In general, the set of gravitational forces exerted on the particles $P_1, \ldots, P_N$ of a system S by a body B not belonging to S cannot be represented in a simple way. However, for the purposes of a particular analysis, it may be adequate to assume that $\mathbf{G}_i$, the gravitational force exerted by B on P_i, is given by

$$\mathbf{G}_i = m_i\mathbf{g} \qquad (i = 1, \ldots, N)$$

where m_i is the mass of P_i and $\mathbf{g}$ is a vectorial "constant of proportionality." Show that under these circumstances $(F_r)_G$, the contribution to F_r of the gravitational forces exerted on S by B, is given by

$$(F_r)_G = m\mathbf{g}\cdot\mathbf{v}_r \qquad (r = 1, \ldots, n)$$

where m is the total mass of S and $\mathbf{v}_r$ is the rth partial velocity of the mass center of S.

4.4 In Fig. P4.4, A designates a sun gear, B one of four identical planet gears, and C a ring gear. The axes of A and C coincide and are fixed in a reference frame D. $\hat{A}$ is the point of A that is in contact with B, $\hat{B}$ is the point of B that is in contact with A, and $\hat{C}$ is the point of C that is in contact with B. The radii of A and B are a and b, respectively.

So long as B rolls on A and C without slipping, two generalized coordinates characterize the configuration in D of the system S formed by A, B, and C. Defining u_1 and u_2 as

$$u_1 \triangleq {}^D\boldsymbol{\omega}^A \cdot \mathbf{n}_3 \qquad u_2 \triangleq {}^D\boldsymbol{\omega}^B \cdot \mathbf{n}_3$$

and expressing the forces exerted by A on B and by B on C as

$$\mathbf{F}^{A/B} = P_1\mathbf{n}_1 + P_2\mathbf{n}_2 + P_3\mathbf{n}_3 \qquad \mathbf{F}^{B/C} = Q_1\mathbf{n}_1 + Q_2\mathbf{n}_2 + Q_3\mathbf{n}_3$$

where $\mathbf{n}_1$, $\mathbf{n}_2$, $\mathbf{n}_3$ are mutually perpendicular unit vectors directed as shown, determine (*a*) the contributions of $\mathbf{F}^{A/B}$ to the generalized active forces F_1 and F_2, (*b*) the contributions of the force exerted by B on A to F_1 and F_2, (*c*) the contributions of $\mathbf{F}^{B/C}$ to F_1 and F_2, and (*d*) the total contributions to F_1 and F_2 of all forces exerted on each other by A and B and by B and C.

Results (*a*) aP_2, 0; (*b*) $-aP_2$, 0; (*c*) aQ_2, $2bQ_2$; (*d*) 0, 0

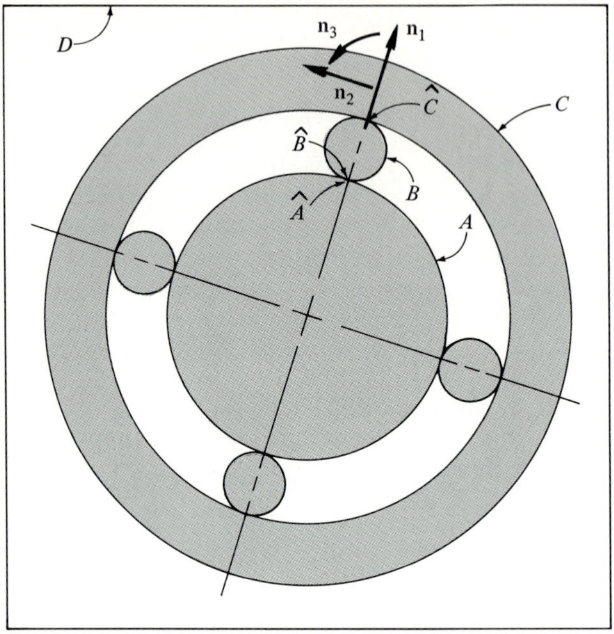

Figure P4.4

4.5 A uniform sphere S of radius R is placed in a spherical cavity of a rigid body B, as indicated in Fig. P4.5. The radius of the cavity is only slightly greater than that of S, and the space between S and B is filled with a viscous fluid. The system formed by S and B moves in a reference frame C in the absence of external forces.

Defining $u_1, \ldots, u_9$ in terms of the angular velocities of B and S in C and the velocity of the center S^* of S in C as

$$u_i \triangleq \begin{cases} \boldsymbol{\omega}^B \cdot \mathbf{b}_i & (i = 1, 2, 3) \\ \boldsymbol{\omega}^S \cdot \mathbf{b}_{i-3} & (i = 4, 5, 6) \\ \mathbf{v}^{S^*} \cdot \mathbf{b}_{i-6} & (i = 7, 8, 9) \end{cases}$$

where $\mathbf{b}_1$, $\mathbf{b}_2$, $\mathbf{b}_3$ are mutually perpendicular unit vectors fixed in B, and assuming that the force $d\mathbf{F}$ exerted on S by the fluid across a differential element of S can be expressed as

$$d\mathbf{F} = -\,c\,{}^B\mathbf{v}^P\,dA$$

where c is a constant, ${}^B\mathbf{v}^P$ is the velocity in B of a point P of S lying within the element under consideration, and dA is the area of the element, determine the generalized active forces F_2 and F_4 for the system in C.

Results $\quad F_2 = \frac{8}{3}\pi c R^4(u_5 - u_2) \qquad F_4 = \frac{8}{3}\pi c R^4(u_1 - u_4)$

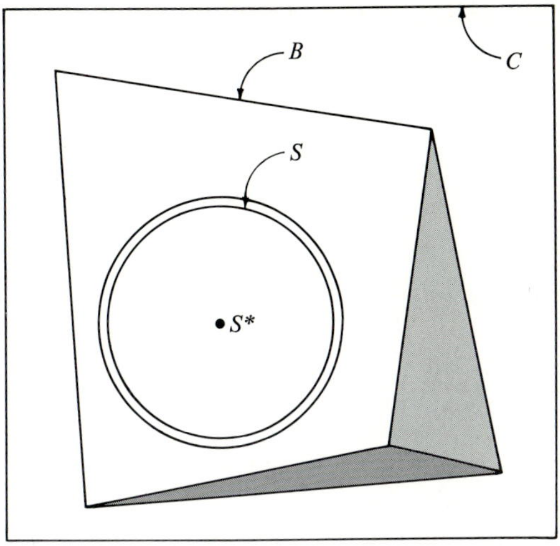

Figure P4.5

4.6 Referring to the example in Sec. 4.2, suppose that a contact force is applied to B at B^* in such a way that the motion of B^* in N is prescribed as a function of time, so that the configuration of B in N can be characterized by a single generalized coordinate. Show that under these circumstances a potential energy of B in N is given by

$$W = -\frac{G\overline{m}mL^2}{16{q_2}^3}\cos 2q_3$$

if the gravitational forces exerted on B by $\overline{P}$ are replaced with a couple of torque $\mathbf{T}$ and a force $\mathbf{R}$ applied at B^*, with (*a*) $\mathbf{T} = \widetilde{\mathbf{M}}$, $\mathbf{R} = \widetilde{\mathbf{F}}$ and (*b*) $\mathbf{T} = \widetilde{\mathbf{M}}$, $\mathbf{R} = \hat{\mathbf{F}}$, where $\widetilde{\mathbf{M}}$, $\widetilde{\mathbf{F}}$, and $\hat{\mathbf{F}}$ are given in Eqs. (2.6.3), (2.3.6), and (2.3.5), respectively.

4.7 Referring to Prob. 4.1, introduce generalized coordinates $q_1, \ldots, q_8$ by letting q_1, q_2, q_3 be body-three orientation angles for B in A (see Sec. 1.7), defining q_4, q_5, q_6 as rectangular Cartesian coordinates of B^* in A, and setting $q_7 = s_1$, $q_8 = s_1 + s_2$. Construct an expression for a potential energy W of S in A, both for u_r $(r = 1, \ldots, 8)$ as defined in Prob. 4.1 and for $u_r \triangleq \dot{q}_r$ $(r = 1, \ldots, 8)$. Show that, if a force $F\mathbf{b}_1$ is applied to B at B^*, with F a constant, then there exists a potential energy of S in A if u_r $(r = 1, \ldots, 8)$ are defined as in Prob. 4.1, but not if $u_r \triangleq \dot{q}_r$ $(r = 1, \ldots, 8)$.

Result $\quad W = \frac{1}{2}[k_1{q_7}^2 + k_2(q_8 - q_7)^2]$

4.8 One point of a rigid body B is fixed in a reference frame A; and, when space-three orientation angles q_1, q_2, q_3 relating mutually perpendicular unit vectors $\mathbf{a}_1$, $\mathbf{a}_2$, $\mathbf{a}_3$ fixed in A to similar unit vectors $\mathbf{b}_1$, $\mathbf{b}_2$, $\mathbf{b}_3$ fixed in B are used as generalized coordinates, and generalized speeds are defined as $u_i \triangleq \dot{q}_i$ $(i = 1, 2, 3)$, then B possesses a potential energy $W(q_1, q_2, q_3)$.

Suppose that generalized speeds are introduced as $u_i \triangleq \boldsymbol{\omega} \cdot \mathbf{b}_i$ $(i = 1, 2, 3)$, where $\boldsymbol{\omega}$ is the angular velocity of B in A. Then the generalized active force F_i associated with u_i can be expressed in terms of $q_1, q_2, q_3, \partial W/\partial q_1, \partial W/\partial q_2, \partial W/\partial q_3$. Letting $s_i \triangleq \sin q_i$ and $c_i \triangleq \cos q_i$ $(i = 1, 2, 3)$, determine F_1, F_2, F_3.

Results

$$F_1 = -\frac{\partial W}{\partial q_1}$$

$$F_2 = -\frac{1}{c_2}\left[c_1 c_2 \frac{\partial W}{\partial q_2} + s_1\left(s_2 \frac{\partial W}{\partial q_1} + \frac{\partial W}{\partial q_3}\right)\right]$$

$$F_3 = -\frac{1}{c_2}\left[c_1\left(s_2 \frac{\partial W}{\partial q_1} + \frac{\partial W}{\partial q_3}\right) - s_1 c_2 \frac{\partial W}{\partial q_2}\right]$$

4.9 Show that $\mathbf{T}^*$ as defined in Eq. (4.3.4) can be expressed as

$$\mathbf{T}^* = -\dot{\mathbf{H}}$$

where $\mathbf{H}$ is the central angular momentum of S in A and the dot denotes time-differentiation in A.

4.10 Referring to Prob. 1.28, let P have a mass m, and A a mass M; regard the linkage as massless; assume that the central principal axes of inertia of A are parallel to $\mathbf{a}_1, \mathbf{a}_2, \mathbf{a}_3$; and let the associated moments of inertia have the values I_1, I_2, I_3. Letting $q_7 = r$ and $q_8 = \theta$, determine the generalized inertia forces F_1^*, F_5^*, and F_8^* for S in N.

Results

$$\begin{aligned} F_1^* = &-m(c - q_7 - bc_8)[\dot{u}_6 + u_1u_5 - u_2u_4 \\ &+ (\dot{u}_1 + u_2u_3)(c - q_7 - bc_8) - (\dot{u}_2 - u_3u_1)bs_8 \\ &+ 2(u_1u_8 - u_2u_7c_8)] - I_1\dot{u}_1 + u_2u_3(I_2 - I_3) \end{aligned}$$

$$\begin{aligned} F_5^* = &-M(\dot{u}_5 + u_3u_4 - u_1u_6) - m[\dot{u}_5 + u_3u_4 - u_1u_6 + \dot{u}_8 \\ &+ (\dot{u}_3 + u_1u_2)bs_8 + 2u_3u_7c_8 - (u_3^2 + u_1^2)(c - q_7 - bc_8)] \end{aligned}$$

$$\begin{aligned} F_8^* = &-m[\dot{u}_5 + u_3u_4 - u_1u_6 + \dot{u}_8 + (\dot{u}_3 + u_1u_2)bs_8 + 2u_3u_7c_8 \\ &- (u_3^2 + u_1^2)(c - q_7 - bc_8)] \end{aligned}$$

4.11 The system considered in Prob. 4.10 is modified by the addition of a particle $\bar{P}$ of mass $\bar{m}$ at point L_2. Determine the contribution of $\bar{P}$ to the generalized inertia force F_7^*.

Result

$$\begin{aligned} &-\bar{m}\{c_8[\dot{u}_4 + \dot{u}_7c_8 - \dot{u}_3(c - bc_8) + u_2u_6 - u_3u_5 + u_1u_2(c - bc_8) \\ &- (u_2^2 + u_3^2)bs_8] + s_8[\dot{u}_5 + \dot{u}_7s_8 + \dot{u}_3bs_8 + u_3u_4 - u_1u_6 \\ &- (u_3^2 + u_1^2)(c - bc_8) + u_1u_2bs_8]\} \end{aligned}$$

4.12 The system considered in Prob. 4.10 is modified as follows: a uniform right-cylindrical rotor B of radius R, made of the same material as A and having a mass

$M/10$, is installed in a cavity in A, the cavity being just big enough to accommodate B. Placing the axis of B parallel to $\mathbf{a}_3$, calling C the portion of A that remains, and assuming that B is driven relative to C by means of a torque motor in such a way that Ω, defined as $\Omega \triangleq {}^C\boldsymbol{\omega}^B \cdot \mathbf{a}_3$, is given by $\Omega = 10(1 + 0.1 \sin 2t)$ rad/sec, determine the generalized inertia forces F_3^* and F_5^* for the system formed by B, C, and P.

Results

$$
\begin{aligned}
F_3^* = &- m\{bs_8[\dot{u}_5 + u_3u_4 - u_1u_6 + \dot{u}_8 + (\dot{u}_3 + u_1u_2)bs_8 \\
&+ 2u_3u_7c_8 - u_1^2(c - q_7 - bc_8)] - (c - q_7 - bc_8)[\dot{u}_4 + u_2u_6 \\
&- u_3u_5 + \dot{u}_7c_8 - (\dot{u}_3 - u_1u_2)(c - q_7 - bc_8) - 2u_3u_8 \\
&- u_7^2s_8/b - u_2^2bs_8]\} - I_3\dot{u}_3 + u_1u_2(I_1 - I_2) - (M/10)R^2 \cos 2t \\
F_5^* = &- M(\dot{u}_5 + u_3u_4 - u_1u_6) - m[\dot{u}_5 + u_3u_4 - u_1u_6 + \dot{u}_8 \\
&+ (\dot{u}_3 + u_1u_2)bs_8 + 2u_3u_7c_8 - (u_3^2 + u_1^2)(c - q_7 - bc_8)]
\end{aligned}
$$

4.13 Referring to Prob. 1.28, let P have a mass m, and A a mass M; regard the linkage as massless; assume that the central principal axes of inertia of A are parallel to $\mathbf{a}_1$, $\mathbf{a}_2$, $\mathbf{a}_3$; and let the associated moments of inertia have the values I_1, I_2, I_3. Introduce generalized coordinates $q_1, \ldots, q_8$ as follows: let q_1, q_2, q_3 be the radian measures of body-three orientation angles (see Sec. 1.7) relating $\mathbf{a}_1$, $\mathbf{a}_2$, $\mathbf{a}_3$ to a dextral set of mutually perpendicular unit vectors $\mathbf{n}_1$, $\mathbf{n}_2$, $\mathbf{n}_3$ fixed in N; let $q_4 = \mathbf{p} \cdot \mathbf{n}_1$, $q_5 = \mathbf{p} \cdot \mathbf{n}_2$, $q_6 = \mathbf{p} \cdot \mathbf{n}_3$, where $\mathbf{p}$ is the position vector of A^* relative to a point fixed in N; and take $q_7 = r$, $q_8 = \theta$ (see Fig. P1.28).

After verifying that the kinetic energy of S in N can be expressed as

$$
\begin{aligned}
K = (M/2)(v_1^2 + v_2^2 + v_3^2) &+ (m/2)\{[v_1 + b\dot{q}_8c_8 - \omega_3(c - q_7 - bc_8)]^2 \\
&+ (v_2 + b\dot{q}_8s_8 - \dot{q}_7 + \omega_3bs_8)^2 + [v_3 + \omega_1(c - q_7 - bc_8) - \omega_2bs_8]^2\} \\
&+ (I_1\omega_1^2 + I_2\omega_2^2 + I_3\omega_3^2)/2
\end{aligned}
$$

where $\omega_i \triangleq {}^N\boldsymbol{\omega}^A \cdot \mathbf{a}_i$ and $v_i \triangleq {}^N\mathbf{v}^{A^*} \cdot \mathbf{a}_i$ $(i = 1, 2, 3)$, use Eq. (4.4.6) to form the generalized inertia force associated with $\dot{q}_1$, assuming that $\dot{q}_1, \ldots, \dot{q}_8$ are used as generalized speeds. Use the notations $s_i \triangleq \sin q_i$, $c_i \triangleq \cos q_i$ $(i = 1, \ldots, 8)$ to express the result, and comment briefly on the amount of labor required to obtain this result, on the one hand, and F_1^* in Prob. 4.10, on the other hand.

Result

$$
\begin{aligned}
&[\dot{q}_6(s_1s_2c_3 + s_3c_1) - \dot{q}_5(-c_1s_2c_3 + s_3s_1)]\{Mv_1 + m[v_1 + b\dot{q}_8c_8 \\
&- \omega_3(c - q_7 - bc_8)]\} + [\dot{q}_6(-s_1s_2s_3 + c_3c_1) \\
&- \dot{q}_5(c_1s_2s_3 + c_3s_1)][Mv_2 + m(v_2 + b\dot{q}_8s_8 - \dot{q}_7 + \omega_3bs_8)] \\
&- (\dot{q}_6s_1 + \dot{q}_5c_1)c_2\{Mv_3 + m[v_3 + \omega_1(c - q_7 - bc_8) - \omega_2bs_8]\} \\
&- m\{-[\dot{v}_1 + b(\ddot{q}_8c_8 - \dot{q}_8^2s_8) - \dot{\omega}_3(c - q_7 - bc_8) \\
&- \omega_3(b\dot{q}_8s_8 - \dot{q}_7)]s_2(c - q_7 - bc_8) - [v_1 + b\dot{q}_8c_8 \\
&- \omega_3(c - q_7 - bc_8)][\dot{q}_2c_2(c - q_7 - bc_8) + s_2(b\dot{q}_8s_8 - \dot{q}_7)]
\end{aligned}
$$

(continued on next page)

$$+ [\dot{v}_2 + b(\ddot{q}_8 s_8 + \dot{q}_8{}^2 c_8) - \ddot{q}_7 + \dot{\omega}_3 b s_8 + \omega_3 b \dot{q}_8 c_8] s_2 b s_8$$
$$+ (v_2 + b\dot{q}_8 s_8 - \dot{q}_7 + \omega_3 b s_8) b(\dot{q}_2 c_2 s_8 + \dot{q}_8 s_2 c_8)$$
$$+ [\dot{v}_3 + \dot{\omega}_1(c - q_7 - bc_8) + \omega_1(b\dot{q}_8 s_8 - \dot{q}_7) - \dot{\omega}_2 b s_8$$
$$- \omega_2 \dot{q}_8 b c_8][c_2 c_3(c - q_7 - bc_8) + c_2 s_3 b s_8] + [v_3 + \omega_1(c - q_7 - bc_8)$$
$$- \omega_2 b s_8][-(\dot{q}_2 s_2 c_3 + \dot{q}_3 c_2 s_3)(c - q_7 - bc_8) + c_2 c_3(b\dot{q}_8 s_8 - \dot{q}_7)$$
$$+ (-\dot{q}_2 s_2 s_3 + \dot{q}_3 c_2 c_3) b s_8 + c_2 s_3 b \dot{q}_8 c_8]\} - I_1[\dot{\omega}_1 c_2 c_3$$
$$- \omega_1(\dot{q}_2 s_2 c_3 + \dot{q}_3 c_2 s_3)] + I_2[\dot{\omega}_2 c_2 s_3 + \omega_2(-\dot{q}_2 s_2 s_3 + \dot{q}_3 c_2 c_3)]$$
$$- I_3(\dot{\omega}_3 s_2 + \omega_3 \dot{q}_2 c_2)$$

4.14 Referring to Probs. 1.28, 4.2, and 4.10, and assuming that N is a Newtonian reference frame, formulate the dynamical equations associated with u_1, u_5, and u_8.

Results

$$T_1 + (c - q_7 - bc_8)R_3 - m(c - q_7 - bc_8)[\dot{u}_6 + u_1 u_5 - u_2 u_4$$
$$+ (\dot{u}_1 + u_2 u_3)(c - q_7 - bc_8) - (\dot{u}_2 - u_3 u_1) b s_8$$
$$+ 2(u_1 u_8 - u_2 u_7 c_8)] - I_1 \dot{u}_1 + u_2 u_3 (I_2 - I_3) = 0$$

$$R_2 + S_2 - M(\dot{u}_5 + u_3 u_4 - u_1 u_6) - m[\dot{u}_5 + u_3 u_4 - u_1 u_6 + \dot{u}_8$$
$$+ (\dot{u}_3 + u_1 u_2) b s_8 + 2 u_3 u_7 c_8 - (u_3^2 + u_1^2)(c - q_7 - bc_8)] = 0$$

$$\sigma + R_2 - m[\dot{u}_5 + u_3 u_4 - u_1 u_6 + \dot{u}_8 + (\dot{u}_3 + u_1 u_2) b s_8$$
$$+ 2 u_3 u_7 c_8 - (u_3^2 + u_1^2)(c - q_7 - bc_8)] = 0$$

4.15 Referring to Probs. 1.28 and 4.2, let P have a mass m, and A a mass M; regard the linkage as massless; assume that the central principal axes of inertia of A are parallel to $\mathbf{a}_1$, $\mathbf{a}_2$, $\mathbf{a}_3$; let the associated moments of inertia have the values I_1, I_2, I_3; and let N be a Newtonian reference frame. Introduce q_7 and q_8 as

$$q_7 \triangleq r - L \qquad q_8 \triangleq \theta - \frac{\pi}{2} \quad \text{rad}$$

where L is the natural length of the spring that connects P to L_2, and assume that the torsional springs at L_1 and L_4 are undeformed when $\theta = \pi/2$ rad. Finally, define $u_1, \ldots, u_8$ as

$$u_i \triangleq \begin{cases} \boldsymbol{\omega} \cdot \mathbf{a}_i & (i = 1, 2, 3) \\ \mathbf{v}^{A^*} \cdot \mathbf{a}_{i-3} & (i = 4, 5, 6) \\ b\dot{q}_8 & (i = 7) \\ b\dot{q}_8 \cos q_8 - \dot{q}_7 & (i = 8) \end{cases}$$

where $\boldsymbol{\omega}$ is the angular velocity of A in N, and $\mathbf{v}^{A^*}$ is the velocity of A^* in N.

In the absence of external forces, the system can remain at rest in N with $u_1 = \cdots = u_8 = q_7 = q_8 = 0$. Derive dynamical equations governing "small" departures from this state of rest, that is, equations linearized in $u_1, \ldots, u_8, q_7$, and q_8, taking $\sigma = -k_1 q_7$ and $\tau = -k_2 q_8$, where k_1 and k_2 are constants.

Results

$$[I_1 + m(c - L)^2]\dot{u}_1 + m(c - L)(\dot{u}_6 - b\dot{u}_2) = 0$$

$$(I_2 + mb^2)\dot{u}_2 - mb[(c - L)\dot{u}_1 + \dot{u}_6] = 0$$

$$[I_3 + m(c - L)^2 + mb^2]\dot{u}_3 + m[b(\dot{u}_5 + \dot{u}_8) - (c - L)\dot{u}_4] = 0$$

$$(M + m)\dot{u}_4 - m(c - L)\dot{u}_3 = 0$$

$$(M + m)\dot{u}_5 + m(\dot{u}_8 + b\dot{u}_3) = 0$$

$$(M + m)\dot{u}_6 + m[(c - L)\dot{u}_1 - b\dot{u}_2] = 0$$

$$k_1 q_7 - 2k_2 q_8/b = 0$$

$$m(\dot{u}_5 + \dot{u}_8 + b\dot{u}_3) + k_1 q_7 = 0$$

4.16 Figure P4.16 shows n identical rigid bars $B_1, \ldots, B_n$ forming an n-tuple pendulum attached to a rigid circularly cylindrical body C of radius R. The bars are connected to each other with pins whose axes all are parallel to the axis of C, and only motions during which these axes remain perpendicular to a plane that is fixed in a Newtonian reference frame N are considered.

Each bar has a length L and mass m; the mass center B_i^* of B_i is at the midpoint of B_i, and the moment of inertia of B_i about a line parallel to the pin axes and passing through B_i is I $(i = 1, \ldots, n)$. C has a mass M and an axial moment of inertia J. A linear torsional spring of modulus σ is located at each joint; when all springs are undeformed, all bars coincide with the line Z passing through the mass center C^* of C and through the point at which B_1 is attached to C.

Introduce mutually perpendicular unit vectors $\mathbf{c}_1, \mathbf{c}_2, \mathbf{c}_3$ as shown in Fig. P4.16. Define generalized speeds u_1, u_2, u_3 as

$$u_i \triangleq {}^N\mathbf{v}^{C^*} \cdot \mathbf{c}_i \qquad (i = 1, 2) \qquad u_3 \triangleq {}^N\boldsymbol{\omega}^C \cdot \mathbf{c}_3$$

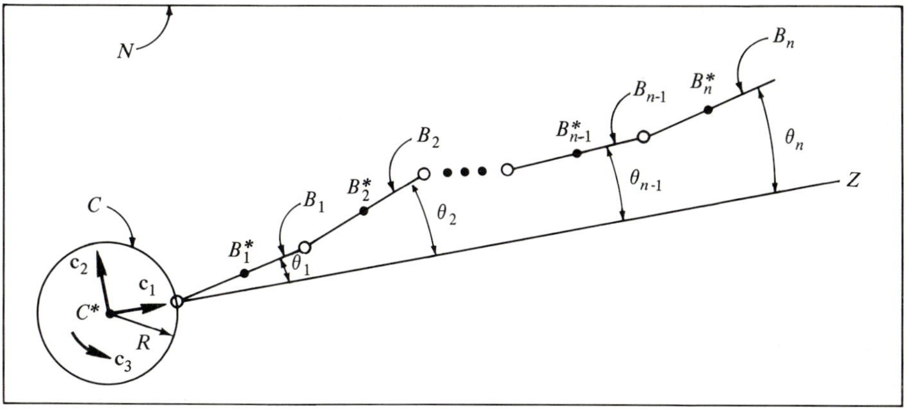

Figure P4.16

Let θ_i denote the angle between B_i and Z, and express θ_i as

$$\theta_i = \sum_{j=1}^{\nu} A_{ij} q_j \qquad (i = 1, \ldots, n)$$

where A_{ij} $(i = 1, \ldots, n; j = 1, \ldots, \nu)$ are constants. With u_{3+j} defined as

$$u_{3+j} \triangleq \dot{q}_j \qquad (j = 1, \ldots, \nu)$$

and assuming that $q_1, \ldots, q_\nu$ remain "small," verify that the dynamical equations governing $u_1, \ldots, u_{3+\nu}$ can be expressed as

$$(M + nm)\dot{u}_1 - m \sum_{i=1}^{n} \beta_i \dot{u}_3 = M u_2 u_3 - m \sum_{i=1}^{n} \gamma_i$$

$$(M + nm)\dot{u}_2 + m \sum_{i=1}^{n} \alpha_i \dot{u}_3 + mL \sum_{k=1}^{\nu} \sum_{i=1}^{n} \left(\sum_{j=1}^{i} A_{jk} - \frac{A_{ik}}{2} \right) \dot{u}_{3+k} = -\left(M u_3 u_1 + m \sum_{i=1}^{n} \delta_i \right)$$

$$m \sum_{i=1}^{n} \beta_i \dot{u}_1 - m \sum_{i=1}^{n} \alpha_i \dot{u}_2 - \left(m \sum_{i=1}^{n} \alpha_i^2 + nI + J \right) \dot{u}_3 - \sum_{k=1}^{\nu} \sum_{i=1}^{n} \left[mL\alpha_i \left(\sum_{j=1}^{i} A_{jk} - \frac{A_{ik}}{2} \right) + IA_{ik} \right] \dot{u}_{3+k} = m \sum_{i=1}^{n} [\beta_i u_3(u_2 + u_3\alpha_i) + \alpha_i \delta_i]$$

$$\begin{aligned} mL \sum_{i=1}^{n} \sum_{j=1}^{i} \left(A_{js}\theta_j - \frac{A_{is}\theta_i}{2} \right) \dot{u}_1 - mL \sum_{i=1}^{n} \left(\sum_{j=1}^{i} A_{js} - \frac{A_{is}}{2} \right) \dot{u}_2 \\ - \sum_{i=1}^{n} \left[mL\alpha_i \left(\sum_{j=1}^{i} A_{js} - \frac{A_{is}}{2} \right) + IA_{is} \right] \dot{u}_3 \\ - \sum_{k=1}^{\nu} \sum_{i=1}^{n} \left[mL^2 \left(\sum_{j=1}^{i} A_{jk} - \frac{A_{ik}}{2} \right) \left(\sum_{l=1}^{i} A_{ls} - \frac{A_{is}}{2} \right) + IA_{is}A_{ik} \right] \dot{u}_{3+k} \\ = mL \sum_{i=1}^{n} \left[u_3(u_2 + u_3\alpha_i) \left(\sum_{j=1}^{i} A_{js}\theta_j - \frac{A_{is}\theta_i}{2} \right) + \delta_i \left(\sum_{j=1}^{i} A_{js} - \frac{A_{is}}{2} \right) \right] \\ + \sigma \left[A_{1s}\theta_1 + \sum_{i=2}^{n} (A_{i-1,s} - A_{is})(\theta_{i-1} - \theta_i) \right] \qquad (s = 1, \ldots, \nu) \end{aligned}$$

where α_i, β_i, γ_i, δ_i are given by

$$\alpha_i = R + iL - \frac{L}{2}$$

$$\beta_i = L \sum_{j=1}^{i} \theta_j - \frac{L\theta_i}{2}$$

$$\gamma_i = -u_3\left(u_2 + u_3\alpha_i - L\sum_{k=1}^{\nu} A_{ik}u_{3+k} + 2L\sum_{j=1}^{i}\sum_{k=1}^{\nu} A_{jk}u_{3+k}\right)$$

$$\delta_i = u_3(u_1 - u_3\beta_i)$$

4.17 Figure P4.17 shows a planar double pendulum consisting of two identical, uniform, pin-connected rods, each of length L and mass m. Equal linear torsional springs of modulus σ are attached as indicated and both springs are undeformed when the angles q_1 and q_2 are equal to zero.

Considering "small" motions during which the pendulum remains in a horizontal plane, (*a*) show that the equations of motion can be written

$$M\ddot{q} + Sq = 0$$

where

$$q \triangleq \begin{bmatrix} q_1 \\ q_2 \end{bmatrix} \qquad M \triangleq \begin{bmatrix} 8 & 3 \\ 3 & 2 \end{bmatrix} \qquad S \triangleq \frac{\sigma}{mL^2}\begin{bmatrix} 12 & -6 \\ -6 & 6 \end{bmatrix}$$

(*b*) form the upper triangular matrix V such that $M = V^T V$; (*c*) letting $W \triangleq (V^{-1})^T S V^{-1}$, determine the eigenvalues of W and show that they are equal to those of $M^{-1}S$; (*d*) verify that if C_1 and C_2 are eigenvectors of W, then $V^{-1}C_1$ and $V^{-1}C_2$ are the corresponding eigenvectors of $M^{-1}S$.

Results (*b*) $\begin{bmatrix} \sqrt{8} & 3/\sqrt{8} \\ 0 & \sqrt{7}/\sqrt{8} \end{bmatrix}$

(*c*) $0.3409\sigma/(mL^2)$, $15.09\sigma/(mL^2)$

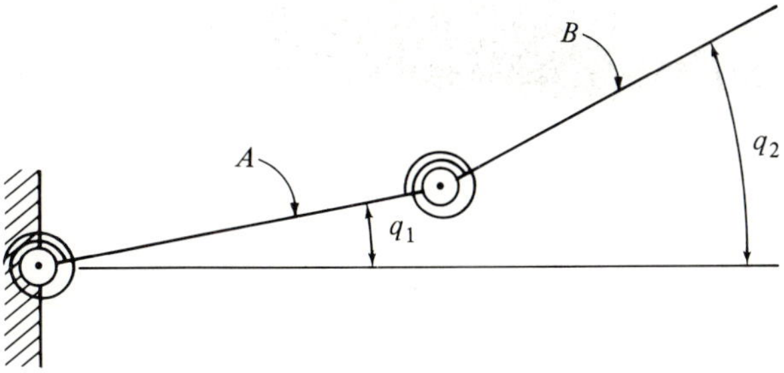

Figure P4.17

4.18 After showing that there exist no values of x_1 and x_2 such that all five of the equations $x_1 + 2x_2 = 1$, $2x_1 + 3x_2 = 2$, $3x_1 + 4x_2 = 1$, $4x_1 + 5x_2 = 2$, and $5x_1 + 6x_2 = 1$ are satisfied simultaneously, (*a*) find values x_1* and x_2* of x_1 and x_2, respectively, representing the "best approximate solution" of the equations, using the method employed to solve Eq. (4.9.156); (*b*) with $S(x_1, x_2)$ defined as

$$S(x_1, x_2) \triangleq (x_1 + 2x_2 - 1)^2 + (2x_1 + 3x_2 - 2)^2 + \cdots + (5x_1 + 6x_2 - 1)^2$$

evaluate $S(x_1{}^*, x_2{}^*)$ and compare the result with $S(\bar{x}_1, \bar{x}_2)$, where $\bar{x}_1$ and $\bar{x}_2$ are any numbers whatsoever.

Results (*a*) -1.4, 1.4
(*b*) 1.2, $S(x_1{}^*, x_2{}^*) < S(\bar{x}_1, \bar{x}_2)$

4.19 Figure P4.19 shows a plane truss T that has n joints $P_1, \ldots, P_n$, and moves in its own plane, which is fixed in a Newtonian reference frame N. Mutually perpendicular lines N_1 and N_2 are fixed in N and intersect at point N^*. Lines R_1 and R_2, intersecting at point R^*, are fixed in an auxiliary reference frame R (not shown) that moves relative to N, this motion being characterized by the distances z_1 and z_2 and the angle θ.

Dynamical equations governing the motion of T in N are to be formulated after T has been discretized as a set of particles placed at $P_1, \ldots, P_n$ and connected by massless, linear springs, the particle at P_i having a mass m_i equal to half the total mass of all truss members meeting at P_i. A force $\mathbf{Q}_i$ is presumed to be applied to T at P_i.

Introduce $3 + \nu$ generalized speeds as

$$u_1 \triangleq \dot{\theta} \qquad u_2 \triangleq {}^N\mathbf{v}^{R^*} \cdot \mathbf{r}_1 \qquad u_3 \triangleq {}^N\mathbf{v}^{R^*} \cdot \mathbf{r}_2 \qquad u_{3+k} \triangleq \dot{q}_k \qquad (k = 1, \ldots, \nu)$$

where ${}^N\mathbf{v}^{R^*}$ is the velocity of R^* in N, $1 \le \nu \le n$, and q_k is a generalized coordinate

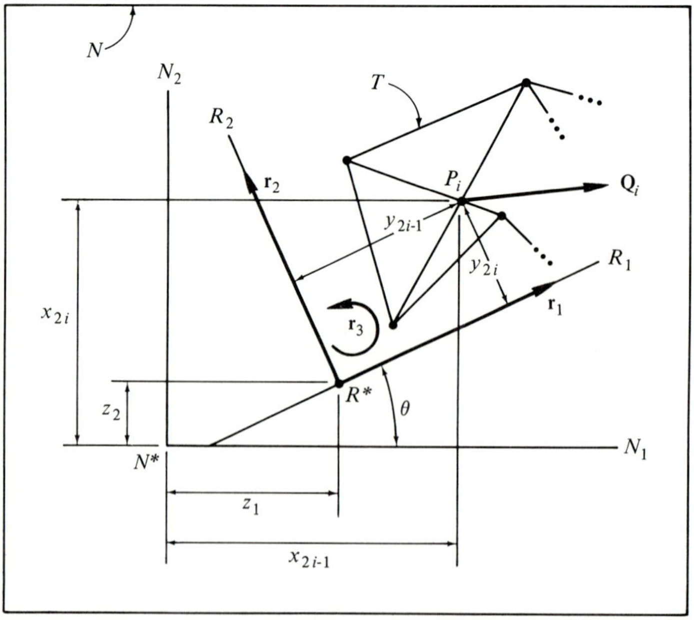

Figure P4.19

brought into the discussion as follows: if x_{2i-1}, x_{2i} and y_{2i-1}, y_{2i} are the Cartesian coordinates of P_i relative to N_1, N_2 and R_1, R_2, respectively (see Fig. P4.19), and $\hat{x}_1, \ldots, \hat{x}_{2n}$ are values of $x_1, \ldots, x_{2n}$, respectively, such that T is undeformed, let $A_{jk}(j = 1, \ldots, 2n;\ k = 1, \ldots, \nu)$ be constants and express y_j as

$$y_j = \hat{x}_j + \sum_{k=1}^{\nu} A_{jk} q_k \qquad (j = 1, \ldots, 2n)$$

Show that if A_{jk} $(j = 1, \ldots, 2n;\ k = 1, \ldots, \nu)$ are formed like their counterparts in Sec. 4.9 [see Eq. (4.9.123)], then the dynamical equations governing $u_1, \ldots, u_{3+\nu}$ can be expressed as

$$I\dot{u}_1 - w_2\dot{u}_2 + w_1\dot{u}_3 + \sum_{j=1}^{\nu}\sum_{i=1}^{n} m_i(y_{2i-1}A_{2i,j} - y_{2i}A_{2i-1,j})\dot{u}_{3+j}$$
$$= \sum_{i=1}^{n} [(Q_{2i} - m_i a_{i2})y_{2i-1} - (Q_{2i-1} - m_i a_{i1})y_{2i}]$$

$$-w_2\dot{u}_1 + \overline{m}\,\dot{u}_2 = \sum_{i=1}^{n} (Q_{2i-1} - m_i a_{i1})$$

$$w_1\dot{u}_1 + \overline{m}\,\dot{u}_3 = \sum_{i=1}^{n} (Q_{2i} - m_i a_{i2})$$

$$-\sum_{i=1}^{n} m_i(y_{2i-1}A_{2i,k} - y_{2i}A_{2i-1,k})\dot{u}_1 - \dot{u}_{3+k}$$
$$= -G_k + \lambda_k q_k + \sum_{i=1}^{n} m_i(A_{2i-1,k}a_{i1} + A_{2i,k}a_{i2}) \qquad (k = 1, \ldots, \nu)$$

where

$$\overline{m} \triangleq \sum_{i=1}^{n} m_i \qquad I \triangleq \sum_{i=1}^{n} m_i(y_{2i-1}^2 + y_{2i}^2)$$

$$w_j \triangleq \sum_{i=1}^{n} m_i y_{2i-2+j} \qquad (j = 1, 2)$$

$$Q_{2i-1} \triangleq \mathbf{Q}_i \cdot \mathbf{r}_1 \qquad Q_{2i} \triangleq \mathbf{Q}_i \cdot \mathbf{r}_2 \qquad (i = 1, \ldots, n)$$

$$a_{i1} \triangleq -u_1\left(2\sum_{k=1}^{\nu} A_{2i,k}u_{3+k} + u_3 + u_1 y_{2i-1}\right) \qquad (i = 1, \ldots, n)$$

$$a_{i2} \triangleq u_1\left(2\sum_{k=1}^{\nu} A_{2i-1,k}u_{3+k} + u_2 - u_1 y_{2i}\right) \qquad (i = 1, \ldots, n)$$

$$G_k \triangleq \sum_{i=1}^{n} (A_{2i-1,k}Q_{2i-1} + A_{2i,k}Q_{2i}) \qquad (k = 1, \ldots, \nu)$$

and where $\lambda_1, \ldots, \lambda_\nu$ are any ν nonzero eigenvalues of $M^{-1/2}\,SM^{-1/2}$, arranged in ascending order, S and M being, respectively, the stiffness matrix of T and the diagonal mass matrix of T (see Sec. 4.9).

4.20 A complete description of the motion of truss T of Prob. 4.19 is said to be in hand when $x_1, \ldots, x_{2n}$ are known as functions of time. To obtain such a description, one can express $x_1, \ldots, x_{2n}$ in terms of z_1, z_2, θ, and $q_1, \ldots, q_\nu$ after determining these quantities by solving simultaneously the differential equations $\dot{q}_k = u_{3+k}$ $(k = 1, \ldots, \nu)$, the $3 + \nu$ dynamical equations developed in Prob. 4.19, and three additional kinematical differential equations that relate θ, z_1, z_2 to u_1, u_2, u_3.

(*a*) Verify that $x_1, \ldots, x_{2n}$ are related to z_1, z_2, θ and $q_1, \ldots, q_\nu$ by the equations

$$x_{2i-1} = z_1 + y_{2i-1} \cos\theta - y_{2i} \sin\theta \qquad (i = 1, \ldots, n)$$

$$x_{2i} = z_2 + y_{2i-1} \sin\theta + y_{2i} \cos\theta \qquad (i = 1, \ldots, n)$$

(*b*) Show that the needed kinematical differential equations are

$$\dot{\theta} = u_1 \qquad \dot{z}_1 = u_2 \cos\theta - u_3 \sin\theta \qquad \dot{z}_2 = u_2 \sin\theta + u_3 \cos\theta$$

(*c*) Assuming that the initial values of x_i and $\dot{x}_i$, denoted by $x_i(0)$ and $\dot{x}_i(0)$ $(i = 1, \ldots, 2n)$, are given and that $\hat{x}_i \approx x_i(0)$ $(i = 1, \ldots, 2n)$, show that $z_1(0)$, $z_2(0)$, $\theta(0)$, and $q_k(0)$ $(k = 1, \ldots, \nu)$ must satisfy the $2n$ linear equations

$$z_1(0) - x_{2i}(0)\,\theta(0) + \sum_{k=1}^{\nu} A_{2i-1,k}\, q_k(0) = x_{2i-1}(0) - \hat{x}_{2i-1} \qquad (i = 1, \ldots, n)$$

$$z_2(0) + x_{2i-1}(0)\,\theta(0) + \sum_{k=1}^{\nu} A_{2i,k}\, q_k(0) = x_{2i}(0) - \hat{x}_{2i} \qquad (i = 1, \ldots, n)$$

while $u_1(0)$, $u_2(0)$, $u_3(0)$, and $u_{3+k}(0)$ $(k = 1, \ldots, \nu)$ are governed by the linear equations

$$-\,y_{2i}(0)u_1(0) + u_2(0) + \sum_{k=1}^{\nu} A_{2i-1,k}\, u_{3+k}(0) = \dot{x}_{2i-1}(0) + \dot{x}_{2i}(0)\,\theta(0) \qquad (i = 1, \ldots, n)$$

$$y_{2i-1}(0)u_1(0) + u_3(0) + \sum_{k=1}^{\nu} A_{2i,k}\, u_{3+k}(0) = -\,\dot{x}_{2i-1}(0)\,\theta(0) + \dot{x}_{2i}(0) \qquad (i = 1, \ldots, n)$$

where $y_j(0)$ is given by

$$y_j(0) = \hat{x}_j + \sum_{k=1}^{\nu} A_{jk}\, q_k(0) \qquad (j = 1, \ldots, 2n)$$

and is, therefore, available as soon as the equations governing $z_1(0)$, $z_2(0)$, $\theta(0)$, and $q_k(0)$ $(k = 1, \ldots, \nu)$ have been solved.

4.21 Figure P4.21(a) shows a triangular truss T that moves in its own plane, which is fixed in a Newtonian reference frame N. Mutually perpendicular lines N_1 and N_2 are fixed in N and intersect at point N^*. Lines R_1 and R_2, intersecting at point R^*, are fixed in an auxiliary reference frame R (not shown) that moves relative to N,

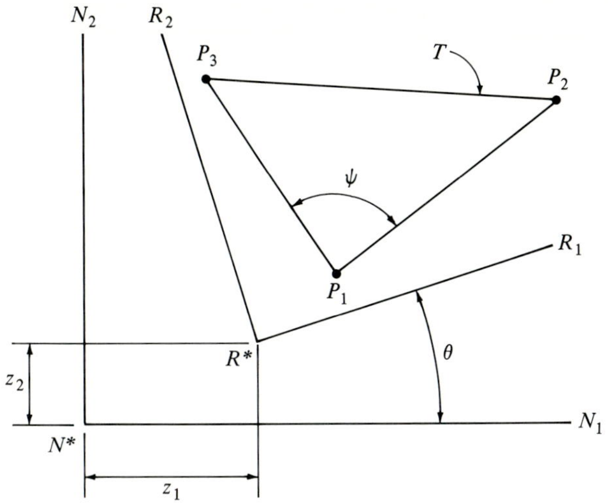

Figure P4.21(a)

this motion being characterized by the distances z_1 and z_2 and the angle θ. Each truss member has a Young's modulus of 10^9 N/m^2, a cross-section area of 10^{-4} m^2, and mass per unit of volume of 10^4 kg/m^3; and, when T is undeformed, the members connecting points P_1 and P_2, P_2 and P_3, and P_3 and P_1 have lengths of 4 m, 5 m, and 3 m, respectively. The N_1, N_2 coordinates of joint P_i are x_{2i-1}, x_{2i}, respectively ($i = 1, 2, 3$), and x_j has the value $\hat{x}_j$ ($j = 1, \ldots, 6$) when T is undeformed.

(*a*) Referring to Prob. 4.20, and taking $\hat{x}_1 = 0$, $\hat{x}_2 = 0$, $\hat{x}_3 = 4$ m, $\hat{x}_4 = 0$, $\hat{x}_5 = 0$, $\hat{x}_6 = 3$ m, verify that the constants A_{ij} ($i = 1, \ldots, 6$; $j = 1, 2, 3$) are given by

$$[A_{ij}] = \begin{bmatrix} -0.285761 & -0.316759 & -0.0767620 \\ -0.183059 & 0.175246 & -0.324885 \\ -0.0506765 & 0.321784 & 0.135214 \\ 0.204701 & -0.0565620 & -0.0566329 \\ 0.307052 & -0.0848430 & -0.0849493 \\ -0.0701124 & -0.0897081 & 0.347987 \end{bmatrix}$$

(*b*) Taking

$$\begin{aligned} x_1(0) &= 0.01 \text{ m} & \dot{x}_1(0) &= 0.01 \text{ m/sec} \\ x_2(0) &= 0.02 \text{ m} & \dot{x}_2(0) &= -0.01 \text{ m/sec} \\ x_3(0) &= 4.03 \text{ m} & \dot{x}_3(0) &= 0.02 \text{ m/sec} \\ x_4(0) &= -0.01 \text{ m} & \dot{x}_4(0) &= -0.02 \text{ m/sec} \\ x_5(0) &= -0.02 \text{ m} & \dot{x}_5(0) &= 0.03 \text{ m/sec} \\ x_6(0) &= 3.04 \text{ m} & \dot{x}_6(0) &= -0.03 \text{ m/sec} \end{aligned}$$

determine the quantities $Z_1, \ldots, Z_{30}$ in Table P4.21.

Table P4.21

	$\nu = 1$	$\nu = 2$	$\nu = 3$
$\theta(0)$ (deg)	Z_1	Z_9	Z_{19}
$z_1(0)$ (m)	Z_2	Z_{10}	Z_{20}
$z_2(0)$ (m)	Z_3	Z_{11}	Z_{21}
$q_1(0)$ (kg$^{1/2}$m)	Z_4	Z_{12}	Z_{22}
$q_2(0)$ (kg$^{1/2}$m)	. . .	Z_{13}	Z_{23}
$q_3(0)$ (kg$^{1/2}$m)	. . .	. . .	Z_{24}
$u_1(0)$ (rad/sec)	Z_5	Z_{14}	Z_{25}
$u_2(0)$ (m/sec)	Z_6	Z_{15}	Z_{26}
$u_3(0)$ (m/sec)	Z_7	Z_{16}	Z_{27}
$u_4(0)$ (kg$^{1/2}$m/sec)	Z_8	Z_{17}	Z_{28}
$u_5(0)$ (kg$^{1/2}$m/sec)	. . .	Z_{18}	Z_{29}
$u_6(0)$ (kg$^{1/2}$m/sec)	. . .	. . .	Z_{30}

(*c*) In Sec. 4.9, linear structural theory is used to characterize truss deformations. Investigate the applicability of this theory to motions during which truss deformations become large, proceeding as follows.

Let forces $\mathbf{Q}_1$, $\mathbf{Q}_2$, $\mathbf{Q}_3$ be applied, respectively, to joints P_1, P_2, P_3 of T, in such a way that $\mathbf{Q}_1$ is directed from P_1 to P_2, $\mathbf{Q}_2$ from P_2 to P_3, and $\mathbf{Q}_3$ from P_3 to P_1. Let the magnitudes of $\mathbf{Q}_1$, $\mathbf{Q}_2$, $\mathbf{Q}_3$ be 4000 N, 5000 N, 3000 N, respectively. (Note that if T were rigid, this system of forces would be a couple.) Referring to Prob. 4.19, verify that $\lambda_1 = 5.87615 \times 10^3$ sec^{-2}, $\lambda_2 = 1.43990 \times 10^4$ sec^{-2}, $\lambda_3 = 1.97248 \times 10^4$ sec^{-2}; take $\nu = 3$, $x_i(0) = \hat{x}_i$ $(i = 1, \ldots, 6)$ [see (*a*) for $\hat{x}_i$], $\dot{x}_i(0) = 0$ $(i = 1, \ldots, 6)$; and perform a numerical integration of the equations of motion of T from $t = 0$ to $t = 0.6$ sec. Use the results of this numerical integration to plot the angle $\phi \triangleq \psi - 90°$ versus t, where ψ is the angle shown in Fig. P4.21(a).

(*d*) Treating T as a system of three particles connected by massless linear springs, verify that exact differential equations of motion of this set of particles can be written as

$$m_1\dot{v}_1 - \frac{[FL_1 + k_1(d_1 - L_1)](x_3 - x_1)}{d_1} + \frac{k_3(d_3 - L_3)(x_1 - x_5)}{d_3} = 0$$

$$m_1\dot{v}_2 - \frac{[FL_1 + k_1(d_1 - L_1)](x_4 - x_2)}{d_1} + \frac{k_3(d_3 - L_3)(x_2 - x_6)}{d_3} = 0$$

$$m_2\dot{v}_3 - \frac{[FL_2 + k_2(d_2 - L_2)](x_5 - x_3)}{d_2} + \frac{k_1(d_1 - L_1)(x_3 - x_1)}{d_1} = 0$$

$$m_2\dot{v}_4 - \frac{[FL_2 + k_2(d_2 - L_2)](x_6 - x_4)}{d_2} + \frac{k_1(d_1 - L_1)(x_4 - x_2)}{d_1} = 0$$

$$m_3\dot{v}_5 - \frac{[FL_3 + k_3(d_3 - L_3)](x_1 - x_5)}{d_3} + \frac{k_2(d_2 - L_2)(x_5 - x_3)}{d_2} = 0$$

$$m_3\dot{v}_6 - \frac{[FL_3 + k_3(d_3 - L_3)](x_2 - x_6)}{d_3} + \frac{k_2(d_2 - L_2)(x_6 - x_4)}{d_2} = 0$$

$$\dot{x}_i = v_i \qquad (i = 1, \ldots, 6)$$

where

$$d_1 \triangleq [(x_3 - x_1)^2 + (x_4 - x_2)^2]^{1/2}$$
$$d_2 \triangleq [(x_5 - x_3)^2 + (x_6 - x_4)^2]^{1/2}$$
$$d_3 \triangleq [(x_1 - x_5)^2 + (x_2 - x_6)^2]^{1/2}$$

while $F = 1000$ N/m, $L_1 = 4$ m, $L_2 = 5$ m, $L_3 = 3$ m, $m_1 = 3.5$ kg, $m_2 = 4.5$ kg, $m_3 = 4.0$ kg, $k_1 = 2.5 \times 10^4$ N/m, $k_2 = 2.0 \times 10^4$ N/m, $k_3 = 3.33333 \times 10^4$ N/m.

(*e*) Take $x_i(0) = \hat{x}_i$ [see (*a*)], $v_i(0) = 0$ $(i = 1, \ldots, 6)$; for $0 \le t \le 0.6$ sec, perform a numerical integration of the equations given in (*d*); use the results of this numerical integration to plot the angle $\overline{\phi} \triangleq \overline{\psi} - 90°$ versus t, where $\overline{\psi}$ is the angle between the line connecting P_1 and P_2 and the line connecting P_1 and P_3. Display this plot and the one generated in (*c*) on the same set of axes.

Results (*b*) -1.06×10^{-1}, 4.09×10^{-3}, 1.80×10^{-2}, -7.04×10^{-2}, -1.70×10^{-3}, 1.85×10^{-2}, -1.74×10^{-2}, 1.82×10^{-2}, -9.69×10^{-2}, 5.23×10^{-3}, 1.74×10^{-2}, -7.19×10^{-2}, 3.73×10^{-2}, -1.61×10^{-3}, 1.91×10^{-2}, -1.77×10^{-2}, 1.75×10^{-2}, 1.86×10^{-2}, -1.05×10^{-1}, 5.64×10^{-3}, 1.82×10^{-2}, -7.47×10^{-2}, 3.97×10^{-2}, 5.78×10^{-2}, -1.55×10^{-3}, 1.89×10^{-2}, -1.81×10^{-2}, 1.87×10^{-2}, 1.75×10^{-2}, -2.60×10^{-2}

(*c*) Figure P4.21(b)

4.22 Transverse vibrations of an unrestrained beam of length L, constant flexural rigidity EI, and constant mass per unit of length ρ are governed by Eq. (4.10.1) and by the boundary conditions $y''(0, t) = y'''(0, t) = y''(L, t) = y'''(L, t) = 0$. Show that the general solution of Eq. (4.10.1) satisfying these boundary conditions can be expressed as

$$y = \sum_{i=1}^{\infty} \overline{\phi}_i \overline{q}_i$$

where

$$\overline{\phi}_i = \cosh\frac{\lambda_i x}{L} + \cos\frac{\lambda_i x}{L} - \frac{\cosh\lambda_i - \cos\lambda_i}{\sinh\lambda_i - \sin\lambda_i}\left(\sinh\frac{\lambda_i x}{L} + \sin\frac{\lambda_i x}{L}\right)$$

$$\overline{q}_i = \alpha_i \cos p_i t + \beta_i \sin p_i t$$

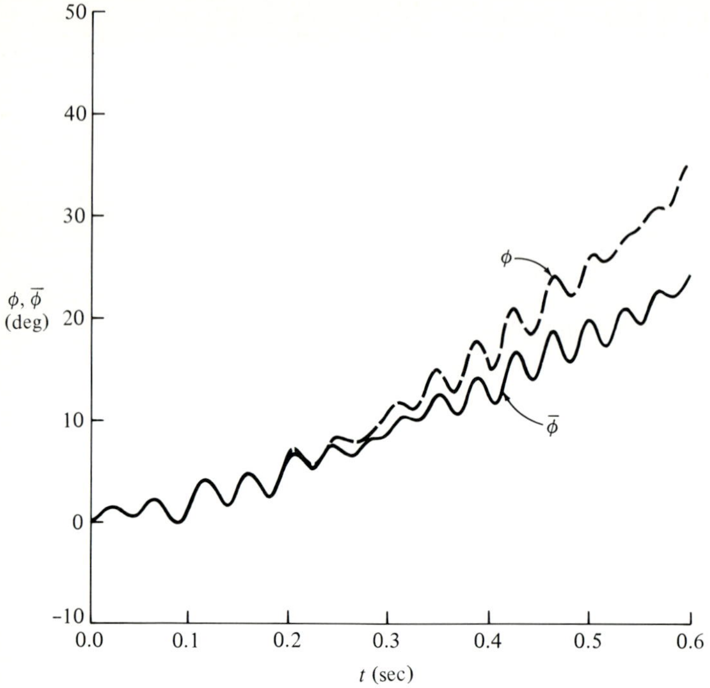

Figure P4.21(b)

and λ_i ($i = 1, \ldots, \infty$) are the consecutive roots of the transcendental equation†

$$\cos \lambda \cosh \lambda - 1 = 0$$

while

$$p_i \triangleq \left(\frac{\lambda_i}{L}\right)^2 \left(\frac{EI}{\rho}\right)^{1/2}$$

and α_i and β_i are constants that depend upon initial conditions.

4.23 Figure P4.23 shows an unrestrained, uniform beam B of length L, constant flexural rigidity EI, and constant mass per unit of length ρ. B moves in a Newtonian reference frame N under the action of gravitational forces exerted by a particle O that has a mass m_O and is fixed in N. Only planar motions of B are to be considered.

Dynamical equations governing the motion of B in N are to be formulated. To this end, an auxiliary reference frame R that moves in N is introduced, as are mutually perpendicular lines R_1 and R_2 fixed in R and intersecting at a point R^*, as shown in Fig. P4.23, where D denotes the distance from O to R^*, θ is the angle between R_1 and the line Z determined by O and R^*, and Q is one of the endpoints of B. R_2 is required to pass through Q at all times.

† The first 30 positive roots of this equation are listed in Table 4.10.1.

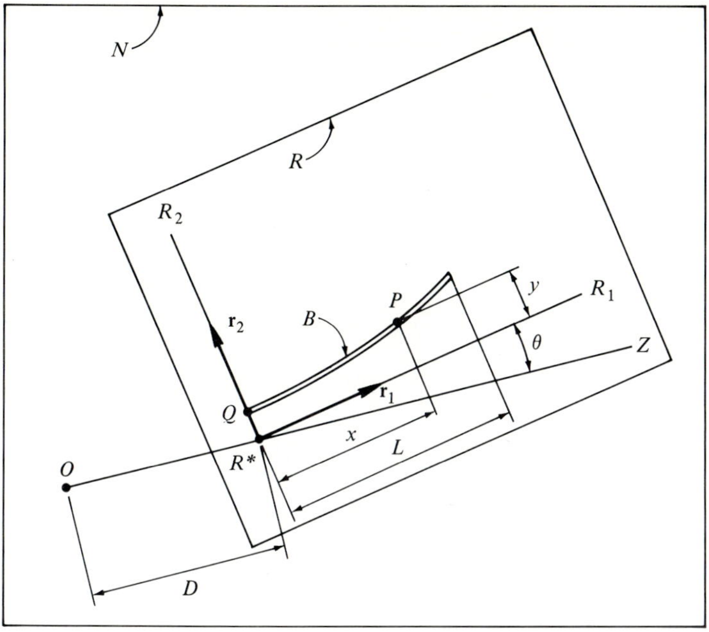

Figure P4.23

Express the distance y from R_1 to a generic point P on the neutral axis of B as

$$y = \sum_{i=1}^{\nu} \phi_i q_i$$

where $\phi_1, \ldots, \phi_\nu$ $(1 \le \nu \le \infty)$ are functions of the distance x from P to R_2, and $q_1, \ldots, q_\nu$ are functions of the time t; let $\gamma \triangleq Gm_O$, where G is the universal gravitational constant; introduce $3 + \nu$ generalized speeds as

$$u_1 \triangleq {}^N\mathbf{v}^{R^*} \cdot \mathbf{r}_1 \qquad u_2 \triangleq {}^N\mathbf{v}^{R^*} \cdot \mathbf{r}_2 \qquad u_3 \triangleq {}^N\boldsymbol{\omega}^R \cdot \mathbf{r}_3$$

$$u_{3+k} \triangleq \dot{q}_k \qquad (k = 1, \ldots, \nu)$$

where ${}^N\mathbf{v}^{R^*}$ is the velocity of R^* in N, ${}^N\boldsymbol{\omega}^R$ is the angular velocity of R in N, $\mathbf{r}_1$ and $\mathbf{r}_2$ are unit vectors directed as shown in Fig. P4.23, and $\mathbf{r}_3 \triangleq \mathbf{r}_1 \times \mathbf{r}_2$.

Making use of Eqs. (4.10.20), (4.10.21), and (4.10.37), assuming that $D >> L$, and proceeding as in the derivation of Eqs. (4.10.48)–(4.10.51) to deal with the gravitational forces exerted on B by O, formulate the dynamical differential equations governing $u_1, \ldots, u_{3+\nu}$.

Results $\dot{u}_1 m_B - \dot{u}_3 \sum_{i=1}^{\nu} q_i E_i = m_B u_2 u_3 + e_B u_3^2 + 2u_3 \sum_{i=1}^{\nu} u_{3+i} E_i$

$$+ \left(\frac{\gamma}{D^2}\right)\left[\frac{e_B(3\cos^2\theta - 1)}{D} - m_B\cos\theta - 3\sin\theta\cos\theta \sum_{i=1}^{\nu} \frac{E_i q_i}{D}\right]$$

$$\dot{u}_2 m_B + \dot{u}_3 e_B + \sum_{i=1}^{\nu} \dot{u}_{3+i} E_i = -m_B u_3 u_1 + u_3^2 \sum_{i=1}^{\nu} q_i E_i$$
$$+ \left(\frac{\gamma}{D^2}\right)\left[m_B \sin\theta - \frac{3e_B \sin\theta\cos\theta}{D} + (3\sin^2\theta - 1)\sum_{i=1}^{\nu}\frac{E_i q_i}{D}\right]$$

$$\dot{u}_1 \sum_{i=1}^{\nu} q_i E_i - \dot{u}_2 e_B - I_B \dot{u}_3 - \sum_{i=1}^{\nu} \dot{u}_{3+i} F_i = e_B u_3 u_1 + u_2 u_3 \sum_{i=1}^{\nu} q_i E_i$$
$$- \left(\frac{\gamma}{D^2}\right)\left\{e_B \sin\theta - \frac{3I_B \sin\theta\cos\theta}{D} + \sum_{i=1}^{\nu}\left[E_i\cos\theta + \frac{3F_i(1 - 2\cos^2\theta)}{D}\right]q_i\right\}$$

$$\dot{u}_2 E_j + \dot{u}_3 F_j + \sum_{i=1}^{\nu} \dot{u}_{3+i} G_{ij} = -u_3 u_1 E_j + \sum_{i=1}^{\nu}(u_3^2 G_{ij} - H_{ij})q_i$$
$$+ \left(\frac{\gamma}{D^2}\right)\left[E_j \sin\theta - \frac{3F_j \sin\theta\cos\theta}{D} + (3\sin^2\theta - 1)\sum_{i=1}^{\nu}\frac{G_{ij} q_i}{D}\right]$$
$$(j = 1, \ldots, \nu)$$

4.24 Using for ϕ_i in Eqs. (4.10.21) and (4.10.37) the function $\overline{\phi}_i$ of Prob. 4.22, show that $E_i = F_i = 0$ $(i = 1, \ldots, \nu)$ while $G_{ij} = m_B\,\delta_{ij}$ and $H_{ij} = p_i^2 m_B\,\delta_{ij}$ $(i, j = 1, \ldots, \nu)$, where p_i is given by Eq. (4.10.7) and δ_{ij} is the Kronecker delta.

4.25 In Prob. 4.23, $3 + \nu$ dynamical differential equations were formulated. Since these equations involve the $5 + 2\nu$ variables $D, \theta, u_1, \ldots, u_{3+\nu}, q_1, \ldots, q_\nu$, they do not suffice for the determination of these variables. Show that the $2 + \nu$ required additional equations are the kinematical relationships $\dot{q}_i = u_{3+i}$ $(i = 1, \ldots, \nu)$, $\dot{D} = u_1\cos\theta - u_2\sin\theta$, $\dot{\theta} = u_3 - (u_1\sin\theta + u_2\cos\theta)/D$.

4.26 Figure P4.26(a) shows a system consisting of a block A, a uniform beam B, and a light linear spring S. A is constrained to move in a guide G, one end of B is clamped to A, and S connects A to a fixed support.

If $w(x, t)$ denotes the displacement of a point P located a distance x from the clamped end of B ($w = 0$ when both B and S are undeformed), EI is the flexural rigidity of B, σ is the modulus of S, L is the length of B, ρ is the mass per unit of length of B, and m_A is the mass of A, then w is governed by the partial differential equation

$$EI\frac{\partial^4 w}{\partial x^4} + \rho\frac{\partial^2 w}{\partial t^2} = 0$$

together with the (time-dependent) boundary conditions

$$\frac{\partial w(0, t)}{\partial x} = \frac{\partial^2 w(L, t)}{\partial x^2} = \frac{\partial^3 w(L, t)}{\partial x^3} = 0$$

$$m_A\frac{\partial^2 w(0, t)}{\partial t^2} + EI\frac{\partial^3 w(0, t)}{\partial x^3} + \sigma w(0, t) = 0$$

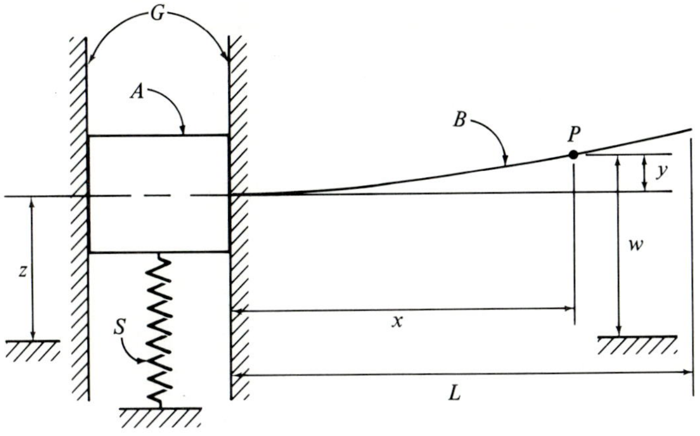

Figure P4.26(a)

and the initial conditions

$$w(x, 0) = g(x) \qquad \frac{\partial w(x, 0)}{\partial t} = h(x)$$

where $g(x)$ and $h(x)$ are specified functions of x. The solution of the partial differential equation can be expressed as

$$w(x, t) = \sum_{i=1}^{\infty} [f_i(x)(B_i \sin \omega_i t + C_i \cos \omega_i t)]$$

where

$$f_i(x) \triangleq \sin \frac{\lambda_i x}{L} + \alpha_i \cos \frac{\lambda_i x}{L} - \sinh \frac{\lambda_i x}{L} + \beta_i \cosh \frac{\lambda_i x}{L}$$

Here λ_i $(i = 1, \ldots, \infty)$ are the roots of the equation

$$\left(\lambda^4 - \frac{\rho\sigma L^4}{m_A EI}\right)\left[e^{-\lambda} + \tfrac{1}{2}(1 + e^{-2\lambda}) \cos \lambda\right] + \frac{\rho L}{m_A} \lambda^3 \left[\tfrac{1}{2}(1 + e^{-2\lambda}) \sin \lambda + \tfrac{1}{2}(1 - e^{-2\lambda}) \cos \lambda\right] = 0$$

and the constants α_i, β_i, ω_i, B_i, and C_i are defined as

$$\alpha_i \triangleq \{e^{-\lambda_i} \sin \lambda_i + \tfrac{1}{2}(1 - e^{-2\lambda_i}) + \rho L(1 + e^{-2\lambda_i})[1 - \rho\sigma L^4/(m_A EI\lambda_i{}^4)]^{-1}(m_A \lambda_i)^{-1}\}/[- e^{-\lambda_i} \cos \lambda_i - \tfrac{1}{2}(1 + e^{-2\lambda_i})]$$

$$\beta_i \triangleq \{2\rho L(m_A \lambda_i)^{-1} e^{-\lambda_i}[1 - \rho\sigma L^4/(m_A EI\lambda_i{}^4)]^{-1} \cos \lambda_i - e^{-\lambda_i} \sin \lambda_i - \tfrac{1}{2}(1 - e^{-2\lambda_i})\}/[- e^{-\lambda_i} \cos \lambda_i - \tfrac{1}{2}(1 + e^{-2\lambda_i})]$$

$$\omega_i \triangleq \left(\frac{\lambda_i}{L}\right)^2 \left(\frac{EI}{\rho}\right)^{1/2}$$

$$B_i \triangleq \frac{\int_0^L h(x) f_i(x)\,dx + (m_A/\rho)\,h(0) f_i(0)}{\omega_i \left[\int_0^L f_i^2(x)\,dx + (m_A/\rho) f_i^2(0)\right]}$$

$$C_i \triangleq \frac{\int_0^L g(x) f_i(x)\,dx + (m_A/\rho)\,g(0) f_i(0)}{\int_0^L f_i^2(x)\,dx + (m_A/\rho) f_i^2(0)}$$

An alternative description of $w(x, t)$, obtained as in Sec. 4.10, can be produced as follows: let y denote the deflection of B at P from its undeformed position [Fig. P4.26(a)], and express y as

$$y = \sum_{i=1}^{\nu} \overline{\phi}_i q_i$$

where $\overline{\phi}_i$ is given by Eq. (4.10.4), q_i is a function of t, and ν is a positive integer. Introduce z as $z \triangleq w - y$ (z is the "stretch" of S), and, after defining generalized speeds $u_1, \ldots, u_{1+\nu}$ as $u_1 \triangleq \dot{z}$, $u_{1+i} \triangleq \dot{q}_i$ $(i = 1, \ldots, \nu)$, solve the dynamical equations of motion governing $u_1, \ldots, u_{1+\nu}$, together with the kinematical equations $\dot{z} = u_1$, $\dot{q}_i = u_{1+i}$ $(i = 1, \ldots, \nu)$. Finally, form w as

$$w = z + \sum_{i=1}^{\nu} \overline{\phi}_i q_i$$

(*a*) Formulate the dynamical differential equations governing $u_1, \ldots, u_{1+\nu}$.

(*b*) Taking $m_A = 10$ kg, $L = 10$ m, $\rho = 1000$ kg/m, $EI = 100$ N $\cdot$ m^2, $\sigma = 394.784$ N/m (when σ has this value, the system formed by A and S alone has a natural vibration period of 1 sec), and, assuming that the system starts from rest with B undeformed and S extended by 0.1 m, make time-plots, for $0 \le t \le 10$ sec, of the displacement $w(L, t)$ of the free end of B, using each of the two above-mentioned methods. Make two sets of plots, first using five terms in the exact solution and $\nu = 4$, and then using nine terms in the exact solution and $\nu = 8$. Comment briefly on the relative amounts of computational effort required to implement each of the two methods.

Results (*a*) $(m_A + \rho L)\dot{u}_1 + \sum_{i=1}^{\nu} E_i \dot{u}_{1+i} + \sigma z = 0$

$$E_j \dot{u}_1 + \rho L \dot{u}_{1+j} + \frac{EI \lambda_j^4 q_j}{L^3} = 0 \qquad (j = 1, \ldots, \nu)$$

where $\lambda_1, \ldots, \lambda_\nu$ appear in the first column of Table 4.10.1, and $E_1, \ldots, E_\nu$ are given by Eq. (4.10.66).

(*b*) Figures P4.26(b) and P4.26(c).

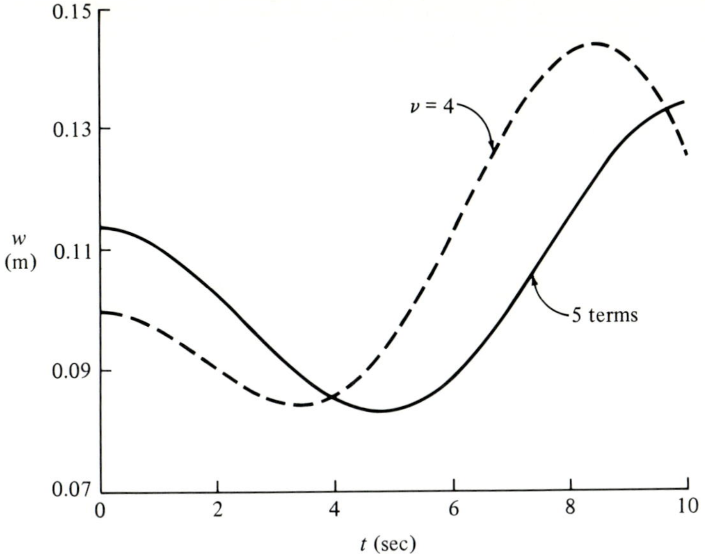

Figure P4.26(b)

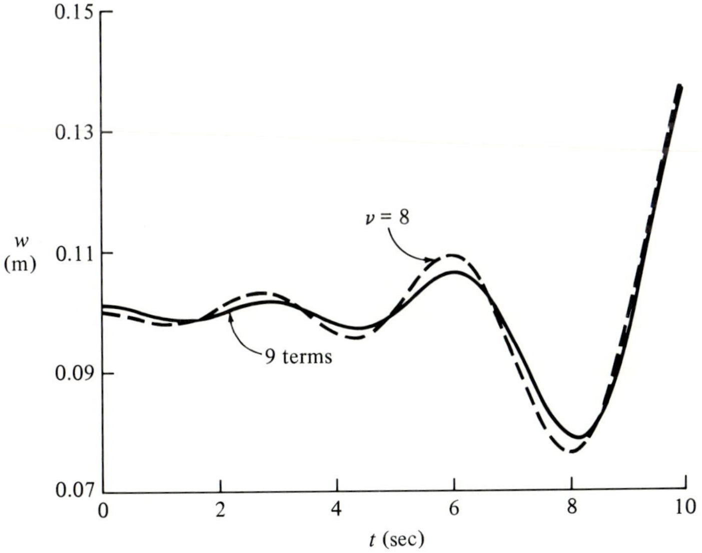

Figure P4.26(c)

4.27 A cantilever beam has a length of 20 m, a flexural rigidity of 5 N · m^2, and a mass per unit length of 0.2 kg/m. Using a single finite element, find the associated global mass and stiffness matrices $\widetilde{M}$ and $\widetilde{S}$, the eigenvalues $\widetilde{\lambda}_1$ and $\widetilde{\lambda}_2$ of $\widetilde{M}^{-1}\widetilde{S}$, and the corresponding eigenvectors $\widetilde{A}_1$ and $\widetilde{A}_2$ of $\widetilde{M}^{-1}\widetilde{S}$, normalized with respect to $\widetilde{M}$.

Results

$$\widetilde{M} = \begin{bmatrix} 1.48571 & -4.19048 \\ -4.19048 & 15.2381 \end{bmatrix}$$

$$\widetilde{S} = \begin{bmatrix} 7.5 \times 10^{-3} & -7.5 \times 10^{-2} \\ -7.5 \times 10^{-2} & 1.0 \end{bmatrix}$$

$$\widetilde{\lambda}_1 = 1.95003 \times 10^{-3}$$

$$\widetilde{\lambda}_2 = 0.189300$$

$$\widetilde{A}_1 = \begin{bmatrix} 1.00976 \\ 6.95473 \times 10^{-2} \end{bmatrix}$$

$$\widetilde{A}_2 = \begin{bmatrix} 1.40726 \\ 5.36342 \times 10^{-1} \end{bmatrix}$$

4.28 The circular frequency p_i associated with the ith mode of free vibrations of a cantilever beam of length L, flexural rigidity EI, and mass per unit of length ρ is given by Eq. (4.10.7), which involves quantities λ_i ($i = 1, \ldots, \infty$), the first two of which have the values $\lambda_1 = 1.87510$, $\lambda_2 = 4.69409$. Approximations to p_i can be found by means of the finite-element method. To test the quality of such approximations, let p_{ij} denote the circular frequency associated with the ith mode when j elements of equal length are employed in the finite-element analysis of the beam considered in Prob. 4.27 and evaluate the ratio p_{ij}/p_i for $i = 1, 2$ and $j = 1, 2, 3$.

Suggestion: Letting M_j and S_j denote, respectively, the global mass matrix and the global stiffness matrix of the beam when j elements are employed, and designating as λ_{ij} ($i = 1, \ldots, 2j$) the consecutive eigenvalues of $M_j^{-1}S_j$, verify that $p_{ij} = \lambda_{ij}^{1/2}$; note that λ_{11} and λ_{21} are available in Prob. 4.27; and use the following as values for λ_{ij} ($i = 1, 2; j = 2, 3$): $\lambda_{12} = 1.93349 \times 10^{-3}$, $\lambda_{22} = 7.71553 \times 10^{-2}$, $\lambda_{13} = 1.93201 \times 10^{-3}$, $\lambda_{23} = 7.63614 \times 10^{-2}$.

Results Table P4.28.

Table P4.28

p_{ij}/p_i	$i = 1$	$i = 2$
$j = 1$	1.005	1.580
$j = 2$	1.000	1.008
$j = 3$	1.000	1.003

Table P4.29(a)

	Mode number i		
	1	2	3
$\lvert\widetilde{\phi}_i\rvert_{x=L/2}$ ($m = 1$) Eq. (4.11.29)	Q_1	Q_4	. . . .
$\lvert\widetilde{\phi}_i\rvert_{x=L/2}$ ($m = 10$) Eq. (4.11.29)	Q_2	Q_5	Q_7
$\lvert\overline{\phi}_i m_B^{-1/2}\rvert_{x=L/2}$ Eq. (4.10.4)	Q_3	Q_6	Q_8

4.29 In the finite-element analysis of vibrations of a cantilever beam, the functions $\widetilde{\phi}_i$ $(i = 1, \ldots, \nu)$ defined in Eqs. (4.11.29) play a role analogous to that played by the functions $\overline{\phi}_i$ $(i = 1, \ldots, \infty)$ defined in Eq. (4.10.4) in connection with the "exact" analysis of such vibrations. To explore the relationship between $\widetilde{\phi}_i$ and $\overline{\phi}_i$, begin by using Eq. (4.11.30), the third of Eqs. (4.10.21), and Eq. (4.11.45) to verify that

$$\int_0^L \widetilde{\phi}_i \widetilde{\phi}_j \rho \, dx = \delta_{ij} \qquad (i, j = 1, \ldots, \nu)$$

Next, show with the aid of Eq. (4.10.8) that if there exists a constant c such that $\widetilde{\phi}_i = c\overline{\phi}_i$ $(i = 1, \ldots, \nu)$, then $c = m_B{}^{1/2}$. Finally, considering the midpoint of the beam of Prob. 4.27, evaluate the quantities $Q_1, \ldots, Q_8$ indicated in Table P4.29(*a*), using $\widetilde{A}_{ij}$ $(i, j = 1, 2)$ as found in Prob. 4.27 when dealing with $m = 1$, and assigning to $\widetilde{A}_{ij}$ $(i = 1, \ldots, 20; j = 1, 2, 3)$ for $m = 10$ the values shown in Table P4.29(*b*).

Results 0.33101, 0.33952, 0.33952, 0.63723, 0.71371, 0.71367, 0.01971, 0.01969

Table P4.29(*b*)

A_{ij}	i	j = 1	2	3
	1	0.0167735	0.0926354	−0.228184
	2	0.0163703	0.0838835	−0.188372
	3	0.0638710	0.301075	−0.604814
	4	0.0303259	0.116210	−0.155986
	5	0.136483	0.526168	−0.756627
	6	0.0418917	0.101759	0.0177607
	7	0.229885	0.683515	−0.526200
	8	0.0511296	0.0505744	0.203096
	9	0.339524	0.713713	−0.0197071
	10	0.0581528	−0.0226585	0.277741
	11	0.461135	0.589515	0.474000
	12	0.0631374	−0.100977	0.189659
	13	0.590877	0.317073	0.657761
	14	0.0663307	−0.168558	−0.0178446
	15	0.725479	−0.0700400	0.395084
	16	0.0680579	−0.214396	−0.236888
	17	0.862401	−0.523786	−0.228612
	18	0.0687259	−0.235492	−0.367110
	19	1.00000	−1.00007	−1.00050
	20	0.0688254	−0.239055	−0.392637

4.30 E_i as given in Eq. (4.11.37) represents an approximation to $E_i\, m_B^{-1/2}$ with E_i given by Eq. (4.10.66); similarly for F_i as in Eq. (4.11.42) and $F_i\, m_B^{-1/2}$ with F_i as in Eq. (4.10.67). To test the quality of these approximations, evaluate the quantities $Q_1, \ldots, Q_8$ indicated in Tables P4.30(*a*) and P4.30(*b*).

Table P4.30(*a*)

	$\lvert E_i \rvert$ $(m = 1)$ Eq. (4.11.37)	$\lvert E_i \rvert$ $(m = 10)$ Eq. (4.11.37)	$E_i\, m_B^{-1/2}$ Eq. (4.10.66)
$i = 1$	Q_1	1.56598	Q_2
$i = 2$	Q_3	0.867871	Q_4

Table P4.30(*b*)

	$\lvert F_i \rvert$ $(m = 1)$ Eq. (4.11.42)	$\lvert F_i \rvert$ $(m = 10)$ Eq. (4.11.42)	$F_i\, m_B^{-1/2}$ Eq. (4.10.67)
$i = 1$	Q_5	22.7530	Q_6
$i = 2$	Q_7	3.63067	Q_8

Results 1.55587, 1.56598, 0.761093, 0.867872, 22.7095, 22.7530, 3.50408, 3.63067

4.31 Equations (4.10.13), (4.10.57)–(4.10.60), (4.10.62), and (4.10.63) form a set of $5 + 2\nu$ equations governing the quantities R, θ, $q_1, \ldots, q_\nu$, and $u_1, \ldots, u_{3+\nu}$, all of whose initial values must be known before one can undertake a solution of the equations. As regards R and θ, this poses no problem, for each of these variables has a readily apparent physical significance; similarly for u_1, u_2, u_3 [see Eqs. (4.10.10)]; but $q_k(0)$ and $u_{3+k}(0)$ $(k = 1, \ldots, \nu)$ generally must be found by making use of given information regarding the initial deflection and initial motion of B relative to A, that is, $y(x, 0)$ and $\partial y(x, 0)/\partial t$.

(*a*) Taking for ϕ_k in Eq. (4.10.12) the function $\overline{\phi}_k$ defined in Eq. (4.10.4), show that

$$q_k(0) = \frac{1}{m_B} \int_0^L y(x, 0)\, \overline{\phi}_k\, \rho\, dx \qquad (k = 1, \ldots, \nu)$$

$$u_{3+k}(0) = \frac{1}{m_B} \int_0^L \frac{\partial y(x, 0)}{\partial t}\, \overline{\phi}_k\, \rho\, dx \qquad (k = 1, \ldots, \nu)$$

(*b*) Referring to Sec. 4.11, show that, for the stepped beam there considered,

$$q_k(0) = \sum_{i=1}^{2m} \sum_{j=1}^{2m} \widetilde{A}_{jk} \widetilde{M}_{ji}\, \widetilde{z}_i(0) \qquad (k = 1, \ldots, \nu)$$

$$u_{3+k}(0) = \sum_{i=1}^{2m} \sum_{j=1}^{2m} \widetilde{A}_{jk} \widetilde{M}_{ji}\, \dot{\widetilde{z}}_i(0) \qquad (k = 1, \ldots, \nu)$$

(*c*) Referring to the Example in Sec. 4.11, suppose that the initial deflection of B is given by

$$y(x, 0) = \left(\frac{x}{20}\right)^2$$

and that B is initially at rest relative to A. Making use of Table P4.31, which contains results obtained from a structural analysis of B, determine $q_5(0)$ and $u_8(0)$.

Table P4.31

i	$\sum_{j=1}^{20} \tilde{A}_{j5} \tilde{M}_{ji}$	i	$\sum_{j=1}^{20} \tilde{A}_{j5} \tilde{M}_{ji}$
1	0.224024	11	0.196386
2	0.0382657	12	0.0476021
3	0.385068	13	0.247772
4	−0.00239025	14	−0.0394262
5	0.157218	15	−0.168500
6	−0.0542421	16	−0.0496868
7	−0.272344	17	−0.159556
8	−0.0516500	18	0.0557229
9	−0.274485	19	0.128851
10	0.0712732	20	−0.0299777

Result 0.001818, 0

APPENDIX

I

DIRECTION COSINES AS FUNCTIONS OF ORIENTATION ANGLES

(see Sec. 1.7)

In this appendix, the direction cosines associated with each of 24 sets of angles describing the orientation of rigid body B in a reference frame A are tabulated. To use the tables, proceed as follows: let $\mathbf{a}_1$, $\mathbf{a}_2$, $\mathbf{a}_3$ be a dextral set of mutually perpendicular unit vectors fixed in the reference frame A, and let $\mathbf{b}_1$, $\mathbf{b}_2$, $\mathbf{b}_3$ be a similar such set fixed in the body B. Regard $\mathbf{b}_i$ as initially aligned with $\mathbf{a}_i$ $(i = 1, 2, 3)$; select the type of rotation sequence of interest (that is, body-three, body-two, space-three, or space-two); letting θ_1, θ_2, θ_3 denote the amounts of the first, the second, and the third rotation, respectively, pick the rotation sequence of interest [for example, 3-1-2 (corresponding to a $\theta_1\mathbf{b}_3$, $\theta_2\mathbf{b}_1$, $\theta_3\mathbf{b}_2$ body-three sequence or a $\theta_1\mathbf{a}_3$, $\theta_2\mathbf{a}_1$, $\theta_3\mathbf{a}_2$ space-three sequence)]; finally, locate the table corresponding to the rotation sequence chosen. The nine entries in the table [with s_i and c_i standing, respectively, for $\sin\theta_i$ and $\cos\theta_i$ $(i = 1, 2, 3)$] are the elements C_{ij} of the associated direction cosine matrix, which elements are defined as $C_{ij} \triangleq \mathbf{a}_i \cdot \mathbf{b}_j$ $(i, j = 1, 2, 3)$. Moreover, by reading a row or column of the table, one can determine how to express any of $\mathbf{a}_1$, $\mathbf{a}_2$, $\mathbf{a}_3$ in terms of $\mathbf{b}_1$, $\mathbf{b}_2$, $\mathbf{b}_3$, or any of $\mathbf{b}_1$, $\mathbf{b}_2$, $\mathbf{b}_3$ in terms of $\mathbf{a}_1$, $\mathbf{a}_2$, $\mathbf{a}_3$. For example, in the table corresponding to the body-three 3-1-2 sequence, the third row reveals that $\mathbf{a}_3 = -c_2s_3\mathbf{b}_1 + s_2\mathbf{b}_2 + c_2c_3\mathbf{b}_3$, while the second column indicates that $\mathbf{b}_2 = -s_1c_2\mathbf{a}_1 + c_1c_2\mathbf{a}_2 + s_2\mathbf{a}_3$.

Body-three: 1-2-3

	$\mathbf{b}_1$	$\mathbf{b}_2$	$\mathbf{b}_3$
$\mathbf{a}_1$	c_2c_3	$-c_2s_3$	s_2
$\mathbf{a}_2$	$s_1s_2c_3 + s_3c_1$	$-s_1s_2s_3 + c_3c_1$	$-s_1c_2$
$\mathbf{a}_3$	$-c_1s_2c_3 + s_3s_1$	$c_1s_2s_3 + c_3s_1$	c_1c_2

Body-three: 2-3-1

	$\mathbf{b}_1$	$\mathbf{b}_2$	$\mathbf{b}_3$
$\mathbf{a}_1$	c_1c_2	$-c_1s_2c_3 + s_3s_1$	$c_1s_2s_3 + c_3s_1$
$\mathbf{a}_2$	s_2	c_2c_3	$-c_2s_3$
$\mathbf{a}_3$	$-s_1c_2$	$s_1s_2c_3 + s_3c_1$	$-s_1s_2s_3 + c_3c_1$

Body-three: 3-1-2

	$\mathbf{b}_1$	$\mathbf{b}_2$	$\mathbf{b}_3$
$\mathbf{a}_1$	$-s_1s_2s_3 + c_3c_1$	$-s_1c_2$	$s_1s_2c_3 + s_3c_1$
$\mathbf{a}_2$	$c_1s_2s_3 + c_3s_1$	c_1c_2	$-c_1s_2c_3 + s_3s_1$
$\mathbf{a}_3$	$-c_2s_3$	s_2	c_2c_3

Body-three: 1-3-2

	$\mathbf{b}_1$	$\mathbf{b}_2$	$\mathbf{b}_3$
$\mathbf{a}_1$	c_2c_3	$-s_2$	c_2s_3
$\mathbf{a}_2$	$c_1s_2c_3 + s_3s_1$	c_1c_2	$c_1s_2s_3 - c_3s_1$
$\mathbf{a}_3$	$s_1s_2c_3 - s_3c_1$	s_1c_2	$s_1s_2s_3 + c_3c_1$

Body-three: 2-1-3

	$\mathbf{b}_1$	$\mathbf{b}_2$	$\mathbf{b}_3$
$\mathbf{a}_1$	$s_1s_2s_3 + c_3c_1$	$s_1s_2c_3 - s_3c_1$	s_1c_2
$\mathbf{a}_2$	c_2s_3	c_2c_3	$-s_2$
$\mathbf{a}_3$	$c_1s_2s_3 - c_3s_1$	$c_1s_2c_3 + s_3s_1$	c_1c_2

Body-three: 3-2-1

	$\mathbf{b}_1$	$\mathbf{b}_2$	$\mathbf{b}_3$
$\mathbf{a}_1$	c_1c_2	$c_1s_2s_3 - c_3s_1$	$c_1s_2c_3 + s_3s_1$
$\mathbf{a}_2$	s_1c_2	$s_1s_2s_3 + c_3c_1$	$s_1s_2c_3 - s_3c_1$
$\mathbf{a}_3$	$-s_2$	c_2s_3	c_2c_3

Body-two: 1-2-1

	$\mathbf{b}_1$	$\mathbf{b}_2$	$\mathbf{b}_3$
$\mathbf{a}_1$	c_2	s_2s_3	s_2c_3
$\mathbf{a}_2$	s_1s_2	$-s_1c_2s_3 + c_3c_1$	$-s_1c_2c_3 - s_3c_1$
$\mathbf{a}_3$	$-c_1s_2$	$c_1c_2s_3 + c_3s_1$	$c_1c_2c_3 - s_3s_1$

Body-two: 1-3-1

	$\mathbf{b}_1$	$\mathbf{b}_2$	$\mathbf{b}_3$
$\mathbf{a}_1$	c_2	$-s_2c_3$	s_2s_3
$\mathbf{a}_2$	c_1s_2	$c_1c_2c_3 - s_3s_1$	$-c_1c_2s_3 - c_3s_1$
$\mathbf{a}_3$	s_1s_2	$s_1c_2c_3 + s_3c_1$	$-s_1c_2s_3 + c_3c_1$

Body-two: 2-1-2

	$\mathbf{b}_1$	$\mathbf{b}_2$	$\mathbf{b}_3$
$\mathbf{a}_1$	$-s_1c_2s_3 + c_3c_1$	s_1s_2	$s_1c_2c_3 + s_3c_1$
$\mathbf{a}_2$	s_2s_3	c_2	$-s_2c_3$
$\mathbf{a}_3$	$-c_1c_2s_3 - c_3s_1$	c_1s_2	$c_1c_2c_3 - s_3s_1$

Body-two: 2-3-2

	$\mathbf{b}_1$	$\mathbf{b}_2$	$\mathbf{b}_3$
$\mathbf{a}_1$	$c_1c_2c_3 - s_3s_1$	$-c_1s_2$	$c_1c_2s_3 + c_3s_1$
$\mathbf{a}_2$	s_2c_3	c_2	s_2s_3
$\mathbf{a}_3$	$-s_1c_2c_3 - s_3c_1$	s_1s_2	$-s_1c_2s_3 + c_3c_1$

Body-two: 3-1-3

	$\mathbf{b}_1$	$\mathbf{b}_2$	$\mathbf{b}_3$
$\mathbf{a}_1$	$-s_1c_2s_3 + c_3c_1$	$-s_1c_2c_3 - s_3c_1$	s_1s_2
$\mathbf{a}_2$	$c_1c_2s_3 + c_3s_1$	$c_1c_2c_3 - s_3s_1$	$-c_1s_2$
$\mathbf{a}_3$	s_2s_3	s_2c_3	c_2

Body-two: 3-2-3

	$\mathbf{b}_1$	$\mathbf{b}_2$	$\mathbf{b}_3$
$\mathbf{a}_1$	$c_1c_2c_3 - s_3s_1$	$-c_1c_2s_3 - c_3s_1$	c_1s_2
$\mathbf{a}_2$	$s_1c_2c_3 + s_3c_1$	$-s_1c_2s_3 + c_3c_1$	s_1s_2
$\mathbf{a}_3$	$-s_2c_3$	s_2s_3	c_2

Space-three: 1-2-3

	$\mathbf{b}_1$	$\mathbf{b}_2$	$\mathbf{b}_3$
$\mathbf{a}_1$	c_2c_3	$s_1s_2c_3 - s_3c_1$	$c_1s_2c_3 + s_3s_1$
$\mathbf{a}_2$	c_2s_3	$s_1s_2s_3 + c_3c_1$	$c_1s_2s_3 - c_3s_1$
$\mathbf{a}_3$	$-s_2$	s_1c_2	c_1c_2

Space-three: 2-3-1

	$\mathbf{b}_1$	$\mathbf{b}_2$	$\mathbf{b}_3$
$\mathbf{a}_1$	c_1c_2	$-s_2$	s_1c_2
$\mathbf{a}_2$	$c_1s_2c_3 + s_3s_1$	c_2c_3	$s_1s_2c_3 - s_3c_1$
$\mathbf{a}_3$	$c_1s_2s_3 - c_3s_1$	c_2s_3	$s_1s_2s_3 + c_3c_1$

Space-three: 3-1-2

	$\mathbf{b}_1$	$\mathbf{b}_2$	$\mathbf{b}_3$
$\mathbf{a}_1$	$s_1s_2s_3 + c_3c_1$	$c_1s_2s_3 - c_3s_1$	c_2s_3
$\mathbf{a}_2$	s_1c_2	c_1c_2	$-s_2$
$\mathbf{a}_3$	$s_1s_2c_3 - s_3c_1$	$c_1s_2c_3 + s_3s_1$	c_2c_3

Space-three: 1-3-2

	$\mathbf{b}_1$	$\mathbf{b}_2$	$\mathbf{b}_3$
$\mathbf{a}_1$	c_2c_3	$-c_1s_2c_3 + s_3s_1$	$s_1s_2c_3 + s_3c_1$
$\mathbf{a}_2$	s_2	c_1c_2	$-s_1c_2$
$\mathbf{a}_3$	$-c_2s_3$	$c_1s_2s_3 + c_3s_1$	$-s_1s_2s_3 + c_3c_1$

Space-three: 2-1-3

	$\mathbf{b}_1$	$\mathbf{b}_2$	$\mathbf{b}_3$
$\mathbf{a}_1$	$-s_1s_2s_3 + c_3c_1$	$-c_2s_3$	$c_1s_2s_3 + c_3s_1$
$\mathbf{a}_2$	$s_1s_2c_3 + s_3c_1$	c_2c_3	$-c_1s_2c_3 + s_3s_1$
$\mathbf{a}_3$	$-s_1c_2$	s_2	c_1c_2

Space-three: 3-2-1

	$\mathbf{b}_1$	$\mathbf{b}_2$	$\mathbf{b}_3$
$\mathbf{a}_1$	c_1c_2	$-s_1c_2$	s_2
$\mathbf{a}_2$	$c_1s_2s_3 + c_3s_1$	$-s_1s_2s_3 + c_3c_1$	$-c_2s_3$
$\mathbf{a}_3$	$-c_1s_2c_3 + s_3s_1$	$s_1s_2c_3 + s_3c_1$	c_2c_3

Space-two: 1-2-1

	$\mathbf{b}_1$	$\mathbf{b}_2$	$\mathbf{b}_3$
$\mathbf{a}_1$	c_2	s_1s_2	c_1s_2
$\mathbf{a}_2$	s_2s_3	$-s_1c_2s_3 + c_3c_1$	$-c_1c_2s_3 - c_3s_1$
$\mathbf{a}_3$	$-s_2c_3$	$s_1c_2c_3 + s_3c_1$	$c_1c_2c_3 - s_3s_1$

Space-two: 1-3-1

	$\mathbf{b}_1$	$\mathbf{b}_2$	$\mathbf{b}_3$
$\mathbf{a}_1$	c_2	$-c_1s_2$	s_1s_2
$\mathbf{a}_2$	s_2c_3	$c_1c_2c_3 - s_3s_1$	$-s_1c_2c_3 - s_3c_1$
$\mathbf{a}_3$	s_2s_3	$c_1c_2s_3 + c_3s_1$	$-s_1c_2s_3 + c_3c_1$

Space-two: 2-1-2

	$\mathbf{b}_1$	$\mathbf{b}_2$	$\mathbf{b}_3$
$\mathbf{a}_1$	$-s_1c_2s_3 + c_3c_1$	s_2s_3	$c_1c_2s_3 + c_3s_1$
$\mathbf{a}_2$	s_1s_2	c_2	$-c_1s_2$
$\mathbf{a}_3$	$-s_1c_2c_3 - s_3c_1$	s_2c_3	$c_1c_2c_3 - s_3s_1$

Space-two: 2-3-2

	$\mathbf{b}_1$	$\mathbf{b}_2$	$\mathbf{b}_3$
$\mathbf{a}_1$	$c_1c_2c_3 - s_3s_1$	$-s_2c_3$	$s_1c_2c_3 + s_3c_1$
$\mathbf{a}_2$	c_1s_2	c_2	s_1s_2
$\mathbf{a}_3$	$-c_1c_2s_3 - c_3s_1$	s_2s_3	$-s_1c_2s_3 + c_3c_1$

Space-two: 3-1-3

	$\mathbf{b}_1$	$\mathbf{b}_2$	$\mathbf{b}_3$
$\mathbf{a}_1$	$-s_1c_2s_3 + c_3c_1$	$-c_1c_2s_3 - c_3s_1$	s_2s_3
$\mathbf{a}_2$	$s_1c_2c_3 + s_3c_1$	$c_1c_2c_3 - s_3s_1$	$-s_2c_3$
$\mathbf{a}_3$	s_1s_2	c_1s_2	c_2

Space-two: 3-2-3

	$\mathbf{b}_1$	$\mathbf{b}_2$	$\mathbf{b}_3$
$\mathbf{a}_1$	$c_1c_2c_3 - s_3s_1$	$-s_1c_2c_3 - s_3c_1$	s_2c_3
$\mathbf{a}_2$	$c_1c_2s_3 + c_3s_1$	$-s_1c_2s_3 + c_3c_1$	s_2s_3
$\mathbf{a}_3$	$-c_1s_2$	s_1s_2	c_2

APPENDIX

II

KINEMATICAL DIFFERENTIAL EQUATIONS IN TERMS OF ORIENTATION ANGLES

(see Sec. 1.7)

In this appendix, the kinematical differential equations associated with each of twenty-four sets of angles describing the orientation of a rigid body B in a reference frame A are tabulated. To use the tables, proceed as follows: let $\mathbf{a}_1$, $\mathbf{a}_2$, $\mathbf{a}_3$ be a dextral set of mutually perpendicular unit vectors fixed in the reference frame A, and let $\mathbf{b}_1$, $\mathbf{b}_2$, $\mathbf{b}_3$ be a similar such set fixed in the body B. Regard $\mathbf{b}_i$ as initially aligned with $\mathbf{a}_i$ $(i = 1, 2, 3)$; select the type of rotation sequence of interest (i.e., body-three, body-two, space-three, or space-two); letting θ_1, θ_2, θ_3 denote the amounts (in radians) of the first, the second, and the third rotation, respectively, pick the rotation sequence of interest [for example, 3-1-2 (corresponding to a $\theta_1\mathbf{b}_3$, $\theta_2\mathbf{b}_1$, $\theta_3\mathbf{b}_2$ body-three sequence or a $\theta_1\mathbf{a}_3$, $\theta_2\mathbf{a}_1$, $\theta_3\mathbf{a}_2$ space-three sequence)]; finally, locate the table corresponding to the rotation sequence chosen. The table contains the relationships between $\dot{\theta}_1$, $\dot{\theta}_2$, $\dot{\theta}_3$ and ω_1, ω_2, ω_3, where $\omega_i \triangleq {}^A\boldsymbol{\omega}^B \cdot \mathbf{b}_i$ $(i = 1, 2, 3)$, ${}^A\boldsymbol{\omega}^B$ being the angular velocity of B in A.

Body-three: 1-2-3

$\omega_1 = \dot{\theta}_1 c_2 c_3 + \dot{\theta}_2 s_3$	$\dot{\theta}_1 = (\omega_1 c_3 - \omega_2 s_3)/c_2$
$\omega_2 = -\dot{\theta}_1 c_2 s_3 + \dot{\theta}_2 c_3$	$\dot{\theta}_2 = \omega_1 s_3 + \omega_2 c_3$
$\omega_3 = \dot{\theta}_1 s_2 + \dot{\theta}_3$	$\dot{\theta}_3 = (-\omega_1 c_3 + \omega_2 s_3) s_2/c_2 + \omega_3$

Body-three: 2-3-1

$\omega_1 = \dot{\theta}_1 s_2 + \dot{\theta}_3$	$\dot{\theta}_1 = (\omega_2 c_3 - \omega_3 s_3)/c_2$
$\omega_2 = \dot{\theta}_1 c_2 c_3 + \dot{\theta}_2 s_3$	$\dot{\theta}_2 = \omega_2 s_3 + \omega_3 c_3$
$\omega_3 = -\dot{\theta}_1 c_2 s_3 + \dot{\theta}_2 c_3$	$\dot{\theta}_3 = \omega_1 + (-\omega_2 c_3 + \omega_3 s_3) s_2/c_2$

Body-three: 3-1-2

$\omega_1 = -\dot{\theta}_1 c_2 s_3 + \dot{\theta}_2 c_3$	$\dot{\theta}_1 = (-\omega_1 s_3 + \omega_3 c_3)/c_2$
$\omega_2 = \dot{\theta}_1 s_2 + \dot{\theta}_3$	$\dot{\theta}_2 = \omega_1 c_3 + \omega_3 s_3$
$\omega_3 = \dot{\theta}_1 c_2 c_3 + \dot{\theta}_2 s_3$	$\dot{\theta}_3 = (\omega_1 s_3 - \omega_3 c_3) s_2/c_2 + \omega_2$

Body-three: 1-3-2

$\omega_1 = \dot{\theta}_1 c_2 c_3 - \dot{\theta}_2 s_3$	$\dot{\theta}_1 = (\omega_1 c_3 + \omega_3 s_3)/c_2$
$\omega_2 = -\dot{\theta}_1 s_2 + \dot{\theta}_3$	$\dot{\theta}_2 = -\omega_1 s_3 + \omega_3 c_3$
$\omega_3 = \dot{\theta}_1 c_2 s_3 + \dot{\theta}_2 c_3$	$\dot{\theta}_3 = (\omega_1 c_3 + \omega_3 s_3) s_2/c_2 + \omega_2$

Body-three: 2-1-3

$\omega_1 = \dot{\theta}_1 c_2 s_3 + \dot{\theta}_2 c_3$	$\dot{\theta}_1 = (\omega_1 s_3 + \omega_2 c_3)/c_2$
$\omega_2 = \dot{\theta}_1 c_2 c_3 - \dot{\theta}_2 s_3$	$\dot{\theta}_2 = \omega_1 c_3 - \omega_2 s_3$
$\omega_3 = -\dot{\theta}_1 s_2 + \dot{\theta}_3$	$\dot{\theta}_3 = (\omega_1 s_3 + \omega_2 c_3) s_2/c_2 + \omega_3$

Body-three: 3-2-1

$\omega_1 = -\dot{\theta}_1 s_2 + \dot{\theta}_3$	$\dot{\theta}_1 = (\omega_2 s_3 + \omega_3 c_3)/c_2$
$\omega_2 = \dot{\theta}_1 c_2 s_3 + \dot{\theta}_2 c_3$	$\dot{\theta}_2 = \omega_2 c_3 - \omega_3 s_3$
$\omega_3 = \dot{\theta}_1 c_2 c_3 - \dot{\theta}_2 s_3$	$\dot{\theta}_3 = \omega_1 + (\omega_2 s_3 + \omega_3 c_3) s_2/c_2$

Body-two: 1-2-1

$\omega_1 = \dot{\theta}_1 c_2 + \dot{\theta}_3$	$\dot{\theta}_1 = (\omega_2 s_3 + \omega_3 c_3)/s_2$
$\omega_2 = \dot{\theta}_1 s_2 s_3 + \dot{\theta}_2 c_3$	$\dot{\theta}_2 = \omega_2 c_3 - \omega_3 s_3$
$\omega_3 = \dot{\theta}_1 s_2 c_3 - \dot{\theta}_2 s_3$	$\dot{\theta}_3 = \omega_1 - (\omega_2 s_3 + \omega_3 c_3) c_2/s_2$

Body-two: 1-3-1

$\omega_1 = \dot{\theta}_1 c_2 + \dot{\theta}_3$	$\dot{\theta}_1 = (-\omega_2 c_3 + \omega_3 s_3)/s_2$
$\omega_2 = -\dot{\theta}_1 s_2 c_3 + \dot{\theta}_2 s_3$	$\dot{\theta}_2 = \omega_2 s_3 + \omega_3 c_3$
$\omega_3 = \dot{\theta}_1 s_2 s_3 + \dot{\theta}_2 c_3$	$\dot{\theta}_3 = \omega_1 + (\omega_2 c_3 - \omega_3 s_3) c_2/s_2$

Body-two: 2-1-2

$\omega_1 = \dot\theta_1 s_2 s_3 + \dot\theta_2 c_3$	$\dot\theta_1 = (\omega_1 s_3 - \omega_3 c_3)/s_2$
$\omega_2 = \dot\theta_1 c_2 + \dot\theta_3$	$\dot\theta_2 = \omega_1 c_3 + \omega_3 s_3$
$\omega_3 = -\dot\theta_1 s_2 c_3 + \dot\theta_2 s_3$	$\dot\theta_3 = (-\omega_1 s_3 + \omega_3 c_3)c_2/s_2 + \omega_2$

Body-two: 2-3-2

$\omega_1 = \dot\theta_1 s_2 c_3 - \dot\theta_2 s_3$	$\dot\theta_1 = (\omega_1 c_3 + \omega_3 s_3)/s_2$
$\omega_2 = \dot\theta_1 c_2 + \dot\theta_3$	$\dot\theta_2 = -\omega_1 s_3 + \omega_3 c_3$
$\omega_3 = \dot\theta_1 s_2 s_3 + \dot\theta_2 c_3$	$\dot\theta_3 = -(\omega_1 c_3 + \omega_3 s_3)c_2/s_2 + \omega_2$

Body-two: 3-1-3

$\omega_1 = \dot\theta_1 s_2 s_3 + \dot\theta_2 c_3$	$\dot\theta_1 = (\omega_1 s_3 + \omega_2 c_3)/s_2$
$\omega_2 = \dot\theta_1 s_2 c_3 - \dot\theta_2 s_3$	$\dot\theta_2 = \omega_1 c_3 - \omega_2 s_3$
$\omega_3 = \dot\theta_1 c_2 + \dot\theta_3$	$\dot\theta_3 = -(\omega_1 s_3 + \omega_2 c_3)c_2/s_2 + \omega_3$

Body-two: 3-2-3

$\omega_1 = -\dot\theta_1 s_2 c_3 + \dot\theta_2 s_3$	$\dot\theta_1 = (-\omega_1 c_3 + \omega_2 s_3)/s_2$
$\omega_2 = \dot\theta_1 s_2 s_3 + \dot\theta_2 c_3$	$\dot\theta_2 = \omega_1 s_3 + \omega_2 c_3$
$\omega_3 = \dot\theta_1 c_2 + \dot\theta_3$	$\dot\theta_3 = (\omega_1 c_3 - \omega_2 s_3)c_2/s_2 + \omega_3$

Space-three: 1-2-3

$\omega_1 = \dot\theta_1 - \dot\theta_3 s_2$	$\dot\theta_1 = \omega_1 + (\omega_2 s_1 + \omega_3 c_1)s_2/c_2$
$\omega_2 = \dot\theta_2 c_1 + \dot\theta_3 s_1 c_2$	$\dot\theta_2 = \omega_2 c_1 - \omega_3 s_1$
$\omega_3 = -\dot\theta_2 s_1 + \dot\theta_3 c_1 c_2$	$\dot\theta_3 = (\omega_2 s_1 + \omega_3 c_1)/c_2$

Space-three: 2-3-1

$\omega_1 = -\dot\theta_2 s_1 + \dot\theta_3 c_1 c_2$	$\dot\theta_1 = (\omega_1 c_1 + \omega_3 s_1)s_2/c_2 + \omega_2$
$\omega_2 = \dot\theta_1 - \dot\theta_3 s_2$	$\dot\theta_2 = -\omega_1 s_1 + \omega_3 c_1$
$\omega_3 = \dot\theta_2 c_1 + \dot\theta_3 s_1 c_2$	$\dot\theta_3 = (\omega_1 c_1 + \omega_3 s_1)/c_2$

Space-three: 3-1-2

$\omega_1 = \dot\theta_2 c_1 + \dot\theta_3 s_1 c_2$	$\dot\theta_1 = (\omega_1 s_1 + \omega_2 c_1)s_2/c_2 + \omega_3$
$\omega_2 = -\dot\theta_2 s_1 + \dot\theta_3 c_1 c_2$	$\dot\theta_2 = \omega_1 c_1 - \omega_2 s_1$
$\omega_3 = \dot\theta_1 - \dot\theta_3 s_2$	$\dot\theta_3 = (\omega_1 s_1 + \omega_2 c_1)/c_2$

Space-three: 1-3-2

$\omega_1 = \dot{\theta}_1 + \dot{\theta}_3 s_2$	$\dot{\theta}_1 = \omega_1 + (-\omega_2 c_1 + \omega_3 s_1) s_2 / c_2$
$\omega_2 = \dot{\theta}_2 s_1 + \dot{\theta}_3 c_1 c_2$	$\dot{\theta}_2 = \omega_2 s_1 + \omega_3 c_1$
$\omega_3 = \dot{\theta}_2 c_1 - \dot{\theta}_3 s_1 c_2$	$\dot{\theta}_3 = (\omega_2 c_1 - \omega_3 s_1)/c_2$

Space-three: 2-1-3

$\omega_1 = \dot{\theta}_2 c_1 - \dot{\theta}_3 s_1 c_2$	$\dot{\theta}_1 = (\omega_1 s_1 - \omega_3 c_1) s_2 / c_2 + \omega_2$
$\omega_2 = \dot{\theta}_1 + \dot{\theta}_3 s_2$	$\dot{\theta}_2 = \omega_1 c_1 + \omega_3 s_1$
$\omega_3 = \dot{\theta}_2 s_1 + \dot{\theta}_3 c_1 c_2$	$\dot{\theta}_3 = (-\omega_1 s_1 + \omega_3 c_1)/c_2$

Space-three: 3-2-1

$\omega_1 = \dot{\theta}_2 s_1 + \dot{\theta}_3 c_1 c_2$	$\dot{\theta}_1 = (-\omega_1 c_1 + \omega_2 s_1) s_2 / c_2 + \omega_3$
$\omega_2 = \dot{\theta}_2 c_1 - \dot{\theta}_3 s_1 c_2$	$\dot{\theta}_2 = \omega_1 s_1 + \omega_2 c_1$
$\omega_3 = \dot{\theta}_1 + \dot{\theta}_3 s_2$	$\dot{\theta}_3 = (\omega_1 c_1 - \omega_2 s_1)/c_2$

Space-two: 1-2-1

$\omega_1 = \dot{\theta}_1 + \dot{\theta}_3 c_2$	$\dot{\theta}_1 = \omega_1 - (\omega_2 s_1 + \omega_3 c_1) c_2 / s_2$
$\omega_2 = \dot{\theta}_2 c_1 + \dot{\theta}_3 s_1 s_2$	$\dot{\theta}_2 = \omega_2 c_1 - \omega_3 s_1$
$\omega_3 = -\dot{\theta}_2 s_1 + \dot{\theta}_3 c_1 s_2$	$\dot{\theta}_3 = (\omega_2 s_1 + \omega_3 c_1)/s_2$

Space-two: 1-3-1

$\omega_1 = \dot{\theta}_1 + \dot{\theta}_3 c_2$	$\dot{\theta}_1 = \omega_1 + (\omega_2 c_1 - \omega_3 s_1) c_2 / s_2$
$\omega_2 = \dot{\theta}_2 s_1 - \dot{\theta}_3 c_1 s_2$	$\dot{\theta}_2 = \omega_2 s_1 + \omega_3 c_1$
$\omega_3 = \dot{\theta}_2 c_1 + \dot{\theta}_3 s_1 s_2$	$\dot{\theta}_3 = (-\omega_2 c_1 + \omega_3 s_1)/s_2$

Space-two: 2-1-2

$\omega_1 = \dot{\theta}_2 c_1 + \dot{\theta}_3 s_1 s_2$	$\dot{\theta}_1 = (-\omega_1 s_1 + \omega_3 c_1) c_2 / s_2 + \omega_2$
$\omega_2 = \dot{\theta}_1 + \dot{\theta}_3 c_2$	$\dot{\theta}_2 = \omega_1 c_1 + \omega_3 s_1$
$\omega_3 = \dot{\theta}_2 s_1 - \dot{\theta}_3 c_1 s_2$	$\dot{\theta}_3 = (\omega_1 s_1 - \omega_3 c_1)/s_2$

Space-two: 2-3-2

$\omega_1 = -\dot{\theta}_2 s_1 + \dot{\theta}_3 c_1 s_2$	$\dot{\theta}_1 = -(\omega_1 c_1 + \omega_3 s_1) c_2 / s_2 + \omega_2$
$\omega_2 = \dot{\theta}_1 + \dot{\theta}_3 c_2$	$\dot{\theta}_2 = -\omega_1 s_1 + \omega_3 c_1$
$\omega_3 = \dot{\theta}_2 c_1 + \dot{\theta}_3 s_1 s_2$	$\dot{\theta}_3 = (\omega_1 c_1 + \omega_3 s_1)/s_2$

Space-two: 3-1-3

$\omega_1 = \dot{\theta}_2 c_1 + \dot{\theta}_3 s_1 s_2$	$\dot{\theta}_1 = -(\omega_1 s_1 + \omega_2 c_1) c_2 / s_2 + \omega_3$
$\omega_2 = -\dot{\theta}_2 s_1 + \dot{\theta}_3 c_1 s_2$	$\dot{\theta}_2 = \omega_1 c_1 - \omega_2 s_1$
$\omega_3 = \dot{\theta}_1 + \dot{\theta}_3 c_2$	$\dot{\theta}_3 = (\omega_1 s_1 + \omega_2 c_1) / s_2$

Space-two: 3-2-3

$\omega_1 = \dot{\theta}_2 s_1 - \dot{\theta}_3 c_1 s_2$	$\dot{\theta}_1 = (\omega_1 c_1 - \omega_2 s_1) c_2 / s_2 + \omega_3$
$\omega_2 = \dot{\theta}_2 c_1 + \dot{\theta}_3 s_1 s_2$	$\dot{\theta}_2 = \omega_1 s_1 + \omega_2 c_1$
$\omega_3 = \dot{\theta}_1 + \dot{\theta}_3 c_2$	$\dot{\theta}_3 = (-\omega_1 c_1 + \omega_2 s_1) / s_2$

INDEX